Nanotechnology for Environmental Management

Editors

Jayanta Kumar Biswas
Department of Ecological Studies and
International Centre for Ecological Engineering
University of Kalyani
Kalyani, Nadia, West Bengal, India

Mahendra Rai
Department of Biotechnology
SGB Amravati University
Amravati, Maharashtra, India

CRC Press
Taylor & Francis Group
Boca Raton London New York

CRC Press is an imprint of the
Taylor & Francis Group, an **informa** business
A SCIENCE PUBLISHERS BOOK

Cover credit: Cover illustrations reproduced by kind courtesy of Elanit Rubalcava Avila and Yossef Rubalcava Avila, Chihuahua City, Mexico.

First edition published 2024
by CRC Press
2385 NW Executive Center Drive, Suite 320, Boca Raton FL 33431

and by CRC Press
4 Park Square, Milton Park, Abingdon, Oxon, OX14 4RN

CRC Press is an imprint of Taylor & Francis Group, LLC

Library of Congress Cataloging-in-Publication Data (applied for)

ISBN: 978-1-032-39696-5 (hbk)
ISBN: 978-1-032-39697-2 (pbk)
ISBN: 978-1-003-35094-1 (ebk)

DOI: 10.1201/9781003350941

Typeset in Times New Roman
by Prime Publishing Services

Foreword

Environmental pollution is a serious global concern due to factors such as rapid population expansion, industrialization, urban development, intensive agriculture and other anthropogenic activities. This pollution encompasses the contamination of soil, surface water and groundwater by potentially toxic metal(loid)s, pesticides, petroleum hydrocarbons, pharmaceuticals, solvents, dyes, radionuclides, explosives, crude oil, etc. Removal of these contaminants employing nanotechnology has emerged as a highly sustainable approach, often called 'green' nanotechnology, which offers numerous advantages over expensive, resource and energy-intensive, and environmentally disruptive conventional technological methods.

The book, 'Nanotechnology for Environmental Management' edited by Profs. Biswas and Rai, contain chapters describing diversity in approaches and applications of nanomaterials and nanotechnologies for environmental remediation, energy conservation, resource management, and sustainable development. It is divided into six Sections—I. Introductory; II. Soil/Land Remediation/ Restoration and Waste Management; III. Water and Wastewater Treatment/Remediation; IV. Agriculture; V. Energy and Conservation; and VI. Toxicity and Risk. The chapters have been contributed by nationally and internationally recognized scientists involved in multidisciplinary nanotechnology research. Their contributions highlight the latest advancements in applying green nanotechnology to address a broad spectrum of environmental challenges.

The book chapters present an array of advanced nanotechnologies in developing cost-effective, high-performance treatment and remediating systems that require not only an extensive advancement of its science and engineering aspects but also the realization of the already documented approaches to real-world scenarios. The chapters effectively illustrate 'lab to land' applications and case studies of environmental nanotechnology, which will assist planners, government and non-government organizations in applying green nanotechnologies for *in-situ* problem-solving. Furthermore, it sheds light on the 'flip side of the coin,' describing ecotoxicity and the ethical dimensions of nanotechnology to make readers aware of the challenges, potentials pitfalls and risks associated with green nanotechnology.

This timely and collaboratively authored book is designed to engage a broad audience, including students, academics, scientists, and a wide range of professionals responsible for regulating, monitoring, and designing environmental facilities. It is particularly relevant for those interested in addressing the removal of contaminants from various environmental mediums and matrices for sustainable development. Engineering consultants, industrial waste managers, and government regulators will also find this book a valuable resource and comprehensive reference guide.

I congratulate the editors, Profs. Biswas and Rai for their dedicated efforts in creating this high-quality book, which will undoubtedly enhance and enrich our current understanding of environmental nanotechnology.

Ravindra Pandey
Professor of Physics and Dean of CSA
Michigan Technological University
Fellow of the American Physical Society, USA

Preface

Exponential increase in human population, unplanned urbanization, extensive industrialization, and intensive agriculture have resulted in all-pervasive pollution in different domains of the environment—soil, water, and air with legacy (e.g., nutrients, potentially toxic metal (loid)s, pesticides, dyes, petroleum hydrocarbons, etc) and emerging contaminants (e.g., pharmaceuticals, endocrine disrupting compounds, microplastics, etc.). Nanotechnology has emerged as one of the most prominent and potential emerging technologies for the clean-up of environmental contaminants. Nanotechnology opens up new avenues of scientific innovations and offers opportunities for the development of new generation of materials, products and chemicals which have a broad spectrum of environmental applications with promoted performance but less material, energy, and carbon footprints, reducing harm to the environment. Nanotech has propelled the progress and practice of novel and cost-effective techniques for catalytic degradation, adsorptive elimination, and sensing contaminants in the environment. Nanomaterials (NMs) and their organic and inorganic variants developed from modifications with polymers, zeolites, activated carbon, clay minerals, biochars, etc. have been tried, tested and trusted for confiscating contaminants with enhanced efficiency.

There is the other side of the nanorevolution story. Taking advantage of this new technology for health, environmental, and sustainability benefits, science needs to examine the environmental and health implications adhered to applications of nanomaterials. The environment is prone to suffer from nanopollution and toxic effects due to the release and build-up of nanoparticles' contaminants during different phases of the lifecycle of nanomaterials including their production, transportation, disposal, entry into ecosystems, accumulation, bioconcentration, trophic transfer and so on. A sensible, integrated, and holistic approach is required to develop best management practices (BMPs) based on the scientific understanding of nanomaterials and nanotechnological tools and techniques.

With the suitable breadth of coverage, this edited book has been designed, written, and organized for the broad-spectrum multi-sectorial audience of researchers, students, professionals, and policymakers. It is expected to be a pointer and reference resource and will cater to the needs of the researchers, professionals as well as the graduate and post-graduate students of nearly every discipline of science and engineering—toxicology, biotechnology, chemistry, physics, biology, medical science, material science and engineering, environmental science, and environmental engineering, who are interested to know the ecotoxicological aspects of the nanomaterial and the potential hazards and risks involved therein.

This book would be an inspirational comprehensive resource for researchers, environmental professionals, and practitioners as well as industrial portfolio holders, who are interested to earn a working knowledge of this fascinating field of Nanotechnology and Nanotoxicology.

This book provides a wide-spectrum exploration and reflection on the state-of-the-art of nanotechnological research and development (R&D) in the domain of environmental science, engineering, and management. This book includes important chapters written by specialists and experts in the concerned topic, and hopefully, it will win wide acceptance among undergraduate and postgraduate students of chemistry, biochemistry, pharmacology, microbiology, biotechnology, and allied subjects.

The edited volume is divided into six Sections—I. Introductory; II. Soil/land Remediation/ Restoration and Waste Management; III. Water and Wastewater Treatment/Remediation; IV. Agriculture; V. Energy and Conservation; and VI. Toxicity and risk.

The book starts with an introductory overview of nanotechnology for environmental management (Chapter 1). The authors (Biswas et al.) present a synoptic overview of the diversity and functionality of environmentally relevant nanomaterials and their applications in environmental management, potential environmental risks, and possible remedies.

Chapters 2 and 3 deal with applications of nanotechnology in soil remediation and restoration of degraded land, while Chapter 4 concerns with waste management.

In Chapter 2, Faizan et al. narrates how applications of nanomaterials have facilitated phytoremediation of contaminated soils removing contaminants, promoting plant growth, and enhancing phytoavailability.

Chapter 3 presents a plethora of nanoadsorbents as a cutting-edge tool for the removal of metal (loid)s from the soil environment, their mechanisms, and associated factors regulating metal (loid)s removal significantly not affecting soil health.

Chapter 4 discusses different aspects and prospects of nanotechnology in sustainable industrial waste management.

Chapters 5 to 12 show how biogenic and chemogenic nanomaterials find wide applications in water and wastewater treatment and remediation

Chapter 5 emphasizes the importance of advanced oxidation processes (AOPs) relying on the photon energy for photocatalytic degradation of contaminants, and sheds mechanistic insights into the contaminants' degradation process involving AOPs combined with ultraviolet irradiation of nanomaterials.

In Chapter 6 Aboagye et al. discuss the state-of-the-art of nanomaterials development and applications and future prospects in this area of environmental remediation of water resources with due consideration of economic feasibility and environmental sustainability.

Chapter 7 deals with purification and desalination of water using nanomaterials where the authors illustrate how an extraordinary assemblage of nanomaterials' properties comprising high surface/volume ratio, tunable pore size, photosensitivity, electrochemical, optical and magnetic properties, surface chemistry, antimicrobial activity, etc. realize clean-up of wide-ranging contaminants comprising heavy metal(loid)s, organic matter, organic dyes, emerging contaminants, and others through adsorption, disinfection, decontamination and high-performance membrane filtration.

In Chapter 8 Skorupska and others review the functionalization of 3D graphene-based materials for the removal of water impurities highlighting the role of surface functional groups as adsorption centers for water contaminants.

Chapter 9 champions biochar-based nanocomposites as a novel and sustainable solution to (waste)water contamination. Here the authors highlight nanocomposites' structural and functional diversity in achieving water decontamination

Chapter 10 presents nanoparticles decorated biochar matrix (biochar-based nanocomposites) as a very efficient hybrid adsorbent for the removal of toxic azo dyes from wastewater.

Chapter 11 deals with applications of nanostructured materials for the removal of microplastics from water. Here Sarker et al. discuss how several nanostructured materials viz. reduced graphene oxide, nano Fe_3O_4, nano TiO_2, etc. are effectively employed in removing or degrading microplastics via adsorption, magnetic separation, and catalytic degradation.

Chapter 12 reviews the significance of diverse adsorbents (viz. carbonaceous material-based nanocomposites, polymeric nanoadsorbents and magnetic nanoparticles) in fluoride removal from contaminated water.

Chapters 13 and 14 cover nanotechnological applications in the field of agriculture and plant protection in promoting plant growth and providing protection against abiotic and biotic stresses.

Chapter 13 showcases the potential of nanobiotechnology nanomaterials controlled release and targeted delivery of agrochemicals, which enhances the yield by optimizing nutrient and water conditions.

Chapter 14 focuses on the role of nano-enabled sensors and probes in detecting and controlling plant pathogens.

Chapters 15 and 16 deal with nanomaterials' application in the generation and storage of clean and green energy while Chapter 17 is about the promise of nanomaterials in material conservation.

Chapter 15 emphasizes the utilization of nanotechnology in enhancing the quality of electrodes and electrolyte material of fuel cells, electrical double-layer capacitors, pseudocapacitors, and hybrid supercapacitors as well as the potential of incorporation of different nanomaterials in developing the new generations of solar cells.

In Chapter 16 Dagar and coauthors highlight TiO_2 as a "magical nanomaterial" capable of harvesting clean energy and ensuring environmental sustainability.

Chapter 17 discusses the use of nanomaterials in conserving concrete materials and adding value to green development.

Chapter 18 and 19 narrate the other side of the story of nanotechnology shaded with apprehension of toxicity and potential risks resulting from nanoparticle pollution in the environment.

Chapter 18 highlights the interactions between plants and nanoparticles particularly focusing on the phytotoxicity and modulation of the proteome profile.

Chapter 19 is focused on different aspects of toxicity, safety, and regulations of the applications of nanomaterials which are used in contaminant remediation, (waste)water treatment, delivery of bioactive compounds, and other forms of environmental management.

We anticipate that this book will be relevant to scientists and researchers working in the domains of nano(bio) technology, toxicology, microbiology, ecology and other related fields, and interested in real-time applications of nanotechnology for environmental management.

We are very grateful to all contributors for their valuable contributions, enduring efforts and hard work in the making of this book. We sincerely thank the Commissioning Editor at CRC Press (Taylor & Francis) for all-round support and cooperation during this project.

Last, but not the least, we deeply acknowledge the steadfast support, love, and sacrifice of our family members.

Jayanta Kumar Biswas and
Mahendra Rai
(Editors)

Contents

Section I

Introductory

Chapter 1

Nanotechnology for Environmental Management: An Introductory Overview

Jayanta Kumar Biswas[1]* and Barsha Roy[1]

[1]Enviromicrobiology, Ecotoxicology and Ecotechnology Research Laboratory (3E-MicroToxTech Lab), Department of Ecological Studies and International Centre for Ecological Engineering, University of Kalyani, Kalyani, Nadia–741235, West Bengal, India

Mahendra Rai[2]

[2]Nanobiotechnology Lab., Department of Biotechnology, SGB Amravati University, Amravati–444602, Maharashtra, India

1.1 INTRODUCTION

Though nanotechnology was already present in nature from the beginning, the term was formally discovered by the American physicist and Nobel Laureate Dr. Richard Phillips Feynman in 1959 (Subramani et al. 2019). The US government defined the term 'nanotechnology' as "research and technology development at the atomic, molecular or macromolecular level in the length scale of approximately 1–100 nm range, to provide a fundamental understanding of phenomena and materials at the nanoscale and to create and use structures, devices, and systems that have novel properties and functions because of their small and/or intermediate size" (US-Government). It was observed that reducing the particle size below 100 nm altered their properties considerably (physicochemical and biological) compared to the bulk material (Khan et al. 2019).

Nanoparticles (NPs) have at least one dimension below 100 nm (Eaton et al. 2017). Though NPs were naturally present from the onset of the Universe, it was only in 1914 that scientist Richard Adolf Zsigmondy first coined the unit 'nanometer' (nm), which denoted 10^{-6} times small dimension than a millimetre (mm) (Hulla et al. 2015). Generally, nanomaterials (NMs)

*Corresponding Author: jkbiswas@klyuniv.ac.in; biswajoy2008@gmail.com

can be categorized into natural and artificial. Various environmental factors create natural NMs, whereas humans manufacture artificial NMs for specific purposes. The NPs differ from their bulk equivalent in physiochemical properties due to their smaller size and larger surface area (Sapsford et al. 2013). Because of their altered physicochemical properties, NPs are used in broad fields like energy production, information technology, food, biomedical applications, biotechnology, agriculture, water treatment techniques etc. (Biswas and Sarkar, 2019). NPs can be categorized into many groups based on size, chemical constitution, and shape. These include carbon-based NPs, ceramic NPs, lipid-based NPs, metal and metal oxide NPs, and polymeric NPs (Khan et al. 2019).

The environment consists of a combination of abiotic and biotic components that have some influence on its organisms (Bashir and Chisti, 2017). Our environment has been polluted with many pollutants, such as heavy metals, oil spills, particulate matter, toxic gases, herbicides, pesticides, fertilizers, etc. Various technologies are available to tackle these contaminants but with many shortcomings, such as high cost and energy intensity, eco-invasiveness, secondary pollution, etc. (Guerra et al. 2018). Nanotechnology has shown immense potential in solving such environmental problems. This field provides enormous opportunities to evolve new strategies to remediate environmental pollution and generate clean energy (Jawad and Hamooshy 2021). Using diverse NPs, nanotechnology has enabled the generation of innovative and cost-effective methods for remediation, detection and monitoring of environmental pollution (U.S. EPA 2007).

The present chapter deals with the details of different environmentally relevant nanomaterials, applications of nanotechnology in environmental management, potential environmental risks, and possible remedies to overcome the risks.

1.2 DIVERSITY OF ENVIRONMENTAL POLLUTANTS

Pollutants are introduced into the environment naturally or anthropogenically and have detrimental effects on environmental health (Xu et al. 2014). Different inorganic, organic, and harmful microorganisms (Figure 1.1) are responsible for the pollution of the environment (Guerra et al. 2018). Other human ventures, like domestic activities, agricultural practices, and unsustainable industrial procedures, increase environmental pollutants. The diversity of poisonous, bio-resistant, and dangerous chemicals emitted into the environment reflects the diversity of ever-growing, extremely polluting industries such as foods and beverages, tannery, textiles, pharmaceuticals, and cosmetics (Pandey 2018, Razack and Duraiarasan 2020). Details of different environmental pollutants are stated below:

1.2.1 Inorganic Pollutants

Heavy metals, metalloids, and nutrients are considered inorganic pollutants. Heavy metals (e.g. Cr, Cu, Ni, Pb) and metalloids (e.g. As, Se) are generally mixed in the aqueous environment emanating from diverse human activities and industrial operations (Liang et al. 2021). In soil, these metals and metalloids come from different sources like mining and smelting, rocks, etc. (Chen et al. 2015). These compounds are not bio-degradable; hence, they can accumulate in the environment and cause damage to animals and humans (Atieh et al. 2017). These heavy metals and metalloids have a density of more than 5 g cm^{-3} and an atomic mass higher than 20. According to this information, 51 elements of the periodic table can be considered heavy metals and metalloids (Ali and Khan 2018). These heavy metals and metalloids are very toxic, and their chronic exposure can harm various organs like the kidney, brain, lungs, bones, and liver (Islam et al. 2018).

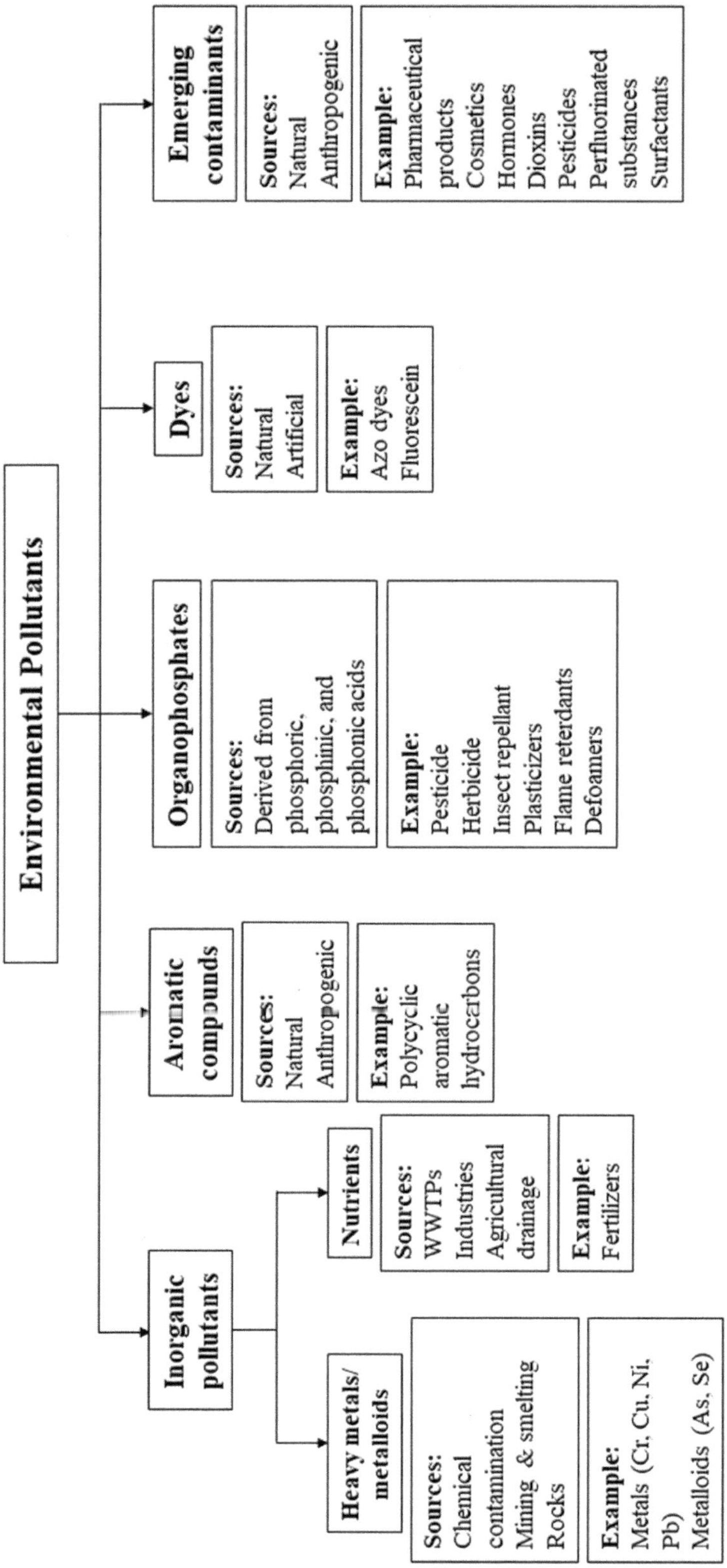

Figure 1.1 Schematic representation of pollutants diversity.

The other type of inorganic pollutant is nutrients. Increased discharge of human-generated nutrients pollutes aquatic ecosystems (Grizzetti et al. 2017). Excessive load of nitrogen (N) and phosphorus (P) from the nutrients results in eutrophication and hypoxia (Lintern et al. 2020). Nutrients can come from various sources like wastewater treatment plants, industries, agricultural drainage, etc. (Carey et al. 2013). Eutrophication leads to a reduction in fish production. Also, eutrophication reduces the biodiversity in aquatic environments (both autotrophs and heterotrophs), enhances the generation of methane, nitrous oxide, and other greenhouse gases, and induces the proliferation of harmful algal blooms (Wurtsbaugh et al. 2019).

1.2.2 Aromatic Compounds

Polycyclic aromatic compounds (PACs) are of significant concern among aromatic compounds. These are the materials where the carbon-carbon bonds in the surroundings strongly support the central molecular structure. These compounds are noticeable for their surprising chemical stability. These PACs are generated upon exposure of the organic substances to high temperatures, a process known as pyrolysis. The final product formed by pyrolysis has higher stability when compared to its precursor compound (Blumer, 1976). PACs can be naturally found in coal, petroleum and fossil fuels. These can also be formed during the ignition of organic compounds and thus are released into the environment. As these PACs can come from various sources and because of their higher persistence, they are considered a predominant environmental pollutant worldwide (Larsson et al. 2018).

1.2.3 Organophosphates

Chemicals that are phosphinic, phosphonic and phosphoric acid derivatives are mainly known as organophosphates (OPs). The first representative OPs were synthesized in the early 19th century, and since then, similar compounds have been used in several products utilized worldwide (Naughton and Terry Jr. 2018). OPs are widely used in pesticides, herbicides, insect repellants, plasticizers, flame retardants, and defoamers (Zhong et al. 2020). These OPs enter human bodies from contaminated soil and aquatic environments (Stasinska et al. 2014, Bekele et al. 2019). These compounds are tough to degrade, and only biological processes are not helpful for their degradation (Silva and Baltrusaitis 2021).

1.2.4 Dyes

Typically, dyes can be classified into two categories–natural and artificial. Parts of plants like leaves, fruits, wood, bark, and roots are the primary source of natural dyes. Synthetic dyes are made from chemicals, petroleum products, and minerals (Adnan et al. 2019, Pavithra and Jaikumar 2019). Various industries like paper, cosmetics, leather, and textiles use artificial dyes for colouring and printing jobs (Lai et al. 2019). Dyes can further be categorized into two types, non-ionic and ionic. According to the charged groups (positive or negative), ionic dyes are either cationic or anionic (León et al. 2018). Dyes have complex molecular structures with high molecular weight. They are water-soluble, non-degradable, and can show potential carcinogenicity and mutagenicity. Also, they can decrease photosynthesis by inhibiting the penetration of sunlight (Jusoh et al. 2013). Additionally, when mixed with wastewater, dyes disrupt the ecological balance of aquatic ecosystems (Lai et al. 2019). Further, organic dyes like fluorescein and azo dyes are difficult to decompose and cause cytotoxicity in mammalian tissues (Li et al. 2012). It was observed that hybrid nanocomposite *n*Ag-TC effectively removed toxic dyes (Shaikh et al. 2022).

1.2.5 Emerging Contaminants

A new category of environmental pollutants, emerging contaminants (ECs), draw significant attention because of their unwanted synergistic effects (Rout et al. 2021). They have comparatively low potency but intense toxicity (Rathi et al. 2021). ECs are mainly organic compounds, and their concentration in the environment is very low (ng L^{-1} to μg L^{-1}) (Rodriguez-Narvaez et al. 2017). The definition of ECs, according to the United States Geological Survey, states "any chemical of synthetic or natural origin or any microorganism that is not commonly monitored in the environment but has the potential to cause adverse ecological and/or human health effects" (USGS 2017). ECs are highly acidic or alkaline (polar) compounds (Rego et al. 2021). Cosmetic and pharmaceutical products, dioxins, hormones, pesticides, perfluorinated substances, surfactants, etc., fall under the category of ECs (Naidu et al. 2016, Tijani et al. 2016). The immeasurable utilization of ECs made them ubiquitous in the environment (Roberts et al. 2016). Because of their hydrophobicity, ECs can be piled up in the organisms' lipid-enriched tissues and damage the endocrine system of animals and humans (Rodriguez-Narvaez et al. 2017).

1.3 DIVERSITY OF NANOMATERIALS

In general, NMs can be categorized into three groups. They are (i) Inorganic NMs, (ii) Carbon-based NMs, and (iii) Polymer-based NMs, respectively. Figure 1.2 represents the detailed schema of the categorization of NMs.

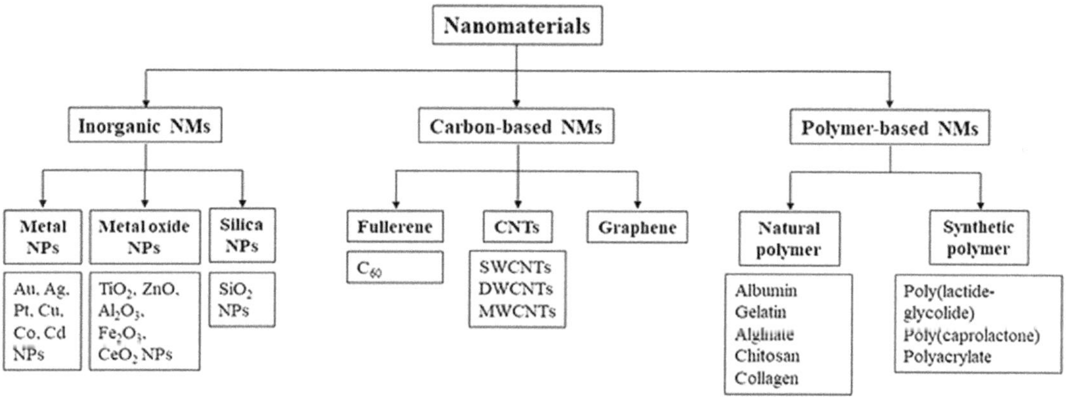

Figure 1.2 Schematic representation of nanoparticle classification.

1.3.1 Inorganic Nanomaterials

NMs that do not have carbon atoms are known as inorganic NMs. Typically, all the metal and metal oxide nanoparticles and silica nanoparticles (SiO_2 NPs) fall under the category of inorganic NMs.

Metal and metal oxide-based nanomaterials: Because of their distinctive features and potential, metal and metal oxide NPs are used widely in present days. In 1857, Michael Faraday first acknowledged the presence of the metal NPs (Nasrollahzadeh et al. 2019). Almost all metals can be used to convert to their nanoforms. NPs from noble metals like gold (Au), silver (Ag), and platinum (Pt) are mainly used in the fields of medicine, agriculture, and energy (Nasrollahzadeh et al. 2019). Some of the other metal NPs that are commonly used are copper (Cu), cadmium (Cd), and cobalt (Co). All these metal NPs exhibit higher reactivity and potency than their bulk forms (Tai et al. 2007). Researchers formed various types of metal oxide NPs from diverse ranges of metals. As these metal oxide NPs have large numbers of geometrical structures and diverse electronic structures, they can be used as conductors, semiconductors, or insulators. Among metal oxide NPs, titania (TiO_2) and zinc oxide (ZnO) NPs are widely used. Besides these two, aluminium oxide (Al_2O_3), iron oxide (Fe_2O_3), and cerium oxide (CeO_2) NPs are also used extensively (Nasrollahzadeh et al. 2019).

Silica nanomaterials: Silica or silicon dioxide (SiO_2) is one of the most abundant and complicated materials available in nature and can be found as a component in various minerals and synthetic compounds like fused-silica, silica-gel, fumed-silica, and aerogels. In almost all the silicates, the Si atom forms a tetrahedral orientation, where four oxygen atoms surround one Si atom in the centre. Silica NPs (SiO_2 NPs) exhibit primary importance because of their characteristic features (high surface area to volume ratios, easy modification of the surface, biocompatibility, and adjustable pore size) (Kim et al. 2015). By interchanging the surfactant during the synthesis of SiO_2 NPs, two different pore-sized NPs (mesoporous and nanoporous) can be obtained (Tang and Cheng 2013). These NPs are considered the non-toxic safest NPs for biomedical applications.

1.3.2 Carbon-based Nanomaterials

These NMs are formed solely with carbon atoms. Fullerene, carbon nanotubes (CNTs), and graphene are examples of carbon-based NMs. These particles have broad applications from biomedical to electronics and mechanics for their multifaceted shapes and properties like thermal and electronic conductivity (Ema et al. 2016, Sardoiwala et al. 2018). Schematic representations of different carbon-based NMs and their super-structured assemblies are provided in Figure 1.3.

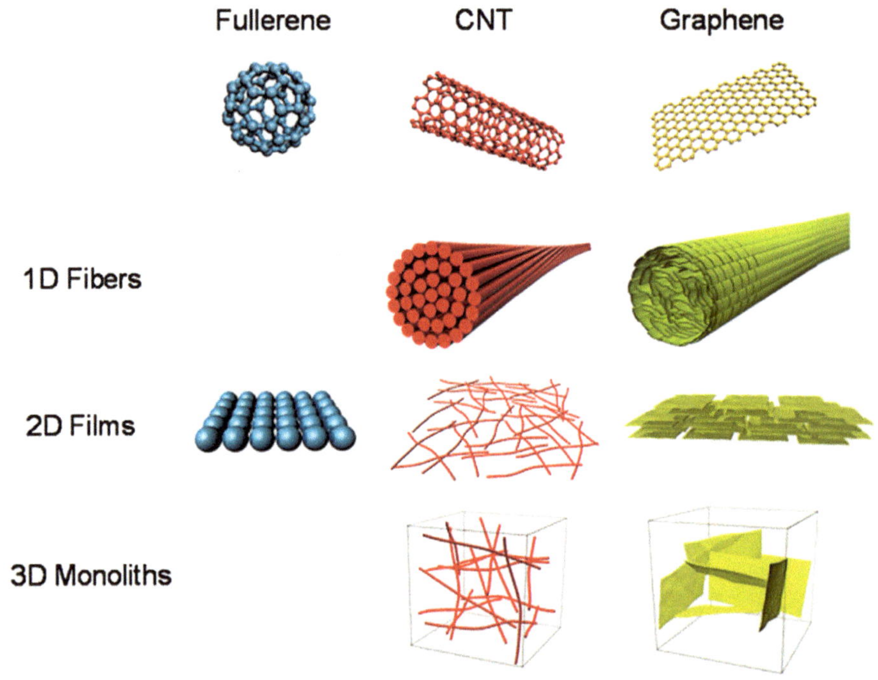

Figure 1.3 Superstructured assemblies of nanocarbons in the macroscopic scale "adapted with permission from Li, Z., Liu, Z., Sun, H. and Gao, C. 2015. Superstructured assembly of nanocarbons: fullerenes, nanotubes, and graphene. *Chemical reviews*, 115(15): 7046–7117. Copyright 2015, American Chemical Society".

1.3.2.1 Fullerene

In 1985, Kroto and Smalley discovered the first fullerene (C60) molecule; for this discovery, they were awarded the Nobel Prize in Chemistry in the same year (Kroto et al. 1985). C_{60} fullerene comprises 60 carbon atoms arranged into 12 pentagons connected with 20 hexagons and thus forms a hollow spherical structure resembling the soccer ball. Therefore, it is also called buckyball or buckminsterfullerene (Kroto et al. 1985, Bethune et al. 1993, Taylor and Walton, 1993). Though fullerenes can comprise of wide ranges of carbon atoms and symmetry ($C_n \geq 20$), C_{60} is the most stable and commonly found structure (Nasrollahzadeh et al. 2019). Fullerenes exhibit various unique properties, including electron transport capacity, electron acceptor and photosensitization, and other physicochemical properties. These allow them to be applied in catalysis, photovoltaics, and biological and medicinal fields (Li et al. 2015).

1.3.2.2 Carbon Nanotubes (CNTs)

The invention of CNTs in 1991 by Iijima paved a new path for nanobiotechnology research in the scientific society (Iijima, 1991). CNTs have 1D structure (cylindrical structure formed with rolled up graphite sheet) with a very high aspect ratio (length: diameter, 10^2–10^7). The size of the CNTs can vary up to several centimetres also. Depending on the number of walls, CNTs can be classified into three types: single-walled CNTs (SWCNTs), double-walled CNTs (DWCNTs), and multiwalled CNTs (MWCNTs) (Li et al. 2015). Because of the high modulus and tensile strength, the CNTs become very stiff. CNTs are used in several fields, including electronics, catalysis, medicine and therapeutics, sensors, data storage devices, etc. (Patel et al. 2019).

1.3.2.3 Graphene

Although the theoretical investigation of graphene NPs started in the 1940s, it was not practically used until the 1970s (Tiwari et al. 2018). Carbon atoms form sp^2-hybridization and are densely packed into a two-dimensional (2D) hexagonal lattice and assemble into a single planar layer of carbon atoms (Novoselov et al. 2004). Placing one carbon atom at each of the vertices of the hexagon makes graphene extremely transparent, strong, flexible, and a good conductor of electricity. It has six times higher electrical conductivity than copper, and its thermal conductivity is also quite high (Nasrollahzadeh et al. 2019). They are used extensively in aerospace industries and optoelectronic and electronic fields. They are also used in biomedical fields as biosensors, photothermal therapy, drug delivery, and bioimaging (Liao et al. 2018).

1.3.3 Polymer-based Nanomaterials

These NPs are harmless colloid particles having size ranges within 1000 nm. They can be spherical, elliptical, rod, or disc-shaped. Biodegradable synthetic polymers like poly-lactide-glycolide (PLGA) copolymers, poly caprolactones, and polyacrylates or natural polymers like collagen, alginate, albumin, gelatin, and chitosan are generally used to synthesize the polymer-based NPs (Wu et al. 2015, Ling et al. 2016, Xu et al. 2016). Based on the requirement of the NPs, either functional or non-function polymer groups can be used for the synthesis. Non-functional polymer groups evade the endosomal recognition system; hence, lysosomes cannot clear and degrade them inside the body (Rathore et al. 2019). On the other hand, charged polymer groups help in systemic drug delivery based on the response of the stimuli (Zhang et al. 2015). Because of their non-toxic nature, these NPs are mainly used in the medicinal field, especially as a career for drug delivery (DeFrates et al. 2018).

1.4 APPLICATIONS OF NANOTECHNOLOGY IN ENVIRONMENTAL MANAGEMENT

The present advent of nanotechnology results in its diverse application in different fields. Because of the unique properties of the nanoparticles, they can effectively use to remediate different environmental pollutants (Biswas et al. 2023). Nanotechnology, mainly green nanotechnology, is essential in defining the present field of environmental engineering and sciences. Hence, this green nanotechnology can be used to increase environmental sustainability, recommends using more environment-friendly nanoproducts instead of old existing products, and reduces the potential risk to human health and the environment from contamination and pollution by utilizing different nanoproducts (Bashir and Chisti 2017). Additional applications of nanotechnology in environmental management are discussed below in detail.

1.4.1 Production of Environmentally Sustainable Products

The application of nanotechnology to agricultural practices may be crucial in developing long-term answers to the global issues of food production, safety, and resource shortages (Pandey et al. 2023). Fertilizers have a crucial role in maintaining and improving the production of crops and fruits. But, it was observed that the use of chemical fertilizers in excess quantity finally affected the quality of the food and also degraded the soil condition (Sharma et al. 2021). It was observed that plants could use only 30 percent of the fertilizers applied, and the remaining parts were susceptible to mineralization, leaching, and bioconversions (Raliya et al. 2016). Nano fertilizers, on the other hand, are made up of nanoformulations and, when applied to plants, resulting in high uptake efficiency. Because of the small size of the NPs than their bulk counterparts, plants find it easy to absorb than bulk components or ionic salts. This phenomenon is a significant advantage of nano fertilizers (Raliya et al. 2017). Nano fertilizers can supply different nutrients to plants, and based on this, they can be classified into macro-nano fertilizers, micro-nano fertilizers, nanocoatings, nano biofertilizers, etc. (Sharma et al. 2021).

1.4.2 Nanosensors for the Detection of Pollutants

NMs are useful as biomarkers to monitor different toxic chemicals in the environment. This monitoring becomes possible because of their altered physicochemical properties. Nanosensors generally work in two ways—first, one is to identify pollutants by detecting the unique surface markers on the pollutants, secondly, they can increase the analytical signal. These features made them reliable in detecting even extremely low concentrations of contaminants when compared to the traditional techniques (Yadav et al. 2021). Nanobiosensors are mainly used to detect pathogens, pesticides, and heavy metals (Lukyanenko et al. 2019). Comparing nanobiosensors to traditional sensors, the former are much more sensitive and trustworthy in biosensing (Bhagat et al. 2023). Various types of NMs, like metallic, carbon-based, and quantum dots, prepare biosensors for detecting different environmental pollutants.

1.4.3 Remediation of environmental pollutants

Different hazardous ingredients (biological, physical, and chemical) are released daily into the environment through various industrial effluents and human activities (Manisalidis et al. 2020). These impart several harmful effects on living organisms, including humans. So, remediating these pollutants from the environment is of utmost necessity. Conventionally, physical, chemical, and biological techniques are involved in the remediation process of contaminants (Yadav et al. 2021). Adsorption, filtration, sedimentation, ion exchange, electro-dialysis flocculation, and coagulation are some of the physical methods used for remediation (Werkneh and Rene 2019). These processes are expensive, tedious, energy-consuming, less sensitive, and have other issues like membrane fouling (Mishra et al. 2019). Coagulation and oxidation are used in chemical remediation, but the chemicals used are costly, making the detection process also expensive (Sharma and Bhattacharya 2017). Bioremediation uses microbes primarily to degrade pollutants (Jacob et al. 2018). Other kinds of bioremediation include phytoremediation, bioaugmentation, biofilter, bio-sparging, biostimulation, land farming, etc. (Sharma 2020). All these techniques are time-consuming, expensive, less sensitive, inefficient, and complicated. To eradicate these problems, scientists invented new and more effective tools using nanotechnology to remediate contaminants in soil and water (Mandal et al. 2018, Ingle et al. 2020, Mukhopadhyay et al. 2021, Rodriguez-Couto et al. 2021). Recent study has shown that nanoparticles in their many forms offer an exceptional opportunity for the removal of different pollutants, including heavy metals, as compared to their bulk equivalents, therefore becoming a subject of interest for researchers in numerous other areas (Inobeme et al. 2023). How different types of NMs are used as adsorbents or catalysts to remediate and filtrate techniques are described below in detail. NMs and NPs can also act as antimicrobial agents and helps protect plants (Rai et al. 2018a). Different approaches involved in the nano bioremediation process are mentioned in Figure 1.4.

1.4.3.1 Nanoadsorbent-mediated Pollutant Remediation

Nanoadsorbents show better adsorption efficiency using active materials (Kyzas and Matis 2015). Nanoadsorbents can comprise a metal oxide, silica, CNTs, clay, etc., in the nanocomposite conformation (Anjum et al. 2019). The potency of the nanoadsorbents relies on several determinants like their size, surface properties, functionalization group, intrinsic composition etc. (Biswas et al. 2022). Different factors like incubation period, temperature, pH, and concentrations also affect the adsorption of the pollutants on the adsorbent (Anjum et al. 2019). CNTs, graphene, TiO_2, ZnO, MnO, MgO, and Fe_2O_3 NPs are mainly used to prepare nano adsorbents (Osagie et al. 2021). Inherent surface and circumferential functionalization are two main factors of nano adsorbents. Configuration of the external surface, outer size, and internal structure determine their material, physical, and chemical properties (Anjum et al. 2019). Factors affecting the adsorption mechanism in an aqueous environment are chemical and adsorption activity, high surface area and surface binding energy, positions of atoms on the surface, and deficiency of internal diffusion resistance (Khajeh et al. 2013). These nano-based procedures are cost-effective and more efficient in the removal of pollutants from aquatic environments and hence, in turn, improve human health and the surrounding environment (Yadav et al. 2021).

1.4.3.2 Nanocatalysis-dependent Pollutant Remediation

Inorganic materials like metal oxide and semiconductors are used as nanocatalysts, the significant component of nanocatalysis. Different nanocatalysts like electrocatalysts, photocatalysts, and Fenton-based catalysts are used to treat wastewater (Dutta et al. 2014). Fenton-based catalysts have antimicrobial properties and enhance the redox reaction for organic contaminants (Ma et al. 2013). Light energy mediates the photocatalytic reaction of metallic NPs, further degrading organic pollutants (Gupta et al. 2020). To degrade detergents, dyes, volatile organic pollutants, and pesticides, semiconductor NPs act as the photocatalyst (Lin et al. 2014). These NPs can also degrade halogenated and non-halogenated organic pollutants (Anjum et al. 2019). Light induces the excitation of the electron in the catalysts, which further results in photocatalysis. TiO_2 NPs are the most used photocatalyst because of their unique consistency and chemical reactivity. TiO_2 NPs can also eliminate pathogenic bacteria from wastewater (Anjum et al. 2019). ZnO NPs are also one strong photocatalyst. Factors like pH, band gap energy, pollutant doses and concentrations, and particle size modulate the catalysis process (Yadav et al. 2021). Researchers sometimes modify

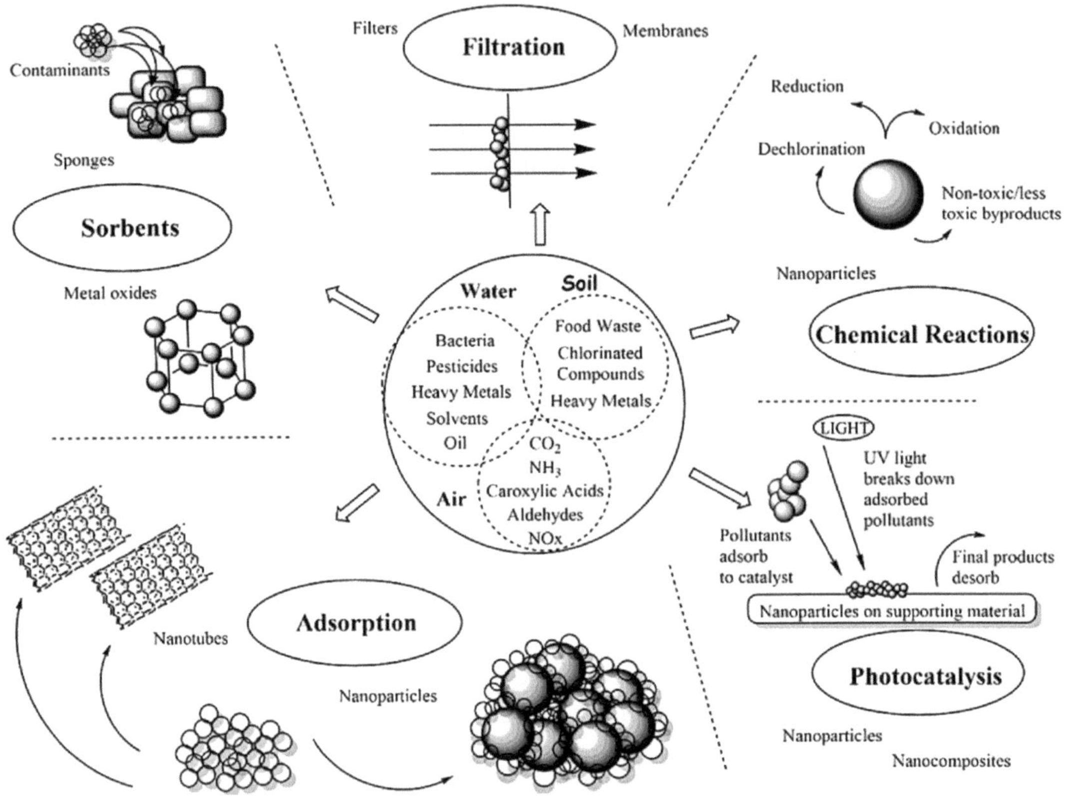

Figure 1.4 Environmental remediation approaches. "adapted from Guerra, F.D., Attia, M.F., Whitehead, D.C. and Alexis, F. 2018. Nanotechnology for environmental remediation: materials and applications. *Molecules*, *23*(7): 1760."

nanocatalysts by doping them with metal impurities, forming hybrid NPs or composites using low band gap semiconductors or anions, or sensitizing them with dye (Anjum et al. 2019). Moreover, to the present date, pathogens are found to produce no resistance against NPs because of the arbitrary nature and various modes of action of the NPs to the microorganisms (Ingle et al. 2022).

1.4.3.3 Remediation of Pollutants by Nanomembrane-based Filtration

The traditional wastewater treatment processes cannot remove the remarkably low concentrations of emerging pollutants present in the wastewater. CNTs can act as membranes, adsorbents, and catalysts and can be used effectively in wastewater treatment (Sarkar et al. 2019). CNTs, nano-fibres, and nano-ribbons are used to synthesize nanomembranes. These nanomembranes can also remove organic pollutants, heavy metals, and dyes from wastewater. Researchers have created various nanomembranes like hybrid nanomembranes, electrospun nanofiber membranes, and carbon nanotube membranes to filter polluted wastewater (Anjum et al. 2019). The usage of nanomembranes has several benefits, like they have better permeability, high sensitivity, and catalytic activity, and are resistant to fouling. Furthermore, water treatment with nanomembranes requires a small space, is cost-effective, and is simple (Guo et al. 2016). In addition, these nanomembranes can also be used to remove heavy metals like chromium, cadmium, copper, and nickel (Anjum et al. 2019). A few NPs, like AgNPs, TiO_2 NPs, and ZnO NPs, are also used to disinfect water from pathogenic microorganisms (Werkneh and Rene 2019).

1.4.4 Renewable Energy Harvesting

To date, fossil fuels (mainly coal and oil) are the primary and most accessible energy source. But, burning fossil fuels releases different types of greenhouse gases (CO_2, SO_2 NO_x) and carbons into the environment (Ahmadi et al. 2019). It is necessary to find other energy sources to prevent environmental pollution from these greenhouse gases (Whitesides 2001). Nanotechnology generates different renewable energies from solar, ocean, and wind. These energies are known as green energy because they allow society to decrease the dependency on petroleum products and thus help solve environmental problems related to the usage of fossil fuels (Khan et al. 2005). How nanotechnology is used to harvest these renewable energies is described below.

Nanotechnology in solar energy: Active solar systems convert solar energy into other forms of usable energy (Mohammadi et al. 2018). These are used for different energy by energy conversion in buildings like cooling, heating, etc. Photocatalysts, solar photovoltaics, solar collectors, and dye-sensitized solar cells (DSSC) are some active solar systems used. Solar collectors and solar photovoltaics were modified using nanofluids (NPs suspended in the carrier fluid) to enhance their performance, as nanofluids show increased thermal conductivity than pure fluids. Though TiO_2 NPs are mainly used as photocatalysts, they are also used in DSSCs (Ahmadi et al. 2019).

1.4.4.1 Nanotechnology in Ocean Energy

As a significant portion of the world consists of water, several locations use the power of waves to generate electricity. Supplementary aids like wave turbines, tidal current turbines, and tidal barrages are some of the most needed devices to generate electricity from waves. These technologies can be further combined with other less advanced technologies depending on salinity gradient energy. Corrosion is a crucial concern in ocean energy, and hence it requires particular focus to produce protective covering material, making it cost-effective (Serrano et al. 2009). Triboelectric nanogenerators (TENGs) are generally used to collect mechanical energies from the environment. The central concept of TENGs is based on contact electrification within two materials, and the electrodes exchange charge because of the electrostatic induction (Wu et al. 2019). The erosion layers are mainly composed of polymer and metal pairs. They are inexpensive, lightweight, easy to make, and come in various materials. Furthermore, TENGs have an energy conversion efficiency of up to 55%. TENGs can collect energy across multiple frequencies, including human movement and ocean waves (Ahmadi et al. 2019).

1.4.4.2 Nanotechnology in Wind Energy

Wind energy can be transformed into a source of beneficial energy. For example, wind energy can be converted into electricity by wind turbines, mechanical power by windmills, and water or drainage pumping by wind pumps. Nanotechnology can be utilized to create coatings, lubricants, and lightweight materials with good durability for application in wind energy transformation. Furthermore, nanotechnology-based sensing materials can monitor stability or potential hazards. A nanocomposite coating for wind turbine blades has been suggested to increase blade durability and guard against sand, rain, and lightning strikes. Another feature of the wind turbine that involves nanotechnology is gearbox lubrication. Higher wear resistance in the gearbox has been possible by utilizing an additive such as nano-colloidal boron nitride (Ahmadi et al. 2019).

1.5 ECOTOXICITY AND ENVIRONMENTAL RISK

Despite the substantial rise in the usage of NMs, there has been little discussion regarding their potential toxicity to the environment and human health. These NPs are anthropogenic and cannot be found in natural environments. Hence, the organisms do not possess any suitable defensive mechanisms to tackle these nano-intruders (Biswas et al. 2018). The intrinsic features of the NPs, such as size, surface area, and agglomeration/ dispersion ability, make them fascinating from the viewpoint of technological applicability and may be disadvantageous if released into the environment. These features of the NPs can expedite their translocation through the ecosystem and induce harm to the food web (Almeida et al. 2020). Due to these constraints, the terminology ‘nanotoxicology’ was coined to describe the field of study examining how nanodevices and NMs affect living beings (Oberdörster et al. 2005). The unique properties of the NPs are frequently associated with increased biological reactivity. The large surface area could also facilitate the generation of free radicals like superoxide anion or hydroxyl radicals. Because of this, oxidative stress may be a significant factor in NPs toxicity, particularly for metal-based NPs (Mosa et al. 2018). It was also observed that some NPs can show toxicity to the exposed organisms even at very low exposure concentrations. Hence, it is necessary to investigate in detail the factors responsible for determining the toxic potential of the NPs (Rai et al. 2018b).

Information on the relationship between dose-response and exposure to humans and the environment will be essential to evaluate the hazards of NMs following inhalation, ingestion, or topical exposure modes. The significance of dose metric, dose rate, dosage, and biokinetics are all critical parameters for assessing the safety of newly engineered NPs.

Inhalation is one of the major entry points of the NPs inside the body. It was observed that lung inflammation was induced by exposure to NPs. NPs had been reported to be distributed throughout the lymphatic and bloodstream system. The other important entry route of the NPs is skin due to accidental contact, use of cosmetics, and topical treatments. The stratum corneum, the outermost part of the epidermis, defends against almost all environmental disturbances, but TiO_2 NPs were found to penetrate this part and hair follicles. They can enter the body (Mosa et al. 2018).

For all these reasons, it is necessary to consider the factors like biodegradability, non-toxicity, recyclability, facile synthesis, green chemistry, and cost-effectiveness while synthesizing new NMs for remediating different environmental pollutants. Further additional steps are required to prevent agglomeration, improve monodispersity, and boost stability (Guerra et al. 2018). It is also essential to deal with the critical concerns of ecotoxicity and the consequences of NPs on the environment and living beings, their vulnerability, threats, etc. (Rai and Biswas 2018).

1.6 AVENUES AHEAD

Creating novel nanotechnology-based tools and materials is imperative for identifying and removing all environmental pollutants. Nanoscale instruments and gadgets should have exceptional sensitivity, consistency, economy, and environmental friendliness. They should also use the least amount of energy and chemicals possible. For superior analytical performances compared to traditional techniques, currently, accessible nanodevices need additional modifications or enhancements. These devices must be simple, reliable, easy to use, portable, and able to perform in challenging environmental circumstances. Environmental nano techniques must also be able to operate constantly without interference and with a minimum amount of technician supervision. Systems for identifying emerging pollutants make it simple enough to gather specimens from the air, water, and soil environment. Since most nano instruments and gadgets are now employed in laboratory settings, they must be commercialized fast for hazardous environmental components to be identified and destroyed on-site to prevent adverse effects on biological systems.

1.7 CONCLUSIONS

Due to the ever-increasing human population growth and industrialization, the environment is polluted with numerous pollutants. If not appropriately treated, these pollutants would cause severe effects on the ecosystem and living beings. The present chapter dealt in detail with different environmental pollutants and their sources. Along with this, a detailed discussion was done on how nanotechnology can help environmental management. Though nanotechnology helps in remediating various contaminants, produces different environment-sustainable products, acts as a biomonitor, and helps harvest renewable energy sources, the NPs used in the process might become harmful to the environment because of their unique physicochemical characteristics. So, performing a risk assessment for the NPs used is necessary. This risk assessment will help us develop more bio-friendly NPs, reducing the burden of environmental pollution and providing a cleaner environment.

References

Adnan, M., Muhd, Julkapli, N., Amir, M.N.I. and Maamor, A. 2019. Effect on different TiO_2 photocatalyst supports on photodecolorization of synthetic dyes: a review. IJEST. 16(1): 547–566.

Ahmadi, M.H., Ghazvini, M., Alhuyi Nazari, M., Ahmadi, M.A., Pourfayaz, F., Lorenzini, G. and Ming, T. 2019. Renewable energy harvesting with the application of nanotechnology: a review. Int. J. Energy Res. 43(4): 1387–1410.

Ali, H. and Khan, E. 2018. What are heavy metals? Long–standing controversy over the scientific use of the term ‘heavy metals’—proposal of a comprehensive definition. Toxicol Environ. Chem. 100(1): 6–19.

Almeida, L., Felzenszwalb, I., Marques, M. and Cruz, C. 2020. Nanotechnology activities: environmental protection regulatory issues data. Heliyon. 6(10): e05303.

Anjum, M., Miandad, R., Waqas, M., Gehany, F. and Barakat, M.A. 2019. Remediation of wastewater using various nanomaterials. Arab. J. Chem. 12(8): 4897–4919.

Atieh, M.A., Ji, Y. and Kochkodan, V. 2017. Metals in the environment: toxic metals removal. Bioinorg. Chem. Appl. 2017: 4309198

Bashir, R. and Chisti, H. 2017. Nanotechnology for environmental control and remediation. pp. 1217–1238. *In*: Materials Science and Engineering: Concepts, Methodologies, Tools, and Applications. Information Resources Management Association. IGI Global.

Bekele, T.G., Zhao, H., Wang, Q. and Chen, J. 2019. Bioaccumulation and trophic transfer of emerging organophosphate flame retardants in the marine food webs of Laizhou Bay, North China. Environ. Sci. Technol. 53(22): 13417–13426.

Bethune, D.S., Johnson, R.D., Salem, J.R., De Vries, M.S. and Yannoni, C.S. 1993. Atoms in carbon cages: the structure and properties of endohedral fullerenes. Nature. 366(6451): 123–128.

Bhagat, R., Ingle, A.P. and Chen, H. 2023. Nanosensors and nanobiosensors for sustainable agriculture. pp. 93–112. *In*: Ingle, A.P. (ed.). Nanotechnology in Agriculture and Agroecosystems. Cover for Nanotechnology in Agriculture and Agroecosystems. Elsevier.

Biswas, J.K., Rai, M., Ingle, A.P., Mondal, M. and Biswas, S. 2018. Nano–bio interactions and ecotoxicity in aquatic environment: plenty of room at the bottom but tyranny at the top!. pp. 19–36. *In*: Rai, M. and Biswas, J. (eds). Nanomaterials: Ecotoxicity, Safety, and Public Perception. Springer. Cham.

Biswas, J.K. and Sarkar, D. 2019. Nanopollution in the aquatic environment and ecotoxicity: no nano issue!. Curr. Pollut. Rep. 5(1): 4–7.

Biswas, J.K., Mitra, A., Shah, M.P. and Rodriguez–Couto, S. 2022. Nanoadsorbents for scavenging emerging contaminants from wastewater. pp. 1–22. *In*: Rodriguez-Couto, S., Shah, M.P. and Biswas, J.K. (eds). Development in Wastewater Treatment Research and Processes: Removal of Emerging Contaminants from Wastewater through Bio-nanotechnology, 1st Ed.. Amsterdam: Elsevier.

Biswas, P., Polash, S.A., Dey, D., Kaium, M.A., Mahmud, A.R., Yasmin, F., Baral, S.K., et al. 2023. Advanced implications of nanotechnology in disease control and environmental perspectives. Biomed. Pharmacother. 158: 114172.

Blumer, M. 1976. Polycyclic aromatic compounds in nature. Sci. Am. 234(3): 34–45.

Carey, R.O., Hochmuth, G.J., Martinez, C.J., Boyer, T.H., Dukes, M.D., Toor, G.S. and Cisar, J.L. 2013. Evaluating nutrient impacts in urban watersheds: Challenges and research opportunities. Environ. Pollut. 173: 138–149.

Chen, M., Xu, P., Zeng, G., Yang, C., Huang, D. and Zhang, J. 2015. Bioremediation of soils contaminated with polycyclic aromatic hydrocarbons, petroleum, pesticides, chlorophenols and heavy metals by composting: applications, microbes and future research needs. Biotechnol. Adv. 33(6): 745–755.

DeFrates, K., Markiewicz, T., Gallo, P., Rack, A., Weyhmiller, A., Jarmusik, B. and Hu, X. 2018. Protein polymer–based nanoparticles: Fabrication and medical applications. Int. J. Mol. Sci. 19(6): 1717.

Dutta, A.K., Maji, S.K. and Adhikary, B. 2014. γ–Fe_2O_3 nanoparticles: an easily recoverable effective photo–catalyst for the degradation of rose bengal and methylene blue dyes in the waste–water treatment plant. Mater. Res. Bull. 49: 28–34.

Eaton, P., Quaresma, P., Soares, C., Neves, C., De Almeida, M.P., Pereira, E. and West, P. 2017. A direct comparison of experimental methods to measure dimensions of synthetic nanoparticles. Ultramicroscopy. 182: 179–190.

Ema, M., Hougaard, K.S., Kishimoto, A. and Honda, K. 2016. Reproductive and developmental toxicity of carbon–based nanomaterials: a literature review. Nanotoxicology. 10(4): 391–412.

Gangwar, R.K., Tomar, G.B., Dhumale, V.A., Zinjarde, S., Sharma, R.B. and Datar, S. 2013. Curcumin conjugated silica nanoparticles for improving bioavailability and its anticancer applications. J. Agric. Food Chem. 61(40): 9632–9637.

Grizzetti, B., Pistocchi, A., Liquete, C., Udias, A., Bouraoui, F. and Van De Bund, W. 2017. Human pressures and ecological status of European rivers. Sci. Rep. 7(1): 1–11.

Guerra, F.D., Attia, M.F., Whitehead, D.C. and Alexis, F. 2018. Nanotechnology for environmental remediation: materials and applications. Molecules, 23(7): 1760.

Guo, J., Zhang, Q., Cai, Z. and Zhao, K. 2016. Preparation and dye filtration property of electrospun polyhydroxybutyrate–calcium alginate/carbon nanotubes composite nanofibrous filtration membrane. Sep. Purif. 161: 69–79.

Gupta, N.K., Ghaffari, Y., Kim, S., Bae, J., Kim, K.S. and Saifuddin, M. 2020. Photocatalytic degradation of organic pollutants over MFe_2O_4 (M=Co, Ni, Cu, Zn) nanoparticles at neutral pH. Sci. Rep. 10(1): 4942.

Hulla, J.E., Sahu, S.C. and Hayes, A.W. 2015. Nanotechnology: history and future. Hum. Exp. Toxicol. 34(12): 1318–1321.

Iijima, S. 1991. Helical microtubules of graphitic carbon. Nature. 354(6348): 56–58.

Ingle, A.P., Wagh, S., Biswas, J., Mondal, M., Feitosa, C.M. and Rai, M. 2020. Phyto–fabrication of different nanoparticles and evaluation of their antibacterial and anti–biofilm efficacy. Curr. Nanosci. 16(6): 1002–1015.

Ingle, P.U., Biswas, J.K., Mondal, M., Rai, M.K., Kumar, P.S. and Gade, A.K. 2022. Assessment of in vitro antimicrobial efficacy of biologically synthesized metal nanoparticles against pathogenic bacteria. Chemosphere. 291: 132676.

Inobeme, A., Mathew, J.T., Adetunji, C.O., Ajai, A.I., Inobeme, J., Maliki, M., Okonkwo, S., et al. 2023. Recent advances in nanotechnology for remediation of heavy metals. Environ. Monit. Assess. 195(1): 1–24.

Islam, M.M., Karim, M.R., Zheng, X. and Li, X. 2018. Heavy metal and metalloid pollution of soil, water and foods in bangladesh: a critical review. Int. J. Environ. Res. Public Health. 15(12): 2825.

Jacob, J.M., Karthik, C., Saratale, R.G., Kumar, S.S., Prabakar, D., Kadirvelu, K. and Pugazhendhi, A. 2018. Biological approaches to tackle heavy metal pollution: a survey of literature. J. Environ. Manage. 217: 56–70.

Jawad, N.K. and Hamooshy, E.A. 2021. Nanotechnology and environment study. Earthline J. Chem. Sci. 6(2): 249–260.

Jusoh, N.W., Jalil, A.A., Triwahyono, S., Setiabudi, H.D., Sapawe, N., Satar, M.A., Karim, A.H., et al. 2013. Sequential desilication–isomorphous substitution route to prepare mesostructured silica nanoparticles loaded with ZnO and their photocatalytic activity. Appl. Catal. A: Gen. 468: 276–287.

Khajeh, M., Laurent, S. and Dastafkan, K. 2013. Nanoadsorbents: classification, preparation, and applications (with emphasis on aqueous media). Chem. Rev. 113(10): 7728–7768.

Khan, F.I., Hawboldt, K. and Iqbal, M.T. 2005. Life cycle analysis of wind–fuel cell integrated system. Renew. Energ. 30(2): 157–177.

Khan, I., Saeed, K. and Khan, I. 2019. Nanoparticles: Properties, applications and toxicities. Arab. J. Chem. 12(7): 908–931.

Kim, I.Y., Joachim, E., Choi, H. and Kim, K. 2015. Toxicity of silica nanoparticles depends on size, dose, and cell type. Nanomed. Nanotechnol. Biol. Med. 11(6): 1407–1416.

Kroto, H.W., Heath, J.R., O'Brien, S.C., Curl, R.F. and Smalley, R.E. 1985. C60: Buckminsterfullerene. Nature. 318(6042): 162–163.

Kyzas, G.Z. and Matis, K.A. 2015. Nanoadsorbents for pollutants removal: a review. J. Mol. Liq. 203: 159–168.

Lai, X., Guo, R., Xiao, H., Lan, J., Jiang, S., Cui, C. and Ren, E. 2019. Rapid microwave–assisted bio–synthesized silver/Dandelion catalyst with superior catalytic performance for dyes degradation. J. Hazard. Mater. 371: 506–512.

Larsson, M., Lam, M.M., van Hees, P., Giesy, J.P. and Engwall, M. 2018. Occurrence and leachability of polycyclic aromatic compounds in contaminated soils: chemical and bioanalytical characterization. Sci. Total Environ. 622: 1476–1484.

León, O., Muñoz–Bonilla, A., Soto, D., Pérez, D., Rangel, M., Colina, M. and Fernández–García, M. 2018. Removal of anionic and cationic dyes with bioadsorbent oxidized chitosans. Carbohydr. Polym. 194: 375–383.

Li, X., Zhu, J. and Li, H. 2012. Comparative study on the mechanism in photocatalytic degradation of different–type organic dyes on SnS_2 and CdS. Appl. Catal. B: Environ. 123: 174–181.

Li, Z., Liu, Z., Sun, H. and Gao, C. 2015. Superstructured assembly of nanocarbons: fullerenes, nanotubes, and graphene. Chem. Rev. 115(15): 7046–7117.

Liang, L., Xi, F., Tan, W., Meng, X., Hu, B. and Wang, X. 2021. Review of organic and inorganic pollutants removal by biochar and biochar–based composites. Biochar. 3: 255–281.

Liao, C., Li, Y. and Tjong, S.C. 2018. Graphene nanomaterials: synthesis, biocompatibility, and cytotoxicity. Int. J. Mol. Sci. 19(11): 3564.

Lin, S.T., Thirumavalavan, M., Jiang, T.Y. and Lee, J.F. 2014. Synthesis of ZnO/Zn nano photocatalyst using modified polysaccharides for photodegradation of dyes. Carbohydr. Polym. 105: 1–9.

Ling, X., Huang, Z., Wang, J., Xie, J., Feng, M., Chen, Y., Abbas, F., et al. 2016. Development of an itraconazole encapsulated polymeric nanoparticle platform for effective antifungal therapy. J. Mater. Chem. B. 4(10): 1787–1796.

Lintern, A., McPhillips, L., Winfrey, B., Duncan, J. and Grady, C. 2020. Best management practices for diffuse nutrient pollution: wicked problems across urban and agricultural watersheds. Environ. Sci. Technol. 54(15): 9159–9174.

Lukyanenko, K.A., Denisov, I.A., Sorokin, V.V., Yakimov, A.S., Esimbekova, E.N. and Belobrov, P.I. 2019. Handheld enzymatic luminescent biosensor for rapid detection of heavy metals in water samples. Chemosensors. 7(1): 16.

Ma, X., Gurung, A. and Deng, Y. 2013. Phytotoxicity and uptake of nanoscale zero–valent iron (nZVI) by two plant species. Sci. Total Environ. 443: 844–849.

Mandal, S., Sarkar, B., Mukhopadhyay, R., Biswas, J.K. and Manjaiah, K.M. 2018. Microparticle–supported nanocomposites for safe environmental applications. pp. 305–317. *In*: Rai, M. and Biswas, J.K. (eds). Nanomaterials: Ecotoxicity, Safety, and Public Perception. Springer, Cham.

Manisalidis, I., Stavropoulou, E., Stavropoulos, A. and Bezirtzoglou, E. 2020. Environmental and health impacts of air pollution: a review. Front. Public Health. 14.

Mishra, S., Chowdhary, P. and Bharagava, R.N. 2019. Conventional methods for the removal of industrial pollutants, their merits and demerits. pp. 1–31. *In*: Emerging and Eco–Friendly Approaches for Waste Management. Springer, Singapore.

Mohammadi, A., Ahmadi, M.H., Bidi, M., Ghazvini, M. and Ming, T. 2018. Exergy and economic analyses of replacing feedwater heaters in a Rankine cycle with parabolic trough collectors. Energy Rep. 4: 243–251.

Mosa, K.A., El–Naggar, M., Ramamoorthy, K., Alawadhi, H., Elnaggar, A., Wartanian, S., Ibrahim, E., et al. 2018. Copper nanoparticles induced genotoxicty, oxidative stress, and changes in superoxide dismutase (SOD) gene expression in cucumber (*Cucumis sativus*) plants. Front. Plant Sci. 9: 872.

Mukhopadhyay, R., Bhaduri, D., Sarkar, B., Rusmin, R., Hou, D., Khanam, R., Sarkar, S., et al. 2021. Clay–polymer nanocomposites: progress and challenges for use in sustainable water treatment. J. Hazard. Mater. 383: 121125.

Naidu, R., Espana, V.A.A., Liu, Y. and Jit, J. 2016. Emerging contaminants in the environment: risk–based analysis for better management. Chemosphere. 154: 350–357.

Nasrollahzadeh, M., Sajadi, S.M., Sajjadi, M. and Issaabadi, Z. 2019. An introduction to nanotechnology. pp. 1–27. *In*: Arthur T. Hubbard (ed.). Interface Science and Technology, Vol. 28. New York: Academic Press Inc.

Naughton, S.X. and Terry Jr., A.V. 2018. Neurotoxicity in acute and repeated organophosphate exposure. Toxicology. 408: 101–112.

Novoselov, K.S., Geim, A.K., Morozov, S.V., Jiang, D.E., Zhang, Y., Dubonos, S.V., Grigorieva, I.V., et al. 2004. Electric field effect in atomically thin carbon films. Science. 306(5696): 666–669.

Oberdörster, G., Oberdörster, E. and Oberdörster, J. 2005. Nanotoxicology: an emerging discipline evolving from studies of ultrafine particles. Environ. Health Perspect. 113(7): 823–839.

Osagie, C., Othmani, A., Ghosh, S., Malloum, A., Esfahani, Z.K. and Ahmadi, S. 2021. Dyes adsorption from aqueous media through the nanotechnology: A review. J. Mater. Res. Technol. 14:2195–2218.

Pandey, G. 2018. Challenges and future prospects of agri–nanotechnology for sustainable agriculture in India. Environ. Technol. Innov. 11: 299–307.

Pandey, G., Tripathi, S., Bajpai, S. and Kamboj, M. 2023. Approaches, challenges, and prospects of nanotechnology for sustainable agriculture. pp. 83–103. *In*: Fernandez-Luqueno, F. and Patra, J.K. (eds). Agricultural and Environmental Nanotechnology: Novel Technologies and their Ecological Impact. Singapore: Springer.

Patel, K.D., Singh, R.K. and Kim, H.W. 2019. Carbon–based nanomaterials as an emerging platform for theranostics. Mater. Horiz. 6(3): 434–469.

Pavithra, K.G. and Jaikumar, V.J.J.O.I. 2019. Removal of colorants from wastewater: A review on sources and treatment strategies. J. Ind. Eng. Chem. 75P: 1–19.

Rai, M. and Biswas, J.K. (eds). 2018. Nanomaterials: Ecotoxicity, Safety, and Public Perception. Cham, Switzerland: Springer International Publishing.

Rai, M., Gupta, I., Ingle, A. P., Biswas, J. K. and Sinitsyna, O.V. 2018a. Nanomaterials: what are they, why they cause ecotoxicity, and how this can be dealt with? pp. 3–18. *In*: Rai, M. and Biswas, J.K. (eds). Nanomaterials: Ecotoxicity, Safety, and Public Perception. Springer. Cham.

Rai, M., Ingle, A.P., Pandit, R., Paralikar, P., Shende, S., Gupta, I., Biswas, J.K., et al. 2018b. Copper and copper nanoparticles: Role in management of insect–pests and pathogenic microbes. Nanotechnol. Rev. 7(4): 303–315.

Raliya, R., Franke, C., Chavalmane, S., Nair, R., Reed, N. and Biswas, P. 2016. Quantitative understanding of nanoparticle uptake in watermelon plants. Front. Plant Sci. 7: 1288.

Raliya, R., Saharan, V., Dimkpa, C. and Biswas, P. 2017. Nanofertilizer for precision and sustainable agriculture: current state and future perspectives. J. Agric. Food Chem. 66(26): 6487–6503.

Rathore, B., Sunwoo, K., Jangili, P., Kim, J., Kim, J.H., Huang, M., Xiong, J., et al. 2019. Nanomaterial designing strategies related to cell lysosome and their biomedical applications: A review. Biomaterials 211: 25–47.

Rathi, B.S., Kumar, P.S., Ponprasath, R., Rohan, K. and Jahnavi, N. 2021. An effective separation of toxic arsenic from aquatic environment using electrochemical ion exchange process. J. Hazard. Mater. 412: 125240.

Razack, S.A. and Duraiarasan, S. 2020. A way to create a sustainable environment: Green nanotechnology–with an emphasis on noble metals. pp. 359–425. *In*: Chaudhery, M.H. (ed.). The ELSI Handbook of Nanotechnology: Risk, Safety, ELSI and Commercialization. Wiley.

Rego, R.M., Kuriya, G., Kurkuri, M.D. and Kigga, M. 2021. MOF based engineered materials in water remediation: Recent trends. J. Hazard. Mater. 403: 123605.

Roberts, J., Kumar, A., Du, J., Hepplewhite, C., Ellis, D.J., Christy, A.G. and Beavis, S.G. 2016. Pharmaceuticals and personal care products (PPCPs) in Australia's largest inland sewage treatment plant, and its contribution to a major Australian river during high and low flow. Sci. Total Environ. 541: 1625–1637.

Rodriguez–Couto, S., Shah, M.P. and Biswas, J.K. (eds). 2021. Development in Wastewater Treatment Research and Processes: Removal of Emerging Contaminants from Wastewater through Bio–nanotechnology. Elsevier.

Rodriguez–Narvaez, O.M., Peralta–Hernandez, J.M., Goonetilleke, A. and Bandala, E.R. 2017. Treatment technologies for emerging contaminants in water: A review. Chem. Eng. J. 323, 361–380.

Rout, P.R., Zhang, T.C., Bhunia, P. and Surampalli, R.Y. 2021. Treatment technologies for emerging contaminants in wastewater treatment plants: a review. Sci. Total Environ. 753: 141990.

Sapsford, K.E., Algar, W.R., Berti, L., Gemmill, K.B., Casey, B.J., Oh, E., Stewart, M.H., et al. 2013. Functionalizing nanoparticles with biological molecules: developing chemistries that facilitate nanotechnology. Chem. Rev. 113(3): 1904–2074.

Sardoiwala, M.N., Kaundal, B. and Choudhury, S.R. 2018. Development of engineered nanoparticles expediting diagnostic and therapeutic applications across blood–brain barrier. pp. 696–709. *In*: Chaudhery, M.H. (ed.). Handbook of Nanomaterials for Industrial Applications: A volume in Micro and Nano Technologies. Elsevier.

Sarkar, B., Mandal, S., Tsang, Y.F., Vithanage, M., Biswas, J.K., Yi, H., Dou, X., et al. 2019. Sustainable sludge management by removing emerging contaminants from urban wastewater using carbon nanotubes. pp. 553–571. *In*: Prasad, M.N.V., de Campos Favas, P.J., Vithanage, M. and Venkata Mohan, S. (eds). Industrial and Municipal Sludge: Emerging Concerns and Scope for Resource Recovery. Butterworth–Heinemann.

Serrano, E., Rus, G. and Garcia–Martinez, J. 2009. Nanotechnology for sustainable energy. Renewable and Sustainable Energy Rev. 13(9): 2373–2384.

Shaikh, W.A., Kumar, A., Chakraborty, S., Islam, R.U., Bhattacharya, T. and Biswas, J.K. 2022. Biochar–based nanocomposite from waste tea leaf for toxic dye removal: from facile fabrication to functional fitness. Chemosphere. 291: 132788.

Sharma, S. and Bhattacharya, A. 2017. Drinking water contamination and treatment techniques. Appl. Water Sci. 7(3): 1043–1067.

Sharma, I. 2020. Bioremediation techniques for polluted environment: concept, advantages, limitations, and prospects. *In*: Mario Alfonso Murillo-Tovar, Hugo Saldarriaga-Noreña and Agnieszka Saeid (eds). Trace Metals in the Environment–New Approaches and Recent Advances. IntechOpen.

Sharma, S., Rana, V.S., Pawar, R., Lakra, J. and Racchapannavar, V. 2021. Nanofertilizers for sustainable fruit production: a review. Environ. Chem. Lett. 19: 1693–1714.

Silva, M. and Baltrusaitis, J. 2021. Destruction of emerging organophosphate contaminants in wastewater using the heterogeneous iron–based photo–Fenton–like process. J. Hazard. Mater. Lett. 2: 100012.

Stasinska, A., Heyworth, J., Reid, A., Callan, A., Odland, J.Ø., Duong, P.T., Ho, Q.V., et al. 2014. Polybrominated diphenyl ether (PBDE) concentrations in plasma of pregnant women from Western Australia. Sci. Total Environ. 493: 554–561.

Subramani, K., Elhissi, A., Subbiah, U. and Ahmed, W. 2019. Introduction to nanotechnology. pp. 3–18. *In*: Nanobiomaterials in Clinical Dentistry. Elsevier.

Tai, C.Y., Tai, C.T., Chang, M.H. and Liu, H.S. 2007. Synthesis of magnesium hydroxide and oxide nanoparticles using a spinning disk reactor. Ind. Eng. Chem. Res. 46(17): 5536–5541.

Tang, L. and Cheng, J. 2013. Nonporous silica nanoparticles for nanomedicine application. Nano Today. 8(3): 290–312.

Taylor, R. and Walton, D.R. 1993. The chemistry of fullerenes. Nature. 363(6431): 685–693.

Tijani, J.O., Fatoba, O.O., Babajide, O.O. and Petrik, L.F. 2016. Pharmaceuticals, endocrine disruptors, personal care products, nanomaterials and perfluorinated pollutants: a review. Environ. Chem. Lett. 14: 27–49.

Tiwari, S.K., Mishra, R.K., Ha, S.K. and Huczko, A. 2018. Evolution of graphene oxide and graphene: from imagination to industrialization. Chem. Nano Mat. 4(7): 598–620.

U.S. EPA. Science Policy Council. 2007. Nanotechnology white paper. U.S. Environmental Protection Agency. Available at: http://www.epa.gov/ncer/nano/publications/whitepaper12022005.pdf

USGS, 2017. Contaminants of Emerging Concern in the Environment. Environmental Health–Toxic Substances Hydrology Program. U.S. Geological Survey.

Werkneh, A.A. and Rene, E.R. 2019. Applications of nanotechnology and biotechnology for sustainable water and wastewater treatment. pp. 405–430. *In*: Bui, X.-T., Chiemchaisri, C., Fujioka, T. and Varjani, S. (eds). Water and Wastewater Treatment Technologies. Singapore: Springer.

Whitesides, G.M. 2001. The once and future nanomachine. Sci. Am. 285(3): 78–83.

Wu, J., Zhao, L., Xu, X., Bertrand, N., Choi, W.I., Yameen, B., Shi, J., et al. 2015. Hydrophobic cysteine poly (disulfide)-based redox-hypersensitive nanoparticle platform for cancer theranostics. Angew. Chem. 127(32): 9350–9355.

Wu, C., Wang, A.C., Ding, W., Guo, H. and Wang, Z.L. 2019. Triboelectric nanogenerator: a foundation of the energy for the new era. Adv. Energy Mater. 9(1): 1802906.

Wurtsbaugh, W.A., Paerl, H.W. and Dodds, W.K. 2019. Nutrients, eutrophication and harmful algal blooms along the freshwater to marine continuum. Wiley Interdiscip. Rev. Water. 6(5): e1373.

Xu, X.D., Cheng, Y.J., Wu, J., Cheng, H., Cheng, S.X., Zhuo, R.X. and Zhang, X.Z. 2016. Smart and hyper–fast responsive polyprodrug nanoplatform for targeted cancer therapy. Biomaterials. 76: 238–249.

Xu, Y., Sun, Q., Yi, L., Yin, X., Wang, A., Li, Y. and Chen, J. 2014. The source of natural and anthropogenic heavy metals in the sediments of the Minjiang River Estuary (SE China): implications for historical pollution. Sci. Total Environ. 493: 729–736.

Yadav, N., Garg, V.K., Chhillar, A.K. and Rana, J.S. 2021. Detection and remediation of pollutants to maintain ecosustainability employing nanotechnology: a review. Chemosphere. 280: 130792.

Zhang, H., Fei, J., Yan, X., Wang, A. and Li, J. 2015. Enzyme-responsive release of doxorubicin from monodisperse dipeptide-based nanocarriers for highly efficient cancer treatment in vitro. Adv. Funct. Mater. 25(8): 1193–1204.

Zhong, M., Tang, J., Guo, X., Guo, C., Li, F. and Wu, H. 2020. Occurrence and spatial distribution of organophosphorus flame retardants and plasticizers in the Bohai, Yellow and East China seas. Sci. Total Environ. 741: 140434.

Zhao, H., Lin, Z.Y., Yildirimer, L., Dhinakar, A., Zhao, X. and Wu, J. 2016. Polymer–based nanoparticles for protein delivery: Design, strategies and applications. J. Mater. Chem. B. 4(23): 4060–4071.

Section II

Soil/Land Remediation/Restoration and Waste Management

Chapter 2

Nanotechnology in the Restoration of Polluted/Degraded Soil

Mohammad Faizan[1] and S. Maqbool Ahmed[1]
[1]Botany Section, School of Sciences, Maulana Azad National Urdu University, Hyderabad-50032, India

Priyadarshani Rajput[2], Vishnu D. Rajput*[2], Sudhir S. Shende[2], Tatiana Minkina[2], Svetlana Sushkova[2] and Saglara Mandzhieva[2]
[2]Academy of Biology and Biotechnology, Southern Federal University, Rostov-on-Don, 344006, Russia

Shamsul Hayat[3] and Anjuman Hussain[3]
[3]Plant Physiology Lab, Department of Botany, Aligarh Muslim University, Aligarh-202002, India

Fangyuan Yu[4]
[4]Collaborative Innovation Centre of Sustainable Forestry in Southern China, College of Forest Science, Nanjing Forestry University, 159 Longpan Road, Nanjing, Jiangsu, 210037, China

2.1 INTRODUCTION

The ecological roles which plants play in their surroundings are numerous, and they also influence the local wildlife. Plants are essential to all living creatures on the planet, directly or indirectly (Cetin 2016). As reported by Cetin et al. (2017) and Guney et al. (2017), the capability of plants to perform their tasks is largely dependent on the presence of suitable environmental and climatic

*Corresponding Author: rajput.vishnu@gmail.com; rvishnu@sfedu.ru

conditions. Soil is a prerequisite for the survival of plants. But upon closer inspection, the soil is revealed to have a composite construction, and the biological and biochemical activities that take place, where it serves as the foundation for terrestrial ecosystems (Guney et al. 2017, Abdalla et al. 2018). In this respect, it is crucial to assess soil health changes and ascertain how they relate to the plant. Additionally, enhancing soil quality through the conservation of soil and remediation techniques and eradicating situations that endanger soil quality are two key factors in determining the sustainability of soils (Fernández-Luqueño et al. 2017).

Soil has been polluted by various types of contaminants; especially in recent years as a result of rapidly ongoing technological developments (Xu et al. 2017). The contaminants added to soils are non-biodegradable, disrupt the soil ecosystem and inhibit the expansion of plants and soil microbes. These pollutants damage the living organisms and prevent the growth of plant roots and leaves. Physical, chemical, and biological techniques are currently used to clean up polluted soil. These procedures are expensive tedious and inefficient. For instance, the microbial remediation approach has limitations because not all bacteria are very environment-adaptive, making it impossible to ensure the clean-up of polluted soil (Chen et al. 2015). As a result, they are not ideal tools for solving the environmental degradation problems which we are currently confronting (Chen et al. 2015).

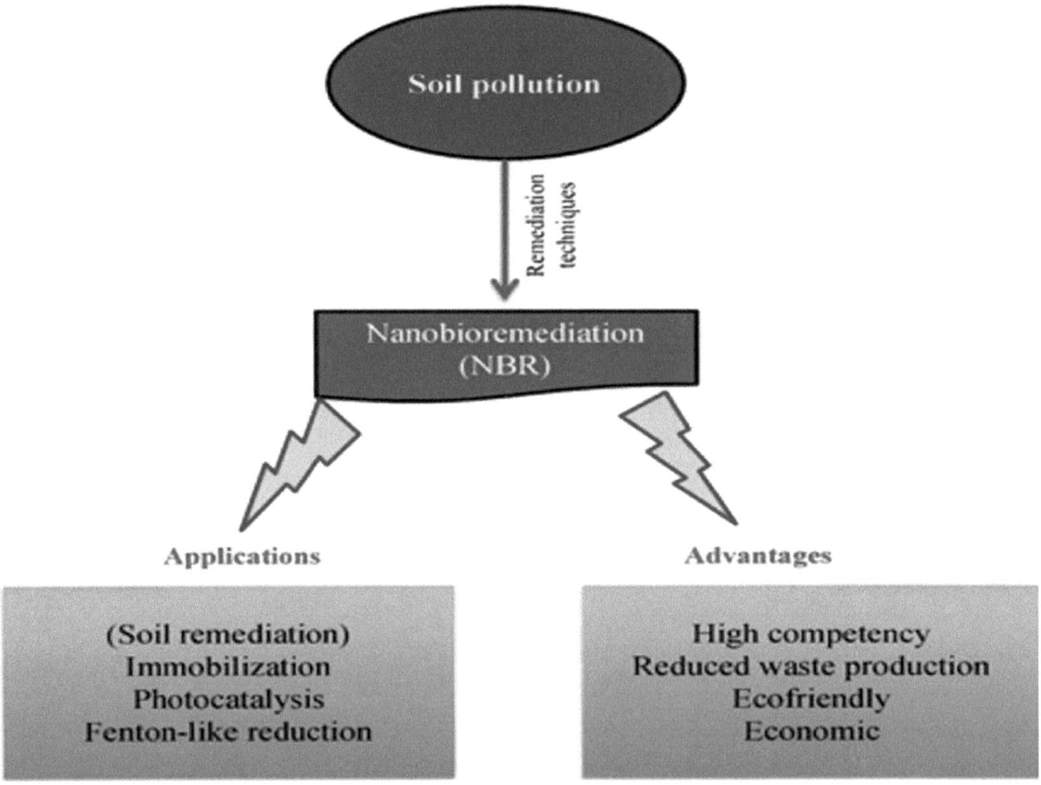

Figure 2.1 The pictorial representation of nanobioremediation.

A breakthrough technique is developing and expanding its applicability across all aspects, i.e., nanotechnology. Outlining, analyzing, and modifying the matter at the atomic, molecular, macromolecular, and micro-molecular scales, so that the final substance has special and wholly different qualities from the source bulk material being offered by nanotechnology (Bhatt and Tripathi 2011, Khan, et al. 2017, Shende et al. 2021, Faizan et al. 2023). The resulting particles are called engineered nanoparticles (ENPs) or nanoparticles (NPs) with one exterior dimension having sizes ranging from 1 and 100 nm (Shi et al. 2011, Thuesombat et al. 2014, Faizan et al. 2020, 2021, Shende et al. 2022). Numerous industries, including agriculture, food manufacture and wrapping, pharmaceuticals, health diagnostics, nano-based encapsulation of pesticides, genetic material release in plants, drug release in humans, cancer cure, and many more use of nanotechnology (Gogos et al. 2012, De Oliveira et al. 2014). Other than this, it's expected that the applications of NPs in soil and water cleanup could be a boon; especially for groundwater cleanup, wastewater treatment, and remediation of contaminated soils (Kuiken 2010, Roh et al. 2010, Shi et al. 2010). Growing utilization and application of nano-based materials for environmental clean-up may result in critical need for a technology that is quicker to deliver output without adding further weight to the cleaning process in the form of environmental endurance (Yan et al. 2013, Cai et al. 2019).

Numerous applications of nanotechnology are being utilized for soil remediation, including (i) using nanotechnology, heavy metals converted into moderate toxic material. (ii) pesticide and POP degradation, (iii) soil pesticide remains finding using nanotechnology, and (iv) bioremediation of contaminated soil with the assistance of nanomaterials. Nanobioremediation (NBR), which uses NPs to speed up natural biodegradation processes, has recently been recognized as a solution for clearing up environmental contaminants.

NBR is a method for removing dangerous chemicals from soils that combines NPs with microbes or plants (Cecchin et al. 2017). Several NBR processes—such as nano-phytoremediation and microbial-enhanced nano remediation—are described and formulated on the type of organism used to remove contaminants (Rajput et al. 2020a,b, Kumari et al. 2022). The costs of chemical and physical processes are increasing, and NBR technologies that are mediated by microorganisms and plants are getting more attention. Figure 2.1 shows the pictorial representation of nanobioremediation.

2.2 NANOPARTICLES USED IN THE SOIL REMEDIATION

Environmental researchers are paying more and more attention to employment of NPs for pollution reduction and environmental cleanup as a result of the development of nanotechnology (Tang et al. 2014). Amongst NPs, carbon-based and metal-based NPs are the subjects of extensive research (Gong et al. 2009, Chen et al. 2017). In both Europe and the United States, numerous field utilization of ENPs for soil and groundwater remediation has been completed successfully (Mueller and Nowack 2010).

NPs have recently been used in research to report on phytoremediation of polluted soil. Incorporating NPs into established phytoremediation methods is promising (Liu et al. 2023). In NBR method, pollutants are first reduced by NPs to a state that is favorable for biodegradation, and then the pollutants are destroyed by living organisms. High competency decreased creation of chemical and biological misuse, selectivity, lack of additional food necessities, bio-sorbent regeneration, likelihood of metal revival, etc. are the key benefits of bioremediation over conventional techniques (Chauhan et al. 2020).

Organic contaminants are common in soil and are primarily produced by human activity. Due to their negative impacts, persistence, and bioaccumulation, these pollutants must be eliminated (Jones and de Voogt 1999, Kang 2014). In soil contaminated with trichloroethylene, endosulfan, and trinitrotoluene, it has been found that adding NPs improved the phytoremediation process (Pillai and Kottekottil 2016). According to Ma and Wang (2010), fullerene NPs may enhance eastern cottonwood's ability to absorb trichloroethylene in the phytoremediation system.

Application of fullerene NPs at concentrations of 2 mg/L and 15 mg/L resulted in increase in trichloroethylene absorption by 26% and 82%, respectively. Fullerene NPs which were added had no immediate harmful effects on the plants. NPs of nZVI are added to the soil, and it significantly eliminates trinitrotoluene by *Panicum maximum* (Jiamjitrpanich et al. 2012). The outcome of recent works suggests that adding nZVI particles could significantly increase the effectiveness of phytoremediation. Table 1 depicts the use of various NPs in the remediation of soil.

Table 2.1 Nanomaterials used in the soil remediation

Nanomaterial	Soil Pollutant	Remediation mechanism	References
Magnetite nanoparticles	As, Polycyclic aromatic hydrocarbons (PAH)	Adsorption processes on the nanoparticles led to immobilization.	Baragaño et al. 2020
Biochar and titanium dioxide nanoparticles	Sb	Sb is becoming more immobile in the soil. TiO_2 NPs significantly improved the Sorghum bicolor capacity to accumulate Sb.	Zand et al. 2020
Iron sulfide nanoparticles	Cr(VI)	Immobilization was caused by adsorption, reduction and co-precipitation.	Wang et al. 2019a
Modified carbon black nanoparticles	Petroleum, Ni, Cd	Following $KMnO_4$-HNO_3 modification of CB nanoparticles, MNCB reduced metal availability in co-contaminated soil and prevented heavy metal transfer from the soil to plants.	Cheng et al. 2019
Magnetite nanoparticles	As	Immobilization procedure by lowering arsenite's bioavailability to humans, plants, release risk, and leachability.	Zialame et al. 2021
Nano-chlorapatite	Cd and Pb	Immobilization via precipitation as metal-phosphate compounds	Wan et al. 2018
Nano-selenium	Hg	The creation of insoluble mercuric selenide, which immobilizes elemental mercury (HgSe)	Wang et al. 2017
Carbon nanotubes	Organochlorine pesticides	Immobilization through adsorption for in situ remediation	Zhang et al. 2017
nZVI	Cr	Stabilization by reductive transformation of Cr(VI) into more stable and less toxic form	Wang et al. 2014
Pd/Fe bimetallic nanoparticles	Polychlorinated biphenyls (PCBs)	Degradation through a process involving Pb/Fe nanoparticles and hydrogenation, dichlorination	Chen et al. 2014

2.3 REMEDIATION OF SOIL WITH IRON NANOPARTICLES

Iron (Fe) is the fourth most common metal on earth, and iron oxide (FeO) nanocrystals with a diameter of 5–100 nm are found in soil. Numerous organic and inorganic materials can be attached because of the surface area's size (Kien and Catherine 2023). Iron oxides have the potential to be used in the preparation of nanomaterials to repair soil because they are environmentally acceptable and have the ability to absorb pollutants. Nano FeO can be applied to contaminated soil at a reasonable cost with little to no secondary contamination. The same study discovered and concluded that using micro FeOs may completely remove all heavy metals from polluted soil. In the meantime, it was discovered that nanohematite and nanomagnetite adsorption were identical (Shipley et al. 2011).

Zero-valent iron (ZVI) nanoparticles have a wide surface area, which enable touch with the contaminant and hence enhance cleansing effectiveness. They also have several hasty sites, which increases their reactivity compared to their bulk part. These characteristics allow ZVI nanoparticles to penetrate contaminated sites. Nanoscale iron particles can be delivered directly into impure soils, sediments, and aquifers because they remain suspended in their colloidal solution. Because of their excellent adsorption and reducing abilities, they can react with heavy metals like mercury (Hg), nickel (Ni), and cadmium (Cd) (Cd). Zero-valent iron NPs are an excellent absorbent for a variety of contaminants and a powerful reducing agent (Cundy et al. 2008). They are frequently used to remove heavy metals from contaminated soil, including Mercury (Hg^{2+}), Nickel (Ni^{2+}), Cadmium (Cd^{2+}), Lead (Pb^{2+}), and Chromium (Cr^{6+}) (Zhang 2003, Caliman et al. 2011). Adsorption and reduction, via which heavy metals are changed into less harmful species, immobilized, or become less accessible in soil, are the main mechanisms involved in heavy metal elimination (Cundy et al. 2008). The effectiveness of ZVI nanoparticles in immobilising arsenic in soil samples has been shown in other studies (Gil-D'az et al. 2016, 2017a,b, V'tkova' et al. 2016). In a different study, it was shown that a decreased As availability in the soil resulted in a decrease in soil phytotoxicity. This is a significant aspect to take into account when deciding whether nanoremediation treatment is an effective way to restore land usage in As-polluted soils. The authors of a study with As solutions treated with nZVI came to the conclusion that As(III) and As(V) may be eliminated using nZVI by inner-sphere surface complexation in a matter of minutes. Additionally, it was discovered that after 90 days, nZVI reduced 25% of As(V) to As(III) (Kanel et al. 2005, 2006). In another experiment, it was found that As(V) was converted to As(III) and As(0) after 24 hours in a spiking solution that had been treated with nZVI. As(III) was reduced and oxidized by the reaction of nZVI with As(III), which resulted in the generation of As(0), As(III), and As(V) (Ramos et al. 2009). In samples of soil and sediment, the sorption mechanisms of nanoscale iron oxides with As were investigated.

Earlier research found that the precipitation of $Zn\text{-}AsO^4$ species boosted the preservation of As when Zn was present. Thus, it has been demonstrated that using nZVI is a successful method for immobilizing As in acid and calcareous soils, (Grafe et al. 2004). Gil-Díaz et al. (2016) found that adding nZVI to soils with chromium (Cr) contamination considerably reduced the availability of the metal in both single-metal and multi-metal contaminated soils. Iron sulphide nanoparticles were used to treat soil and sediment samples for Hg contamination, and the availability of Hg was significantly reduced (Van Koetsem et al. 2016). The submission of nZVI reduced accessibility of both elements in two areas of brownfield that were considered according to experiment and types of soil samples taken from brownfield which were contaminated with As and Hg. However, there were no clear effect on soil pH and electrical conductivity (Gil-Díaz et al. 2017a,b).

2.4 REMEDIATION OF SOIL WITH CARBON NANOPARTICLES

New NPs, known as carbon nanotubes (CNTs) have the capacity to remove both organic and inorganic pollutants (Wang et al. 2019b). CNTs are unique carbon-based nanomaterials because of their mechanical capability, and high thermal and electrical conductivity (Ibrahim 2013, Veeman et al. 2021). Additionally, CNTs have outstanding chemical characteristics, that coupled with its other features enable it to achieve greater success in the environmental cleanup process (Wang et al. 2019). Carbon nanotubes have the ability to adsorb pollutants, immobilizing them similar to phytostabilization.

It has been shown that CNTs are quite effective at adsorbing a variety of pollutants (Kang et al. 2018). Interactions between CNTs and organic contaminants involve hydrophobic contact, electrostatic attraction, and p-p bonding (Song et al. 2018). The co-existence of many interactions during the adsorption process keeps the mixture of pollutants and CNTs generally stable. Single-wall carbon nanotubes (SWCNTs) and multi-wall carbon nanotubes (MWCNTs) are examples of carbon NPs that have great potential and promise in a variety of research domains, including the bioremediation of soil and groundwater (Thines et al. 2017). They are effective in removing nonpolar inorganic pollutants from the environment, such as dioxin and PAHs, and pure CNTs are about three times more effective at doing so than activated carbon (Ahmed et al. 2018).

Polar contaminants cannot be bound to CNTs because of their naturally nonpolar surface; however, they can be changed to have polar chemical groups on both their outer and inner surfaces, which makes it easier for them to do so. In truth, it is not economical to incorporate a lot of CNTs into the soil. A small expansion reduces costs while preventing subsequent environmental contamination effects. Due to their hydrophobicity and nonpolar nature, initial pure CNTs agglomerate when added to water. However, this can be avoided by altering the CNTs' surface or adding surfactants to lower surface tension. A repulsive force close enough to avoid aggregation is anticipated to be provided by surface modification (Hyung et al. 2007, Jaisi et al. 2008). According to one study, natural organic matter accounts for the stabilizing capacity of MWCNTs, itbeing almost the same in both in vitro and in vivo experiments (Hyung et al. 2007). The transport behavior of SWCNTs and fullerene was investigated and compared. Fullerene is more effective in moving through the soil than SWCNT. However, it is also known that fullerene is more sensitive than SWCNT to changes in the ionic strength of the environment (Jaisi et al. 2008). According to the literature (Liu et al. 2009), even MWCNT has limited mobility. When CNT health risks are assessed, respiratory disorders are said to be in the lead (Chen et al. 2006).

2.5 REMEDIATION OF SOIL WITH METAL OXIDE NANOPARTICLES

It was previously known that MnO_2 was useful in the breakdown of pharmaceutical compounds that had been discharged into water bodies, particularly groundwater. But its relative significance and possibility in cleaning up pollutants that have been absorbed into the soil were never investigated. Han et al. (2015) created these MnO_2 NPs in a stabilized form for the first time in 2015 by employing carboxymethyl celluloses (CMC) as a stabilizer. They next put them to the test for degradation of both soil- and aqueous-sorbed estradiol. This provided unequivocal evidence that MnO_2 NPs might be employed as a viable source for soil remediation. The researchers found that, at a normal aquatic pH of 6–7, CMC-stabilized MnO_2 displayed a higher rate of oxidative destruction of 17–estradiol than unsterilized MnO_2. When used to address estradiol that has been absorbed into the soil, the reactive activity becomes considerably more prominent. This is mostly due to CMC's capacity to combine with metal ions, which prevents the reactive sites from binding soil components and reduces their ability to impede activity. For MnO_2 that has been stabilized by CMC, the oxidation kinetics might be explained by a retarded first-order rate model.

Numerous conditions have the tendency to reduce these NPs' ability to oxidize. Some of them are desorption rate, soil-NPs contact, soil-released metals, and reductants. It is believed that iron-cerium oxide in mixed oxide states or as a multi-ferrite composite may be able to remove such contaminants through affinity-based binding, which would then remove heavy metals from the soil (Yi et al. 2015). Another justification for such a mixed oxide formulation is to give these nanostructures stability (Vivekananthan et al. 2014). Due to their natural semi-conductivity, cerium oxide NPs exhibit photo-catalysis. This behavior is crucial for the breakdown of soil-polluting dyes (Korsvik et al. 2007).

2.6 SOIL REMEDIATION WITH PHOSPHATE-BASED NANOPARTICLES

Phosphate and heavy metals can combine to generate very insoluble compounds that have a high degree of stability in polluted soil. Phosphate-based NPs can be used to treat substances that are heavily laden with lead. Common lead products like PbO_4, PbS, and $PbCO_3$ often have solubilities between 10 ppm or above, whereas lead phosphate compounds like pyromorphites have solubilities under 10 ppm. This demonstrates unequivocally that the soil water is less soluble in lead phosphates. When phosphate is added to Pb-polluted soils, stable complexes are created that stop the Pb from spreading further, creating immobilized species. Phosphate can be used to treat Cu, Zn, Cd, Co, Ba, Cr, Eu, and U, and its complexes can be created using H_3PO_4, NaH_2PO_4, $(NH_4)_2H_2PO_4$, apatite, natural rock phosphate, and biologically produced materials such animal bones (Basta and McGowen 2004, Raicevic et al. 2006). The soil can become too acidic by adding too much phosphoric acid and ammonium phosphate (Basta and McGowen 2004).

It is essential to be concerned about the dose interacting with the soil, and certain rules and laws must be established in this regard. The mobility of phosphate in the soil decreases as the size of the phosphate source, such as appetite particles, increases. The reduced size is therefore preferable. The problem is successfully solved while reducing secondary environmental damage caused by soil treatment for heavy metals. In addition, these produced NPs can function as fertilizers that include P (DeRosa et al. 2010, Liu 2011).

2.7 BIOREMEDIATION OF SOIL WITH ENGINEERED POLYMERIC NANOPARTICLES

PAHs, a hydrophobic contaminant, have the potential to have a remarkable effect on the environment. They last a long period in the soil because they are insoluble in water. Created poly (ethylene) glycol-altered urethane acrylate (PMUA) NPs as one method to boost the effectiveness of bioremediation. Natural NPs present in nature might improve their ability to prevent the soil from liberating phenanthrene, a hydrophobic organic pollutant that is a part of PAH. They capture PAHs, and spread them in the groundwater, and a study also showed that they had the ability to release the captured PAHs. With the formulation, the rate of mineralization also rose.

The conjugated NPs may be consumed by microbes like bacteria, which aided in accelerating in-situ bioremediation. It has been discovered that PMUA NPs are more stable in an environment with a variety of bacteria. Even after bacteria have had an impact on the prepared NPs, they may still be used again. Even in bioreactors, this repair and recycling process is possible (Tungittiplakorn et al. 2005). We now have novel technologies to address and recognize a wide extent of environmental applications, including sorbents, high-flux membranes, depth filters, antimicrobial agents, environmental sensors, renewable energy technologies, and pollution control measures, thanks to the brilliant properties of carbon-based NPs like nanocrystals and CNTs (Mauter and Elimelech 2008, Rizwan et al. 2014). For instance, CNT-modified equivalents and CD-co-hexamethylene/toluene diisocyanate polyurethanes have been created and used successfully to remove organic pollutants from soil at a very low level (Li et al. 2010). Amphiphilic polyurethane (APU) NPs are employed to treat soil that has been contaminated with polynuclear PAH. Polyurethane acrylate anion omer (UAA) or poly urea is used to make the particles (ethylene glycol). APU can move through soil more easily because they are sturdy and have an exterior that is hydrophilic and their interior is hydrophobic.

According to recent studies conducted on plants growing in mining pools; *Gundelia tournefortii, Centaurea virgata, Reseda lutea, Scariola orientalis, Eleagnum angustifolia, and Noaea mucronata* have a significant capacity to acquire heavy metals from soil, including lead, copper, nickel, zinc, and others. *Noaea mucronata* was found to have the greatest capacity to absorb lead and other heavy metals among these plants. These plant sources were employed to create NPs, which are now being used in the process of NBR of the environment (Mohsenzadeh and Rad 2012).

2.8 CONCLUSION AND FUTURE PROSPECTS

There are currently several traditional and innovative technologies accessible for soil remediation. Practically speaking, each has disadvantages. An eco-friendly approach, for instance, might not be economical. Therefore, before using these techniques, it is crucial to weigh the advantages and disadvantages. Comparing the conventional and new technologies, it is clear that novel technologies, specifically soil remediation boosted by nanotechnology, are more cost-effective and efficient overall. A lot of emphases have been paid to the use of NPs in the remediation of polluted media. These materials' traits—high surface area, improved reactivity, and often emerging features at the nanoscale level—open up promising possibilities for applications involving the removal of pollutants.

As in the chapter discussed, applied nanotechnology methods have shown to be more successful and affordable. Regardless of the steps used to clean up the contaminants in the soil, it is imperative to make sure sustainability of the soil and the environment in the interim. Even in later phases, the process must not affect the soil at its conclusion. Eco-friendly NPs are more practicable to use in soil remediation than traditional approaches. In situ soil remediation benefits, more from these components. To ensure the successful and ongoing use of natural resources by all living things, future research must be focused on the creation of biodegradable, effective, and economical nanomaterials.

2.9 ACKNOWLEDGMENT

This work is supported by the laboratory «Soil Health» of the Southern Federal University with the financial support of the Ministry of Science and Higher Education of the Russian Federation, agreement No. 075-15-2022-1122.

References

Abdalla, M., Hasting, A., Chadwick, D.R., Jones, D.L., Evans, C.D., Jones, M.B., Rees, R.M., et al. 2018. Critical review of the impacts of grazing intensity on soil organic carbon storage and other soil quality indicators in extensively managed grasslands. Agric Ecosyst Environ. 253: 62.

Ahmed, W., Elhissi, A., Dhanak, V. and Subramani, K. 2018. Carbon nanotubes: applications in cancer therapy and drug delivery research. pp. 371–389. *In*: Subramani, K. and Ahmed, W. (eds). Emerging Nanotechnologies in Dentistry, 2nd Ed. Micro and Nano Technologies. Elsevier Available from: https://doi.org/10.1016/B978-0-12-812291-4.00018-2.

Baragaño, D., Alonso, J., Gallego, J.R., Lobo, M.C. and Gil-Díaz, M. 2020. Magnetite nanoparticles for the remediation of soils co-contaminated with As and PAHs. Chem. Eng. J. 399: 125809.

Basta, N.T. and McGowen, S.L. 2004. Evaluation of chemical immobilization treatments for reducing heavy metal transport in a smelter-contaminated soil. Environ. Pollut. 127(1): 73–82.

Bhatt, I. and Tripathi, B.N. 2011. Interaction of engineered nanoparticles with various components of the environment and possible strategies for their risk assessment. Chemosphere. 82(3): 308–317.

Cai, C., Zhao, M., Yu, Z., Rong, H. and Zhang, C. 2019. Utilization of nanomaterials for in-situ remediation of heavy metal (loid) contaminated sediments: a review. Sci. Total Environ. 662: 205–217.

Caliman, F.A., Robu, B.M., Smaranda, C., Pavel, V.L. and Gavrilescu, M. 2011. Soil and groundwater cleanup: benefits and limits of emerging technologies. Clean Techn. Environ. Policy. 13: 241–268.

Cecchin, I., Reddy, K.R., Thome, A., Tessaro, E.F. and Schnaid, F. 2017. Nanobioremediation: integration of nanoparticles and bioremediation for sustainable remediation of chlorinated organic contaminants in soils. Int. Biodeter. Biodegr.119: 419–428.

Cetin, M. 2016. Sustainability of urban coastal area management: a case study on Cide. J. Sustain. Forest. 35(7): 527–541.

Cetin, M., Sevik, H. and Saat, A. 2017. Indoor air quality: the samples of safranbolu bulak Mencilis cave. Fresenius Environ. Bull. 26(10): 5965–5970.

Chauhan, R., Yadav, H.O.S. and Sehrawat, N. 2020. Nanobioremediation: a new and a versatile tool for sustainable environmental clean-up-overview. J. Mater. Environ. Sci. 11(4): 564–573.

Chen, X., Tam, U.C., Czlapinski, J.L., Lee, G.S., Rabuka, D., Zettl, A. and Bertozzi, C.R. 2006. Interfacing carbon nanotubes with living cells. J. Am. Chem. Soc. 128(19): 6292–6293.

Chen, X., Yao, X., Yu, C., Su, X., Shen, C., Chen, C., Huang, R. and Xu, X. 2014. Hydrodechlorination of polychlorinated biphenyls in contaminated soil from an e-waste recycling area, using nanoscale zerovalent iron and Pd/Fe bimetallic nanoparticles. Environ. Sci. Pollut. Res. 21(7): 5201–5210.

Chen, M., Xu, P., Zeng, G., Yang, C., Huang, D. and Zhang, J. 2015. Bioremediation of soils contaminated with polycyclic aromatic hydrocarbons, petroleum, pesticides, chlorophenols and heavy metals by composting: applications, microbes and future research needs. Biotechnol. Adv. 33: 745–755.

Chen, M., Zeng, G., Xu, P., Zhang, Y., Jiang, D. and Zhou, S. 2017. Understanding enzymatic degradation of single-walled carbon nanotubes triggered by functionalization using molecular dynamics simulation. Environ. Sci. Nano. 4(3): 720–727.

Cheng, J., Sun, Z., Yu, Y., Li, X. and Li, T. 2019. Effects of modified carbon black nanoparticles on plant-microbe remediation of petroleum and heavy metal co-contaminated soils. Int. J. Phytoremediation. 21(7): 634–642.

Cundy, A.B., Hopkinson, L. and Whitby, R.L.D. 2008. Use of iron-based technologies in contaminated land and groundwater remediation: a review. Sci Total Environ. 400: 42–51.

De Oliveira, J.L., Campos, E.V.R., Bakshi, M., Abhilash, P.C. and Fraceto, L.F. 2014. Application of nanotechnology for the encapsulation of botanical insecticides for sustainable agriculture: prospects and promises. Biotechnol. Advances. 32(8): 1550–1561.

DeRosa, M.C., Monreal, C., Schnitzer, M., Walsh, R. and Sultan, Y. 2010. Nanotechnology in fertilizers. Nat. Nanotechnol. 5(2): 91–91.

Faizan, M., Faraz, A., Mir, A.R. and Hayat, S. 2020. Role of zinc oxide nanoparticles in countering negative effects generated by cadmium in *Lycopersicon esculentum*. J. Plant Growth Regul. 40. 101–115.

Faizan, M., Sehar, S., Rajput, V.D., Faraz, A., Afzal, S., Minkina, T., Sushkova, S., et al. 2021. Modulation of cellular redox status and antioxidant defense system after synergistic application of zinc oxide nanoparticles and salicylic acid in rice (*Oryza sativa*) plant under arsenic stress. Plants. 10: 2254.

Faizan, M., Karabulut, F., Alam, P., Yusuf, M., Tonny, S.H., Adil, M.F., Sehar, S., et al. 2023. Nanobionics: a sustainable agricultural approach towards understanding plant response to heavy metals, drought, and salt stress. Nanomaterials. 13: 974.

Fernández-Luqueño, F., Lopez-Valdez, F., Pérez-Morales, C., García-Mayagoitia, S., Sarabia-Castillo, C.R. and Perez-Rios, S.R. 2017. Enhancing decontamination of PAHs-polluted soils: role of organic and mineral amendments. pp. 339–368. *In*: Anjum, N.A., Gill, S.S. and Tuteja N. (eds). Enhancing Cleanup of Environmental Pollutants,Vol. 2: Non-Biological Approache. Springer. Switzerland.

Gil-Díaz, M., Diez-Pascual, S., González, A., Alonso, J., Rodríguez-Valdés, E., Gallego, J.R. and Lobo, M.C. 2016. A nanoremediation strategy for the recovery of an As-polluted soil. Chemosphere. 149: 137–145.

Gil-Díaz, M., Alosno, J., Rodríguez-Valdés, E., Gallego, J.R. and Lobo, M.C. 2017a. Comparing different commercial zero valent iron nanoparticles to immobilize As and Hg in brownfield soil. Sci. Total Environ. 15: 1324–1332.

Gil-Díaz, M., Pinilla, P., Alonso, J. and Lobo, C. 2017b. Viability of a nanoremediation process in single or multimetal(loid) contaminated soils. J. Hazard. Mater. 321: 812–819.

Gogos, A., Knauer, K. and Bucheli, T.D. 2012. Nanomaterials in plant protection and fertilization: current state, foreseen applications, and research priorities. J. Agric. Food Chem. 60(39): 9781–9792.

Gong, J.L., Wang, B., Zeng, G.M., Yang, C.P., Niu, C.G., Niu, Q.Y., Zhou, W.J., et al. 2009. Removal of cationic dyes from aqueous solution using magnetic multi-wall carbon nanotube nanocomposite as adsorbent. J. Hazard Mater. 164(2–3): 1517–1522.

Gräfe, M., Nachtegaal, M. and Sparks, D.L. 2004. Formation of metal-arsenate precipitates at the goethite water interface. Environ. Sci. Technol. 38(24): 6561–6570.

Guney, K., Cetin, M., Guney, K.B. and Melekoglu, A. 2017. The effects of some hormone applications on *Lilium martagon* L. germination and morpholgical characters. Pol. J. Environ. Stud. 26(6): 2533–2538.

Han, B., Zhang, M., Hao, D. and Feng, Y. 2015. Degradation of aqueous and soil-sorbed estradiol using a new class of stabilized manganeseoxide nanoparticles. Water Res. 70: 288–299.

Hyung, H., Fortner, J.D., Hughes, J.B. and Kim, J.H. 2007. Natural organic matter stabilizes carbon nanotubes in the aqueous phase. Environ. Sci. Technol. 41(1): 179–184.

Ibrahim, K.S. 2013. Carbon nanotubes-properties and applications: a review. Carbon Lett. 14(3): 131–144.

Jaisi, D.P., Saleh, N.B., Blake, R.E. and Elimelech, M. 2008. Transport of single-walled carbon nanotubes in porous media: filtration mechanisms and reversibility. Environ. Sci. Technol. 42(22): 8317–8323.

Jiamjitrpanich, W., Parkpian, P., Polprasert, C. and Kosanlavit, R. 2012. Enhanced phytoremediation efficiency of TNT-contaminated soil by nanoscale zero valent iron. 2nd International Conference on Environment and Industrial Innovation. 35: 82–86. Singapore. IACSIT Press.

Jones, K.C. and de Voogt, P. 1999. Persistent organic pollutants (POPs): state of the science. Environ. Pollut. 100(1–3): 209–221.

Kanel, S.R., Manning, B., Charlet, L. and Choi, H. 2005. Removal of arsenic-(III) from groundwater by nano scale zero-valent iron. Environ. Sci. Technol. 39: 1291–1298.

Kanel, S.R., Greneche, J.M. and Choi, H. 2006. Arsenic(V) removal from groundwater using nano scale zerovalent iron as a colloidal reactive barrier material. Environ. Sci. Technol. 40: 2045–2050.

Kang J.W. 2014. Removing environmental organic pollutants with bioremediation and phytoremediation. Biotechnol. Lett. 36: 1129–1139.

Kang, J., Duan, X., Wang, C., Sun, H., Tan, X., Tade, M.O. and Wang, S. 2018. Nitrogen-doped bamboo-like carbon nanotubes with Ni encapsulation for persulfate activation to remove emerging contaminants with excellent catalytic stability. Chem. Eng. J. 332: 398–408.

Khan, M.N., Mobin, M., Abbas, Z.K., AlMutairi, K.A. and Siddiqui, Z.H. 2017. Role of nanomaterials in plants under challenging environments. Plant. Physiol. Biochem. 110: 194–209.

Kien A. Vu. and Catherine N. Mulligan. 2023. Remediation of organic contaminated soil by Fe-based nanoparticles and surfactants: a review. Environ. Tech. Rev. 12: 1, 60–82

Korsvik, C., Patil, S., Seal, S. and Self, W.T. 2007. Superoxide dismutase mimetic properties exhibited by vacancy engineered ceria nanoparticles. Chem. Commun. 10: 1056–1058.

Kuiken, T. 2010. Cleaning up contaminated waste sites: is nanotechnology the answer? Nano Today. 5(1): 6–8.

Kumari, A., Kumari, P., Rajput, V.D., Sushkova, S.N. and Minkina, T. 2022. Metal (loid) nanosorbents in restoration of polluted soils: geochemical, ecotoxicological, and remediation perspectives. Environ. Geochem. Health. 44(1): 235–246.

Li, Y., Liu, F., Xia, B., Du, Q., Zhang, P., Wang, D., Wang, Z., et al. 2010. Removal of copper from aqueous solution by carbon nanotube/calcium alginate composites. J. Hazard Mater. 177: 876–880.

Liu, X., O'Carroll, D.M., Petersen, E.J., Huang, Q. and Anderson, C.L. 2009. Mobility of multiwalled carbon nanotubes in porous media. Environ. Sci. Technol. 43(21): 8153– 8158.

Liu, R. 2011. *In-situ* lead remediation in a shoot-range soil using stabilized apatite nanoparticles. Coll. Sur. Sci. Symp. Canada. 85.

Liu, Z., Faizan, M., Zheng, L., Cui, L., Han, C., Chen, H. and Yu, F. 2023. Nanoparticles enhance plant resistance to abiotic stresses: a bibliometric statistic. Agronomy. 13(3): 729.

Ma, X. and Wang, C. 2010. Fullerene nanoparticles affect the fate and uptake of trichloroethylene in phytoremediation system. Environ. Eng. Sci. 27(11): 989–992.

Mauter, M.S. and Elimelech, M. 2008. Environmental applications of carbon-based nanomaterials. Environ. Sci. Technol. 42(16): 5843–5859.

Mohsenzadeh, F. and Rad, A.C. 2012. Bioremediation of heavy metal pollution by nano-particles of Noaeamucronat. Int. J. Biosci. Biochem. Bioinformatics. 2: 85–89.

Mueller, N.C. and Nowack, B. 2010. Nano zero valent iron—the solution for water and soil remediation? ObservatoryNANO Focus Report. 1–34.

Pillai, H.P. and Kottekottil, J. 2016. Nano-phytotechnological remediation of endosulfan using zero valent iron nanoparticles. J. Environ. Prot. 7: 734–744.

Raicevic, S., Wright, J., Veljkovic, V. and Conca, J. 2006. Theoretical stability assessment of uranyl phosphates and apatites: selection of amendments for in situ remediation of uranium. Sci. Total Environ. 355(1–3): 13–24.

Rajput, V.D., Minkina, T., Fedorenko, A., Chernikova, N., Hassan, T., Mandzhieva, S., Sushkova, S., et al. 2021a. Effects of zinc oxide nanoparticles on physiological and anatomical indices in spring barley tissues. Nanomater. 11(7): 1722.

Rajput, V.D., Singh, A., Singh, V.K., Minkina, T.M. and Sushkova, S. 2021b. Impact of nanoparticles on soil resource. pp. 65–85. *In*: Amrane, A., Mohan, D., Nguyen, T.A., Assadi, A.A. and Yasin, G. (eds). Nanomaterials for Soil Remediation. Elsevier.

Ramos, M.A.V., Yan, W., Li, X.Q., Koel, B.E. and Zhang, W.X. 2009. Simultaneous oxidation and reduction of arsenic by zero-valent iron nanoparticles: understanding the significance of the Core_Shell structure. J. Phys. Chem. C. 113: 14591–14594.

Rizwan, M., Singh, M., Mitra, C.K. and Morve, R.K. 2014. Ecofriendly application of nanomaterials: nanobioremediation. J. Nanoparticles. 1–7.

Roh, J., Y., Park, Y.K., Park, K. and Choi, J. 2010. Ecotoxicological investigation of CeO_2 and TiO_2 nanoparticles on the soil nematode *Caenorhabditis elegans* using gene expression, growth, fertility, and survival as endpoints. Environ Toxicol Pharm. 29(2): 167–172.

Shende, S.S., Ingle, P.U., Rajput, V.D., Gade, A.K., Minkina, T.M., Sushkova, S.N., Mandzhieva, S.S., et al. 2021. Biogenic synthesis of silver and gold nanoparticles by microbes. pp. 47–78. *In*: Shende, S.S., Rajput, V.D., Gorovtsov, A.V., Minkina, T.M. and Sushkova, S.N. (eds). Microbial Synthesis of Nanomaterials. Nova Science Publishers Inc. USA.

Shende, S., Rajput, V., Gade, A., Minkina, T., Sushkova, S.N., Mandzhieva, S.S. and Boldyreva, V.E. 2022. Metal-based green synthesized nanoparticles: boon for sustainable agriculture and food security. IEEE Transactions on Nano Biosci. 21(1): 44–54. https://doi.org/10.1109/TNB.2021.3089773

Shi, J., Votruba, A.R., Farokhzad, O.C. and Langer, R. 2010. Nanotechnology in drug delivery and tissue engineering: From discovery to applications. Nano Lett. 10(9): 3223–3230.

Shi, J., Abid, A.D., Kennedy, I.M., Hristova, K.R. and Silk, W.K. 2011. To duckweeds (*Landoltia punctata*), nanoparticulate copper oxide is more inhibitory than the soluble copper in the bulk solution. Environ. Pollut. 159(5): 1277–1282.

Shipley, H.J., Engates, K.E. and Guettner, A.M. 2011. Study of iron oxide nanoparticles in soil for remediation of arsenic. J. Nanopart. Res. 13(6): 2387–2397.

Song, B., Xu, P., Zeng, G., Gong, J., Zhang, P., Feng, H., Liu, Y., et al. 2018. Carbon nanotube-based environmental technologies: The adopted properties, primary mechanisms, and challenges. Rev. Environ. Sci. Biotechnol. 17: 571–590.

Tang, W.W., Zeng, G.M., Gong, J.L., Liang, J., Xu, P., Zhang, C. and Huang, B.B. 2014. Impact of humic/fulvic acid on the removal of heavy metals from aqueous solutions using nanomaterials: a review. Sci. Total Environ. 468: 1014–1027.

Thines, R.K., Mubarak, N.M., Nizamuddin, S., Sahu, J.N., Abdullah, E.C. and Ganesan, P. 2017. Application potential of carbon nanomaterials in water and wastewater treatment: a review. J. Taiwan Inst. Chem. Eng.72: 116–133.

Thuesombat, P., Hannongbua, S., Akasit, S. and Chadchawan, S. 2014. Effect of silver nanoparticles on rice (*Oryza sativa* L. cv. KDML 105) seed germination and seedling growth. Ecotoxicol. Environ. Safe. 104: 302–309.

Tungittiplakorn, W., Cohen, C. and Lion, L.W. 2005. Engineered polymeric nanoparticles for bioremediation of hydrophobic contaminants. Environ. Sci. Technol. 39(5): 1354–1358.

Van Koetsem, F., Van Havere, L. and Laing, G.D. 2016. Impact of carboxymethyl cellulose coating on iron sulphide nanoparticles stability, transport, and mobilization potential of trace metals present in soils and sediment. J. Environ. Manage. 168: 210–218.

Veeman, D., Shree, M.V., Sureshkumar, P., Jagadeesha, T., Natrayan, L., Ravichandran, M. and Paramasivam, P. 2021. Sustainable development of carbon nanocomposites: synthesis and classification for environmental remediation. J. Nanomater. 5840645.

Vítková, M., Rákosová, S., Michálková, Z. and Komárek, M. 2016. Metal(loid)s behaviour in soils amended with nano zero-valent iron as a function of pH and time. J. Environ. Manage. 186: 268–276.

Vivekananthan, V., Selvapriya, A., Janani, D. and Narendhar, C. 2014. Synthesis of mixed oxides of cerium-iron nanostructures for effective removal of heavy metals from waste water. Res. J. Recent. Sci. 212–217.

Wan, J., Zeng, G., Huang, D., Hu, L., Xu, P., Huang, C., Deng, R., et al. 2018. Rhamnolipid stabilized nano-chlorapatite: Synthesis and enhancement effect on Pb-and Cd-immobilization in polluted sediment. J. Hazard Mater. 343: 332–339.

Wang, C., Liu, H., Chen, J., Tian, Y., Shi, J., Li, D., Guo, C., et al. 2014. Carboxylated multiwalled carbon nanotubes aggravated biochemical and subcellular damages in leaves of broad bean (*Vicia faba* L.) seedlings under combined stress of lead and cadmium. J. Hazard Mat. 274: 404–412.

Wang, X., Zhang, D., Pan, X., Lee, D.J., Al-Misned, F.A., Mortuza, M.G. and Gadd, G.M. 2017. Aerobic and anaerobic biosynthesis of nano-selenium for remediation of mercury contaminated soil. Chemosphere. 170: 266–273.

Wang, Y., Pan, C., Chu, W., Vipin, A.K. and Sun, L. 2019a. Environmental remediation applications of carbon nanotubes and graphene oxide: adsorption and catalysis. Nanomater. 9(3): 439.

Wang, T., Liu, Y., Wang, J., Wang, X., Liu, B. and Wang, Y. 2019b. In-situ remediation of hexavalent chromium contaminated groundwater and saturated soil using stabilized iron sulfide nanoparticles. J. Environ. Manage. 231: 679–686.

Xu, X., Pliego, G., Zazo, J.A., Sun, S., García-Muñoz, P., He, L., Casas, J.A., et al. 2017. An overview on the application of advanced oxidation processes for the removal of naphthenic acids from water. Crit. Rev. Environ. Sci. Technol. 47(15): 1337–1370.

Yi, P., Pignatello, J.J., Uchimiya, M. and White, J.C. 2015. Hetero aggregation of cerium oxide nanoparticles and nanoparticles of pyrolyzed biomass. Environ. Sci. Technol. 49(22): 13294–13303.

Yan, W., Lien, H.L., Koel, B.E. and Zhang, W.X. 2013. Iron nanoparticles for environmental clean-up: Recent developments and future outlook. Environ. Sci. Process Impacts. 15(1): 63–77.

Zand, A.D., Tabrizi, A.M. and Heir, A.V. 2020. Co-application of biochar and titanium dioxide nanoparticles to promote remediation of antimony from soil by *Sorghum bicolor:* Metal uptake and plant response. Heliyon. 6(8): p.e04669.

Zhang, W.X. 2003. Nanoscale iron particles for environmental remediation: an overview. J. Nanoparticle Res. 5: 323–332.

Zialame, A., Jamshidi-Zanjani, A. and Darban, A.K. 2021. Stabilized magnetite nanoparticles for the remediation of arsenic contaminated soil. J. Environ. Chem. Eng. 9(2): 104821.

Chapter 3

Nanoadsorbents as a Cutting-edge Tool for Removal of Metal(loid)s from the Environment

Pedda Ghouse Peera Sheikh Kulsum[1]

[1]C.V. Raman Global University, Odisha, India

Rubina Khanam*[2]

[2]ICAR—National Rice Research Institute, Cuttack-753006, Odisha, India

Jayanta Kumar Biswas[3]

[3]Department of Ecological Studies and International Centre for Ecological Engineering, University of Kalyani, Kalyani, Nadia- 741235, West Bengal, INDIA

3.1 INTRODUCTION

During the past few decades, several methods have been devised to handle heavy metal pollution, including physical (adsorption, agglomeration, evaporation and filtration); chemical (precipitation, ion exchange and electrochemical changes) and biological approaches (Siddiqui et al. 2019). Despite the wide range of methods, due to potential demerits like installation cost, and complexity in procedures, a significant research gap was observed in nanoadsorbent research. Among all, adsorption is a relatively better method widely used because of its low cost, and high removal efficiency (Gupta et al. 2021a).

In recent years, the development of innovative approaches for efficient and cost-effective removal of metal(loid)s from the environment has gained considerable attention. Among these approaches, nanotechnology has emerged as a cutting-edge tool, offering promising solutions for remediation processes. In particular, nanoadsorbents have demonstrated great potential in tackling the challenges associated with metal(loid) contamination.

---

*Corresponding Author: erinakhanam@gmail.com

Selection of the most suitable adsorption material is vital to achieving high removal efficiency. Currently, the most commonly used adsorbents are biochar, activated carbon, carbon film, biopolymers, clay materials, and nanomaterials (Kaushal and Singh 2017).

Nanomaterials (1–100 nm) are promising adsorbents and catalysts for use in soil and environmental heavy metal remediation due to more surface, chemical reactivity and low-temperature modification (Gu et al. 2020). They also have distinct redox characteristics that are useful in removing heavy metals sensitive to redox conditions. However, some of the commonly used nanomaterials do have limitations, including high costs, potential toxicity, difficulty in recycling, and easy interaction with other media. Although nanoadsorbents and their use in heavy metal remediation have been widely reported by several researchers, a research gap was identified to address their use in a soil environment. Therefore, this book chapter covers the nanoadsorbents associated mechanisms, regulating factors and potential qualities of nanoadsorbents for efficient removal of heavy metals from soil and environment. This knowledge is essential for the design of efficient nanoadsorbents with capacities to remove heavy metals at levels prevailing in the environment.

This chapter will delve into the recent advancements and applications of nanoadsorbents in the field of environmental remediation. We will explore their effectiveness in the removal of specific metal(loid) contaminants from soil, water, and air, presenting case studies and highlighting the challenges and opportunities associated with their implementation on a larger scale. Moreover, we will discuss the potential synergistic effects of combining nanoadsorbents with other remediation techniques, such as photocatalysis and electrochemical processes, to further enhance metal(loid) removal efficiency.

By comprehensively examining the role of nanoadsorbents in metal(loid) removal, this chapter aims to provide researchers, environmental engineers, and policymakers with a comprehensive understanding of the potential and limitations of nanoadsorbents as a cutting-edge tool for addressing the pressing environmental challenge of metal(loid) contamination. The insights gained from this chapter can pave the way for the development of sustainable and effective strategies to mitigate the adverse impacts of metal(loid)s and safeguard the environment for future generations.

3.2 MECHANISMS OF NANOADSORBENTS

At the nanoscale (1–100 nm), materials show a wide range of properties compared to the microscale, due to two effects (i) surface properties (more surface atoms, free surface energy thereby more chemical reactions), (ii) volume effects (due to lower wavelength, super-paramagnetism, a blue shift of atoms for optical absorption spectra).

3.3 SURFACE CHARACTERISTICS

Assessing the nature and location of active sites on nano-adsorbent surfaces is a key to achieving efficiency in heavy metal sorption. The potential surface accessibility, number of active sites and accordingly surface energy to targeted heavy metals are affected by factors like the location of the atoms in the surface, high surface area, high reactivity, high adsorptive capacity, high surface binding energy, low internal resistance for diffusion. For example, metal NPs (MNPs), metal oxide NPs and CNMs are excellent adsorbents due to their more internal surface area and electron configuration.

3.3.1 Crosstalk with Functional Groups

Surface coupling with functional groups prevents NPs from agglomeration and facilitates great selectivity. They also improve adsorption capacity, affinity and faster desorption. It is achieved by reacting NMs surface with organic donor atoms which further enhance the interaction of NPs with targeted heavy metal. For example, (i) $-NH_2$ and –O groups form electrostatic bonding with transition metals, (ii) –COOH groups forms amide bonding with multiwalled carbon nanotubes (MWCNTs), triggers Cd^{2+} adsorption from soil water. When the pH of the solution is more than isoelectric point (IEP) of nano-adsorbents, cation exchange occurs facilitating cation adsorption. Similarly, pH less than IEP attracts anions from solution to adsorbent surface.

Higher zeta potential will impart stability to nanoparticles in suspension due to repulsion among the particles whereas low zeta potential leads to coagulation. As a general rule, the stability as per ZP (in milli volts) is given as:

$$-30 \text{ mV} > \text{ZP} > +30 \text{ mV}$$

This suggests that the value of ZP can attain either positive or negative values when the solution attains more positive ions (H^+) or more negative ions (OH^-), respectively. This is where pH comes into picture. It plays a crucial role in maintaining the zeta potential and hence the stability of the nanoparticles. Therefore, a wise person would always report the pH corresponding to the measured ZP.

3.4 NANOADSORBENTS FOR REMOVAL OF HEAVY METALS IN A SOIL ENVIRONMENT

Adsorption has long been commonly used for heavy metal removal due to its low cost, efficiency, and simplicity. The specificity of a particular hazardous chemical, the adsorption efficiency, and the benefit-cost ratio are important factors in the selection of adsorbents. In the recent past, different nanomaterials have been synthesized (carbon analogues, carbonaceous nanostructured composites, nanomagnetic materials, microporous glasses, adsorbent sieves ceramics, clay polymers and so on) (Kumar et al. 2020). In comparison to the materials utilized historically and commercially as adsorbents, they have been found to have a high adsorption capacity for heavy metals from soil water. Table 3.1 summarizes the nano-adsorbents vital in heavy metal removal in a complex matrix-like soil environment and is further discussed in the following sections.

3.4.1 Metal Oxide Nanoparticles and Metal Nanoparticles (MNPs)

Metal oxides, such as Fe, Cu, Ti, Mn, Mg, Zn, Si and Al can be used to synthesize nanoparticles. Due to metal-ligand precipitation or the generation of ternary ligands, metal oxide nanoparticles have a greater degree of adsorption than regular-sized oxide (Stietiya and Wang, 2014). Metal oxide-based nanomaterials with very well structural, crystalline, and surface properties act as semiconductors with large band gaps. They also have advantageous properties such as non-toxicity and excellent water stability. Metal oxides such as ferric oxides, titanium (Ti), cerium (Ce), and zinc (Zn) provide cost-effective adsorbent. The adsorption capacity of metal-based nanoparticles is pH-dependent. Heavy metals adsorption from the media gets enhanced with pH due to increase in electrostatic interactions and formation of ionic or co-valent bonds.

A study was conducted to evaluate how nano-TiO_2 (2–6 nm) can reduce the effect of wastewater (containing heavy metals) on the development of maize seedlings. During in vitro experiments, varying concentrations of nano-TiO_2 suspension were introduced into either autoclaved wastewater or deionized water. Nano-TiO_2 at a concentration of 25 mg L^{-1} significantly reduced the negative effects of wastewater on maize development indices (Yaqoob et al. 2018). Alumina nanoadsorbent

Table 3.1 Characteristics of Nanoadsorbents used for heavy metal removal in soil environment

Adsorbent	Characteristics	pH_{PZC}	Disadvantages	Reference
1. Nanoparticles (NPs)				
1.1. Metal oxide nanoparticles				
(a) ZnO NPs	Highly efficient catalyst, very high porosity, high surface area and pore volume, inexpensive, Non-toxic for solid waste disposal	8.75	Phyto-toxic at higher levels	Gu et al. 2020, Akhtar et al. 2021a
(b) Al_2O_3 NPs	Degree of surface reactivity, mechanical strength, high surface binding energy, resistance to pH changes	7.9 9.5	(i) Selective adsorbent, (ii) Organic molecules are not adsorbed, (iii) hydrophilic, (iv) surface alteration is required, (v) sensitive to the agglomeration	Borji et al. 2020, Gupta et al. 2021a
(c) TiO_2 NPs	Non-toxic, photochemically stable, high specific surface area	4.8–6.2	(i) Precipitates at higher pH, (ii) Alteration required for higher surface adsorption and selectivity	Zand et al. 2020, Liu et al. 2020
1.2. Metallic nanoparticles (MNPs)	Higher electrocatalytic activity, Compatible with functional groups including –SH and $–NH_2$	—	(i) Very limited work available, (ii) High cost	Raffa et al. 2021, Gupta et al. 2021b
1.3. Nanostructured mixed oxides (Fe-Ti mixed oxides)	Special magnetic and electrical properties, pH dependent anion and cation exchange and desorption,	5.5–6.0	More selectivity for polar molecules.	Siddiqui et al. 2019
1.4. Magnetic nanoparticles (MNPs)	Chemical composition, type and degree of defectiveness of the crystal lattice, no agglomeration, high efficiency, low cost	7.0	(i) Not suitable for complex soil matrix, (ii) Sophisticated coating is required to use, (iii) Not target specific	Zhu et al. 2020, Khan et al. 2020a
2. Carbon Nanotubes (CNTs)	Higher sorption-desorption, surface area 140–3200 $m^2\ g^{-1}$,	2.2–11.8	Chemical inertness and strong van der waal interactions	Oloumi et al. 2018, Hoang et al. 2022
3. Silicon Nanomaterials (SiNMs)				
3.1. Silica Nanotubes (SiNTs)	More polarizable than carbon, stable cross-linked structure	—	Not suitable as nano-adsorbent material	Hussain et al. 2019, Khan et al. 2020b

Adsorbent	Characteristics	pH_{PZC}	Disadvantages	Reference
3.2. SiO_2 Nanoparticles (SiO_2 NPs)	High surface area, low cost and easy accessibility,	2.0–3.0	Extraction of organic pollutants not possible	
4. Nanofibers (NFs)	High surface to volume ratio, length to diameter ration, more active sites for adsorption,	—	—	Kumar et al. 2020
5. Nanoclays	High adsorption rate, more specific surface, more stable and more dispersive with no agglomeration; 1:1 layer with 0.7 mm; 2:1 layer with 1 mm and expandable up to 3 mm.	—	Intercalated structures show exfoliation with time and pretreatment required	Baigorria et al. 2021
6. Polymer based Nanoadsorbents	Adsorption capacity up to 17.0 mg kg^{-1}. Three types based on surface area (i) 1000 m^2 g^{-1}, (ii) > 2000 m^2 g^{-1} (iii) 3000–4500 m^2 g^{-1}	6.86	—	Emenike et al. 2022
7. Aerogels	Larger specific surface area (1000 m^2 g^{-1}), high porosity, internal pore volume (16-18 cm^3 g^{-1})	9.4	—	Vareda and Durães 2019

produced through solution combination synthesis method was reported to be highly efficient in removing Zn(II) from wastewater.

3.4.2 Magnetic Nanoparticles (MNPs)

Nasar et al. 2012 showed 36 mg g^{-1} absorption capacity of Fe_3O_4 nanoadsorbents for removal of Pb(II) ions. Moreover, ultrafine magnesium ferrite ($MgO.27Fe_2.5O_4$) nanoadsorbent synthesized by Tang et al. (2013) act as good adsorbent for As from contaminated water.

Nanosized iron oxide particles have super-paramagnetism, which distinguishes them from other oxide nanoparticles. It imparts higher surface area, are biocompatible, less hazardous, chemically inert, have a low diffusion resistance, and their surface may be changed with organic molecules, inorganic ions, or functional groups, making them ideal surfaces for absorbing heavy metals. Fe_3O_4 superparamagnetic nanoadsorbents with diameter less than 10nm and coated with ascorbic acid were highly efficient to remove As. These nanoadsorbent adsorptions had the maximum adsorption capacity of 16.56 mg g^{-1} and 46.06 mg g^{-1} for As(III) and As(V) respectively (Maghsoudi et al. 2021). Further, Fe_2O_4 MNPs coated with oxide shells of Mn-Co is an excellent adsorber of Pb(II) (481.2 mg g^{-1}), Cu(II) (386.2 mg g^{-1}) and Cd(II) (345.5 mg g^{-1}) (Ma and Wang 2015).

The effects of nano-Fe_3O_4 in reducing heavy metal toxicity in wheat seedlings were investigated by Konate et al. 2017. The use of nano-Fe_3O_4 (@2000 mg L^{-1}) in a 1 Mm solution dramatically reduced the negative impact of heavy metals on growth in wheat seedlings. The beneficial impacts of nano-Fe_3O_4 under heavy metal stress might be due to an increase in antioxidant enzyme activity (Konate et al. 2017). The effectiveness of iron oxide nanoparticles (Fe_3O_4 NP) in dropping detrimental effects of As in *Brassica juncea* was demonstrated by Konate et al. (2018). They also reported application of Fe_3O_4 NP further decrease the stress in plant induced by As contamination. Under cadmium stress, zinc oxide nanoparticles were found to improve wheat chlorophyll content, gas exchange characteristics, antioxidant enzymes, zinc absorption, and yield while lowering Cd concentration.

3.4.3 SiO_2 NPs and SiNPs

Silica nanoparticles are also believed to be highly efficient to remove heavy metals. Silica nanoparticles along with graphite oxide were used to remove Zn, Ni, Cr, Pb and Cd. When metal oxides are employed as adsorbents without any supporting elements (viz., clays or zeolites), they may have limited adsorption capacity, smaller surface area, and release hazardous metals into the environment.

Siddiqui and Al-Whaibi (2014) discovered that 8 g/L of 12 nm nano-silicon oxide increased seed germination in *Lycopersicum esculentum*. It has also been observed that nanostructured silicon dioxide aids in the reduction of plant transpiration rates, as well as the improvement of plant green colour and shoot growth. Application of nano-SiO_2 reduced As uptake in rice seedlings though improving pectin synthesis and the mechanical force of the cell wall. Recently, various iron nano-materials have been employed to reduce heavy metal uptake in plants. In the modern era, application of nanoparticles not only reduce the uptake of heavy metals in plants but also decontaminate the environment through different mechanisms. In the subsequent section, we deal with the different applications of nanomaterial (viz., photocatalyst, nanoabsorbents, nanoclay polymers etc.) to reduce the heavy metal burden in the environment.

3.5 CARBON NANOMATERIALS (CNMs)

Among the nanoabsorbents, carbon nanomaterials are most extensively used for removal of heavy metals. The carbon nanomaterials can be classified as (i) zero-dimensional (having all the three dimensions less than 100 nm), e.g., fullerene and quantum dots. (ii) one-dimensional (1 D) (having only one dimension larger than 100 nm and two dimensions smaller than 100 nm), e.g., nanotubes of carbon and titanium. (iii) two-dimensional nanomaterials (having two dimensions greater than 100 nm), e.g., graphene. (iii) three dimensional (3-D) (having all dimensions greater than 100 nm) e.g., graphite and some nano-composites. The highly efficient nano absorbents are discussed in the subsequent sections (Table 3.2).

3.5.1 Carbon Nanotubes (CNTs)

Carbon nanotubes (CNTs) have great internal sites, interstitial channels, grooves, and outer surfaces acting as adsorption sites. When compared to internal sites, exterior sites attain equilibrium faster. The adsorption process is largely driven by chemical interactions between the surface functional groups of CNTs and metal ions. CNTs often include –OH, –C=O, and –COOH groups, depending on the synthesis and purifying procedure. However, some other functional groups can also be added to CNTs through oxidation using Pd, Ni, or Pt, or removed from CNTs through heat treatment at 2200°C. Cu, Zn, and Pb were removed using CNTs manufactured using the microwave irradiation technique. Kaushal and Singh, 2017 reported removal efficiency of 111 mg g^{-1}, 71 mg g^{-1}, and 85 mg g^{-1} for Cu (II), Pb(II), and Zn(II) respectively by these type of CNTs.

Carbon nanotubes can be classified based on their dimensions into zero-dimensional (0-D, e.g., fullerene and quantum dots), one dimensional (1-D e.g., nanotubes and titanium), two dimensional (2-D, e.g., graphene) or multi-dimensional.

3.6 SINGLE WALLED CNTs (SWCNTs)

The one-dimensional single wall CNTs (SWCNTs) are very promising for their nano-sized porous structure, high surface area and easy surface functionalization. Thus, SWCNTs are widely used for removal of metal pollutants. Researchers demonstrated SWCNT-based nanocomposite can efficiently adsorb Hg (more than 99.56% of within 7 min). In an experiment conducted by Anitha et al. 2015 to evaluate the adsorption capacity of unadorned SWCNTs and their functionalized SWCNTs (viz., SWCNTs–OH, SWCNTs–NH_2, and SWCNTs–COOH) to remove meta(loids). It has been observed that SWCNTs-COOH have highest adsorption capacities (150–230%) compared to unadorned SWCNTs. However, SWCNTs–OH and SWCNTs–NH had poor adsorption capacity. SWCNTs–COOH have also showed adsorption capacity of 96, 77, and 56 mg g^{-1} for Pb^{2+}, Cu^{2+}, and Cd^{2+}, respectively.

3.7 MULTIWALLED CNTs (MWCNTs)

Zhang and Chen 2014 created MWCNTs–TiO_2 and used them to photodegrade methylene blue. Zhang and Chen 2014 reported MWCNTs based nanoadsorbent can efficiently remove Pb (II) and Hg (II) to the extent of 65.40 mg g^{-1} and 65.52 mg g^{-1}, respectively. Ge et al. 2014 synthesized sulfonated multi-walled CNTs (s-MWCNTs) through p-MWCNTs with strong sulfuric acid at high temperatures to adsorb Cu (II). However, since their discovery in the early 1990s, CNT toxicity has been a major issue. Another disadvantage of employing CNTs is their low removal effectiveness and limited selectivity. Furthermore, because the walls of carbon nanotubes are not reactive, the sorption capacities of metal ions are relatively low (Mukhopadhyay et al. 2020).

3.7.1 Clay-polymer Nanocomposites

The clay polymer adsorbents include zeolites (activated and natural), natural clay minerals, modified clay minerals etc. (Mukhopadhyay et al. 2020). Naturally, available clay minerals

Table 3.2 Removal efficiency of different CNTs, MNPs and metal oxide based nano-absorbents

Metal based nano-adsorbent	Metal removal	Efficiency (mg/g)	References
Fe_3O_4 nano-adsorbents	Pb(II)	36	Nassar 2012
Anatase oxide nano-adsorbent	Pb, Cu, As	31.25, 23.74,16.98	Kocabas-Atkali and Yurum 2013
Magnesium ferrite nano-adsorbent	As	127.4 (As^{+3}) and 83.2 (As^{+5})	Tang et al. 2013
Manganese feroxyhyte nano-adsorbent	As	11.7 μg/mg (As^{+3}), 6.7 μg/mg (As^{+5})	Tresintsi et. al. 2013
Fe–La composite oxide	As	58.2	Zhang et al. 2014
Alumina nano-adsorbent	Zn	1047.83	Bhargavi et al. 2015
Fe_3O_4 magnetic nanoparticles	Hg	—	Nassar 2012
Mercapto-functionalized nano-Fe_3O_4 magnetic nano-adsorbent (SH–Fe_3O_4–NMPs)	Hg	—	Chalasani and Vasudevan 2012
Fe_3O_4 superparamagnetic nano-adsorbent coated with ascorbic acid	As	16.56 (As^{+3}) and 46.06 (As^{+5})	Nassar 2012
$MnFe_2O_4$ nano-adsorbent	Pb, Cu and Cd	481.2 for Pb^{+2}, 386.2 for Cu^{+2} and 345.5 for Cd^{+2}	Ma and Wang 2015
SWCNTs	Hg	41.66	Alijani and Shariatinia 2018
SWCNTs–Fe_3O_4–CoS	Hg	1666	Alijani and Shariatinia 2018
MWCNTs	Cr	1.26	Dehghani et al. 2015
Al_2O_3–MWCNTs	Pb	—	Gupta 2017
Porous graphen	As	—	Tabish et al. 2018
Activated carbon	Pb, Cd, Cu, Ni	238.1, 96.2, 87.7, 52.4	Wang et al. 2023

are low cost, easily available and can be intensively used. However, their low surface area, less effective for micro pollutants and lack of standard protocols for regeneration are the major limitations. Recently, researchers have started exploring clay-polymer nanocomposites (CPNs) as a combined adsorbent, integrating both clay minerals and resins into a single material for enhanced adsorption capabilities. The nanocomposites have shown immense potential for removal of major pollutants in water such as Cu, Cr, As, Pb, Ni and Cd. With development of nanoscience, several clay polymer nanocomposites are synthesized such as, epoxy-clay nanocomposites, polystyrene-clay nanocomposites, polyurethane-clay nanocomposites, magnetic clay polymer nanocomposites, polyimide-clay hybrids etc. (Table 3.3). Recently, magnetized-CPN was developed using bentonite clay, iron oxide nanoparticles, and monomer methyl methacrylate for removal of Cr (VI) (Sundaram et al. 2018). The application of these CPN had the adsorption capacity of 113 mg g^{-1}. Whereas, chitosan-Al-pillared-montmorillonite nanocomposite can adsorb Cr (VI) 15.67 mg g^{-1} (Sakulthaew et al. 2017). Recently, chitosan-based clay composite showed high adsorption of Cu (II) in the tune of 96.0 and 99.5%, respectively, at a pH of 6.5 (Sakulthaew et al. 2017). In addition, intercalation polymerization generated organo-bentonite and polyacrylonitrile composite showed the removal of Cd, Zn, and Cu at the tune of 52.6, 65.4, and 77.4 mg g^{-1}, respectively (Mukhopadhyay et al. 2020).

Table 3.3 Efficiency of Clay-polymer nanocomposites in metal removal

Clay-polymer nanocomposites	Metal removed	Efficiency (mg/g)	References
Alginate–montmorillonite nanocomposite	Pb	238.1	Shawky 2011
Polyacrylic acid/bentonite	Ni	270.27	Mukhopadhyay et al. 2020
Polyacrylic acid/bentonite	Pb	1666.67	
Chitosan-Al-pillared-montmorillonite nanocomposite	Pb	—	Mukhopadhyay et al. 2020
Chitosan immobilized on bentonite	Ni	15.82	Mukhopadhyay et al. 2020
Polyacrylic acid/bentonite	Cd	416.67	Mukhopadhyay et al. 2020
Modifed bentonite clay composits	Zn	—	Mukhopadhyay et al. 2020
Polyetherimide/porous activated bentonite	Cd	—	
Chitosan-Al-pillared-montmorillonite nanocomposite	Cu	—	Pereira et al. 2003
Chitosan:clay (0.45:1)	Cu		Pereira et al. 2003
Montmorillonite and chitosan	Se	18.4	
Pillared bentonite by $MnCl_2$	Pb	12.6	Mukhopadhyay et al. 2020
Chitosan-bentonite clay	Ni, Cd	—	Mukhopadhyay et al. 2020
Alginate–montmorillonite nanocomposite	Fe		Mukhopadhyay et al. 2020
Bentonite/thiourea-formaldehyde composite	Mn	4.81	Mukhopadhyay et al. 2020
Na-montmorillonite/cellulose	Cr	22.2	Mukhopadhyay et al. 2020
Chitosan–clay composites and oxides	Se		Bleiman and Mishael 2010

3.8 FACTORS AFFECTING EFFICIENCY OF NANOADSORBENTS IN SOIL ENVIRONMENT

The behaviour of nanoadsorbents in soil is largely undefined. To study the reactions in nanopores within soil matrix, sophisticated methodologies and tools must be devised.

The NPs that are present in soils may undergo transformations, such as growth, dissolution, aggregation, and aging, changing the micro or nanoenvironment surrounding the individual soil NPs. The NPs may interact in different ways with an array of minerals of the soil solid phase, and a variety of soil solution aqueous species.

3.8.1 Dissolution

Some types of metal-based NPs are thermodynamically unstable and undergo chemical dissolution if kinetic restrictions are absent. Soil components such as clay particles or humic molecules, themselves having charged surfaces, will influence the association of NPs with the solid phase.

3.8.2 Ionic Strength

An increase in the ionic strength will lead to a decrease in the electrical double layer thickness, which favours particle association and leads to increasing agglomeration. The effects of ionic strength on the aggregation of Ag–NPs (uncoated, citrate, and sodium borohydride–coated particles) have been observed for suspensions with pH higher than 7. This demonstrates that here the particles were stabilized by electrostatic repulsion, caused by the predominance of negative forms (i.e., anions) in the medium and the negative charge of the particles.

3.8.3 Stability

The stability of NPs is a function of their surface energy and NPs are more stable when they possess low surface energy. Soil is a complex matrix and challenging for understanding the physicochemical behaviour of manufactured NPs.

3.8.4 Surface Coating

In the absence of a surface coating, metal-based NPs have charged surfaces resulting from the presence of hydroxyl (–OH) groups that can take up and release protons, and can take up dissolved chemical species such as metal ions and ligands. A suspension of homogenously charged particles will be stable when the magnitude of the zeta potential is greater than 30 mV. The chemistry of the medium will influence the electrostatic surface charge of the particles. Classical colloid DLVO theory of repulsive-attractive forces can be applied to metal and metal oxide NPs.

3.8.5 pH, Point of Zero Charge (PZC), Zeta Potential

pH is mainly responsible for governing the zeta potential of colloids. The colloidal system exhibits minimum stability or maximum coagulation, when pH is at point of zero charge (pzc). The colloid surface is positively charged when the pH is lower than the pzc value and the zeta potential will increase with decreasing pH below the pzc. Conversely, the colloid surface possesses positive charge at pH above pzc and the zeta potential will be more negative with increasing pH. Nanoparticles with low zeta potential tend to coagulate or flocculate, whereas higher value of zeta potential will impart stability to the nano- particle suspension. The solubility of metal-based nanoparticles and their surface charges are governed by solution pH and consequently the electrostatic interactions between nanoparticles and between nanoparticles and porous media. Dissolved counter ions in solution determine the long-range electrostatic interactions for electronically stabilized ENPs and hence decrease the stability and transport of ENPs in the porous media.

3.8.6 Natural Organic Matter

Natural organic matter has the ability to stabilize engineered nanoparticles (ENPs). Fullerene was found to be stabilized in the presence of dissolved organic matter (DOM) in suspension. Generally, after release of the ENPs in soil solution, they may interact with DOM in particulate form and be suspended and transported rather than aggregate and deposit quickly. The DOM adheres to the surface of ENPs and modifies physicochemical properties of their surface and enhances their stability.

3.9 FUTURE RESEARCH NEEDS

The ultra-small sizes of nanoparticles put them into massive use in soil and environmental health monitoring. Conventional methods of heavy metal removal have limitations like leaching, photodegradation, hydrolysis and bio-instability in environment. Nanoadsorbents are one of the convincing and cost-effective tools to reduce the heavy metal load in soil environment. Incorporating clay at appropriate dosages with a variety of inorganic polymers and biopolymers (or nanoclay) for the development of nanocomposites enhanced the effectiveness of nanoabsorbents for pollution removal. However, on non-judicious use, the same properties have equally adverse effects and have the potential to enter into food chain.

A few vividly reported negative effects on crops were plugging of stomata, sealing on stigma, impaired nutrient and water transport through vascular vessels. Longer persistence and high toxicity of NPs also fuel human health ailments like inflammation and protein fibrillation in lungs and induced genotoxicity. Use of B2 biosafety cabinet, maintaining dust concentration of not more than 10^4 m^{-2}, planned experimentation, training and educating on handling of nanoadsorbents are few precautionary measures. Approaches are needed for the synthesis of nanomaterials that are more environmentally friendly and sustainable. Further, advanced tools for ecotoxicological evaluation and safety measures should be developed. Establishment of regulatory agencies at the national and international levels is essential to monitor the diffusion of nanoparticles in environment.

3.10 CONCLUSION

By harnessing the potential of nanoadsorbents, researchers, environmental engineers, and policymakers can develop sustainable and effective strategies to address the pressing environmental challenge of metal(loid) contamination. The insights gained from this chapter provide a comprehensive understanding of the capabilities and limitations of nanoadsorbents, offering a pathway for the development of innovative solutions to mitigate the adverse impacts of metal(loid) s on ecosystems and human health. However, it is important to acknowledge that there are still challenges to overcome in the field of nanoadsorbents. Issues such as scalability, cost-effectiveness, and long-term stability need to be addressed to facilitate their widespread adoption. Additionally, the potential environmental implications of nanoadsorbents themselves, such as their fate and potential release into the environment, require careful consideration. In conclusion, the utilization of nanoadsorbents as a cutting-edge tool for the removal of metal(loid)s from the environment holds great promise. By further advancing research, promoting interdisciplinary collaborations, and addressing the challenges associated with their implementation, nanoadsorbents can play a pivotal role in achieving sustainable and efficient metal(loid) remediation practices. Through these efforts, we can safeguard the environment, protect human health, and pave the way for a cleaner and healthier future.

References

Akhtar, N., Khan, S., Rehman, S.U., Rehman, Z.U., Khatoon, A., Rha, E.S. and Jamil, M. 2021. Synergistic effects of zinc oxide nanoparticles and bacteria reduce heavy metals toxicity in rice (*Oryza sativa* L.) plant. Toxics. 9(5): 113.

Alijani, H. and Shariatinia, Z. 2018. Synthesis of high growth rate SWCNTs and their magnetite cobalt sulfide nanohybrid as super-adsorbent for mercury removal. Chem. Eng. Res. Des. 129: 132–149.

Anitha, K., Namsani, S. and Singh, J.K. 2015. Removal of heavy metal ions using a functionalized single-walled carbon nanotube: a molecular dynamics study. J. Phys. Chem. A. 119(30): 8349–8358. doi: 10.1021/acs.jpca.5b03352.

Baigorria, E., Cano, L. and Alvarez, V.A. 2021. Nanoclays as eco-friendly adsorbents of arsenic for water purification. pp. 455–470. *In*: Kharissova, O.V., Torres-Martínez, L.M. and Kharisov, B.I. (eds). Handbook of Nanomaterials and Nanocomposites for Energy and Environmental Applications. Springer, Cham.

Bhargavi, R.J., Maheshwari, U. and Gupta, S. 2015. Synthesis and use of alumina nanoparticles as an adsorbent for the removal of Zn(II) and CBG dye from wastewater. Int. J. Ind. Chem. 6: 31–41.

Bleiman, N. and Mishael, Y.G. 2010. Selenium removal from drinking water by adsorption to chitosan—clay composites and oxides: batch and columns tests. J. Hazard. Mater. 183(1–3): 590–595.

Borji, H., Ayoub, G.M., Bilbeisi, R., Nassar, N. and Malaeb, L. 2020. How effective are nanomaterials for the removal of heavy metals from water and wastewater? Water Air Soil Pollut. 231(7): 1–35.

Chalasani, R. and Vasudevan, S. 2012. Cyclodextrin functionalized magnetic iron oxide nanocrystal: a host-carrier for magnetic separation of non -polar molecules and arsenic from aqueous media. J. Mater. Chem. 22: 14925–14931.

Dehghani, M.H., Taher, M.M., Bajpai, A.K., Heibati, B., Tyagi, I., Asif, M., Agarwal, S., et al. 2015. Removal of noxious Cr (VI) ions using single-walled carbon nanotubes and multi-walled carbon nanotubes. Chem. Eng. J. 279: 344–352.

Emenike, E.C., Adeniyi, A.G., Omuku, P.E., Okwu, K.C. and Iwuozor, K.O. 2022. Recent advances in nano-adsorbents for the sequestration of copper from water. J. Water Process. Eng. 47: 102715.

Ge, Y., Li, Z., Xiao, D., Xiong, P. and Ye, N. 2014. Sulfonated multi-walled carbon nanotubes for the removal of copper (II) from aqueous solutions. J. Ind. Eng. Chem. 20(4): 1765–1771.

Gu, M., Hao, L., Wang, Y., Li, X., Chen, Y., Li, W. and Jiang, L. 2020. The selective heavy metal ions adsorption of zinc oxide nanoparticles from dental wastewater. Chem. Phys. 534: 110750.

Gupta R.K. 2017. Oil/water separation techniques: a review of recent progresses and future directions. J. Mater Chem A. 5(31): 16025–16058

Gupta, K., Joshi, P., Gusain, R. and Khatri, O.P. 2021a. Recent advances in adsorptive removal of heavy metal and metalloid ions by metal oxide-based nanomaterials. Coord. Chem. Rev. 445: 214100.

Gupta, A.D., Rene, E.R., Giri, B.S., Pandey, A. and Singh, H. 2021b. Adsorptive and photocatalytic properties of metal oxides towards arsenic remediation from water: a review. J. Environ. Chem. Eng. 9(6): 106376.

Hoang, A.T., Nižetić, S., Cheng, C.K., Luque, R., Thomas, S., Banh, T.L. and Nguyen, X.P. 2022. Heavy metal removal by biomass-derived carbon nanotubes as a greener environmental remediation: a comprehensive review. Chemosphere. 287: 131959.

Hussain, A., Rizwan, M., Ali, Q. and Ali, S. 2019. Seed priming with silicon nanoparticles improved the biomass and yield while reduced the oxidative stress and cadmium concentration in wheat grains. Environ. Sci. Pollut. Res. 26(8): 7579–7588.

Kaushal, A. and Singh, S.K. 2017. Removal of heavy metals by nanoadsorbents: a review. Environ. Biotechnol. 6(1): 96–104.

Khan, F.S.A., Mubarak, N.M., Khalid, M., Walvekar, R., Abdullah, E.C., Mazari, S.A. and Karri, R.R. 2020a. Magnetic nanoadsorbents' potential route for heavy metals removal—a review. Environ. Sci. Pollut. Res. 27(19): 24342–24356.

Khan, Z.S., Rizwan, M., Hafeez, M., Ali, S., Adrees, M., Qayyum, M.F., Khalid, S. et al. 2020b. Effects of silicon nanoparticles on growth and physiology of wheat in cadmium contaminated soil under different soil moisture levels. Environ. Sci. Pollut. Res. 27(5): 4958–4968.

Khanam R.A., Kumar, A.K., Nayak, M., Shahid, R., Tripathi, S., Vijayakumar, D., Bhaduri, U., et al. 2020. Metal (loid) s (As, Hg, Se, Pb and Cd) in paddy soil: bioavailability and potential risk to human health. Sci. Total Environ. 699:134330.

Kocabas-Atakli, Z.Ö. and Yürüm, Y. 2013. Synthesis and characterization of anatase nanoadsorbent and application in removal of lead, copper and arsenic from water. Chem. Eng. J. 225: 625–635. http://dx.doi.org/10.1016/j.cej.2013.03.106

Konate A., He, X., Zhang, Z., Ma, Y., Zhang, P., Alugongo, G.M. and Rui. Y. 2017. Magnetic (Fe_3O_4) nanoparticles reduce heavy metals uptake and mitigate their toxicity in wheat seedling. Sustainability. 9: 790.

Konate, A., Wang, Y., He, X., Adeel, M., Zhang, P., Ma, Y. and Zhang, Z. 2018. Comparative effects of nano and bulk-Fe_3O_4 on the growth of cucumber (*Cucumis sativus*). Ecotoxicol. Environ. Saf. 165: 547–554.

Kumar V. and Guleria, P. 2020. Application of DNA-Nanosensor for Environmental Monitoring: Recent Advances and Perspectives. Curr. Pollut. Rep. 12: 1–21.

Kumar, A., Gahoi, P. and Verma, N. 2020. Simultaneous scavenging of Cr (VI) from soil and facilitation of nutrient uptake in plant using a mixture of carbon microfibers and nanofibers. Chemosphere. 239: 124760.

Liu, J., Lv, G., Gu, W., Li, Z., Tang, A. and Mei, L. 2017. A novel luminescence probe based on layered double hydroxides loaded with quantum dots for simultaneous detection of heavy metal ions in water. J. Mater. Chem. 5(20): 5024–5030.

Liu, J., Jiang, J., Meng, Y., Aihemaiti, A., Xu, Y., Xiang, H. and Chen, X. 2020. Preparation, environmental application and prospect of biochar-supported metal nanoparticles: a review. J. Hazard. Mater. 388: 122026.

Ma, H. and Wang, C.N. 2015. Microwave-assisted optimization of platinum-nickel nanoalloys for catalytic water treatment. Appl. Catal. B: Environ. 163: 198–204.

Maghsoudi A.S., Hassani, S., Mirnia, K., and Abdollahi, M. 2021. Recent advances in nanotechnology-based biosensors development for detection of arsenic, lead, mercury, and cadmium. Int. J. Nanomedicine. 16: 803.

Mukhopadhyay, R., Bhaduri, D., Sarkar, B., Rusmin, R., Hou, D., Khanam, R., Sarkar, S., et al. 2020. Clay–polymer nanocomposites: Progress and challenges for use in sustainable water treatment. J. Hazard. Mater. 383: 121125.

Nassar, N.N. 2012. Iron oxide nanoadsorbents for removal of various pollutants from wastewater: an overview. J. Environ. Chem. Eng. 81–118.

Oloumi, H., Mousavi, E.A. and Nejad, R.M. 2018. Multi-wall carbon nanotubes effects on plant seedlings growth and cadmium/lead uptake in vitro. Russ. J. Plant Physiol. 65(2): 260–268.

Pereira, M F.R., Soares, S.F., Órfão, J.J. and Figueiredo, J.L. 2003. Adsorption of dyes on activated carbons: influence of surface chemical groups. Carbon. 41(4): 811–821.

Pereira, F.A., Sousa, K.S., Cavalcanti, G.R., Fonseca, M.G., de Souza, A.G. and Alves. A.P. 2013. Chitosan-montmorillonite biocomposite as an adsorbent for copper (II) cations from aqueous solutions. Int. J. Biol. Macromol. 61: 471–478.

Raffa, C.M., Chiampo, F. and Shanthakumar, S. 2021. Remediation of metal/metalloid-polluted soils: a short review. Appl. Sci. 11(9): 4134.

Sakulthaew, C., Chokejaroenrat, C., Poapolathep, A., Satapanajaru T. and Poapolathep. S. 2017. Hexavalent chromium adsorption from aqueous solution using carbon nano onions (CNOs). Chemosphere. 184: 1168–1174.

Shawky H.A. 2011. Improvement of water quality using alginate/montmorillonite composite beads. J. Appl. Polym. 119(4): 2371–2378.

Siddiqui, M.H. and Al-Whaibi, M.H. 2014. Role of nano-SiO_2 in germination of tomato (*Lycopersicum esculentum* seeds Mill.). Saudi J Biol Sci. 21: 13–17.

Siddiqui, S.I., Naushad, M. and Chaudhry, S.A. 2019. Promising prospects of nanomaterials for arsenic water remediation: A comprehensive review. Process Saf. Environ. Prot. 126: 60–97.

Stietiya, M.H. and J.J. Wang. 2014. Zinc and cadmium adsorption to aluminum oxide nanoparticles affected by naturally occurring ligands. J. Environ. Qual. 43(2): 498–506.

Sundaram, R.M., Sekiguchi, A., Sekiya, M., Yamada T. and Hata, K. 2018. Copper/carbon nanotube composites: research trends and outlook. Royal Soc. Open Sci. 5(11): 180814.

Tabish, T.A., Memon, F.A., Gomez, D.E., Horsell D.W. and Zhang. S. 2018. A facile synthesis of porous graphene for efficient water and wastewater treatment. Sci. Rep. 8(1): 1–14.

Tang, W., Su, Y., Li, Q., Gao S. and Shang, J.K. 2013. Superparamagnetic magnesium ferrite nanoadsorbent for effective arsenic (III, V) removal and easy magnetic separation. Water Res. 47(11): 3624–3634.

Tresintsi, S., Simeonidis, K., Estradé, S., Martinez-Boubeta, C., Vourlias, G., Pinakidou, F., Katsikini, M., et al. 2013. Tetravalent manganese feroxyhyte: a novel nanoadsorbent equally selective for As(III) and As(V) removal from drinking water. Environ. Sci. Technol. 47(17): 9699–9705.

Vareda, J.P. and Durães, L. 2019. Efficient adsorption of multiple heavy metals with tailored silica aerogel-like materials. Environ. Technol. 40(4): 529–541.

Wang, B., Lan, J., Bo, C., Gong, B. and Ou, J. 2023. Adsorption of heavy metal onto biomass-derived activated carbon. RSC Adv. 13(7): 4275–4302.

Yaqoob, S., Ullah, F., Mehmood, S., Mahmood, T., Ullah, M., Khattak, A. and Zeb, M.A. 2018. Effect of waste water treated with TiO_2 nanoparticles on early seedling growth of *Zea mays* L. J. Water Reuse Desaline. 8(3): 424–431.

Zand, A.D., Tabrizi, A.M. and Heir, A.V. 2020. Co-application of biochar and titanium dioxide nanoparticles to promote remediation of antimony from soil by Sorghum bicolor: metal uptake and plant response. Heliyon. 6(8): e04669.

Zhang, R. and Chen, W. 2014. Nitrogen-doped carbon quantum dots: facile synthesis and application as a "turn-off" fluorescent probe for detection of Hg^{2+} ions. Biosens Bioelectron. 55: 83–90.

Zhu, N., Zhang, B. and Yu, Q. 2020. Genetic engineering-facilitated coassembly of synthetic bacterial cells and magnetic nanoparticles for efficient heavy metal removal. ACS Appl. Mater. Interfaces. 12(20): 22948–22957.

Chapter 4

Nano-Technological Interventions for the Sustainable Management of Industrial Solid and Liquid Waste

Oroos Arshi[1]
[1]Department of Cyber Security and Forensics, School of Computer Science, University of Petroleum and Energy Studies, Dehradun

Soumitra K. Gupta[2] **and Surajit Mondal**[2*]
[2]Department of Electrical and Electronics Engineering, School of Engineering, University of Petroleum and Energy Studies, Dehradun

4.1 INTRODUCTION

Photocatalysis is a process that utilizes a substance called a photocatalyst to accelerate chemical reactions when exposed to light. This phenomenon has garnered significant attention due to its potential applications in various fields. The concept of photocatalysis dates back to the early 20th century, with initial studies conducted on the decomposition of oxalic acid and the production of formaldehyde through photocatalytic reactions. However, it wasn't until the discovery of titanium dioxide (TiO_2) as an effective photocatalyst in the 1960s that the field gained momentum. Nanotechnology, on the other hand, has emerged as a powerful tool for manipulating matter at the nanoscale, enabling scientists to design and engineer materials with unique properties. By exploiting the advantages of Nanomaterials, such as their large surface area-to-volume ratio and enhanced reactivity, nanotechnology has revolutionized various scientific disciplines. In the context of photocatalysis, nanotechnology has opened up new avenues for improving the efficiency and effectiveness of photocatalytic reactions. The role of nanotechnology in wastewater treatment has been a subject of intensive research, as the removal of pollutants from industrial liquid waste is a pressing environmental challenge. Industrial pollutants pose severe risks to ecosystems and human health, necessitating the development of advanced and sustainable solutions (Chen and Mao 2007). Nanotechnology offers a range of opportunities to address these challenges, including the use of Nanomaterials as adsorbents and the development of nanofiltration membranes for efficient pollutant removal.

---

*Corresponding Author: surajitmondalee@gmail.com

Moreover, the combination of nanotechnology with microbial processes has demonstrated promising results in wastewater treatment. Microorganisms, either naturally occurring or engineered, can be employed in conjunction with Nanomaterials to enhance pollutant degradation and promote the transformation of waste into valuable resources. The synergy between nanotechnology and enzyme technology has also emerged as a potent approach for the enzymatic degradation of pollutants, providing a greener and more efficient alternative to conventional treatment methods (Zhang and Chen 2011)

In the realm of waste management, nanotechnology has the potential to revolutionize the way we deal with industrial waste. Nanotechnologies offer innovative solutions for waste reduction, recycling, and the conversion of waste into valuable products. From the production of clinker, biogas, and biomolecules to the development of highly efficient adsorbents, nanotechnology plays a pivotal role in achieving higher production rates and promoting sustainable waste-to-resource conversion (Rana and Yadav 2017). With the growing interest and advancements in nanotechnology, researchers and scientists from diverse fields are actively exploring its applications in wastewater treatment and waste management. This paper aims to provide an overview of the role of nanotechnology in these domains, highlighting the potential of Nanomaterials, microbial-assisted nanotechnology, and enzyme technology for effective waste management. By harnessing the unique properties and capabilities of nanotechnology, we can pave the way for a more sustainable and environmentally friendly approach to wastewater treatment and waste management (Kim and Choi 2014).

4.2 Nanomaterials as Photocatalysts

Photocatalysis is a process that involves the use of a photocatalyst to activate a reaction by absorbing photons of light. The photocatalyst can be any material that can absorb light and transfer it to a reactant. Nanomaterials have shown great potential as photocatalysts due to their unique physical and chemical properties (Bhatkhande et al. 2002). The small size of the nanoparticles results in a high surface area to volume ratio, which increases the efficiency of the photocatalytic reaction. The use of Nanomaterials as photocatalysts also results in the reduction of energy consumption and reaction time. Figure. 4.1 illustrates the impact of utilizing Nanomaterials as photocatalysts on energy reduction.

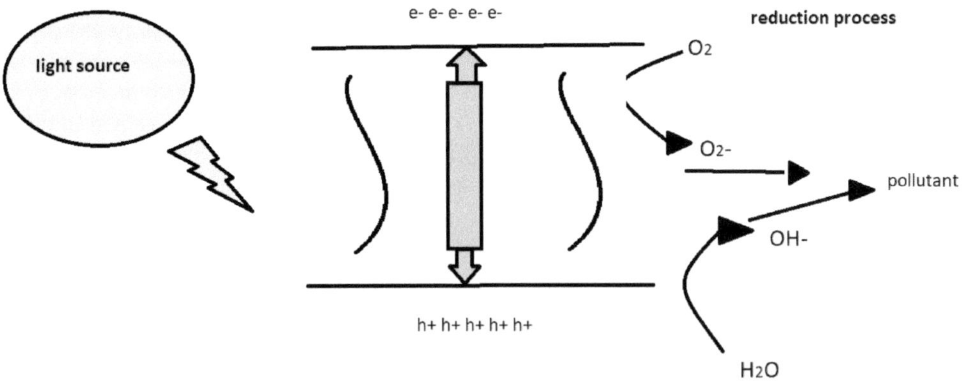

Figure 4.1 Impact of nano materials as photocatalysts on energy reduction.

Various types of Nanomaterials have been used as photocatalysts, including metal oxides, metal sulfides, carbon-based materials, and metal-organic frameworks (MOFs). Among these, metal oxide-based photocatalysts such as titanium dioxide (TiO_2), zinc oxide (ZnO), and tungsten oxide (WO_3) have been extensively studied due to their high efficiency and stability (Hu, X and Wang 2013). The use of TiO_2 as a photocatalyst has been widely investigated due to its high photocatalytic activity and chemical stability. Table 4.1 shows the key aspects and their descriptions related to the utilization of Nanomaterials and photocatalysts.

Table 4.1 Key aspects of nanomaterials in photocatalysis

Aspect	Description
Nanomaterials	Utilization of nanoscale materials as photocatalysts with enhanced surface area and reactivity.
Photocatalytic Activity	The ability of nanomaterials to initiate and accelerate chemical reactions through the absorption of light.
Bandgap Engineering	Tailoring nanomaterials' bandgap enables the absorption of specific wavelengths for efficient photocatalysis.
Reactivity and Selectivity	Nanomaterials' high reactivity and selectivity in driving specific chemical transformations in photocatalytic reactions.
Energy Conversion Efficiency	Enhanced efficiency of energy conversion in photocatalytic reactions facilitated by nanomaterials.
Catalyst Stability and Durability	Improved stability and durability of nanocatalysts, allowing for long-term and repeated use in photocatalysis.
Photogenerated Charge Separation	Nanomaterials promote efficient separation of electron-hole pairs generated during photocatalysis.
Surface Modification	Surface engineering of nanomaterials to enhance their photocatalytic properties, stability, and selectivity.
Heterogeneous Catalysis	Nanomaterials act as solid catalysts, enabling heterogeneous photocatalysis in various chemical reactions.
Environmental Remediation	Application of nanotechnology in the removal and degradation of pollutants, toxins, and contaminants in water and air.
Solar Energy Harvesting	Utilization of nanomaterials to harness solar energy and convert it into chemical energy through photocatalytic processes.
Water Splitting and Hydrogen Production	Nanomaterial-based photocatalysis for water splitting produces clean hydrogen as a renewable energy source.
Degradation of Organic Pollutants	Nanocatalysts effectively degrade organic pollutants and transform them into harmless byproducts through photocatalysis.
Self-Cleaning Surfaces	Nanomaterial-coated surfaces with self-cleaning properties, achieved through photocatalytic degradation of organic matter.
Water Purification and Disinfection	Nanotechnology-based photocatalysis for water purification and disinfection, eliminating harmful pathogens and contaminants.
Air Pollution Control	Use of nanomaterials for photocatalytic degradation of air pollutants, improving air quality, and reducing environmental impact.
Catalyst Design and Optimization	Nanotechnology allows for the design and optimization of catalyst structures and properties for enhanced photocatalytic performance.
Scale-up and Commercialization	Challenges and considerations in scaling up nanotechnology-based photocatalysis for practical and commercial applications.
Future Prospects and Research	Exploration of emerging trends, advancements, and future research directions in nanotechnology for photocatalysis.

4.3 APPLICATIONS OF PHOTOCATALYSTS

Photocatalysis has a wide range of applications in environmental remediation due to its ability to degrade and remove toxic pollutants from various sources. One of the most promising applications of photocatalysts is in the treatment of wastewater. Industrial wastewater is a major source of pollution that contains various organic and inorganic compounds that are harmful to the environment. Photocatalysis can be used to degrade these pollutants into harmless products (Tong et al. 2012). TiO_2-based photocatalysts have been extensively used for the treatment of wastewater due to their high efficiency and stability. The use of ZnO-based photocatalysts has also shown great potential in the treatment of wastewater due to their ability to remove various organic pollutants.

Photocatalysts can also be used for air purification. Air pollution is a major environmental concern that has led to various respiratory diseases (Zhang et al. 2013). The use of photocatalysts for air purification involves the degradation of volatile organic compounds (VOCs) and nitrogen oxides (NOx) into harmless products (6). TiO_2-based photocatalysts have been widely used for air purification due to their high efficiency and stability. The use of MOF-based photocatalysts has also shown great potential in air purification due to their high surface area (Li et al. 2005).

Another promising application of photocatalysts is in the remediation of hazardous waste. Hazardous waste is a major environmental concern that contains various toxic substances that are harmful to human health (Zhang et al. 2015) The use of photocatalysts for hazardous waste remediation involves the degradation of hazardous substances into harmless products. TiO_2-based photocatalysts have been used for the remediation of hazardous waste such as polychlorinated biphenyls (PCBs) and pesticides. The use of ZnO-based photocatalysts has also shown great potential in the remediation of hazardous waste such as dyes and heavy metals (Zhou et al. 2011).

The use of photocatalysts has also been investigated in other areas such as self-cleaning surfaces, antibacterial applications, and energy conversion. The self-cleaning surfaces involve the use of photocatalysts to degrade organic compounds and keep the surface clean. Antibacterial applications involve the use of photocatalysts to kill bacteria and other microorganisms (Wu et al. 2010). Energy conversion involves the use of photocatalysts to convert solar energy into electrical or chemical energy (13). Figure 4.2 illustrates the mechanism of hydrogen production by photocatalyst.

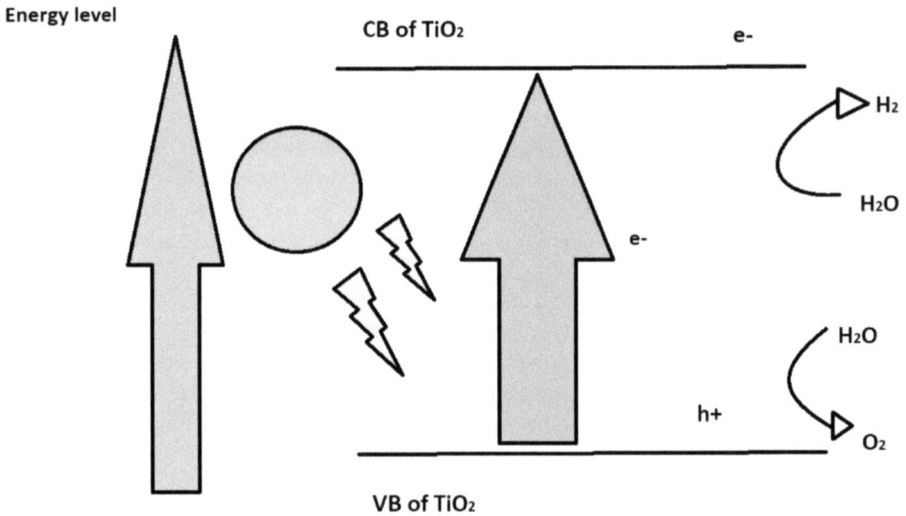

Figure 4.2 Diagrammatic representation of Hydrogen production mechanism by photocatalyst.

4.4 RECOVERY MECHANISM

The recovery mechanism involves the recovery of valuable resources from industrial solid and liquid waste. The use of Nanomaterials such as photocatalysts has shown great potential in the recovery mechanism of various resources such as metals, dyes, and organic compounds. The recovery of metals from industrial waste is one of the most promising areas of recovery mechanism (Amith et al. 2021). The use of photocatalysts can help in the recovery of metals such as gold, silver, and copper by reducing them into their elemental form. The use of TiO_2-based photocatalysts has shown great potential in the recovery of gold and silver from electronic waste. The use of ZnO-based photocatalysts has also been investigated for the recovery of copper from wastewater.

Dye recovery is another area where the use of photocatalysts has shown great potential. Dyes are widely used in various industries such as textile, paper, and leather. The discharge of dye-containing wastewater is a major environmental concern due to its toxicity and color (Huang et al. 2014). The use of photocatalysts can help in the degradation of dyes into harmless products and also in the recovery of valuable resources such as carbon and nitrogen (Pan et al. 2013). TiO_2-based photocatalysts have been extensively used for the degradation of dyes such as reactive black 5 and methyl orange (Liu et al. 2012). The use of MOF-based photocatalysts has also shown great potential in the recovery of carbon and nitrogen from dye-containing wastewater (Jin et al. 2013).

The recovery mechanism of organic compounds from industrial waste is another promising area of research. Organic compounds such as phenol, benzene, and toluene are widely used in various industries such as petrochemical, pharmaceutical, and food. The discharge of organic-containing wastewater is a major environmental concern due to its toxicity and impact on aquatic life (Roy et al. 2011). The use of photocatalysts can help in the degradation of organic compounds into harmless products and also in the recovery of valuable resources such as hydrogen and carbon dioxide. TiO_2-based photocatalysts have been extensively used for the degradation of organic compounds such as phenol and toluene (Li et al. 2015). The use of MOF-based photocatalysts has also shown great potential in the recovery of hydrogen and carbon dioxide from organic-containing wastewater (Asahi et al. 2001).

4.5 THE ROLE OF NANOTECHNOLOGY IN PHOTOCATALYSIS

Photocatalysis has emerged as a promising technology for various applications, including wastewater treatment, air treatment, disinfection, and the degradation of industrial waste. Nanotechnology plays a crucial role in enhancing the efficiency and effectiveness of photocatalytic processes (Linsebigler et al. 1995). This paper explores the utilization of nanotechnology in photocatalysis, focusing on wastewater treatment, the general mechanism of photocatalysis, the electron-hole separation ratio, catalyst structures, incident light intensity and temperature, air treatment, disinfection, and the chemical composition of industrial waste (Peng et al. 2011). The advancements in Nanomaterials and their impact on improving photocatalytic performance are discussed, highlighting the potential for addressing environmental challenges and promoting sustainable development. Photocatalysis has gained significant attention due to its potential in addressing environmental pollution and resource management challenges. By utilizing semiconductor materials, photocatalysis harnesses light energy to initiate chemical reactions and facilitate the degradation of various pollutants (Zou et al. 2001). The advancements in Nanomaterials and their impact on improving photocatalytic performance are discussed, highlighting the potential for addressing environmental challenges and promoting sustainable development. Photocatalysis has gained significant attention due to its potential in addressing environmental pollution and resource management challenges (Zhang et al. 2014). By utilizing semiconductor materials, photocatalysis harnesses light energy to initiate chemical reactions and facilitate the degradation of various pollutants (Liu et al. 2011).

4.6 WASTEWATER TREATMENT

4.6.1 Mechanisms of Photocatalytic Wastewater Treatment

Photocatalysis can effectively degrade organic pollutants present in wastewater through oxidation or reduction reactions (Li and Somorjai 2010). The semiconductor material, typically titanium dioxide (TiO_2), acts as a catalyst to accelerate the breakdown of pollutants. Table 4.2 shows an overview of some aspects and their description related to the utilization of Nanomaterials and photocatalysts in the wastewater treatment process (Chen et al. 2010)

Table 4.2 An overview of key aspects and descriptions related to the utilization of Nanomaterials and photocatalysis in wastewater treatment processes.

Aspect	Description
Nanomaterials	Utilization of nanoscale materials as catalysts, adsorbents, and membranes in wastewater treatment processes.
Photocatalysis	Implementation of nanomaterial-based photocatalysts for degradation of organic pollutants under light irradiation.
Adsorption	Application of nanomaterials with high surface area for adsorptive removal of contaminants from wastewater.
Membrane Technology	Use of nanofiltration and ultrafiltration membranes for selective separation and removal of pollutants.
Antibacterial Properties	Incorporation of antimicrobial nanoparticles to inhibit the growth of bacteria and pathogens in wastewater.
Contaminant Degradation	Nanomaterials' ability to catalyze the breakdown of toxic compounds into less harmful substances in wastewater.
Sensor Technology	Development of nanosensors for real-time monitoring of water quality parameters and contaminants in wastewater.
Filtration Efficiency	Enhanced filtration efficiency through the use of nanomaterial-based filters, reducing suspended solids in water.
Heavy Metal Removal	Nanomaterials' affinity for heavy metal ions, facilitates their removal from wastewater through adsorption processes.
Membrane Fouling Mitigation	Surface modification of membranes with nanoparticles to reduce fouling and improve long-term performance.
Environmental Sustainability	Nanotechnology offers greener and more sustainable approaches to wastewater treatment, minimizing environmental impact.
Energy Efficiency	Nanomaterials-based processes can achieve high energy efficiency, reducing the energy footprint of wastewater treatment.
Water Reuse and Resource Recovery	Nanotechnology facilitates the recovery of valuable resources from wastewater, such as nutrients and clean water.
Scale-up Challenges	Challenges in scaling up nanotechnology-based wastewater treatment processes for practical and large-scale applications.
Cost-effectiveness	Evaluation of the cost-effectiveness and economic viability of implementing nanotechnology in wastewater treatment.
Regulatory Considerations	Addressing regulatory aspects and safety concerns associated with the use of nanomaterials in wastewater treatment.
Future Research Directions	Identification of research gaps and future directions to further enhance the efficiency and effectiveness of nanotechnology in wastewater treatment.

4.6.2 Nanomaterials for Enhanced Wastewater Treatment

The use of Nanomaterials in photocatalysis has shown significant improvements in the efficiency of wastewater treatment (Milis and Hunte 1997). Nanostructured catalysts, such as TiO_2 nanoparticles and nanotubes, exhibit larger surface areas and enhanced adsorption capacities, leading to improved pollutant removal (Fujishima and Zhang 2006).

4.7 GENERAL MECHANISM OF PHOTOCATALYSIS

The photocatalytic process involves the generation of electron-hole pairs upon absorption of photons by the semiconductor material (Chong et al. 2010). The excited electrons and holes participate in redox reactions, leading to the degradation of pollutants.

4.8 ELECTRON-HOLE SEPARATION RATIO

Efficient electron-hole separation is crucial for achieving high photocatalytic activity (Zhang and Wong 2019). Nanostructured catalysts, such as doped or co-catalyst-modified semiconductors, promote effective charge separation, minimizing electron-hole recombination and enhancing reaction rates (Chen et al. 2010). Figure 4.3 represents the process of Electron-Hole separation using nanomaterial photocatalysis.

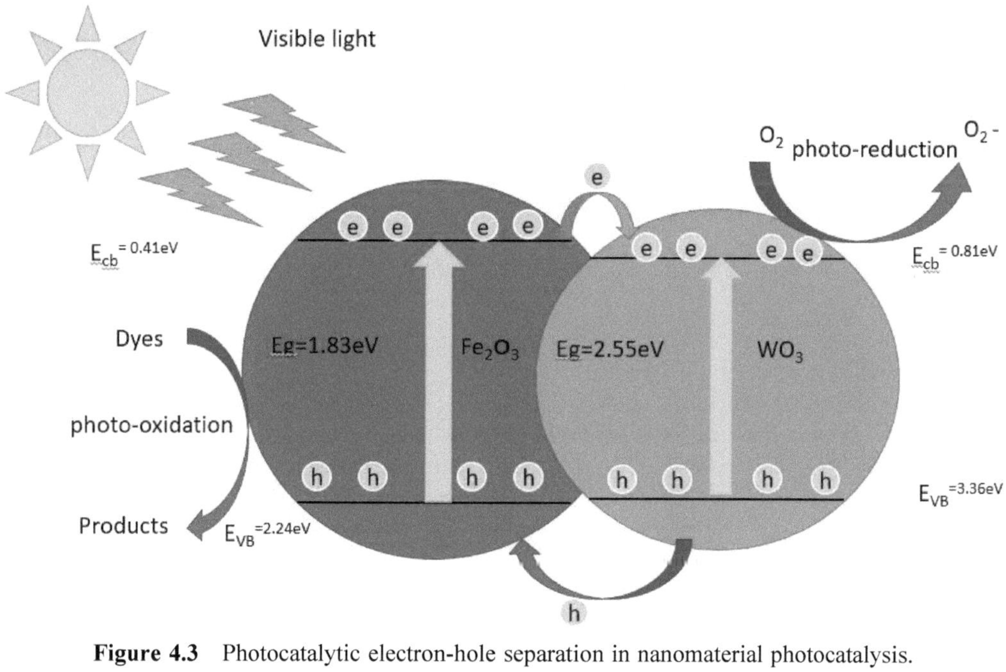

Figure 4.3 Photocatalytic electron-hole separation in nanomaterial photocatalysis.

4.9 CATALYST STRUCTURES

The structure and morphology of catalysts significantly influence photocatalytic performance (Hoffmann et al. 1995). Various nanostructures, including nanoparticles, nanowires, and nanosheets, provide increased surface areas, improved light absorption, and efficient charge transport (Chen et al. 2012).

4.10 INCIDENT LIGHT INTENSITY AND TEMPERATURE

The incident light intensity and temperature play vital roles in photocatalysis (Zhu et al. 2014). Optimal light intensity ensures sufficient photon absorption, while temperature affects reaction kinetics and adsorption-desorption processes (Chen et al. 2012).

4.11 AIR TREATMENT

Photocatalysis offers a sustainable approach to air purification by decomposing harmful gases and volatile organic compounds (VOCs) (Li et al. 2019). Nanomaterials, such as zinc oxide (ZnO) and tungsten trioxide (WO_3) (Gaya and Abdullah 2008), exhibit enhanced photocatalytic activity and selectivity for air treatment applications (Jain et al. 2021). Figure 4.4 illustrates the diagrammatic representation of the photocatalytic Air Treatment Process.

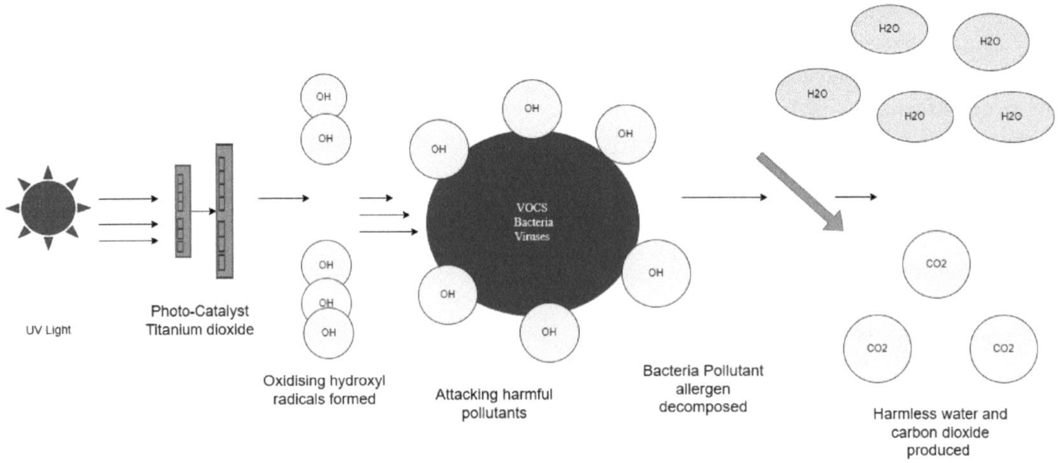

Figure 4.4 Schematic diagram of the photocatalytic air treatment process.

4.12 DISINFECTION

Photocatalytic disinfection utilizes light energy to inactivate microorganisms, offering an environmentally friendly alternative to traditional disinfection methods (Thanh 2014). Nanostructured photocatalysts exhibit enhanced disinfection efficiency, contributing to water and air quality improvement.

4.13 CHEMICAL COMPOSITION ON INDUSTRIAL WASTE

The chemical composition of industrial waste influences the photocatalytic degradation process. Nanotechnology enables the design and synthesis of tailored photocatalysts for specific waste types, optimizing degradation efficiency (Manikandan et al. 2022)

4.14 NANOTECHNOLOGY IN WASTEWATER TREATMENT

Nanotechnology has emerged as a promising approach for enhancing wastewater treatment processes. This paper explores the applications of nanotechnology in wastewater treatment, focusing on nano-adsorbents and nanofiltration membranes, microorganisms-assisted nanotechnology, nanotechnology and enzyme technology, and nano-technologies for waste management (Javaid et al. 2021). The use of Nanomaterials and nanodevices offer numerous advantages, including enhanced adsorption capacities, improved filtration efficiency, targeted pollutant degradation, and efficient waste management. This paper discusses the potential of nanotechnology in revolutionizing wastewater treatment and addressing the challenges associated with water pollution and resource conservation.

Wastewater treatment plays a crucial role in ensuring environmental sustainability and human well-being (Dong et al. 2021). The integration of nanotechnology into wastewater treatment processes offers new possibilities for improving treatment efficiency and achieving higher water quality standards (Gomaa 2022).

4.15 NANO-ADSORBENTS AND NANOFILTRATION MEMBRANES

Microorganisms-assisted nanotechnology is an emerging and rapidly growing field that combines the capabilities of microorganisms with nanotechnology to achieve advanced applications (Hairon et al. 2021). By leveraging the unique characteristics of microorganisms, such as bacteria and fungi, scientists can produce nanoparticles using biomineralization processes, providing a more sustainable and economical alternative to conventional chemical synthesis techniques. Furthermore, microorganisms serve as a source of inspiration for the development of bio-inspired Nanomaterials that mimic natural processes and structures (Chen et al. 2020). This integration of microorganisms and nanotechnology presents significant prospects for creating novel materials, devices, and processes with improved properties and functionalities (Zhang et al. 2019)

4.16 GREEN NANOTECHNOLOGY FOR SUSTAINABLE MANAGEMENT OF WASTAGE

Green/biogenic nanoparticles present promising solutions for waste management, utilizing their unique properties derived from natural sources such as plants or microorganisms. These nanoparticles can be effectively utilized for environmental remediation, particularly in addressing contaminated sites (Jin et al. 2013). Their ability to selectively bind or break down pollutants, including heavy metals and organic compounds, contributes to the detoxification of the environment. Furthermore, these nanoparticles find application in wastewater treatment by serving as efficient catalysts or adsorbents, facilitating the removal of pollutants and improving the purification process. By adopting green nanoparticles, waste management can be approached in a sustainable and environmentally friendly manner, offering viable solutions to environmental challenges and paving the way for a cleaner and healthier future (Serrao Sausa et al. 2017).

4.17 NANO-ADSORBENTS FOR CONTAMINANT REMOVAL

Nanomaterials such as carbon nanotubes, graphene oxide, and metal-organic frameworks have shown exceptional adsorption capacities for various contaminants. These nano-adsorbents can efficiently remove heavy metals, organic pollutants, and emerging contaminants from wastewater (Donaldson and Poland 2013).

4.18 NANOFILTRATION MEMBRANES FOR SELECTIVE FILTRATION

Nanofiltration membranes with nanopores exhibit selective filtration properties, allowing the separation of specific contaminants from wastewater (Zhang et al. 2021). These membranes offer high permeability and rejection rates, contributing to the removal of dissolved pollutants and salts (Enesca and Andronic 2010).

4.19 MICROORGANISMS ASSISTED NANOTECHNOLOGY

4.19.1 Microbial Fuel Cells for Energy Generation

Microbial fuel cells utilize microorganisms to convert organic matter in wastewater into electricity (Zhang et al. 2019). Nanotechnology plays a vital role in improving the performance of microbial fuel cells through the development of efficient electrode materials and nanostructured biofilms (Yakoob et al. 2020).

4.19.2 Microbial Nano-sensors for Monitoring

Microorganisms can be engineered to act as nano-sensors for monitoring water quality parameters. These microbial nano-sensors can detect and quantify specific contaminants, providing real-time information for efficient wastewater treatment (Liang and Liu 2019).

4.20 NANOTECHNOLOGY AND ENZYME TECHNOLOGY

4.20.1 Nano-Enzymes for Enhanced Degradation

Nano-enzymes, consisting of enzymes immobilized on Nanomaterials, exhibit improved stability and catalytic activity. These nano-enzymes can efficiently degrade pollutants in wastewater, offering a more sustainable and efficient treatment approach (Murica et al. 2019).

4.20.2 Enzyme-Mimicking Nanomaterials

Enzyme-mimicking Nanomaterials, such as metal-organic frameworks and nanocatalysts, mimic the catalytic properties of natural enzymes. These Nanomaterials can effectively degrade pollutants and provide a reusable and cost-effective solution for wastewater treatment (Yang et al. 2019).

4.21 NANO-TECHNOLOGIES FOR WASTE MANAGEMENT

4.21.1 Nano-sensors for Pollution Monitoring

Nano-sensors offer precise and real-time monitoring of pollutants in wastewater. These small-scale devices can detect low concentrations of contaminants, enabling efficient waste management and targeted remediation strategies (Wafi et al. 2020).

4.21.2 Nanomaterials for Sludge Treatment

Nanotechnology-based approaches, including nano-adsorbents and photocatalysts, show potential in treating sludge generated during wastewater treatment (Wafi et al. 2020. These Nanomaterials can effectively adsorb heavy metals and degrade organic compounds, reducing the volume and toxicity of sludge. Figure 4.5 illustrates the application of Nanomaterials in the treatment of sludge and highlights the role of Nanomaterials in the process (Belet et al. 2019).

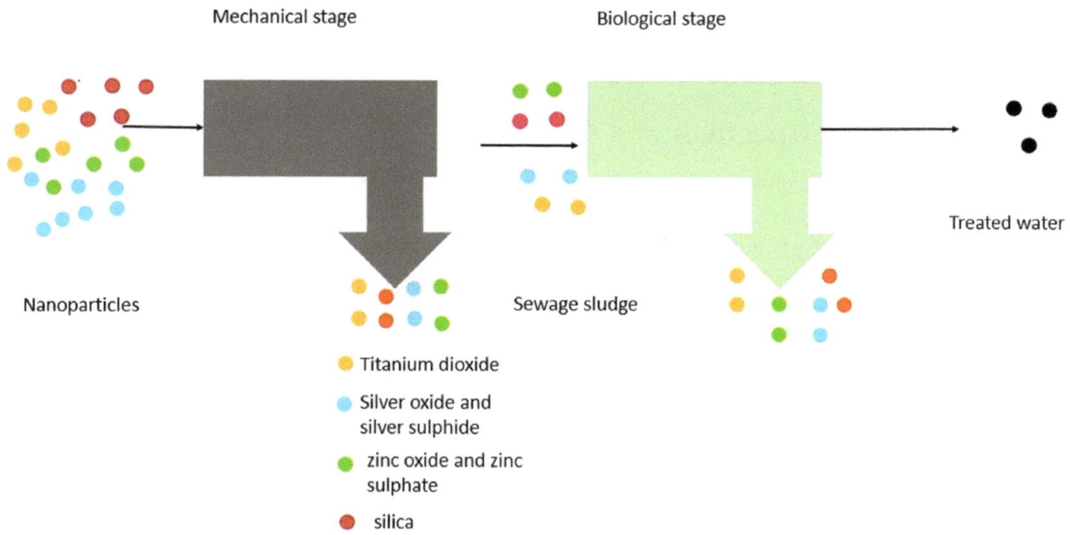

Figure 4.5 An overview of the nanomaterial-assisted sludge treatment process.

4.22 NANO-TECHNOLOGIES FOR WATER REUSE

4.22.1 Advanced Oxidation Processes (AOPs)

Nano-enabled AOPs, such as photocatalysis and electrochemical oxidation, offer efficient degradation of persistent organic pollutants in wastewater, enabling water reuse (Patel et al. 2020). These technologies utilize Nanomaterials to enhance reaction rates and promote complete pollutant removal (Pham et al. 2020).

4.22.2 Nanofiltration and Reverse Osmosis

Nanotechnology-based membranes, including nanofiltration and reverse osmosis membranes, enable the removal of contaminants, salts, and microorganisms, producing high-quality water for various applications (Hernandez et al. 2020). These membranes exhibit excellent rejection capabilities and can contribute to sustainable water reuse practices.

4.23 ENVIRONMENTAL IMPLICATIONS AND SAFETY OF NANOTECHNOLOGY

4.23.1 Nanoparticle Release and Fate

Understanding the fate and behavior of nanoparticles in wastewater treatment systems is crucial for assessing their environmental implications. Research on the release, transport, and potential accumulation of nanoparticles helps in ensuring the safe and responsible application of nanotechnology in wastewater treatment (Luo et al. 2020).

4.23.2 Toxicity and Ecotoxicity Assessment

The potential toxicity of Nanomaterials needs to be evaluated to ensure their safe implementation. Assessing the ecotoxicity of nanoparticles helps in understanding their impact on aquatic organisms and ecosystems, aiding in the development of sustainable wastewater treatment strategies (Qu, Alvarez and Li 2013).

4.23.3 Future Perspectives and Conclusion

The utilization of nanotechnology in wastewater treatment holds immense promise, but several challenges must be addressed. These challenges include scaling up nanotechnology-based processes, ensuring cost-effectiveness, optimizing material synthesis, and addressing environmental and safety concerns (Dona et al. 2021). In conclusion, photocatalysis has emerged as a powerful technology with diverse applications in addressing pollution and energy challenges. The use of Nanomaterials as photocatalysts has revolutionized the field by offering enhanced efficiency and versatility. Semiconductor-based Nanomaterials, with their excellent photocatalytic activity and wide bandgap, have demonstrated superior performance in various photocatalytic reactions. The utilization of photocatalytic technology provides a cost-effective and environmental friendly approach to combating organic and inorganic pollutants, heavy metals, and microbial contaminants. This technology shows great promise in tackling pollution-related problems by efficiently degrading and removing pollutants from wastewater and industrial liquid waste. The ability of photocatalytic reactions to operate under ambient conditions and utilize light as an energy source makes them highly sustainable and energy-efficient. Several parameters influence the photocatalytic activity, including the electron-hole pair separation/recombination rate, temperature, and pH. Fine-tuning these parameters can optimize the efficiency of the photocatalytic process and enhance pollutant removal. The integration of nanotechnology with microorganisms has further expanded the potential of photocatalytic technology. Microorganisms-assisted nanotechnology offers a green and sustainable approach to wastewater treatment by harnessing the synergistic effects of microbial processes and Nanomaterials. This approach has shown promising results in bioremediation, waste valorization, and the production of biohydrogen and bioelectricity from industrial waste. To fully realize the potential of photocatalytic technology, large-scale implementation, and further research and development are essential. Continued efforts should focus on optimizing catalyst structures, improving electron-hole pair separation efficiency, and exploring new Nanomaterials with enhanced photocatalytic properties. Additionally, the development of innovative approaches for scaling up photocatalytic processes and integrating them into existing wastewater treatment systems is crucial. The advancements in photocatalysis and nanotechnology offer immense opportunities for sustainable waste management, pollution control, and the production of clean energy. By harnessing the power of Nanomaterials and exploiting their unique properties, we can create a cleaner and more sustainable future for wastewater treatment and waste management. Overall, photocatalytic technology, coupled with nanotechnology, holds great promise in addressing global environmental challenges and promoting a more sustainable and greener approach to wastewater treatment and waste management.

References

Ajith, M., Aswathi, M., Priyadarshini, E. and Rajamani, P. 2021. Recent innovations of nanotechnology in water treatment: a comprehensive review. Bioresour. Technol. 342: 126000. https://doi.org/10.1016/j.biortech.2021.126000.

Asahi, R., Morikawa, T., Ohwaki, T., Aoki, K. and Taga, Y. 2001. Visible-light photocatalysis in nitrogen-doped titanium oxides. Sci. 293(5528): 269–271.

Belet, A., Wolfs, C., Mahy, J.G., Poelman, D., Vreuls, C., Gillard, N. and Lambert, S.D. 2019. Sol-gel Syntheses of photocatalysts for the removal of pharmaceutical products in water. Nanomaterials. 9: 126.

Bhatkhande, D.S., Pangarkar, V.G. and Beenackers, A.A. 2002. Photocatalytic degradation for environmental applications. J. Chem. Technol. Biotechnol. 77(1): 102–116.

Chen, X. and Mao, S.S. 2007. Titanium dioxide nanomaterials: synthesis, properties, modifications, and applications. Chem. Rev. 107(7): 2891–2959.

Chen, X., Shen, S., Guo, L. and Mao, S.S. 2010. Semiconductor-based photocatalytic hydrogen generation. Chem. Rev. 110(11): 6503–6570.

Chen, X., Li, C., Grätzel, M., Kostecki, R. and Mao, S.S. 2012. Nanomaterials for renewable energy production and storage. Chem. Soc. Rev. 41(23): 7909–7937.

Chen, T., Zhao. Y, Qiu. X, Zhu. X, Liu. X, Yin. J., Shen, D., et al. 2020. Economics analysis of food waste treatment in China and its influencing factors. Front. Environ. Sci. Eng. 15: 33 (2021). https://doi.org/10.1007/s11783-020-1325-y.

Chong, M.N., Jin, B., Chow, C.W. and Saint, C. 2010. Recent developments in photocatalytic water treatment technology: A review. Water. Res. 44(10): 2997–3027.

Donaldson, K. and Poland, C.A. 2013. Nanotoxicity: challenging the myth of nano-specific toxicity. Curr. Opin. Biotechnol. 24(4): 724–734. https://doi.org/10.1016/j.copbio.2013.05.003.

Doña-Rodríguez, J.M. and Pulido Melián, E. 2021. Nano-Photocatalytic materials: possibilities and challenges. Nanomaterials. 11: 688. https://doi.org/10.3390/nano11030688.

Enesca, A. and Andronic, L. 2010. The influence of photoactive heterostructures on the photocatalytic removal of dyes and pharmaceutical active compounds: a mini-review. Nanomaterials. 10: 1766.

Fujishima, A. and Zhang, X. 2006. Titanium dioxide photocatalysis: present situation and future approaches. C.R. Chim. 9(5–6): 750–760.

Gaya, U.I. and Abdullah, A.H. 2008. Heterogeneous photocatalytic degradation of organic contaminants over titanium dioxide: a review of fundamentals, progress and problems. J. Photochem. Photobiol. C. Photochem. Rev. 9(1): 1–12.

Giwa, A., Yusuf, A., Balogun, H.A., Sambudi, N.S., Bilad, M.R., Adeyemi, I., Chakraborty, S., et al. 2021. Recent advances in advanced oxidation processes for removal of contaminants from water: a comprehensive review. Process Saf. Environ. Prot. 146: 220–256. https://doi.org/10.1016/j.psep.2020.08.015.

Gomaa, E.Z. 2022. Nanozymes: a promising horizon for medical and environmental applications. J. Clust. Sci. 33: 1275–1297. https://doi.org/10.1007/s10876-021-02079-4.

Hairom, N.H.H., Soon, C.F., Mohamed, R.M.S.R., Morsin, M., Zainal, N., Nayan, N., Zulkifli, C.Z. et al. 2021. A review of nanotechnological applications to detect and control surface water pollution. Environ. Tech. Innovation. 24: 102032. https://doi.org/10.1016/j.eti.2021.102032.

Hernández Rodríguez, M.J., Pulido Melián, E., Araña, J., Navío, J.A., González Díaz, O.M., Santiago, D.E. and Doña Rodríguez, J.M. 2020. Influence of Water on the Oxidation of NO on Pd/ TiO_2 Photocatalysts. Nanomaterials. 10: 2354.

Hoffmann, M.R., Martin, S.T., Choi, W. and Bahnemann, D.W. 1995. Environmental applications of semiconductor photocatalysis. Chem. Rev. 95(1): 69–96.

Hu, X. and Wang, Y. 2013. Photocatalytic degradation of volatile organic compounds in gas streams: J. Hazard. Mater. 258: 235–246.

Jain. K., Patel, A.S, Pardhi, V.P and Flora, S.J.S. 2021. Nanotechnology in wastewater management: a new paradigm towards wastewater treatment. Molecules. 26(6): 1797. https://doi.org/10.3390/molecules26061797

Javaid, M., Haleem, A., Singh, R.P., Rab, S. and Suman, R. 2020. Exploring the potential of nanosensors: a brief overview. Sens. Int. 2: 100130. https://doi.org/10.1016/j.sintl.2021.100130.

Jin, Z., Li, Z., Guo, L. and Shi, J. 2013. Reduced graphene oxide-coated TiO_2 nanotube arrays for enhanced photoelectrochemical water splitting. J. Mater. Chem. 1(21): 6304–6311.

Kim, T.W. and Choi, W. 2014. Nanoparticle-assisted photocatalytic water treatment using light-emitting diodes. J. Chem. Eng. 251: 34–45.

Li, D., Haneda, H. and Hishita, S. 2005. Preparation and photocatalytic properties of mesoporous Bi_2O_3 powders. J. Appl. Catal. 288(1–2): 143–151.

Li, Y. and Somorjai, G.A. 2010. Nanoscale advances in catalysis and energy applications. or Nano Lett. 10(7): 2289–2295.

Li, W., Chen, J. and Zhao, J. 2015. Visible light–driven photocatalysis on graphene-based nanocomposites. Analyst. 140(10): 3382–3395.

Li, X., Chen, Q. and Yang, S.T. 2019. Photocatalytic disinfection of pathogens using carbon-based Nanomaterials: a review. J. Hazard. Mater. 368: 500–513.

Liang, Y.-C. and Liu, Y.-C. 2019. Microstructures and photodegradation performance toward methylene orange of sputtering-assisted decoration of $ZnFe_2O_4$ crystallites onto TiO_2 nanorods. Nanomaterials. 9: 205.

Linsebigler, A.L., Lu, G. and Yates Jr, J.T. 1995. Photocatalysis on TiO_2 surfaces: principles, mechanisms, and selected results. Chem. Rev. 95(3): 735–758.

Liu, Y., Yu, L. and Zhang, N. 2011. Effects of light intensity on photocatalytic degradation of gaseous formaldehyde by nano–TiO_2 films. J. Hazard. Mater. 190(1–3): 80–86.

Liu, X., Li, Y., Liu, Y., Yang, J., Zhang, Y. and Sun, L. 2012. Photocatalytic hydrogen evolution over graphene-modified TiO_2 nanosheets with exposed (001) facets under visible light. Appl. Catal. B. 111: 87–94.

Luo, G., Liu, H., Li, W. and Lyu, X. 2020. Automobile exhaust removal performance of pervious concrete with nano TiO_2 under photocatalysis. Nanomaterials. 10: 2088.

Manikandan, S., Subbaiya, R., Saravanan, M., Ponraj, M., Selvam, M. and Pugazhendhi, A. 2022. A critical review of advanced nanotechnology and hybrid membrane based water recycling, reuse, and wastewater treatment processes. Chemosphere. 289: 132867. https://doi.org/10.1016/j.chemosphere.2021.132867.

Mills, A. and Le Hunte, S. 1997. An overview of semiconductor photocatalysis. J. Photochem. Photobiol. A. 108(1): 1–35.

Murcia, J.J., Ávila-Martínez, E.G., Rojas, H., Cubillos, J., Ivanova, S., Penkova, A. and Laguna, O.H. 2019. Powder and nanotubes titania modified by dye sensitization as photocatalysts for the organic pollutants elimination. Nanomaterials. 9: 517.

Pan, J., Xu, Y., Li, J., Yu, H., Zhang, Y., Liu, B. and Wang, L. 2013. Enhanced photocatalytic hydrogen evolution of ZnO nanosheets modified by graphene quantum dots under visible light. Appl. Catal. B. 142: 638–644.

Patel, J.; Singh, A.K.; Carabineiro, S.A.C. 2020. Assessing the photocatalytic degradation of fluoroquinolone norfloxacin by Mn:ZnS quantum dots: kinetic study, degradation pathway and influencing factors. Nanomaterials. 10: 964.

Peng, S., Li, L., Hu, Y. and Wang, X. 2011. Shape-controlled synthesis of CdS nanocrystals in an aqueous system with tunable band gap energy. J. Mater. Chem. 21(27): 9910–9916.

Pham, P., Rashid, M., Cai, Y., Yoshinaga, M., Dionysiou, D.D. and O'Shea, K. 2020. Removal of As(III) from water using the adsorptive and photocatalytic properties of humic acid-coated magnetite nanoparticles. Nanomaterials. 10: 1604.

Poornima, S., Manikandan, S., Karthik, V., Balachandar, R., Subbaiya, R., Saravanan, M., Lan Chi, N.T., et al. 2022. Emerging nanotechnology based advanced techniques for wastewater treatment. Chemosphere. 303: 135050. https://doi.org/10.1016/j.chemosphere.2022.135050.

Qu, X., Alvarez, P.J. and Li, Q. 2013. Applications of nanotechnology in water and wastewater treatment. Water Res. 47(12): 3931–3946. https://doi.org/10.1016/j.watres.2012.09.058.

Rana, A.K. and Yadav, R.N. 2017. Photocatalytic wastewater treatment by using nanomaterials. Int. J. Environ. Sci. Nat. Resource. 5(1): 1–5.

Roy, P., Berger, S. and Schmuki, P. 2011. TiO_2 nanotubes: synthesis and applications. Angew. Chem. Int. Ed. 50(13): 2904–2939.

Serrão Sousa, V., Corniciuc, C. and Ribau Teixeira, M. 2017. The effect of TiO_2 nanoparticles removal on drinking water quality produced by conventional treatment C/F/S. Water Res. 109: 1–12. https://doi.org/10.1016/j.watres.2016.11.030.

Thanh, N.T.K. 2014. Magnetic nanoparticle-based approaches to the detection and treatment of diseases. Int. J. Mol. Sci. 15(10): 17275–17297.

Tong, H., Ouyang, S., Bi, Y., Umezawa, N., Oshikiri, M., Ye, J. 2012 Nano-photocatalytic materials: possibilities and challenges. Adv. Mater. 24(2): 229–251. doi: 10.1002/adma.201102752. Epub 2011 Oct 4. PMID: 21972044.

Wafi, A., Szabó-Bárdos, E., Horváth, O., Pósfai, M., Makó, É., Juzsakova, T. and Fónagy, O. 2020. The photocatalytic and antibacterial performance of nitrogen-doped TiO_2: surface-structure dependence and silver-deposition effect. Nanomaterials. 10: 2261.

Wei, Z., Endo-Kimura, M., Wang, K., Colbeau-Justin, C. and Kowalska, E. 2019. Influence of semiconductor morphology on photocatalytic activity of plasmonic photocatalysts: titanate nanowires and octahedral anatase nanoparticles. Nanomaterials. 9: 1447.

Wu, J.C., Chen, C.Y., Chen, Y.L. and Tseng, R.L. 2010. Innovative nanocomposite membrane of functionalized carbon nanotubes/polyvinylidene fluoride for efficient separation of oil/water emulsion. J. Hazard. Mater. 181(1–3): 998–1005.

Yang, J., Du, J., Li, X., Liu, Y., Jiang, C., Qi, W., Zhang, K., et al. 2019. Highly hydrophilic TiO_2 nanotubes network by alkaline hydrothermal method for photocatalysis degradation of methyl orange. Nanomaterials. 9: 526.

Yaqoob, A.A., Md. Noor, N.H.B., Serrà, A. and Md. Ibrahim, M.N. 2020. Advances and challenges in developing efficient graphene oxide-based ZnO photocatalysts for dye photo-oxidation. Nanomaterials. 10: 932.

Zhang, H. and Chen, G. 2011. Potentials and challenges of nanomaterials for wastewater treatment. Nano. Today. 6(2): 78–89.

Zhang, J., Liu, S., Wu, J., Guo, Y. and Li, M. 2012. Photocatalytic Cr (VI) reduction by TiO_2 nanoparticles: Mechanism and kinetic study. J. Hazard. Mater. 213: 439–446.

Zhang, H., Chen, G. and Bahnemann, D.W. 2013. Photoelectrocatalytic materials for environmental applications. J. Mater. Chem. A. 1(23): 6691–6709.

Zhang, X., Zhang, Y. and Chen, Y. 2014. Synthesis and visible light photocatalytic activity of graphene oxide-modified TiO_2 nanocomposites. J. Mater. Chem. 2(39): 16534–16542.

Zhang, J., Zhang, M., Zhang, J. and Wang, Y. 2015. Recent advances in photocatalysis for environmental applications. J. Environ. Sci. 28: 123–131.

Zhang, L. and Wong, K.H. 2019. Advanced photocatalytic Nanomaterials for wastewater treatment. *In:* Advanced Environmental Analysis. Springer, Cham. 163–191.

Zhang, H., Yu, M. and Qin, X. 2019. Photocatalytic activity of TiO_2 nanofibers: the surface crystalline phase matters. Nanomaterials. 9: 535.

Zhang, H., Li, J., He, X. and Liu, B. 2020. Preparation of a g-C_3N_4/UiO-66-NH_2/CdS Photocatalyst with enhanced visible light photocatalytic activity for tetracycline degradation. Nanomaterials. 10(9): 1824.

Zhou, M., Chen, X. and Chen, Y. 2011. Environmental applications of semiconductor photocatalysis. Chem. Soc. Rev. 40(3): 1308–1324.

Zhu, Y., Guo, C., Wang, Q., Xu, Y. and Wu, D. 2014. Enhanced photocatalytic activity of ZnO hierarchical microspheres for the degradation of organic dyes. J. Alloys Compd. 583: 177–183.

Zou, Z., Ye, J., Sayama, K. and Arakawa, H. 2001. Direct splitting of water under visible light irradiation with an oxide semiconductor photocatalyst. Nature. 414(6864): 625–627.

Section III

Water and Wastewater Treatment/Remediation

Chapter 5

Photodegradation of Environmental Contaminants by Nanomaterials

Amitava Mukherjee[1]*
[1]Centre of Nanobiotechnology, VIT, Vellore 632014, India

Namrata Roy[1,2] and Krishnan Kannabiran[2]
[2]School of Biosciences and Technology, VIT, Vellore 632014, India

5.1 INTRODUCTION

Over the decades, photocatalysis is considered a promising alternative to the existing water purification technologies (Tsang et al. 2019, Shume et al. 2020, Vaya and Surolia 2020, Ouyang et al. 2021). The process of photocatalytic reactions uses a heterogeneous catalyst that possesses the properties of light absorption. The catalyst interacts with the target compounds present in a liquid or gas phase (Kanan et al. 2020). In a typical mechanism of photodegradation, the irradiation of incident photons excites the electrons (e^-) in the conduction band whereas holes (h^+) are formed in the valence band of the catalyst. These photons generated e^- and h^+ undergo recombination to generate thermal energy or react with the adsorbed compounds on the photocatalyst surface (Peiris, de Silva et al. 2021). Furthermore, radical species like superoxide ($O_2^{\bullet -}$) and hydroxyl ($OH^\bullet$) radicals are generated that are highly responsive and decompose organic compounds (pollutants) on a non-selective basis (Chen et al. 2021). The generation of these reactive radicals is influenced by band positioning and utilization of the charge carrier which directly impacts the photodegradation performance. Therefore, to facilitate the generation of reactive radicals, photocatalysts with steric structure and known irradiation intensity are employed (Samsudin et al. 2018). To satisfy the above parameters, the photocatalyst with improved photocatalytic activity is proposed. In recent years, variety of semiconductors have been applied for pollutant degradation like TiO_2, WO_3, ZnO, CeO_2, BiO_2, g–C_3N_4, and CdS. The inexpensive, non-toxic, thermally and chemically stable, and light harvesting properties have contributed to their recent applications for wastewater treatment (Okpara et al. 2023).

---

*Corresponding Author: amitav@vit.ac.in, amit.mookerjea@gmail.com

Furthermore, the main aim of employing photocatalysts in water treatment techniques is to refine the traditional process of water purification. The available traditional methods suffer from the drawbacks of low efficiency, the requirement for higher energies, and complicated process (Al-Mamun et al. 2019). These pitfalls convinced researchers to conduct experiments for developing nanomaterials that can be naturally activated against contaminants.

5.2 PHOTOCATALYST AS A PROMISING TOOL FOR WATER CONTAMINANT REMOVAL

Fujishima and Honda discovered photocatalyst in 1972, Titanium dioxide (TiO_2) was the first recognized photocatalyst and since then has been explored for water splitting and hydrogen generation in an electrochemical cell (Guo et al. 2019). Later, in 1991, Bahnemann and co-workers discovered that TiO_2 has application in the treatment of wastewater by the process of photocatalysis (Shayegan et al. 2018). As TiO_2 was explored more, its application extended to the industrial fields for air cleaning, self-purification, and anti-corrosion. The influence of light source and thermal and pH change was reported to enhance the potential of TiO_2 photocatalyst (Arun et al. 2023). Photocatalytic research is a solar energy-driven process that harvests solar energy to convert into chemical energy to initiate a chemical reaction. The role of a photocatalyst is to harvest solar power for photon reactions. An ideal photocatalyst needs to satisfy two conditions: (i) the photocatalyst acts indirectly in a reaction and should not be consumed completely; and (ii) should have diverse mechanism routes apart from available photo reactions and fast reaction rates (Natarajan et al. 2018). The other forms of photocatalysts are represented in Figure 5.1.

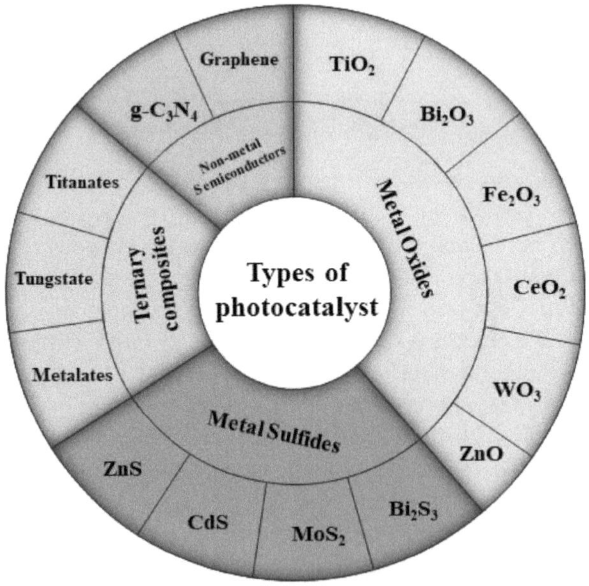

Figure 5.1 Different types of photocatalyst.

5.2.1 Water contaminants

5.2.1.1 Pharmaceuticals and Personal Care Products

The pharmaceuticals and personal care products (PPCPs) have been in the environment for an extended period as compared to the other classes of contaminants (Ghanbari et al. 2021).

However, the PPCPs have gained recognition as one of the emerging contaminants in the past two decades owing to their persistence in the environment (Kumar et al. 2023). These pollutants identify as bioactive compounds, which can have long-term as well as short-term toxic effects on the environment. The treatment of human and animal illnesses is dependent upon pharmaceutical products, the personal care products are applied in the cosmetic industries. The presence of these compounds in the aquatic systems is in few ngL^{-1} to gL^{-1} to be considered micropollutants in the environment (Chen 2020). The detection and degradation of the compounds is difficult in the various levels of water bodies, increasing the genotoxic effects on the aquatic organisms. The widely detected pharmaceuticals in water bodies are penicillin, fluoroquinolones, macrolides, and β-lactams and personal care products compounds are surfactants, plasticizers, parabens, and preservatives (Adhikari et al. 2019, Khalaf et al. 2020, Moradi et al. 2020).

5.2.1.2 Dyes

Dyes are an essential part of various manufacturing industries such as textiles, tannery, printing, and food production houses. Around 100,000 commercially available dyes are water-soluble, non-biodegradable, and damaging to the ecosystem (carcinogenic) (Roy et al. 2022). 10–15% of dyes are estimated to be released in the dying process in the form of effluent (Rafiq et al. 2021). The dyes can be categorized based on their chemical nature like azo, disperse, reactive, substantive, mordant, and acid dyes namely Rhodamine B (RhB), methylene blue (MB), Congo red, malachite green, and 4-nitrophenol (Anwer et al. 2019, Chiu et al. 2019, Su et al. 2021). The textile industries contribute to the release of aromatic amines in the water bodies that cause detrimental effects and produce biotoxicity (Anwer et al. 2019). The reactive dyes cause allergies in humans like skin irritation, respiratory ailment, and infection of the eyes and some have mutagenic effects. Traditional methods of remediation are proving ineffective in cleaning-up dyes and increasing the toxic effects. Hence, the demand for advances in dye degradation methods has increased (Shandilya et al. 2021). A study demonstrated that TiO_2 nanoparticles were stabilized using Pickering emulsion in the form of oil-in-water, wherein the wastewater containing dye and organic matter (insoluble) were effectively treated with high photoactivity of TiO_2 (Nawaz et al. 2017).

5.2.1.3 Heavy Metals

Heavy metal ions are an essential part of metabolism. The sources of heavy metals are metallurgy industries processing raw materials. Some of the heavy metals are released from mining, industrial smoke, waste disposal, and smelting (Kadi et al. 2020). Heavy metal ions discharge can have adverse effects like nephritic syndrome, hepatitis, anemia, and kidney damage (Ren et al. 2021). The commonly found heavy metal ions such as lead are discharged into the environment from acid lead batteries whereas, cadmium is released from electroplating industries, photovoltaic cells, and fabric factories (Gao and Meng 2021). Skin allergies are also developed from nickel ions that are leached from the jewelry (Wang et al. 2023). Therefore, treatment of heavy metals is of primary importance to overcome problematic health issues.

5.2.1.4 Other Pollutants

The other forms of pollutants found in water systems are due to various human activities. Pollutants like arsenic, benzene, fluoride, trichloroethylene, trihalomethanes, mercury, methoxychlore, and vinyl chloride are found through the human application. The sources of such pollutants are residues from chlorination at water treatment plants, pesticide residues, metal-degreaser, leaking fuel tanks, insecticides, and so on (Rani and Shanker 2018, Bratovčić 2019, Bhat and Gogate 2021). These pollutants are known to cause adverse side effects to humans in the form of the damaged central nervous system, reproductive systems, chronic liver damage, cancer in the form of leukemia and anemia, allergenic, laxative action, nasal mucous ulcer, nephritis, and lung cancer (Walker et al. 2019).

5.3 TYPES OF NANOPARTICLES INVOLVED IN THE CONTAMINANT TREATMENT

The semiconductor particles in the size range of nanometers are applied as nano-photocatalysts (Atabaev and Molkenova 2019). A variety of semiconductor metal oxide nanoparticles such as ZnO and TiO_2 have demonstrated higher efficiency in the decomposition of water pollutants. The attraction toward the semiconductor particles is due to their distinctive properties of photo-physical and catalytic as compared to the bulk materials (Chenab et al. 2020). The charge-carrier dynamics depend on the physical form of the materials and are associated with photo-efficiencies of the catalysts. This phenomenon is supported by the observation between quantum-TiO_2 (Q–TiO_2) and Degussa P25 where the interfacial electron transfer is faster in the case of P25–TiO_2 (20–35 nm) (Yahya et al. 2018). To overcome, the con of large recombination volume of charge-carrier in bulk material, pros of reduction in size with an increase in surface area and active sites can improve the rate of charge-carrier transfer with higher photonic efficiency (Ismail et al. 2023). Therefore, an optimum nano-size particle can provide higher photocatalytic efficiency.

5.3.1 Metal and Oxide Nanoparticles

The generally studied metal nanoparticles such as gold (Au), palladium (Pd), copper (Cu), and silver (Ag) in the size range of 10–200 nm for wastewater management are considered for the high surface area and effective photocatalytic activity (Weidner and Ciesielczyk 2019). The identification and size analysis of nanomaterials is done using scanning electron microscopy (SEM) (Figure 5.2). For example, Zn nanoparticles are effective in antimicrobial activity against various deactivations of microorganisms present in wastewater (Serrà, Zhang et al. 2019). Further, iron (Fe) with Pd, Cu, and nickel have been applied to degrade tetracycline from pharmaceutical water. Similarly, metal-oxide nanoparticles such as Fe_3O_4 and ZnO have been used for the degradation of pharmaceutical wastewater pollutants (ciprofloxacin and sulfamethoxazole) owing to their better photocatalytic action (Arora et al. 2016).

5.3.2 Carbon-supported Nanomaterials

The commercially available activated carbon as an adsorbent has been widely used in various fields as well as in the adsorptive removal of dyes, pesticides, and heavy metals (Raja et al. 2020). The removal of organic pollutants with increased adsorption capacity, thermal stability, and electrical conductivity makes it a potential sorbent material (Safaei, Ullah et al. 2018). The leaching of heavy metals such as Cu(II), Zn(II), and Cr(III) from acid mines and the electroplating industries (Baby et al. 2019, Fiyadh et al. 2019). Moreover, the capacity to treat heavy metals in ppb (parts-per-billion) concentrations has gained recognition in recent years. The nanomaterial with comparable properties of adsorption-photocatalytic degradation was developed in the form of titanate nanotubes (TNT@GO) for the removal of methylene blue (MB) dye. The TNT@GO synthesized from the thermal and alkaline treatment of TiO_2 increased the photocatalytic performance of TiO_2. Whereas, the semiconductive properties of graphene can inhibit charge recombination of e^-/h^+ pairs via photosensitization activity (Ali et al. 2019).

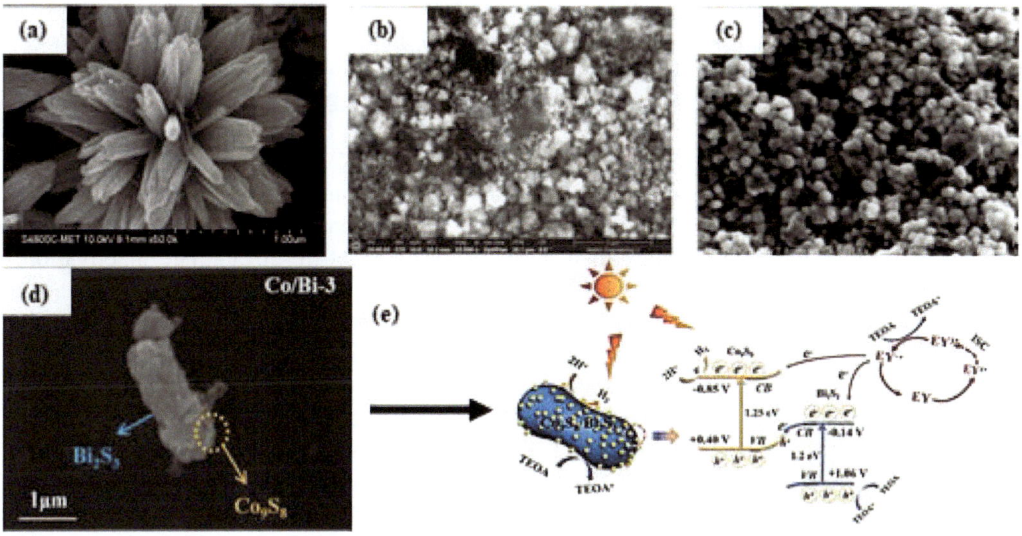

Figure 5.2 SEM image of (a) one individual ZnO flower, (b) TiO_2 NPs, (c) FE-SEM image of silver nanoparticles (d) Co/Bi-3 and graphical representation of photocatalytic hydrogen evolution (e) Mechanism of photocatalytic activity in Co/Bi-3 catalyst. Reprinted from Journal of Colloid and Interface Science, Vol 592, Li, Junke, and Zhiliang Jin, Rational design of a cobalt sulfide/bismuth sulfide S-scheme heterojunction for efficient photocatalytic hydrogen evolution, 237–248, Copyright (2021); Current Research in Green and Sustainable Chemistry, Vol 4, Jaast, Swati, and Anita Grewal, Green synthesis of silver nanoparticles, characterization and evaluation of their photocatalytic dye degradation activity, 100195, Copyright (2021); J. of Molecular Liquids, Vol 316, Chand, K., Cao, D., Fouad, D.E., Shah, A.H., Lakhan, M.N., Dayo, A.Q., Sagar, H.J., Zhu, K. and Mohamed, A.M.A., Photocatalytic and antimicrobial activity of biosynthesized silver and titanium dioxide nanoparticles: a comparative study, 113821, Copyright (2020) with permission from Elsevier (Saikia et al. 2015, Chand et al. 2020, Jaast and Grewal 2021, Li and Jin 2021).

5.3.3 Zeolite-Supported Nanomaterials

Zeolite nanoparticles have been reported to exhibit excellent ion exchange properties, this ability has been applied in water purification as a medium for ion exchange. The residual toxic compounds cause clogging and fouling; therefore, an external pretreatment is applied such as ion exchange and membrane filtration techniques (Iazdani and Nezamzadeh-Ejhieh 2021). The zeolite-supported nZVI particles have been applied in the treatment of nitrate, reducing it to ammonium. The material has demonstrated a high surface area with a uniform dispersion of nZVI particles and prevents aggregation of nanoparticles (Eljamal et al. 2022). Therefore, the low-cost natural clay mineral can function as a stabilizer, dispersant, and support material. Zeolite-magnetite material has been applied in the removal of oxyanions from arsenic and molybdenum contaminated drinking water, using the combined action of the adsorption-photocatalytic process (Weidner and Ciesielczyk 2019).

5.3.4 Clay-supported nanomaterials

As a highly explored photocatalyst, the TiO_2 (Degussa P25) faces certain demerits such as low surface area (50 m^2/g) and difficulty in separation from reaction solutions. Hence, to increase the usability of TiO_2, the clays have been applied as a support material for the catalyst (Mishra et al. 2018). The clays as support frames have gained immense attention as economically feasible, mechanical, chemical, and thermally stable. TiO_2 in combination with clay provides greater surface area and a higher number of surface-active sites, making it a highly active photocatalyst in the nanocomposite form (Alfred et al. 2020).

5.3.5 Advanced Oxidation Process

The treatment of drinkable water by the process of Advanced Oxidation Process (AOPs) was first proposed in the 1980s, the process is defined by the sufficient generation of hydroxyl radicals ($OH^{\bullet}$) that affect the oxidative purification of water (Cardoso et al. 2021). The concept of AOP was later extended to the involvement of sulfate radicals ($SO_4^{\bullet -}$) in the oxidative process. In contrast to the commonly used oxidants such as chlorine and ozone that possess the dual nature of decontamination and disinfection, the application of AOP is primarily focused on the degradation of organic and inorganic contaminants in water systems (Yan et al. 2021). The inactivation of pathogens by the application of AOP is rarely applied for disinfection due to the short life of the radicals (in the order of microseconds). Since the required disinfection detention time is long-term, pathogen removal using AOPs is minimal because of extremely low radical concentrations. In wastewater treatment under the influence of AOPs, the radicals are expected to degrade and mineralize pollutants sufficiently and produce by-products that are less or non-toxic.

The AOPs are broadly divided based on the different generated species such as the H_2O_2 and superoxide ($O_2^{\bullet -}$) for hydroxyl radicals, O_3 in ozone-based reactions, the reaction of H_2O_2 with Fe^{2+} generate reactive species, such as the hydroxyl radicals and ferryl ions in the Fenton process and formation of $OH^{\bullet}$ in the presence of oxidants initiated by photons from surface of heterogenous photocatalyst under UV-based AOPs (Ede et al. 2020). Other processes include electrochemical oxidation, using the water electrolysis for the generation of $OH^{\bullet}$ radicals, sonochemical oxidation applying ultrasound (water sonolysis), and sonoelectrochemistry where the combination of electric current and ultrasound act as the source of radical generation (Zhang et al. 2023).

5.3.6 Overview and Mechanism of Photocatalysis

Amongst the techniques mentioned above, the photocatalysis process is recognized as the most extensively explored area for contaminant treatment in wastewater. To improve the degradation performance, alternatives to the process operating conditions have been proposed such as the application of catalysts like titanium oxide (TiO_2), graphene and its composite forms, carbon nitride, and other oxides such as bismuth, zinc, copper, and nickel (Sharma et al. 2019, Wetchakun et al. 2019). However, the precipitation of toxic metals increases the risk of contamination during the treatment. Therefore, the synthesis of nanometer-size catalysts has been proposed as a feasible alternative to prevent these potential problems. The unique properties of the nanoparticles (large surface area, exposed crystal planes, well dispersion in solution, and low concentration) further enhances photocatalytic capabilities (Lee et al. 2019).

The basic principle of the photocatalytic process involves light photons ($h\upsilon$) adsorption and the electron (e^-) holes (h^+) formation. The active agents generally involved are $OH^{\bullet}$, $O_2^{\bullet -}$, and h^+ (Cardoso et al. 2021). The process is a series of reactions, described as follows:

$$\text{Catalyst} + h\upsilon \rightarrow \text{Catalyst}\ ((e^-_{CB}) + (h^+_{VB})) \tag{1}$$

$$h^+_{VB} + H_2O \rightarrow H^+ + \cdot OH \tag{2}$$

$$2h^+_{VB} + 2H_2O \rightarrow H^+ + H_2O_2 \tag{3}$$

$$H_2O_2 \rightarrow 2\cdot OH \tag{4}$$

$$e^-_{CB} + O_2 \rightarrow O_2^{\bullet -} \tag{5}$$

$$O_2^- + 2\cdot OH + H^+ \rightarrow H_2O_2 + O_2 \tag{6}$$

$$H_2O_2 \rightarrow 2\cdot OH \tag{7}$$

$$\text{Contaminant} + \cdot OH \rightarrow \text{Degradation of contaminant} \tag{8}$$

Nanomaterials are recent highlights due to their distinct attributes namely, thermal stabilities, chemical reactivities, charge separation, increased light harvesting, and selective oxidation and reduction (Figure 5.3). Moreover, nano-catalysts reduce operational costs since they can be used in multiple reaction cycles. There are two different modes of the function of photocatalysis (activation by photons):

1. energy transfer mechanism: the energy from the photon-activated catalyst (C) transfers to the reactant/substrate (S) which in turn stimulates its oxidation. It is difficult to attain the ground state of the substrate/reactant. The catalyst and substrate react to form

$$C \xrightarrow{hv} C^* \Rightarrow S \tag{9}$$

$$C + S^* \Rightarrow P \tag{10}$$

2. electron transfer mechanism: the photon-activated catalyst acts as an acceptor or donor of electrons. In this case, the reaction equilibrium is achieved by oxidation and reduction of reactant/substrate by the activated photocatalysts

$$C \xrightarrow{hv} C^* \Rightarrow S \tag{11}$$

$$C^* + S \Rightarrow C^{*+} \rightarrow S^{*-} \tag{12}$$

$$C + P \tag{13}$$

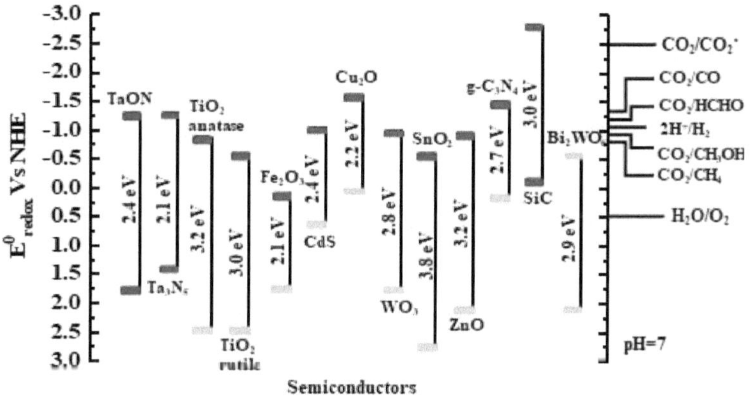

Figure 5.3 Bandgap energies of different photocatalysts.

5.4 OUTLOOK ON THE MECHANISM OF PHOTOCATALYTIC ACTION OF NANOPARTICLES

The nature of a photocatalyst is to activate upon exposure to light energy (UV-Vis) in the form of visible or fluorescent light and ultraviolet irradiation. This is the photoexcitation state of semiconductors. The energy from the light excites the e^- from the valence band (VB) of the catalyst and transfers to the conduction band (CB). This forms a zone of negative charge (e^-) at the CB and positive charge (h^+) at the VB of the photocatalyst. The energy difference between the two bands (CB and VB) is termed the 'bandgap' or otherwise 'forbidden energy zone' (Al-Mamun et al. 2019). The generation of e^-/h^+ pairs by photon energy reacts with the aqueous or atmospheric oxygen (O_2) to form strong radicals such as $OH^{\bullet}$ and $O_2^{\bullet-}$ However, to generate $OH^{\bullet}$ radicals, the valence band must attain sufficient positive holes (positive redox potential). On the contrary, the saturation of negative charges at the valence band that is e^-s (negative redox potential) generates $O_2^{\bullet-}$ in excess (Figure 5.4) (Zhu and Zhou 2019).

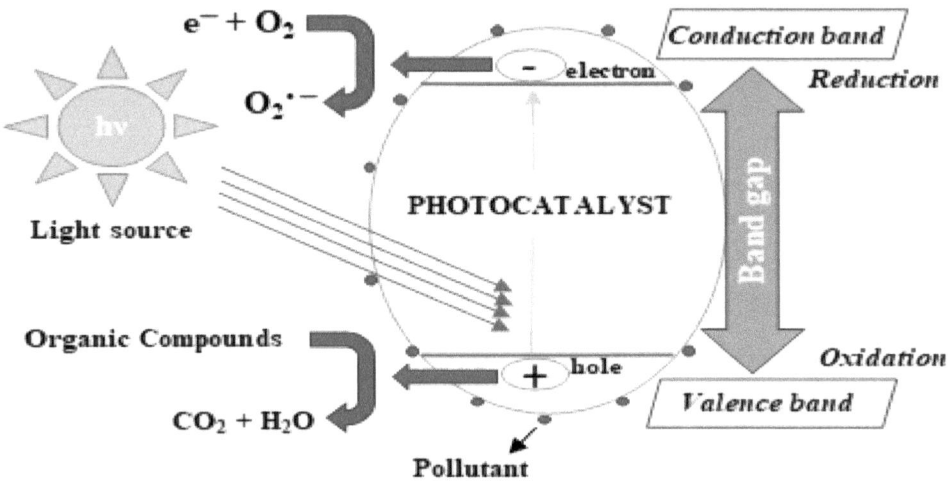

Figure 5.4 Mechanism of photocatalysis activity.

These nanomaterial-based AOPs are extensively studied in the treatment of water contaminants such as dyes, pharmaceuticals and personal care products (PPCPs), pesticides, surfactants, and heavy metals (Duan, Zhou et al. 2021). Certain parameters are also involved in the effective functioning of a photocatalyst and are represented in Figure 5.5.

Furthermore, the photocatalyst in the form of nanomaterials that are coupled with certain UV/AOP techniques is briefly explained.

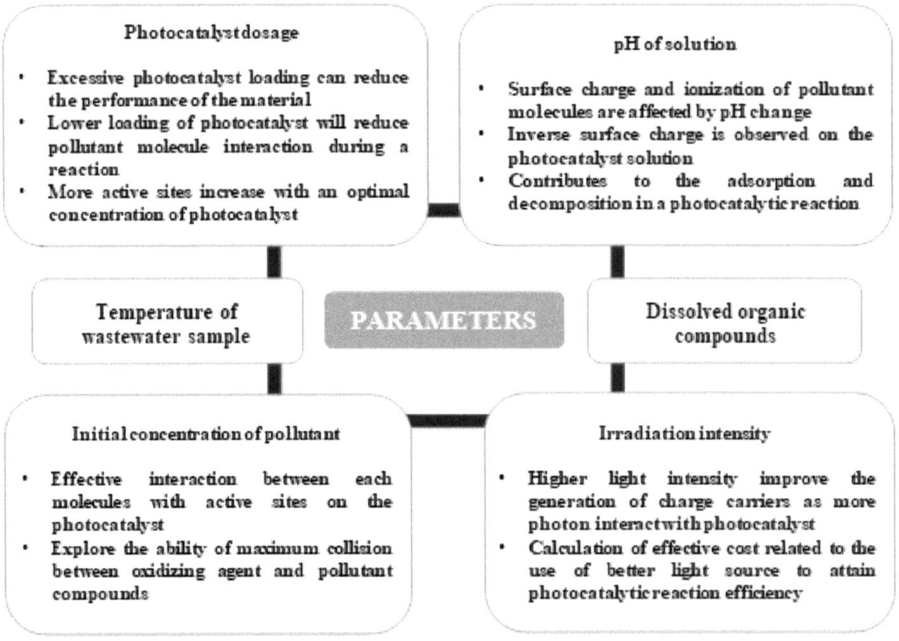

Figure 5.5 Important parameters for effective photocatalysis activity.

5.4.1 UV/hydrogen peroxide Process

A wide spectrum of organic contaminants can be degraded using the UV/hydrogen peroxide (UV/H_2O_2) process which is a promising technique when coupled with nanomaterials (Soni et al. 2020).

The coupling of small size particles having more surface area and exposure with reagent allows increased reactions to take place (Sun et al. 2023). The process can be described as follows

$$H_2O_2 + h\upsilon \rightarrow 2HO^{\bullet} \quad (14)$$

$$H_2O_2 + HO^{\bullet} \rightarrow HO_2^{\bullet} + H_2O \quad (15)$$

$$H_2O_2 + HO_2^{\bullet} \rightarrow HO^{\bullet} + H_2O + O_2 \quad (16)$$

$$2HO^{\bullet} \rightarrow H_2O_2 \quad (17)$$

$$2HO_2^{\bullet} \rightarrow H_2O_2 + O_2 \quad (18)$$

$$HO^{\bullet} + HO_2^{\bullet} \rightarrow H_2O + O_2 \quad (19)$$

The TiO_2 nanoparticle has been widely studied in this process for the removal of dyes and antibiotics, the property of semiconductors with a wide bandgap (3.2 eV) helps in the formation of hydroxyl radicals. The irradiation of TiO_2 nanoparticles in the UV region generates charged e^-/h^+ pair which further reacts with H_2O (Sinha and Ghosal 2023).

$$TiO_2 + h\upsilon \rightarrow e^-CB + h^+ \quad (20)$$

$$e^-_{CB} + O_2 \rightarrow O_2^{\bullet-} \quad (21)$$

$$h^+ + H_2O \rightarrow H^+ + HO^{\bullet} \quad (22)$$

The combination of titanium peroxide semiconductor yield valence band holes (h^+_{VB}) and conduction band electrons (e^-_{CB}). The molecular oxygen present reacts with the e^-_{CB} and produces $O_2^{\bullet-}$ (anion superoxide), whereas the interaction with holes generates hydroxyl radical (Kurian 2021). The zinc oxide (ZnO) when coupled with UV/H_2O_2 has also been shown to increase the process effectiveness in the removal of target contaminants (Jamil, Bokhari et al. 2020).

5.4.2 Persulfate-Based AOP

Sulfate radicals are proven as an effective alternative to the hydroxyl radicals, in remediating organic pollutants present in aqueous solutions. The sulfate activation can be achieved in a cost-effective manner using UV irradiation. The UV-activated persulfate to generate sulfate radicals is shown (Zhou, Shi et al. 2020)

$$S_2O_8^{2-} + h\upsilon \rightarrow 2SO_4^{\bullet-} \quad (23)$$

The radical formation from UV irradiation provides energy to persulfate anion to cleave the peroxide bond resulting in the formation of two sulfate radicals (Yang et al. 2019).

The degradation of organic recalcitrant pollutants, like p-nitroaniline, can be achieved by magnetic iron oxide NPs as these can effectively generate sulfate free radicals due to their structural, catalytic, and magnetic properties (Lin et al. 2022). The activation of persulfate is also possible through one e^- transfer using nanomaterials such as silver, cobalt, copper, and manganese (Lee et al. 2020). The electrophilic nature of sulfate radicals supports faster reaction rates in the case of electron-donating species as compared to electron-withdrawing species (Ding et al. 2020). Moreover, the charged e^-s and peroxide reduce molecular dissolve O_2 to $O_2^{\bullet-}$ participates in the photodegradation of contaminants (Lee, Wang et al. 2020). The reaction is explained below

$$M + h\upsilon \rightarrow e^- + h^+ \quad (24)$$

$$M(_e^-) + O_2 \rightarrow O_2^{\bullet-} \quad (25)$$

$$M(_e^-) + S_2O_8^{2-} \rightarrow [S_2O_8^{2-}]^- \quad (26)$$

$$[S_2O_8^{2-}]^- + O_2 \rightarrow S_2O_8^{2-} + O_2^{\bullet -} \quad (27)$$

$$O_2^{\bullet -} + \text{contaminant} \rightarrow CO_2 + H_2O \quad (28)$$

$$M + \text{contaminant} \rightarrow CO_2 + H_2O \quad (29)$$

where M represents the metal

5.4.3 Fenton-Based AOPs

The process of Fenton reactions is a promising technique for the removal of toxic organic pollutants in wastewater (Li et al. 2020). The heterogeneous Fenton reactions using nano-zero-valent iron (nZVI) are more advantageous than the homogenous analogs due to activity in the vast pH range. The other nano-catalysts applied in a typical Fenton oxidation process are copper oxide, cadmium sulfide, manganese dioxide, and tungsten trioxide (Yu et al. 2019, Xu et al. 2020). The process is divided into four phases such as oxidation, neutralization, flocculation, and phase separation (Gou et al. 2021). The high molecular weight organic molecules are removed in the flocculation phase which is determined by the precipitation of ferric oxyhydroxides (Brillas 2020, Mahtab et al. 2021). The generation of carbon dioxide (CO_2) and ferrous ions takes place by the iron complexes' photodecarboxylation. As an alternative to the use of Fe^{2+} in the classic Fenton reaction for the oxidative activation of H_2O_2, the ferric ion (Fe^{3+}) is applied. The oxidative mechanism forms the $Fe^{III}(HO_2)^{2+}$ unstable complex that decomposes to Fe^{2+}. Further, the Fe^{2+} produces $OH^{\bullet}$ and HO_2 (hydroperoxyl radical) (Ameta et al. 2018).

$$Fe^{2+} + H_2O_2 \rightarrow Fe^{3+} + OH + OH^- \quad (30)$$

$$OH + H_2O_2 \rightarrow HO_2 + H_2O \quad (31)$$

$$Fe^{2+} + OH \rightarrow Fe^{3+} + OH^- \quad (32)$$

$$Fe^{3+} + HO_2 \rightarrow Fe^{2+} + O_2 + H^+ \quad (33)$$

$$OH + OH \rightarrow H_2O_2 \quad (34)$$

Among the Fenton reactions, the process of photoelectron-Fenton is an interesting approach wherein the process is combined with UV irradiation or sunlight (Ameta et al. 2018). The regeneration of ferrous ions and $OH^{\bullet}$ radicals take place as described below:

$$Fe^{3+} + H_2O \rightarrow Fe(OH)^{2+} + H^+ \quad (35)$$

$$Fe(OH)^{2+} + h\upsilon \rightarrow Fe^{2+} + HO^{\bullet} \quad (36)$$

$$Fe(OOCR)^{2+} + h\upsilon \rightarrow Fe^{2+} + CO_2 + R^{\bullet} \quad (37)$$

In a reported study, Fe_3O_4/CdS yolk-shell structured nanoparticles were designed for reaction based on the photo-Fenton process for the degradation of dye-based pollutants (Shi et al. 2016), and the mechanism followed in the process is given below

$$M + h\upsilon \rightarrow e^-(M) + h^+(M) \quad (38)$$

$$H_2O_2 + e^- \rightarrow OH^- + OH^{\bullet} \quad (39)$$

$$Fe^{3+} + e^- \rightarrow Fe^{2+} \quad (40)$$

$$Fe^{2+} + H_2O_2 \rightarrow Fe^{3+} + OH + OH^{\bullet} \quad (41)$$

$$H_2O + h^+ \rightarrow H^+ + OH^{\bullet} \quad (42)$$

$$\text{Contaminant} + h^+ \rightarrow \text{Contaminant}^+ \rightarrow \text{degradation product} \quad (43)$$

5.4.4 Ozone/UV Process

The elimination of toxic organic compounds using convenient and strong oxidants such as ozone (O_3) is another form of AOP. The slow reaction rate of ozone is effective for long-term organic compounds like inactivated aromatics. Therefore, the catalytic ozonation process has lately advanced in the oxidative reduction of persistent organic compounds (Rekhate and Srivastava 2020). The ability of transition metal ions for ozone decomposition is termed homogenous catalytic ozonation (Wang and Chen 2020). The nanomaterials as heterogeneous catalysts in the O_3/H_2O_2 process such as ZnO, TiO_2, and Al_2O_3 nanoparticles for the photons-generated oxidizing species are ecologically considered in the treatment of wastewater in large volumes (Joseph et al. 2021, Raj et al. 2023). The vast range of nanomaterials as heterogeneous catalysts have been studied in the ozonation process, for example, zeolite/metals, Co_3O_4, Fe_3O_4/carbon nanotubes, MgO, $NiFe_2O_4$, CeO_2, and Mn/γ-Al_2O_3 (Gong and Chu 2016, Chen and Wang 2019, Wu et al. 2019, Norabadi et al. 2020). The general mechanism followed in the O_3 oxidation process for the generation of $OH^•$ radicals presence of Fe(II) ions is represented below: (Chaudhary et al. 2023).

$$Fe^{2+} + O_3 \rightarrow FeO^{2+} + O_2 \tag{44}$$

$$FeO^{2+} + H_2O \rightarrow Fe^{3+} + OH^• + OH^- \tag{45}$$

However, the high concentration of Fe(II) can oxide FeO^{2+} to Fe(III) inhibiting the formation of intermediate radical species

$$FeO^{2+} + Fe^{2+} + 2H^+ \rightarrow 2Fe^{3+} + H_2O \tag{46}$$

A study reported, showed the photochemical reduction of Fe^{3+} produced is important for the generation of $OH^•$ radicals (Mecha, Onyango et al. 2016). Exposure to UVA light leads to the following reaction:

$$Fe^{3+} + H_2O + h\upsilon \rightarrow Fe^{2+} + OH^• + H^+ \tag{47}$$

Hence, the catalytic ozonation systems are dependent on the oxidation/reduction reactions as an important parameter in the presence of transition metals.

5.5 CONCLUSIONS

In conclusion, this chapter summarizes the various sub-division techniques of the advanced oxidation process in water treatment and the mechanisms followed. The synergistic effects of photocatalyst in different forms for improvement in photocatalytic performance were listed. The chapter clarifies the steps involved in the photocatalysis process like excitation and separation of charges, adsorption, and semiconductor surface reactions that significantly impact the photocatalytic efficiency. Therefore, taking into consideration the above-mentioned factors a carefully optimized multifunctional photocatalyst can be designed. Furthermore, analyzing the mechanism of the photocatalysis process depending on the type of pollutant can improve the catalyst fabrication as well as the treatment process. The catalytic activity is highly influenced by the structural and chemical properties of materials, the lower dosage of catalyst along with faster kinetics can reduce the processing cost. Moreover, certain bottlenecks in the terms of powder forms of photocatalyst for practical application in water disinfection, such as particle aggregation, difficulty in catalyst separation post water treatment, and factors considered in the treatment are based on laboratory level with a single pollutant system in the simulated wastewater. The treatment application at the industrial or natural level involves higher degrees of complexity which are yet to be considered. The rational designing of a photocatalytic system is important in terms of reactor design with optimized separation efficacy, physicochemical optimization of photocatalyst, and recycling of the catalyst.

References

Adhikari, S., Mandal, S. and Kim, D.-H. 2019. Z-scheme 2D/1D MoS_2 nanosheet-decorated $Ag_2Mo_2O_7$ microrods for efficient catalytic oxidation of levofloxacin. Chem. Eng. J. 373: 31–43.

Al-Mamun, M., Kader, S., Islam, M. and Khan, M. 2019. Photocatalytic activity improvement and application of UV-TiO_2 photocatalysis in textile wastewater treatment: a review. J. Environ. Chem. Eng. 7(5): 103248.

Alfred, M.O., Omorogie, M.O., Bodede, O., Moodley, R., Ogunlaja, A., Adeyemi, O.G., Günter, C., et al. 2020. Solar-active clay-TiO_2 nanocomposites prepared via biomass assisted synthesis: Efficient removal of ampicillin, sulfamethoxazole and artemether from water. Chem. Eng. J. 398: 125544.

Ali, I., Park, K., Kim, S.-R. and Kim, J.-O. 2019. Lectrochemical anodization of graphite oxide-TiO_2 nanotube composite for enhanced visible light photocatalytic activity. Environ. Sci. Pollut. Res. 26(2): 1072–1081.

Ameta, R., Chohadia, A.K., Jain, A. and Punjabi, P.B. 2018. Fenton and photo-Fenton processes. Advanced oxidation processes for waste water treatment. Elsevier: 49–87.

Anwer, H., Mahmood, A., Lee, J., Kim, K.-H., Park, J.-W. and Yip, A.C. 2019. Photocatalysts for degradation of dyes in industrial effluents: opportunities and challenges. Nano Res. 12(5): 955–972.

Arora, A.K., Jaswal, V.S., Singh, K. and Singh, R. 2016. Applications of metal/mixed metal oxides as photocatalyst: (A review). Orient. J. Chem. 32(4): 2035.

Arun, J., Nachiappan, S., Rangarajan, G., Alagappan, R.P., Gopinath, K. and Lichtfouse, E. 2023. Synthesis and application of titanium dioxide photocatalysis for energy, decontamination and viral disinfection: a review. Environ. Chem. Lett. 21(1): 339–362.

Atabaev, T.S. and Molkenova, A. 2019. Upconversion optical nanomaterials applied for photocatalysis and photovoltaics: recent advances and perspectives. Front. Mater. Sci. 13(4): 335–341.

Baby, R., Saifullah, B. and Hussein, M.Z. 2019. Carbon nanomaterials for the treatment of heavy metal-contaminated water and environmental remediation. Nanoscale Res. Lett. 14(1): 1–17.

Bhat, A.P. and Gogate, P.R. 2021. Degradation of nitrogen-containing hazardous compounds using advanced oxidation processes: a review on aliphatic and aromatic amines, dyes, and pesticides. J. Hazard. Mater. 403: 123657.

Bratovčić, A. 2019. Photocatalytic degradation of organic compounds in wastewaters. Technol. Acta: Sci./Prof. J. Chem. and Technol. 11(2): 17–23.

Brillas, E. 2020. A review on the photoelectro-Fenton process as efficient electrochemical advanced oxidation for wastewater remediation. Treatment with UV light, sunlight, and coupling with conventional and other photo-assisted advanced technologies. Chemosphere. 250: 126198.

Cardoso, I.M., Cardoso, R.M. and da Silva, J.C.E. 2021. Advanced oxidation processes coupled with nanomaterials for water treatment. Nanomaterials. 11(8): 2045.

Chand, K., Cao, D., Fouad, D.E., Shah, A.H., Lakhan, M.N., Dayo, A.Q., Sagar, H.J., et al. 2020. Photocatalytic and antimicrobial activity of biosynthesized silver and titanium dioxide nanoparticles: a comparative study. J. Mol. Liq. 316: 113821.

Chaudhary, P., Ahamad, L., Chaudhary, A., Kumar, G., Chen, W.-J. and Chen, S. 2023. Nanoparticle-mediated bioremediation as a powerful weapon in the removal of environmental pollutants. J. Environ. Chem. Eng. 109591.

Chen, G., Yu, Y., Liang, L., Duan, X., Li, R., Lu, X., Yan, B., et al. 2021. Remediation of antibiotic wastewater by coupled photocatalytic and persulfate oxidation system: a critical review. J. Hazard. Mater. 408: 124461.

Chen, H. and Wang, J. 2019. Catalytic ozonation of sulfamethoxazole over Fe_3O4/Co_3O_4 composites. Chemosphere 234: 14–24.

Chen, Y. 2020. Photodegradation of pharmaceutical waste by nano-materials as photocatalysts. pp. 143–152. *In*: Singh, P., Borthakur, A., Mishra, P.K. and Tiwary, D. (eds). Nano-Materials as Photocatalysts for Degradation of Environmental Pollutants: Challenges and Possibilities. Elsevier. 143–152.

Chenab, K.K., Sohrabi, B., Jafari, A. and Ramakrishna, S. 2020. Water treatment: functional nanomaterials and applications from adsorption to photodegradation. Mater. Today Chem. 16: 100262.

Chiu, Y.-H., Chang, T.-F.M., Chen, C.-Y., Sone, M. and Hsu, Y.-J. 2019. Mechanistic insights into photodegradation of organic dyes using heterostructure photocatalysts. Catalysts. 9(5): 430.

Ding, X., Gutierrez, L., Croue, J.-P., Li, M., Wang, L. and Wang, Y. 2020. Hydroxyl and sulfate radical-based oxidation of RhB dye in UV/H_2O_2 and UV/persulfate systems: kinetics, mechanisms, and comparison. Chemosphere. 253: 126655.

Duan, X., Zhou, X., Wang, R., Wang, S., Ren, N.Q. and Ho, S.-H. 2021. Advanced oxidation processes for water disinfection: features, mechanisms and prospects. Chem. Eng. J. 409: 128207.

Ede, J.D., Lobaskin, V., Vogel, U., Lynch, I., Halappanavar, S., Doak, S.H., Roberts, M.G., et al. 2020. Translating scientific advances in the AOP framework to decision making for nanomaterials. Nanomaterials. 10(6): 1229.

Eljamal, O., Eljamal, R., Maamoun, I., Khalil, A.M., Shubair, T., Falyouna, O. and Sugihara, Y. 2022. Efficient treatment of ammonia-nitrogen contaminated waters by nano zero-valent iron/zeolite composite. Chemosphere. 287: 131990.

Fiyadh, S.S., AlSaadi, M.A., Jaafar, W.Z., AlOmar, M.K., Fayaed, S.S., Mohd, N.S., Hin, L.S., et al. 2019. Review on heavy metal adsorption processes by carbon nanotubes. J. Cleaner Prod. 230: 783–793.

Gao, X. and Meng, X. 2021. Photocatalysis for heavy metal treatment: a review. Processes. 9(10): 1729.

Ghanbari, F., Hassani, A., Wacławek, S., Wang, Z., Matyszczak, G., Lin, K.-Y.A. and Dolatabadi, M. 2021. Insights into paracetamol degradation in aqueous solutions by ultrasound-assisted heterogeneous electro-Fenton process: key operating parameters, mineralization and toxicity assessment. Sep. Purif. Technol. 266: 118533.

Gong, H. and Chu, W. 2016. Determination and toxicity evaluation of the generated products in sulfamethoxazole degradation by UV/$CoFe_2O4$/TiO_2. J. Hazard. Mater. 314: 197–203.

Guo, Q., Zhou, C., Ma, Z. and Yang, X. 2019. Fundamentals of TiO_2 photocatalysis: concepts, mechanisms, and challenges. Adv. Mater. 31(50): 1901997.

Gou, Y., Chen, P., Yang, L., Li, S., Peng, L., Song, S. and Xu, Y. 2021. Degradation of fluoroquinolones in homogeneous and heterogeneous photo-Fenton processes: a review. Chemosphere. 270: 129481.

Iazdani, F. and Nezamzadeh-Ejhieh, A. 2021. The photocatalytic rate of ZnO supported onto natural zeolite nanoparticles in the photodegradation of an aromatic amine. Environ. Sci. Pollut. Res. 28(38): 53314–53327.

Ismail, A.A., Al-Hajji, L., Azad, I., Al-Yaqoot, A., Habibi, N., Alseidi, M. and Ahmed, S. 2023. Self-cleaning application of mesoporous ZnO, TiO_2 and Fe_2O_3 films with the accommodation of silver nanoparticles for antibacterial activity. J. Taiwan Inst. Chem. Eng. 142: 104627.

Jaast, S. and Grewal, A. 2021. Green synthesis of silver nanoparticles, characterization and evaluation of their photocatalytic dye degradation activity. Curr. Res. Green Sustainable Chem. 4: 100195.

Jamil, A., Bokhari, T.H., Iqbal, M., Zuber, M. and Bukhari, I.H. 2020. ZnO/UV/H_2O_2 based advanced oxidation of disperse red dye. Zeitschrift für Physikalische Chemie. 234(1): 129–143.

Joseph, C.G., Farm, Y.Y., Taufiq-Yap, Y.H., Pang, C.K., Nga, J.L. and Puma, G.L. 2021. Ozonation treatment processes for the remediation of detergent wastewater: a comprehensive review. J. Environ. Chem. Eng. 9(5): 106099.

Kadi, M.W., Mohamed, R.M., Ismail, A.A. and Bahnemann, D.W. 2020. Decoration of g-C_3N_4 nanosheets by mesoporous $CoFe_2O_4$ nanoparticles for promoting visible-light photocatalytic Hg (II) reduction. Colloids Surf., A: Physicochemical and Engineering Aspects. 603: 125206.

Kanan, S., Moyet, M.A., Arthur, R.B. and Patterson, H.H. 2020. Recent advances on TiO_2-based photocatalysts toward the degradation of pesticides and major organic pollutants from water bodies. Catal. Rev. 62(1): 1–65.

Khalaf, S., Shoqeir, J.H., Lelario, F., Bufo, S.A., Karaman, R. and Scrano, L. 2020. TiO_2 and active coated glass photodegradation of ibuprofen. Catalysts. 10(5): 560.

Kumar, M., Sridharan, S., Sawarkar, A.D., Shakeel, A., Anerao, P., Mannina, G., Sharma, P., et al. 2023. Current research trends on emerging contaminants pharmaceutical and personal care products (PPCPs): A comprehensive review. Sci. Total Environ. 859: 160031.

Kurian, M. 2021. Advanced oxidation processes and nanomaterials-a review. Cleaner Eng. Technol. 2: 100090.

Lee, K., Yoon, H., Ahn, C., Park, J. and Jeon, S. 2019. Strategies to improve the photocatalytic activity of TiO_2: 3D nanostructuring and heterostructuring with graphitic carbon nanomaterials. Nanoscale. 11(15): 7025–7040.

Lee, M.-Y., Wang, W.-L., Du, Y., Hu, H.-Y., Huang, N., Xu, Z.-B., Wu, Q.-Y., et al. 2020. Enhancement effect among a UV, persulfate, and copper (UV/PS/Cu^{2+}) system on the degradation of nonoxidizing biocide: the kinetics, radical species, and degradation pathway. Chem. Eng. J. 382: 122312.

Li, X., Li, J., Shi, W., Bao, J. and Yang, X. 2020. A Fenton-like nanocatalyst based on easily separated magnetic nanorings for oxidation and degradation of dye pollutant. Materials. 13(2): 332.

Li, J. and Jin, Z. 2021. Rational design of a cobalt sulfide/bismuth sulfide S-scheme heterojunction for efficient photocatalytic hydrogen evolution. J. Colloid Interface Sci. 592: 237–248.

Lin, Y., Zhang, Y. and Li, G. 2022. Promotion of sulfameter degradation by coupling persulfate and photocatalytic advanced oxidation processes with Fe-doped MOFs. Sep. Purif. Technol. 282: 119632.

Mahtab, M.S., Farooqi, I.H. and Khursheed, A. 2021. Zero Fenton sludge discharge: a review on reuse approach during wastewater treatment by the advanced oxidation process. Int. J. Environ. Sci. Technol. 1–14.

Mecha, A.C., Onyango, M.S., Ochieng, A., Fourie, C.J. and Momba, M.N. 2016. Synergistic effect of UV–vis and solar photocatalytic ozonation on the degradation of phenol in municipal wastewater: a comparative study. J. Catal. 341: 116–125.

Mishra, A., Mehta, A. and Basu, S. 2018. Clay supported TiO_2 nanoparticles for photocatalytic degradation of environmental pollutants: a review. J. Environ. Chem. Eng. 6(5): 6088–6107.

Moradi, S., Sobhgol, S.A., Hayati, F., Isari, A.A., Kakavandi, B., Bashardoust, P. and Anvaripour, B. 2020. Performance and reaction mechanism of MgO/ZnO/Graphene ternary nanocomposite in coupling with LED and ultrasound waves for the degradation of sulfamethoxazole and pharmaceutical wastewater. Sep. Purif. Technol. 251: 117373.

Natarajan, S., Bajaj, H.C. and Tayade, R.J. 2018. Recent advances based on the synergetic effect of adsorption for removal of dyes from waste water using photocatalytic process. J. Environ. Sci. 65: 201–222.

Nawaz, M., Miran, W., Jang, J. and Lee, D.S. 2017. Stabilization of pickering emulsion with surface-modified titanium dioxide for enhanced photocatalytic degradation of direct Red 80. Catalysis Today. 282: 38–47.

Norabadi, E., Panahi, A.H., Ghanbari, R., Meshkinian, A., Kamani, H. and Ashrafi, S.D. 2020. Optimizing the parameters of amoxicillin removal in a photocatalysis/ozonation process using Box-Behnken response surface methodology. Desalin Water Treat. 192(192): 234–240.

Okpara, E.C., Olatunde, O.C., Wojuola, O.B. and Onwudiwe, D.C. 2023. Applications of transition metal oxides and chalcogenides and their composites in water treatment: a review. Environ. Adv. 100341.

Ouyang, Z., Yang, Y., Zhang, C., Zhu, S., Qin, L., Wang, W., He, D., et al. 2021. Recent advances in photocatalytic degradation of plastics and plastic-derived chemicals. J. Mater. Chem. A. 9(23): 13402–13441.

Peiris, S., de Silva, H.B., Ranasinghe, K.N., Bandara, S.V. and Perera, I.R. 2021. Recent development and future prospects of TiO_2 photocatalysis. J. Chin. Chem. Soc. 68(5): 738–769.

Rafiq, A., Ikram, M., Ali, S., Niaz, F., Khan, M., Khan, Q. and Maqbool, M. 2021. Photocatalytic degradation of dyes using semiconductor photocatalysts to clean industrial water pollution. J. Ind. Eng. Chem. 97: 111–128.

Raj, P., Biju, P., Deepanraj, B. and Menachery, N. 2023. A systematic review on characterization of hybrid aluminium nanocomposites. Materials Today: Proceedings. 72: 2139–2150.

Raja, A., Rajasekaran, P., Selvakumar, K., Arunpandian, M., Kaviyarasu, K., Bahadur, S.A. and Swaminathan, M. 2020. Visible active reduced graphene oxide-$BiVO_4$-ZnO ternary photocatalyst for efficient removal of ciprofloxacin. Sep. Purif. Technol. 233: 115996.

Rani, M. and Shanker, U. 2018. Photocatalytic degradation of toxic phenols from water using bimetallic metal oxide nanostructures. Colloids Surf., A: Physicochemical and Engineering Aspects. 553: 546–561.

Rekhate, C.V. and Srivastava, J. 2020. Recent advances in ozone-based advanced oxidation processes for treatment of wastewater—a review. Chem. Eng. J. Adv. 3: 100031.

Ren, G., Han, H., Wang, Y., Liu, S., Zhao, J., Meng, X. and Li, Z. 2021. Recent advances of photocatalytic application in water treatment: a review. Nanomaterials. 11(7): 1804.

Roy, N., Alex, S.A., Chandrasekaran, N., Kannabiran, K. and Mukherjee, A. 2022. Studies on the removal of acid violet 7 dye from aqueous solutions by green $ZnO@Fe_3O_4$ chitosan–alginate nanocomposite synthesized using *Camellia sinensis* extract. J. Environ. Manage. 303: 114128.

Safaei, J., Ullah, H., Mohamed, N.A., Noh, M.F.M., Soh, M.F., Tahir, A.A., Ludin, N.A., et al. 2018. Enhanced photoelectrochemical performance of Z-scheme $g\text{-}C_3N_4/BiVO_4$ photocatalyst. Appl. Catal., B: Environ. 234: 296–310.

Saikia, L., Bhuyan, D., Saikia, M., Malakar, B., Dutta, D.K. and Sengupta, P. 2015. Photocatalytic performance of ZnO nanomaterials for self sensitized degradation of malachite green dye under solar light. Appl. Catal., A: General. 490: 42–49.

Samsudin, M.F.R., Sufian, S. and Hameed, B.H. 2018. Epigrammatic progress and perspective on the photocatalytic properties of BiVO4-based photocatalyst in photocatalytic water treatment technology: a review. J. Mol. Liq. 268: 438–459.

Serrà, A., Zhang, Y., Sepúlveda, B., Gómez, E., Nogués, J., Michler, J. and Philippe, L. 2019. Highly active ZnO-based biomimetic fern-like microleaves for photocatalytic water decontamination using sunlight. Appl. Catal., B: Environ. 248: 129–146.

Shandilya, P., Raizada, P. and Singh, P. 2021. Photocatalytic degradation of azo dyes in water. pp. 119–146. *In*: Inamuddin, Ahamed, M.I. and Lichtfouse, E. (eds). Water Pollution and Remediation: Photocatalysis. Environmental Chemistry for a Sustainable World, Vol 57. Springer, Cham.

Sharma, K., Dutta, V., Sharma, S., Raizada, P., Hosseini-Bandegharaei, A., Thakur, P. and Singh, P. 2019. Recent advances in enhanced photocatalytic activity of bismuth oxyhalides for efficient photocatalysis of organic pollutants in water: a review. J. Ind. Eng. Chem. 78: 1–20.

Shayegan, Z., Lee, C.-S. and Haghighat, F. 2018. TiO_2 photocatalyst for removal of volatile organic compounds in gas phase–a review. Chem. Eng. J. 334: 2408–2439.

Shi, W., Du, D., Shen, B., Cui, C., Lu, L., Wang, L. and Zhang, J. 2016. Synthesis of yolk–shell structured Fe_3O_4@ void@ CdS nanoparticles: a general and effective structure design for photo-fenton reaction. ACS Appl. Mater. Interfaces. 8(32): 20831–20838.

Shume, W.M., Murthy, H. and Zereffa, E.A. 2020. A review on synthesis and characterization of Ag_2O nanoparticles for photocatalytic applications. J. Chem. 2020.

Sinha, R. and Ghosal, P.S. 2023. A comprehensive appraisal on status and management of remediation of DBPs by TiO_2 based-photocatalysts: Insights of technology, performance and energy efficiency. J. Environ. Manage. 328: 117011.

Soni, R., Pal, A.K., Tripathi, P., Lal, J.A., Kesari, K. and Tripathi, V. 2020. An overview of nanoscale materials on the removal of wastewater contaminants. Appl. Water Sci. 10(8): 1–9.

Su, F., Li, P., Huang, J., Gu, M., Liu, Z. and Xu, Y. 2021. Photocatalytic degradation of organic dye and tetracycline by ternary $Ag_2O/AgBr–CeO_2$ photocatalyst under visible-light irradiation. Sci. Rep. 11(1): 1–13.

Sun, P., Xing, Z., Li, Z. and Zhou, W. 2023. Recent Advances in Quantum Dots Photocatalysts. Chem. Eng. J. 141399.

Tsang, C.H.A., Li, K., Zeng, Y., Zhao, W., Zhang, T., Zhan, Y., Xie, R., et al. 2019. Titanium oxide based photocatalytic materials development and their role of in the air pollutants degradation: overview and forecast. Environ. Int. 125: 200–228.

Vaya, D. and Surolia, P.K. 2020. Semiconductor based photocatalytic degradation of pesticides: an overview. Environ. Technol. Innovation. 20: 101128.

Walker, D., Baumgartner, D., Gerba, C. and Fitzsimmons, K. 2019. Surface water pollution. pp. 261–292. *In*: Brusseau, M.L., Pepper, I.L. and Gerba, C.P. (eds). Environmental and Pollution Science. Elsevier.

Wang, J. and Chen, H. 2020. Catalytic ozonation for water and wastewater treatment: recent advances and perspective. Sci. Total Environ. 704: 135249.

Wang, C.C., Zhang, Q.-C., Kang, S.-G., Li, M.-Y., Zhang, M.-Y., Xu, W.-M., Xiang, P., et al. 2023. Heavy metal(loid)s in agricultural soil from main grain production regions of China: bioaccessibility and health risks to humans. Sci. Total Environ. 858: 159819.

Weidner, E. and Ciesielczyk, F. 2019. Removal of hazardous oxyanions from the environment using metal-oxide-based materials. Materials. 12(6): 927.

Wetchakun, K., Wetchakun, N. and Sakulsermsuk, S. 2019. An overview of solar/visible light-driven heterogeneous photocatalysis for water purification: TiO_2-and ZnO-based photocatalysts used in suspension photoreactors. J. Ind. Eng. Chem.. 71: 19–49.

Wu, M., Zhang, Y., Szeto, W., Pan, W., Huang, H. and Leung, D.Y. 2019. Vacuum ultraviolet (VUV)-based photocatalytic oxidation for toluene degradation over pure CeO_2. Chem. Eng. Sci. 200: 203–213.

Xu, M., Wu, C. and Zhou, Y. 2020. Advancements in the Fenton process for wastewater treatment. Adv. Oxid. Process. 61.

Yahya, N., Aziz, F., Jamaludin, N., Mutalib, M., Ismail, A., Salleh, W., Jaafar, J., et al. 2018. A review of integrated photocatalyst adsorbents for wastewater treatment. J. Environ. Chem. Eng. 6(6): 7411–7425.

Yan, H., Lai, C., Wang, D., Liu, S., Li, X., Zhou, X., Yi, H., et al. 2021. In situ chemical oxidation: peroxide or persulfate coupled with membrane technology for wastewater treatment. J. Mater. Chem. A. 9(20): 11944–11960.

Yang, L., Bai, X., Shi, J., Du, X., Xu, L. and Jin, P. 2019. Quasi-full-visible-light absorption by D35-TiO_2/g-C_3N_4 for synergistic persulfate activation towards efficient photodegradation of micropollutants. Appl. Catal., B: Environ 256: 117759.

Yu, X., Lin, X., Feng, W. and Li, W. 2019. Enhanced catalytic performance of a bio-templated TiO_2 UV-Fenton system on the degradation of tetracycline. Appl. Surf. Sci. 465: 223–231.

Zhang, Y., Zhang, X. and Wang, S. 2023. Recent advances in the removal of emerging contaminants from water by novel molecularly imprinted materials in advanced oxidation processes—a review. Sci. Total Environ. 163702.

Zhou, Y., Shi, J., Li, Y., Long, T., Zhu, X., Sun, C. and Wang, Y. 2020. Degradation kinetics and mechanism of bis (2-chloroethyl) ether by electromagnetic induction electrodeless lamp activated persulfate. Chemosphere. 261: 127709.

Zhu, D. and Zhou, Q. 2019. Action and mechanism of semiconductor photocatalysis on degradation of organic pollutants in water treatment: a review. Environ. Nanotechnol. Monit. Manage. 12: 100255.

Chapter 6

Green Synthesized Bio-nanomaterials for Pollutant Remediation

Ebenezer Agyei Aboagye[1], Mikaele Lorrany Siqueira Santos[1], Pollyanna Vanessa dos Santos Lins[1] and Lucas Meili[1]

[1]Laboratório de Processos, Centro de Tecnologia, Universidade Federal de Alagoas, UFAL, Maceió, Brazil

Dison Stracke Pfingsten Franco[2] and Jordana Georgin[2]

[2]Department of Civil and EnvironmentalEngineering, Universidad de la Costa, CUC, Barranquilla, Atlántico, Colombia

Rangabhashiyam Selvasembian*[3]

[3]Department of Environmental Science and Engineering, School of Engineering and Sciences, SRM University-AP, Amaravati, Andhra Pradesh, 522240, India

6.1 INTRODUCTION

In recent years, nanomaterials have come to have strong participation in various areas of society, and with that science and technology have started to invest in this field, providing a strong expansion of nanoscience (Singh et al. 2021, Nayak et al. 2021). Due to the improvement in synthesis and manipulation, new compounds have emerged with diameters and dimensions that are favorable for certain purposes, an example of which is nanoparticles (Auffan et al. 2009, Mercante et al. 2021). Because they have specific characteristics, these materials can be used for different purposes, including wastewater treatment (Patel et al. 2022). Studies have already proven the high efficiency of nanoparticles based on nanofibers and carbon nanotubes in the treatment of several emerging contaminants (Brescia 2022).

The use of these materials encompasses several technologies for wastewater treatment. Seeking the development of more sustainable materials, bionanomaterials emerged from biological or partially biological sources (Singh 2011, Mubarak et al. 2020). Depending on the purpose of

*Corresponding Author: rambhashiyam@gmail.com

application of the bionanomaterial, its composition and synthesis can be varied, the purposes ranging from environmental to biomedical areas (Paquin et al. 2015). Synthesis directly affects the properties and classification of the bionanomaterial, hence its purpose (Tautzenberger et al. 2012). In the environmental area, in addition to being applied in the removal of contaminants, they have favorable properties for detection and are used as sensors (Suehiro et al. 2003)

The ideal synthesis method aimed at cleaner, sustainable, and economical processes has made biological techniques become more studied and improved (Chaudhry et al. 2018, Nasrollahzadeh et al. 2021). An example is a synthesis through plant extracts (Sheny et al. 2011, Chaudhry et al. 2018), and the synthesis of nanoparticles by microorganisms (Nadaroğlu et al. 2017). In the environmental area, nanoparticles can be applied in a photocatalysis process, where nano photocatalysts are also used in water purification (Chen et al. 2019, Yaqoob et al. 2020). Nanomembranes composed of nanofibers are also applied in the removal of unwanted nanoparticles from the aqueous phase or as a pre-treatment process by reverse osmosis (Jhaveri and Murthy 2016). The synthesis may employ the incorporation of nanostructured catalytic materials such as iron-based free radicals and enzymatic catalysis within the membrane (Abdullah et al. 2019). Finally, the application of nanomaterials as adsorbents has been the subject of constant study because the properties are highly favorable to adsorption (high surface area, high porosity, high affinity with the adsorbate, and fast kinetic time) and allow the adsorption of different chemical compounds such as drugs, dyes, herbicides, metals, and hormones (Khajeh et al. 2013).

The high adsorption capacity for different target pollutants is extremely important since in full-scale treatment the compounds are in the form of a mixture. Fast kinetics, on the other hand, allows the processing time to be made viable, reducing costs (Peng et al. 2017). The most reported nanoadsorbents are polymeric, metallic, metallic oxides, and especially those based on carbon, such as graphene oxide (Yu et al. 2017). Physical adsorption allows the material to be regenerated through desorption, for which there are some regeneration methods depending on the adsorbent, these methods must still be improved in order to achieve greater economic viability in the process (Wongcharee et al. 2017, Manikam et al. 2019).

Therefore, the current and future points are studies that aim to improve the synthesis and applications of these materials. This chapter seeks to provide the current status of nanomaterials development and applications and future prospects in this area. With a main focus on the environment and the issues of emerging contaminants present mainly in water resources. Many of these studies mainly take into account economic feasibility and environmental sustainability. Considering the great variety of materials and consequently of chemical compounds, this area still has much to be explored. It is also observed that in the environmental area, many studies need to break the barrier of the laboratory scale and reach the real space. In this case, the technological areas of society must be present in this scientific environment, investing in the application of these technologies.

6.2 NANOMATERIALS

At the end of the 20th century, Gleiter and Skorokhod made the first classification of nanomaterials. Nonetheless, carbon-based materials and structures with zero, one, two, and three dimensions were not considered in the first classification. However, several years later, Pokropivny and Skorokhod reported a new classification of nanomaterials in which these low-dimensional nanomaterials are included (Garg et al. 2022). Recent research has revealed that nanomaterials can be classified into several categories based on their materials: semiconductors, inorganic oxidants, carbon-based nanomaterials, and organic nanomaterials and compounds (Singh et al. 2021).

There has been a rapid expansion of nanoscience, mainly due to the availability of new methods for synthesizing nanomaterials, as well as tools for their characterization and manipulation.

Low-dimensional materials caught the attention of scientific researchers. Nanoparticles are experimental particles that are often identified by a characteristic structural dimension between 1 and 100 nanometers (Auffan et al. 2009, Mercante et al. 2021). Semiconductors, computer chips, encapsulating agents, carriers, and contaminant removal are used in a wide variety of applications in everyday life (Chen et al. 2011). Also, nanomaterials can be used as nanoadsorbents, nanocatalysts and nanofibers (Jovanović 2021).

An important characteristic that distinguishes nanoparticles from other materials is that they have a higher specific surface area per volume. Due to their extremely high surface area, nanoparticles react extremely comfortably with larger particles (Madeira 2020). Therefore, it is evident that research on nanomaterials and their processes, the development of experimental and theoretical techniques, as well as innovative technologies based on this type of material, will lead to important scientific advances (Wang et al. 2019). Many technologies, for example, catalysis, have been using nanoscale processes for several decades, and the concept of nanoscience is totally new. However, the competition to synthesize specific sizes, and organize and customize materials at the nanoscale is of very recent origin. The size-dependent electrical, optical, and magnetic properties of individual nanostructures made of semiconductors, metals, and other materials are better understood in this decade (Rao and Cheetham 2001).

In addition to well-established techniques of crystallography, electron microscopy, and spectroscopy, scanning probe microscopes have provided robust data for the study of nanoparticles. The size of the nanostructures is small enough to accommodate significant treatment-relevant particles, making them ideal for computer-aided modeling and simulation. The immediate goals of nanomaterials science focus on mastering the synthesis of isolated nanostructures, their unions and assemblies with desired properties, the exploration and establishment of concepts, and the refinement of established research methodologies while discovering better tools to characterize nanostructures (Roy et al. 2021).

Organic and inorganic contaminants in contaminated water have established the demand for new technologies to remove these, which are in evidence and are not effectively removed by physical-chemical treatment methods (Nascimento et al. 2014). Traditional wastewater treatment methods are expensive because they involve heavy materials; a more modern method is the application of nanomaterials for wastewater treatment (Patel et al. 2022). Currently, the possibility of integrating the properties of two materials through the synthesis of nanomaterials has considerably increased the efficiency of various materials in the treatment of water and effluents (Khamlich et al. 2013).

In this respect, nanotechnology has played a crucial role in the creation of effective adsorbents and catalytic materials. The benefits of using nanomaterials for environmental remediation have been established in many other analyses. Such as nanoparticles made with nanofibers, and carbon nanotubes (Brescia 2022). Nanoparticles are used in a variety of approaches, including ultrafiltration membranes, osmosis, filtration, nanofiltration degradation, advanced oxidation, and nanomaterial-based infection control and water remediation processes. The effectiveness of the nanoparticle in removing contaminants from wastewater depends on the physical and chemical properties of the nanomaterial, the contaminant, and the wastewater itself.

6.3 BIONANOMATERIALS DEFINITION

Bionanomaterials are defined as nanomaterials that are wholly or partially composed of biological sources such as nucleic acid (DNA or RNA), lipids, proteins, peptides, and carbohydrates, for example (Singh 2011). The application of bionanomaterials ranged from biomedicine, catalysis, pharmaceuticals, fuel conversion, energy storage, and environmental application. Depending on the nature of the bionanomaterial, different applications are explored. Bionanomaterials are based on nucleic acid or DNA and are the most used in the administration of drugs, chemical and biological sensors, analysis of molecules, and as nanoconstruction material (Paquin et al. 2015). Research on lipid molecules/particles is also related to drug release or capture, and lipid coating of inorganic nanoparticles to increase aggregation and decrease cytotoxic effects (Mashaghi et al. 2013, Luchini and Vitiello 2019). Protein-based bionanomaterials have been investigated in different applications such as catalysis, bioelectronics, bioimaging, biosensing, and cell regeneration (Zhang and Wang 2019, Beloqui and Cortajarena 2020). Peptides are often mixed with proteins to perform similar tasks such as drug delivery, the building blocks for other materials, and even gene therapy (Briggs and Knecht 2012, Jeong et al. 2018). Carbohydrates are a more special group since they present biomedical applications such as those already mentioned and other applications in the development of materials for the solution of environmental problems in the form of adsorbents or catalysts. This fact is mainly due to the high availability of the cellulosic base material (Ansori 2022).

6.3.1 Bionanomaterials Properties

As mentioned above, the applications of each bionanomaterial are different depending on the origin, which directly affects its properties. Materials are typically characterized according to their physical, chemical, biological, mechanical, optical, electrical, and catalytic properties (Cui and Zhao 2019). Bionanomaterials also share characteristics with their nanomaterial counterparts, including size, shape, specific surface area, state of aggregation, particular size distribution, surface morphology, defined structural formula (including purity and impurities), and chemical characteristics of the surface (composition, charge, reactive sites, potential for catalysis) (European Commission 2006). Due to these properties, the bionanomaterial has different applications beyond the biomedical field, including catalysis (for the environment and traditional conversion), and pharmaceutical and environmental applications (adsorption and photocatalysis). However, the main property that separates the application in the biomedical field and traditional environmental engineering is biodegradability and non-toxicity. Without these properties, the application in the human body, for example, is not possible. On the other hand, for environmental applications, the material should have lower biodegradability (avoid changing the material frequently), high selectivity, and high specific surface area to ensure mass transfer and possible surface reactions (Tautzenberger et al. 2012).

6.3.2 Bionanomaterials Classification

The classification of bionanomaterials is shown in Figure 6.1, which are classified according to the nature and mixture of each bionanomaterial. Metal bionanomaterials and metal oxides are easy to synthesize and characterize as they have excellent properties. They can be easily synthesized by chemical, physical, or biological routes, but in general, biologically synthesized bionanomaterials attract attention because biological synthesis is cost-effective, free of chemical toxins, and have no side effects on the environment or life (Fernandes et al. 2020, Singh et al. 2020). Organic sources are frequently used to create mesoporous silica nanoparticles, which are then further changed via surface modification. For this, different molecules can be used, such as Aminopropyltriethoxysilane (APTES), or even simpler molecules such as hydroxyl (Porrang et al. 2022). Polymeric nanoparticles are often colloidal systems made up of natural or synthetic polymers, which present significant stability of polymeric nanoparticles in biological fluids along with the opportunity to functionalize their surfaces. In general, polymeric bionanoparticles can be prepared from emulsification, solvent evaporation, nanoprecipitation, supercritical antisolvent method, and salinization (Gagliardi et al. 2021). Plants/algae are often used as a source of metabolites and extractives, often carried out in the green synthesis process, where phenols and flavonoids are used as starting points to combine with metal ions, such as copper, silver, or even gold (Letchumanan et al. 2021).

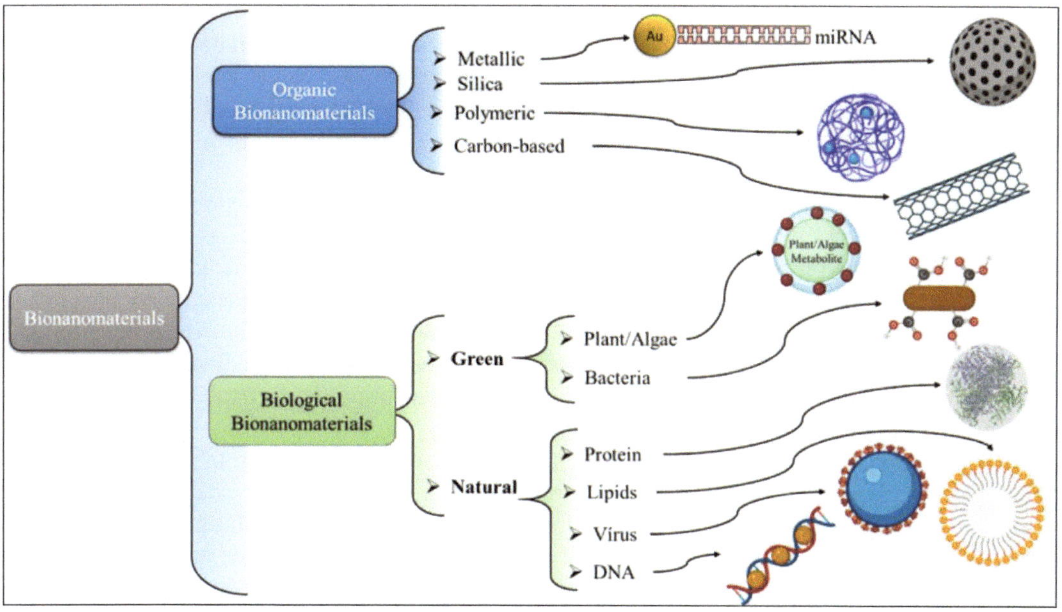

Figure 6.1 Classification of the bionanomaterials according to the origin and mixture.

Bacteria are another source of bionanomaterials, However, in this case, they are often used as a support for further surface modification, through the possibility of covalent bonding, adsorption (bacteria have negative surfaces), biomineralization (production of minerals in bacteria) (Ye et al. 2021). Protein-based bionanoparticles are the next category, This type of material has been much more investigated, presenting different methodologies such as desolvation, self-assembly, emulsification, biomineralization, and covalent/non-covalent interactions. A variety of proteins have frequently been employed to encapsulate or conjugate chemotherapeutic medications, imaging agents, radioactive or photoactive compounds, and functional ligands for diverse diagnostic and therapeutic applications (Iqbal et al. 2021). Lipids are another class of bionanomaterials, often called lipid-based nanocarriers, due to their main application, to carry drugs or target molecules. They are often found in the form of solid lipid bionanoparticles, which constitute liposomes, the structure that the lipid bilayer forms spontaneously in water (Tenchov et al. 2021).

Virus bionanoparticles are a category in which the material relies exclusively on different viruses as a building material or carrier. Depending on the geometry of the viruses, different modifications can be made, However, in general, the viruses can be modified on the surface, through bioconjugation or coupling of molecules, or filled internally through infusion, biomineralization, or through the self-assembly method [twenty-one]. The last category is DNA-based bionanomaterials, which are often used as a building block for other bionanomaterials or in conjunction with gold particles, for example. In particular, DNA has the possibility to form bioconjugates, which is the central mechanism for the construction of bionanoparticles (Zahid et al. 2013).

6.3.3 Bionanomaterials' Environmental Applications

Despite the important properties for biomedical applications, bionanomaterials can also be used in the environmental field, especially in the form of sensors (García-Mayagoitia et al. 2020), elimination (by catalysis) (Kumar et al. 2020), and adsorption (Dhali et al. 2021, Norrrahim et al. 2021). Bionanosensors are used to detect organic compounds, biochemical oxygen demand (BOD), soil contaminants such as heavy metals, insecticides, etc., and to detect toxins and contaminants and their concentration in soil, water, and air. Recently, carbon nanotube-based sensors have gained much attention as they exhibit exceptionally good chemical stability, electrical conductivity, high surface area, and mechanical stiffness. Special carbon nanotubes, including single-wall and multi-wall carbon nanotubes, have been studied for application as sensors (Zhou et al. 2019). An example is the application of a carbon nanotube sensor as a gas sensor for ammonia determination, where the sensors were able to determine ammonia concentrations at mg L^{-1} levels (Suehiro et al. 2003).

Another application is the catalysis of organic substances and the reduction of metal ions. In this case, two different origins are common 1) using carbon-based material as a support, and 2) extracting plant material to serve as an ionic bonding reagent. Nanocellulose coupled with Fe_3O_4 and Pd has been used for the reduction of organic dyes (methylene blue, methylene orange, rhodamine B, and Congo red) in less than 2 min (Kalantari et al. 2021). More recently, the combination of support materials (carbon-based, silica-based, metal-based, or polymer-based) with the immobilization of enzymes that bind through covalent and non-covalent bonds has been explored (Reshmy et al. 2021).

The last common application is adsorption. In this case, the application of carbon and silica-based materials is more common. Special carbon-based material has been extensively studied in adsorption, this group includes carbon nanotubes, graphene, and chitosan/chitin. These materials' exceptional adsorption abilities are mostly a result of their unique surface areas or functional groups, like amides in the case of chitosan and chitin. In addition to that, different studies have investigated the application of chitosan/chitin in different forms of nanoparticles, such as nanowhiskers, nanofilms, and nanogels for the adsorption of different molecules, from dyes to emerging contaminants (Araki et al. 2012, Kim et al. 2020). In addition, different studies have investigated the surface modifications of chitosan by functionalization of the amide group with tyrosine, carboxymethyl, sulfonyl, and many others (Francis et al. 2021). Chitosan and chitin have been combined with different materials (proteins, DNA, lipids), however, research on adsorption application is still lacking (Amaduzzi et al. 2014, Baghdan et al. 2018, Campos et al. 2017).

6.4 TYPES OF SYNTHESIS

Green, biological, or biogenic synthesis are terms commonly used for the ecological synthesis of nanoparticles. Nanoparticles can generally be created through chemical, physical, and biological processes. Although the physical and chemical creation of nanoparticles is increasingly common, the use of harmful chemicals restricts their usage in agriculture, the food industry, and health-related applications. In addition, physical and chemical syntheses are energy-intensive, while biological techniques are more cost-effective, clean, non-toxic, environmentally friendly, and have good polydispersity, dimension, and stability (Chaudhry et al. 2018, Nasrollahzadeh et al. 2021).

Figure 6.2 illustrates how much progress has been made in the field of nanotechnology due to the characteristics of nanomaterials and the products that are being employed in a variety of fields. For the manufacture of nanomaterials, conventional physicochemical strategies involve the participation of mostly hazardous and volatile materials, leading researchers to design and develop biogenic strategies that are ecologically correct and safer for the development of new and efficient adsorbents that can be used to degrade various contaminants in water (Nasrollahzadeh et al. 2021). The biological method that has been presented as an alternative, provides metallic nanoparticles with various shapes, sizes, content, and physicochemical properties.In addition, they are nanoparticles free of chemical contaminants for medical and biological applications where the purity of the nanoparticles is a major concern. The synthesis of nanoparticles can be done extracellularly or intracellularly and there are a large number of microorganisms with the potential to synthesize (Nadaroğlu et al. 2017, Chaudhry et al. 2018).

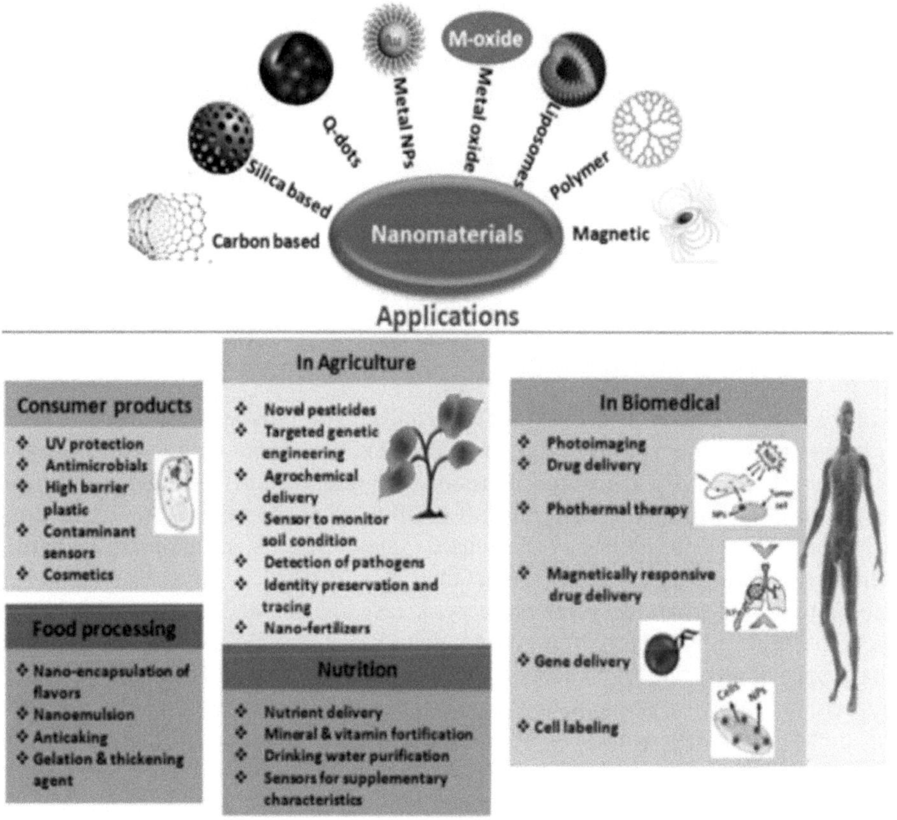

Figure 6.2 Applications of nanomaterials in agriculture, food industry, and biomedicine. *Source:* (Chaudhry et al. 2018).

There are ecological and biological routes for the biogenic fabrication of nanometric materials using plants and microorganisms such as algae, bacteria, fungi, viruses, and yeasts. The creation of sustainable nanostructures, such as nanoflowers, nanowires, nanorods, nanotubes, and nanoparticles, requires the stabilizing and lowering effects of the flavonoids, proteins, vitamins, phenolic acid, glycosides, carbohydrates, and polymers present in such sources. (Nasrollahzadeh et al. 2021). There are many environmental uses for nanoparticles produced through biosynthesis employing living animals and plants, such as the production of Fe and Cu-based adsorbent nanoparticles that demonstrate Malachite for aqueous adsorptive ability. (Zhang et al. 2018b).

Synthesis can be done at one stage using biological organisms such as bacteria, yeast, algae, and plants. Molecules in plants and microorganisms, such as proteins, enzymes, amines, alkaloids, and pigments, carry out a synthesis of nanoparticles by reduction (Nadaroğlu et al. 2017). Biogenic reduction bears resemblance to chemical reduction but involves substituting a biological reducing agent with an extract from a natural product possessing inherent traits such as stabilization, growth inhibition, and capping. This process differs from traditional chemical reduction in that it combines two materials with varying concentrations alongside organic reducing agents, thereby exerting influence over the size and shape of the resulting nanoparticles (Hussain et al. 2016).

The exact mechanism for the synthesis of metal nanoparticles is still not understood. It is believed that some enzymes play a role in the bioreduction of metal ions, with the aim of forming them as metal nanoparticles. A negatively charged bacterial cell wall interacts electrostatically with positively charged metal ions, or the bio-reduction process can be catalyzed by enzymes within or on the cell surface (Barabadi et al. 2017).

6.4.1 Synthesis of Nanoparticles Through Vegetable Extracts

The synthesis of nanoparticles through the plants attracted the attention of two researchers because it occurs in a single stage. The components of the plants can be used (roots, leafs, shoots, fruits, and shell) to obtain the extract used for synthesis of nanoparticles. The major advantage is that the plants components are free of toxic substances and also present several reducing agents and coatings that are necessary to the synthesis and stabilization of the nanoparticles (Chaudhry et al. 2018, Sheny et al. 2011). There are many reports in literature on plant-based green synthesis, producing gold and silver nanoparticles from extracts of *aloe vera*, *geranium*, and *Mirabilis jalapa* leaf, among others (Robinson et al. 2002, Shankar et al. 2004, Hussain et al. 2016).

6.4.2 Synthesis of Nanoparticles by Microorganisms

Reductases are known to allow microorganisms to collect and reduce metal particles into smaller, less polydisperse metal nanoparticles. Using proteins, peptides, enzymes, and cofactors as reducing agents or cofactors to provide the buffer necessary for the synthesis of nanoparticles as well as to prevent aggregation and increase the overall stability of nanoparticles, some microorganisms, including bacteria, fungi, and yeasts, have been used to synthesize metal nanoparticles intracellularly or extracellularly. (Nadaroğlu et al. 2017, Chaudhry et al. 2018).

Due to rapid development, personalized access, easy control, and manipulation of the growth environment, bacteria are involved in the production of nanoparticles. Know that some species of bacteria have special mechanisms to suppress the toxicity of two metals. In this way, through the use of biochemical pathways and reducing agents, such as proteins and enzymes that are present in bacteria, metal ions can be reduced and precipitated by the production of nanoparticles (Hussain et al. 2016). According to the literature, several bacteria can be used for the synthesis of silver and gold nanoparticles, depending on the enzyme nitrate reductase (Zhang et al. 2011, Soni and Prakash 2015).

A synthesis of nanoparticles using fungi consists of a simple method, producing stable and monodisperse materials, which have not been widely explored. The majority of fungi species have valuable metabolites with a high potential for bioaccumulation that can be used to create effective and affordable nanoparticles. The study by Mishra et al. demonstrates its synthesis using Trichoderma species, resulting in a material with properties that promote plant growth and control plant pathogens and can be used both as a biocatalyst and as a microbial agent (Mishra et al. 2014, Zhang et al. 2018a).

Yeasts also have the ability to accumulate toxic metals from their surroundings to adapt to the toxic environment of metals, for which they use different detoxification mechanisms, including chelation, bioprecipitation, biosorption, and extracellular sequestration (Chaudhry et al. 2018). In addition to microorganisms, biological particles like viruses, proteins, peptides, and enzymes also contribute to the creation of nanoparticles (Chaudhry et al. 2018). The mechanism of biofabrication of nanoparticles mediated by extracellular microbes occurs due to a microbial nitrate reductase responsible for the reduction of metal ions in metal nanoparticles. No matter how photosynthesized, phytochemicals such as proteins, flavonoids, phenols, and essential oils presented in plant extracts play a major role in the bioreduction of metal ions, transforming them into metal nanoparticles and also coating the synthesized nanoparticles for stabilization (Barabadi et al. 2017).

The biological synthesis pathway has been carried out, above all, through the use of bacteria, fungi, plants, cyanobacteria, and actinomycetes. There are many reports of mycosynthesis (synthesis via fungi) of metallic nanoparticles using different *Fusarium* species, named *Fusarium oxysporum*, *Fusarium solani*, *Fusarium acuminatum*, *Fusarium culmorum*, *Aspergillus niger*, *Aspergillus tamari*, *Phoma glomerata*, *Alternaria alternata*, and Endophytic Fungi *Pestalotia* sp. Many plants have been used such as coriander (*Coriandrum sativum*), papaya (*Carica papaya*), prickly pear (*Opuntia ficusindica*), datura or thorny apple (*Datura metel*), curry tree (*Murraya koenigii*) and Skunkvine (*Paederia foetida*) for the synthesis of nanoparticles. The phytosynthesis method (synthesis through plants) is fast, ecologically correct, and profitable, with simpler post-processing than mycosynthesis. Compared with phytosynthesis, mycosynthesis takes more time and the biomass is difficult to process. Therefore, phytosynthesis attracts a lot of attention from researchers around the world. There are three key factors in the synthesis of these metallic nanoparticles: the reducing agent, the reaction medium, and the stabilizing agent. In general, biosynthesized nanoparticles are spherical (Nasrollahzadeh et al. 2021).

6.5 POLLUTANTS REMOVAL METHODS

Nanotechnology consists of a large area of research that has grown in recent years, presenting very attractive options for the treatment of sewage and contaminated water (Figure 6.3). Due to their characteristics, including high catalytic activity, high physical-chemical and thermal stability, high surface area, and strong electron exchange capacity, among others, nanomaterials can be used in the degradation and remediation of organic or inorganic contaminants (Zhang et al. 2014, Xu et al. 2019). The nanomaterials and nanoparticles, we see being applied in the treatment of water pollutants, monitoring, and environmental detection are considered an excellent option, once their reactive structure possesses qualities that make them more efficient to remove dangerous pollutants, or at the same time, converting pollutants in toxic-free substances (Nasrollahzadeh et al. 2021).

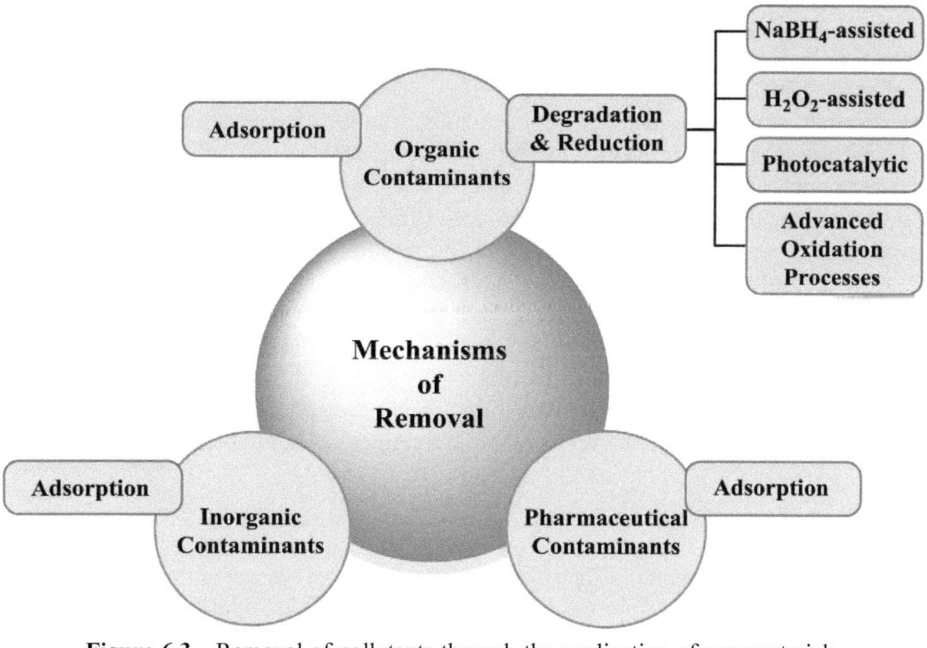

Figure 6.3 Removal of pollutants through the application of nanomaterials.

6.5.1 Nanophotocatalysts

The photocatalysis process entails either running to compost breakdown in the presence of light, activating or stimulating the usage of a light substance (UV, visible, or solar), in either case or sending the activated photocatalyst through photons of light energy. Additionally, nanocatalysts are frequently utilized to clean wastewater, and their increased surface area ratio and shape-dependent properties can boost catalyst reactivity. (Chen et al. 2019, Yaqoob et al. 2020). Due to the efficient formation of oxidizing species on the material's surface, nanophotocatalysts can increase the oxidation capacity, which contributes to the effective degradation of water pollutants. (Gómez-Pastora et al. 2017).

In photocatalysis, two types of processes occur, mineralization and degradation of organic pollutants. There is no degradation process, organic pollutants are broken down into various products, as for mineralization, organic pollutants in water, carbon dioxide, and inorganic gases are completely destroyed. The mechanism of contaminant degradation in the presence of light basically occurs as follows: a semiconductor, such as TiO_2, absorbs light that is greater than or equal to the length of the semiconductor space, creating electron-hole pairs. Leading to the interaction at the surface of the nanocatalyst, wich leads to interaction of the adsorbed species or promotion of the reduction-oxidation reactions (Wang et al. 2014, Nasrollahzadeh et al. 2021).

6.5.2 Nanomembranes

The nanomembranes are composed of different nanofibers and can be used to remove unwanted nanoparticles from the aqueous phase, or as a pre-treatment process for reverse osmosis (Jhaveri and Murthy 2016), due to their technical nature, or removal process that occurs through a high speed. In the case of drinking water, nanomembranes are used to separate toxic particles from water resources, as well as, nanomembrane filters can measure the level of water safety (Yaqoob et al. 2020). There are many studies on nanotechnology, focused on producing multifunctional membranes using different substances, in polymer-based membranes. A membrane containing porous support with a composted layer, normally a composite layer composed of a carbon-based material, such as graphene oxide, for example, dispersed in a polymer matrix, requires significant progress in resistance to fouling and non-transport of water (Zahid et al. 2018, Bassyouni et al. 2019).

Antimicrobial materials, such as silver metals, can be doped with polymer, to produce polymeric membranes that prevent the ingestion of bacteria and inhibit the production of biofilm (Kochkodan and Hilal 2015, Saleh et al. 2019). Some methods, including the insertion of nanostructured catalytic materials like iron-based free radicals and enzymatic catalysis within the membrane, can be taken into consideration in order to synthesize this kind of material and generate it with multifunctional resources. Consequently, it is possible to perform oxidative processes for the removal of contaminants and water purification without using any harmful materials. Furthermore, alongside advancements in nanotechnology, it has played a pivotal role in the creation of numerous catalytic membranes distinguished by their exceptional selectivity, permeability, and resistance to fouling. The most recent techniques in wastewater treatment use hybrid techniques to increase their multifunctional qualities. (Abdullah et al. 2019).

6.5.3 Nanoadsorbents

Nanomaterials, as environmentally friendly adsorbents, have garnered significant interest for their physical and chemical attributes, making them suitable for application without the need for wastewater treatment. The nanomaterials posses material that different from other traditional adsorbents, mainly related with the higher specific surface area, which leads to higher adsorption rates (Khajeh et al. 2013). The percentage of removal of contaminants or other organic pollutants exceeds 90% in most cases (Peng et al. 2017). Difficult to find commercially, nanoadsorbents have received the attention of two researchers and specialists, so that production occurs in greater quantity and at a commercial level (Salim and Ho 2015). The most reported nanoadsorbents are polymeric, metal oxides, metal, and mainly those based on carbon, such as graphene oxide (Yu et al. 2017).

There are many methods to regenerate nanoadsorbents, such as ionic exchange, cleaning agents, and magnetic forces, among others, making these materials even more interesting from an economic point of view since they can be used more than once (Wongcharee et al. 2017, Manikam et al. 2019). In addition, bioadsorbents have properties of high biodegradability, good biocompatibility, and non-toxicity, being able to be replaced by chemically synthesized materials (Yaqoob et al. 2020). At the moment, nanomaterials are highly appealing and are widely used in many industries, including power generation, agriculture, electronics, medicine, and wastewater treatment.

6.6 FUTURE PERSPECTIVES

A number of methods to eliminate the various environmental toxins have been developed and improved by the scientific community in recent years. When analyzing the application of bionanomaterials in this science, we observe that there is still much to be studied and explored. And this fact can be enhanced if we compare the progress in the synthesis of these materials in the field of medicine, which is much further ahead compared to environmental sciences. For this, it is important to break the barrier between areas, so that information can cross, enabling the transfer of knowledge. In this way, new materials can be developed and applied to different target pollutants. The majority of research is focused on the remediation of synthetic dyes, but while creating new biomaterials, attention may also be paid to other organic pollutants, heavy metals, and toxic gases that have a significant impact on aquatic life and the environment.

Recent studies have proven the excellent photocatalytic potential of nanocelluloses to remove dyes. The removal of other contaminants such as food additives, pesticides, volatile organic compounds (VOCs), different heavy metals such as mercury, lead, and copper, and harmful gases such as SO_x and NO_x should therefore be used as model pollutants in future researches using these materials in combination with other remediation techniques. The synthesis of new bionanomaterials can contribute to the optimization and versatility of environmental remediation worldwide since all these water and air pollutants are accumulated in the environment and need to be treated before being released to natural sources. The great future environmental challenge is the application of these new compounds in real conditions and on an industrial scale.

6.7 CONCLUSIONS

In conclusion, the development and classification of nanomaterials over time has resulted in a rapid expansion of the field of nanoscience. Due to their large specific surface area per volume and distinctive features, nanoparticles are highly reactive and well-suited for a variety of applications. Nanoadsorbents, nanocatalysts, and nanofibers with applications in catalysis, water treatment, and environmental remediation have all been developed thanks to the field of nanomaterials. Biomedicine, catalysis, medicines, fuel conversion, energy storage, and environmental applications are just a few of the areas where bionanomaterials, or nanoparticles made from biological sources such nucleic acids, lipids, proteins, peptides, and carbohydrates, have attracted interest. Depending on where they come from, bionanomaterials have unique qualities, but they also have traits in common with other nanomaterials, including size, shape, surface area, and chemical makeup. For biomedical applications, biodegradability and non-toxicity are essential qualities, while selectivity and large surface area are vital for environmental applications.

Depending on their composition and nature, bionanomaterials can be categorized as being made of metal, organic, polymeric, plant/algal, bacterial, protein, lipid, virus, or DNA. There are synthesis techniques and applications for each group. Bionanomaterials are employed in sensors, catalysis, and adsorption; materials based on carbon and silica are frequently used because of their outstanding adsorption properties. Due to their affordability, eco-friendliness, and non-toxicity, green or biological synthesis methods for nanoparticles have become increasingly popular. The advantages of biological synthesis over physical and chemical processes make it more appropriate for use in domains including agriculture, the food industry, and medicine. Overall, substantial breakthroughs in numerous scientific and technological fields have been made possible by the development made in the subject of nanotechnology, particularly in the areas of nanomaterials and bionanomaterials. Further advancements and discoveries in a wide range of applications are anticipated from further research and development in this area, which will advance science and help solve problems in areas including healthcare, environmental sustainability, and materials science.

References

Abdullah, N., Yusof, N., Lau, W.J., Jaafar, J. and Ismail, A.F. 2019. Recent trends of heavy metal removal from water/wastewater by membrane technologies. J. Ind. Eng. Chem. 76: 17–38. https://doi.org/10.1016/j.jiec.2019.03.029.

Amaduzzi, F., Bomboi, F., Bonincontro, A., Bordi, F., Casciardi, S., Chronopoulou, L. and Sennato, S. 2014. Chitosan-DNA complexes: Charge inversion and DNA condensation. Colloids Surf., B: Biointerfaces. 114, 1–10. https://doi.org/10.1016/j.colsurfb.2013.09.029.

Araki, J., Yamanaka, Y. and Ohkawa, K. 2012. Chitin-chitosan nanocomposite gels: reinforcement of chitosan hydrogels with rod-like chitin nanowhiskers. Polym. J. 44(7): 713–717. https://doi.org/10.1038/pj.2012.11.

Auffan, M., Rose, J., Bottero, J.Y., Lowry, G.V., Jolivet, J.P. and Wiesner, M.R. 2009. Towards a definition of inorganic nanoparticles from an environmental, health and safety perspective. Nat. Nanotechnol. 4(10): 634–641. https://doi.org/10.1038/nnano.2009.242

Baghdan, E., Pinnapireddy, S.R., Strehlow, B., Engelhardt, K.H., Schäfer, J. and Bakowsky, U. 2018. Lipid coated chitosan-DNA nanoparticles for enhanced gene delivery. Int. J. Pharm. 535(1–2): 473–479. https://doi.org/10.1016/j.ijpharm.2017.11.045.

Barabadi, H., Ovais, M., Shinwari, Z.K. and Saravanan, M. 2017. Anti-cancer green bionanomaterials: present status and future prospects. Green Chem. Lett. Rev. 10(4): 285–314. https://doi.org/10.1080/17518253.2017.1385856

Bassyouni, M., Abdel-Aziz, M.H., Zoromba, M.S., Abdel-Hamid, S.M.S. and Drioli, E. 2019. A review of polymeric nanocomposite membranes for water purification. J. Ind. Eng. Chem. 73: 19–46. https://doi.org/10.1016/j.jiec.2019.01.045.

Beloqui, A. and Cortajarena, A.L. 2020. Protein-based functional hybrid bionanomaterials by bottom-up approaches. Curr. Opin. Struct. Biol. 63: 74–81. https://doi.org/10.1016/j.sbi.2020.04.005.

Brescia, E.A. 2022. Study of thermodynamic relationship in nanostructured materials applied to adsorption and systems with curved interfaces. Thesis, Federal University of Minas Gerais, Brazil.

Briggs, B.D. and Knecht, M.R. 2012. Nanotechnology meets biology: peptide-based methods for the fabrication of functional materials. J. Phys. Chem. Lett. 3(3): 405–418. https://doi.org/10.1021/jz2016473.

Campos, J., Varas-Godoy, M. and Haidar, Z.S. 2017. Physicochemical characterization of chitosan-hyaluronan-coated solid lipid nanoparticles for the targeted delivery of paclitaxel: a proof-of-concept study in breast cancer cells. Nanomedicine. 12(5): 473–490. https://doi.org/10.2217/nnm-2016-0371.

Chaudhry, N., Dwivedi, S., Chaudhry, V., Singh, A., Saquib, Q., Azam, A. and Musarrat, J. 2018. Bio-inspired nanomaterials in agriculture and food: current status, foreseen applications and challenges. Microb. Pathogen. 123: 196–200. https://doi.org/10.1016/j.micpath.2018.07.013.

Chen, J., Xiu, Z., Lowry, G.V. and Alvarez, P.J.J. 2011. Effect of natural organic matter on toxicity and reactivity of nano-scale zero-valent iron. Water Research. 45(5): 1995–2001. https://doi.org/10.1016/j.watres.2010.11.036.

Chen, W., Liu, Q., Tian, S. and Zhao, X. 2019. Exposed facet dependent stability of ZnO micro/nano crystals as a photocatalyst. Appl. Surf. Sci. 470: 807–816. https://doi.org/10.1016/j.apsusc.2018.11.206.

Cui, W. and Zhao, X. (eds). 2019. Theranostic Bionanomaterials. Amsterdam, Netherlands, Elsevier.

Dhali, K., Ghasemlou, M., Daver, F., Cass, P. and Adhikari, B. 2021. A review of nanocellulose as a new material towards environmental sustainability. Sci. Total Environ. 775: 145871. https://doi.org/10.1016/j.scitotenv.2021.145871.

European Commission. 2006. The appropriateness of existing methodologies to assess the potential risks associated with engineered and adventitious products of nanotechnologies. Scenihr/002/05. Scientific committee on emerging and newly identified health risks (Scenihr).

Fernandes, C.L.F., Volcaõ, L.M., Ramires, P.F., Moura, R.R.D. and Da Silva Júnior, F.M.R. 2020. Distribution of pesticides in agricultural and urban soils of Brazil: a critical review. Environ. Sci. Processes Impacts. 22(2): 256–270. https://doi.org/10.1039/C9EM00433E.

Francis, A.O., Zaini, M.A.A., Muhammad, I.M., Abdulsalam, S. and El-Nafaty, U.A. 2021. Physicochemical modification of chitosan adsorbent: a perspective. Biomass Convers. Biorefin. (0123456789). https://doi.org/10.1007/s13399-021-01599-3.

Gagliardi, A., Giuliano, E., Venkateswararao, E., Fresta, M., Bulotta, S., Awasthi, V. and Cosco, D. 2021. Biodegradable polymeric nanoparticles for drug delivery to solid tumors. Front. Pharmacol. 12: 1–24. https://doi.org/10.3389/fphar.2021.601626.

García-Mayagoitia, S., Pérez-Hernández, H., Medina-Pérez, G., Campos-Montiel, R.G. and Fernández-Luqueño, F. 2020. Bio-nanomaterials in the air pollution treatment. Nanomater. Air Rem. 227–248. https://doi.org/10.1016/B978-0-12-818821-7.00012-9.

Garg, R., Rani, P., Garg, R., Khan, M.A., Khan, N.A., Khan, A.H. and Américo-Pinheiro, J.H.P. 2022. Biomedical and catalytic applications of agri-based biosynthesized silver nanoparticles. Environ. Pollut. 310: 119830. https://doi.org/10.1016/j.envpol.2022.119830.

Gómez-Pastora, J., Dominguez, S., Bringas, E., Rivero, M.J., Ortiz, I. and Dionysiou, D.D. 2017. Review and perspectives on the use of magnetic nanophotocatalysts (MNPCs) in water treatment. Chem. Eng. J. 310: 407–427. https://doi.org/10.1016/j.cej.2016.04.140.

Hussain, I., Singh, N.B., Singh, A., Singh, H. and Singh, S.C. 2016. Green synthesis of nanoparticles and its potential application. Biotechnol. Lett. 38(4): 545–560. https://doi.org/10.1007/s10529-015-2026-7.

Iqbal, H., Yang, T., Li, T., Zhang, M., Ke, H., Ding, D., and Chen, H. 2021. Serum protein-based nanoparticles for cancer diagnosis and treatment. J. Controlled Release. 329(October 2020): 997–1022. https://doi.org/10.1016/j.jconrel.2020.10.030.

Jeong, W.J., Bu, J., Kubiatowicz, L.J., Chen, S.S., Kim, Y.S. and Hong, S. 2018. Peptide–nanoparticle conjugates: a next generation of diagnostic and therapeutic platforms? Nano Convergence. 5(1). 1–18. https://doi.org/10.1186/s40580-018-0170-1.

Jhaveri, J.H. and Murthy, Z.V.P. 2016. A comprehensive review on anti-fouling nanocomposite membranes for pressure driven membrane separation processes. Desalination. 379: 137–154. https://doi.org/10.1016/j.desal.2015.11.009.

Jovanović, D.J. 2021. Low-dimensional nanomaterials: Syntheses, physicochemical properties, and their role in wastewater treatment. pp. 27–58. *In:* Bhanvase, B., Sonawane, S., Pawad. V. and Pandit A. (eds). Handbook of Nanomaterials for Wastewater Treatment. Elsevier. https://doi.org/https://doi.org/10.1016/B978-0-12-821496-1.00028-3.

Kalantari, E., Khalilzadeh, M.A. and Zareyee, D. 2021. Effective reduction of Cr(VI) and organic dyes using Pd NPs/Fe_3O_4@nanocellulose as a recoverable catalyst in aqueous media. J. Inorg. Organomet. Polym. Mater. 31(1): 319–330. https://doi.org/10.1007/s10904-020-01784-3.

Khajeh, M., Laurent, S. and Dastafkan, K. 2013. Nanoadsorbents: classification, preparation, and applications (with emphasis on aqueous media). Chem. Rev. 113(10): 7728–7768. https://doi.org/10.1021/cr400086v.

Khamlich, S., Bello, A., Fabiane, M., Ngom, B.D. and Manyala, N. 2013. Hydrothermal synthesis of simonkolleite microplatelets on nickel foam-graphene for electrochemical supercapacitors. J. Solid State Electrochem. 17(11): 2879–2886. https://doi.org/10.1007/s10008-013-2206-0.

Kim, S.J., Hong, B.M. and Park, W.H. 2020. The effects of chitin/chitosan nanowhiskers on the thermal, mechanical and dye adsorption properties of electrospun PVA nanofibrous membranes. Cellulose. 27(10): 5771–5783. https://doi.org/10.1007/s10570-020-03191-w.

Kochkodan, V. and Hilal, N. 2015. A comprehensive review on surface modified polymer membranes for biofouling mitigation. Desalination. 356: 187–207. https://doi.org/10.1016/j.desal.2014.09.015.

Kumar, A., Sharma, G., Naushad, M., Al-Muhtaseb, A.H., García-Peñas, A., Mola, G.T. and Stadler, F.J. 2020. Bio-inspired and biomaterials-based hybrid photocatalysts for environmental detoxification: a review. Chem. Eng. J. 382: 122937. https://doi.org/10.1016/j.cej.2019.122937.

Letchumanan, D., Sok, S.P.M., Ibrahim, S., Nagoor, N.H. and Arshad, N.M. 2021. Plant-based biosynthesis of copper/copper oxide nanoparticles: An update on their applications in biomedicine, mechanisms, and toxicity. Biomolecules. 11(4): 564. https://doi.org/10.3390/biom11040564.

Luchini, A. and Vitiello, G. 2019. Understanding the nano-bio interfaces: Lipid-coatings for inorganic nanoparticles as promising strategy for biomedical applications. Front. Chem. 7: 1–16. https://doi.org/10.3389/fchem.2019.00343.

Madeira, J.M.C. 2020. Nanoparticles as protein transport systems. Thesis, Universidade de Lisboa.

Manikam, M.K., Halim, A.A., Hanafiah, M.M. and Krishnamoorthy, R.R. 2019. Removal of ammonia nitrogen, nitrate, phosphorus and cod from sewage wastewater using palm oil boiler ash composite adsorbent. Desalin. Water Treat. 149: 23–30. https://doi.org/10.5004/dwt.2019.23842.

Mashaghi, S., Jadidi, T., Koenderink, G. and Mashaghi, A. 2013. Lipid nanotechnology. Int. J. Mol. Sci. 14(2): 4242–4282. https://doi.org/10.3390/ijms14024242.

Mercante, L., Andre, R., Macedo, J., Pavinatto, A. and Correa, D. 2021. Nanofibras Eletrofiadas E Suas Aplicações: Avanços Na Última Década. Química Nova. 44(6): 717–736. https://doi.org/10.21577/0100-4042.20170721.

Mishra, A., Kumari, M., Pandey, S., Chaudhry, V., Gupta, K.C. and Nautiyal, C.S. 2014. Biocatalytic and antimicrobial activities of gold nanoparticles synthesized by Trichoderma sp. Bioresour. Technol. 166: 235–242. https://doi.org/10.1016/j.biortech.2014.04.085

Mubarak, N.S.A., Sabar, S. and Jawad, A.H. 2020. The study of commercial titanium dioxide (TiO_2) Degussa P25 for the Adsorption of Acidic Dye. Sci. Lett. 14: 68–83. https://doi.org/10.24191/sl.v14i1.10607.

Nadaroğlu, H., Alayli Güngör, A. and İnce, S. 2017. Synthesis of Nanoparticles by Green Synthesis Method. INJIRR 1(1): 6–9. Retrieved from http://www.injirr.com/article/view/4.

Nascimento, R.F. do, Lima, A.C.A. de, Vidal, B.C., Melo, D. de Q. and Raulino, G.S.C. 2014. Adsorption: theoretical aspects and environmental applications. https://doi.org/10.13140/RG.2.1.4340.1041.

Nasrollahzadeh, M., Sajjadi, M., Iravani, S. and Varma, R.S. (2021). Green-synthesized nanocatalysts and nanomaterials for water treatment: Current challenges and future perspectives. J. Hazard. Mater. 401(May 2020): 123401. https://doi.org/10.1016/j.jhazmat.2020.123401.

Norrrahim, M.N.F., Md. Kasim, N.A., Knight, V.F., Md. Misenan, M.S., Janudin, N., Ahmad Shah, N.A. and Zin Wan Yunus, W.M. 2021. Nanocellulose: a bioadsorbent for chemical contaminant remediation. RSC Adv. 11(13): 7347–7368. https://doi.org/10.1039/D0RA08005E.

Paquin, F., Rivnay, J., Salleo, A., Stingelin, N. and Silva, C. 2015. Multi-phase semicrystalline microstructures drive exciton dissociation in neat plastic semiconductors. J. Mater. Chem. C. 3: 10715–10722. https://doi.org/10.1039/b000000x.

Patel, N., Srivastav, A.L., Patel, A., Singh, A., Singh, S.K., Chaudhary, V.K., and Bhunia, B. 2022. Nitrate contamination in water resources, human health risks and its remediation through adsorption: a focused review. Environ. Sci. Pollut. Res. 29(46): 69137–69152. https://doi.org/10.1007/s11356-022-22377-2.

Peng, F., Xu, T., Wu, F., Ma, C., Liu, Y., Li, J., and Mao, C. 2017. Novel biomimetic enzyme for sensitive detection of superoxide anions. Talanta. 174: 82–91. https://doi.org/10.1016/j.talanta.2017.05.028.

Porrang, S., Davaran, S., Rahemi, N., Allahyari, S. and Mostafavi, E. 2022. How advancing are mesoporous silica nanoparticles? A comprehensive review of the literature. Int. J. Nanomed. 17: 1803–1827. https://doi.org/10.2147/IJN.S353349.

Rao, C.N.R. and Cheetham, A.K. 2001. Science and technology of nanomaterials: current status and future prospects. J. Mater. Chem. 11(12): 2887–2894. https://doi.org/10.1039/b105058n.

Reshmy, R., Philip, E., Sirohi, R., Tarafdar, A., Arun, K.B., Madhavan, A. and Sindhu, R. 2021. Nanobiocatalysts: advancements and applications in enzyme technology. Bioresour. Technol. 337: 125491. https://doi.org/10.1016/j.biortech.2021.125491

Robinson, T., Chandran, B. and Nigam, P. 2002. Removal of dyes from a synthetic textile dye effluent by biosorption on apple pomace and wheat straw. Water Res. 36: 2824–2830. https://doi.org/10.1016/S0043-1354(01)00521-8.

Roy, A., Sharma, A., Yadav, S., Jule, L.T. and Krishnaraj, R. 2021. Nanomaterials for remediation of environmental pollutants. Bioinorg. Chem. Appl. 2021. https://doi.org/10.1155/2021/1764647.

Saleh, T.A., Parthasarathy, P. and Irfan, M. 2019. Advanced functional polymer nanocomposites and their use in water ultra-purification. Trends Environ. Anal. Chem. 24: e00067. https://doi.org/10.1016/j.teac.2019.e00067.

Salim, W. and Ho, W.W. 2015. Recent developments on nanostructured polymer-based membranes. Curr. Opin. Chem. Eng. 8: 76–82. https://doi.org/10.1016/j.coche.2015.03.003.

Shankar, S.S., Rai, A., Ahmad, A. and Sastry, M. 2004. Rapid synthesis of Au, Ag, and bimetallic Au core-Ag shell nanoparticles using Neem (*Azadirachta indica*) leaf broth. J. Colloid Interface Sci. 275(2): 496–502. https://doi.org/10.1016/j.jcis.2004.03.003.

Sheny, D.S., Mathew, J. and Philip, D. 2011. Phytosynthesis of Au, Ag and Au-Ag bimetallic nanoparticles using aqueous extract and dried leaf of *Anacardium occidentale*. Spectrochim. Acta, Part A: Molecular and Biomolecular Spectroscopy. 79(1): 254–262. https://doi.org/10.1016/j.saa.2011.02.051

Singh, R.P. 2011. Prospects of nanobiomaterials for biosensing. Int. J. Electrochem. 2011: 1–30. https://doi.org/10.4061/2011/125487.

Singh, K.R.B., Nayak, V., Sarkar, T. and Singh, R.P. 2020. Cerium oxide nanoparticles: properties, biosynthesis and biomedical application. RSC Advances. 10(45): 27194–27214. https://doi.org/10.1039/d0ra04736h.

Singh, K.R., Nayak, V., Singh, J., Adetunji, C.O. and Singh, R.P. 2021. Introduction: potentialities of bionanomaterials towards the environmental and agricultural domain. pp. 1–19. *In*: Singh, R.P. and Singh, Kshitij RB. (eds). Bionanomaterials for Environmental and Agricultural Applications. IOP Publishing. https://doi.org/10.1088/978-0-7503-3863-9ch1.

Soni, N. and Prakash, S. 2015. Antimicrobial and mosquitocidal activity of microbial synthesized silver nanoparticles. Parasitol. Res. 114(3): 1023–1030. https://doi.org/10.1007/s00436-014-4268-z.

Suehiro, J., Zhou, G. and Hara, M. 2003. Fabrication of a carbon nanotube-based gas sensor using dielectrophoresis and its application for ammonia detection by impedance spectroscopy. J. Phys. D: Appl. Phys. 36(21): L109–L114. https://doi.org/10.1088/0022-3727/36/21/L01.

Tautzenberger, A., Kovtun, A. and Ignatius, A. 2012. Nanoparticles and their potential for application in bone. Int. J. Nanomed. 7: 4545–4557. https://doi.org/10.2147/IJN.S34127.

Tenchov, R., Bird, R., Curtze, A.E. and Zhou, Q. 2021. Lipid nanoparticles from liposomes to mRNA vaccine delivery, a landscape of research diversity and advancement. ACS Nano. 15(11): 16982–17015. https://doi.org/10.1021/acsnano.1c04996.

Wang, N., Zhang, X., Wang, Y., Yu, W. and Chan, H.L.W. 2014. Microfluidic reactors for photocatalytic water purification. Lab Chip. 14(6): 1074–1082. https://doi.org/10.1039/c3lc51233a.

Wang, Y., Guo, L., Qi, P., Liu, X. and Wei, G. 2019. Synthesis of three-dimensional graphene-based hybrid materials for water purification: a review. Nanomaterials. 9(8): 1123. https://doi.org/10.3390/nano9081123.

Wongcharee, S., Aravinthan, V., Erdei, L. and Sanongraj, W. 2017. Use of macadamia nut shell residues as magnetic nanosorbents. Int. Biodeterior. Biodegrad. 124: 276–287. https://doi.org/10.1016/j.ibiod.2017.04.004.

Xu, C., Nasrollahzadeh, M., Sajjadi, M., Maham, M., Luque, R. and Puente-Santiago, A.R. 2019. Benign-by-design nature-inspired nanosystems in biofuels production and catalytic applications. Renewable Sustainable Energy Rev. 112: 195–252. https://doi.org/10.1016/j.rser.2019.03.062.

Yaqoob, A.A., Parveen, T., Umar, K. and Ibrahim, M.N.M. 2020. Role of nanomaterials in the treatment of wastewater: a review. Water (Switzerland). 12(2): 495. https://doi.org/10.3390/w12020495.

Ye, Z., Liang, L., Lu, H., Shen, Y., Zhou, W. and Li, Y. 2021. Nanotechnology-employed bacteria-based delivery strategy for enhanced anticancer therapy. Int. J. Nanomed. 16: 8069–8086. https://doi.org/10.2147/IJN.S329855.

Yu, L., Ruan, S., Xu, X., Zou, R. and Hu, J. 2017. One-dimensional nanomaterial-assembled macroscopic membranes for water treatment. Nano Today. 17: 79–95. https://doi.org/10.1016/j.nantod.2017.10.012.

Zahid, M., Kim, B., Hussain, R., Amin, R. and Park, S.H. 2013. DNA nanotechnology: a future perspective. Nanoscale Res. Lett. 8(1): 1–13. https://doi.org/10.1186/1556-276x-8-119.

Zahid, Muhammad, Rashid, A., Akram, S., Rehan, Z.A. and Razzaq, W. 2018. A comprehensive review on polymeric nano-composite membranes for water treatment. J. Membr. Sci. Technol. 08(01): 1–20. https://doi.org/10.4172/2155-9589.1000179.

Zhang, X., Yan, S., Tyagi, R.D. and Surampalli, R.Y. 2011. Synthesis of nanoparticles by microorganisms and their application in enhancing microbiological reaction rates. Chemosphere. 82(4): 489–494. https://doi.org/10.1016/j.chemosphere.2010.10.023.

Zhang, X., Lin, M., Lin, X., Zhang, C., Wei, H., Zhang, H. and Yang, B. 2014. Polypyrrole-enveloped Pd and Fe_3O_4 nanoparticle binary hollow and bowl-like superstructures as recyclable catalysts for industrial wastewater treatment. ACS Appl. Mater. Interfaces. 6(1): 450–458. https://doi.org/10.1021/am404440d.

Zhang, M., Liu, W., Zhou, Y., Li, Y., Qin, Y. and Xu, Y. 2018a. Neurodevelopmental toxicity induced by maternal PM2.5 exposure and protective effects of quercetin and vitamin C. Chemosphere. 213: 182–196. https://doi.org/10.1016/j.chemosphere.2018.09.009.

Zhang, P., Hou, D., O'Connor, D., Li, X., Pehkonen, S., Varma, R.S. and Wang, X. 2018b. Green and size-specific synthesis of stable Fe-Cu oxides as earth-abundant adsorbents for malachite green removal. ACS Sustainable Chem. Eng. 6(7): 9229–9236. https://doi.org/10.1021/acssuschemeng.8b01547.

Zhang, D. and Wang, Y. 2019. Functional protein-based bioinspired nanomaterials: From coupled proteins, synthetic approaches, nanostructures to applications. Int. J. Mol. Sci. 20(12): 3054. https://doi.org/10.3390/ijms20123054.

Zhou, Y., Fang, Y. and Ramasamy, R.P. 2019. Non-covalent functionalization of carbon nanotubes for electrochemical biosensor development. Sensors (Switzerland). 19(2): 392. https://doi.org/10.3390/s19020392

Chapter 7

Nanomaterials for Purification and Desalination of Water

Koyel Biswas, Sushil Kumar Mandal*
and Jayanta Kumar Biswas

Department of Ecological Studies and International Centre for Ecological Engineering, University of Kalyani, Kalyani, Nadia-741235, West Bengal, India.

7.1 INTRODUCTION

Water is one of the most important assets of human civilization and potable water supply is a basic human necessity. As the world's population exponentially increases, the global demand for freshwater consumption also increases (Kunduru et al. 2017). Demands for purified water have led to improvements in the quality. Only 2.5% of the world's oceans and seas harness fresh water, 70% of which is frozen. Only <1% fresh water can be used as drinking water. Public and environmental health is threatened by a wide variety of water pollutants, including naturally occurring contaminants and synthetic chemicals. The provision of potable water to match human needs is a huge challenge of the 21st century. Reports show that water-borne pathogens are infecting 250 million people per year resulting in 10 to 20 million deaths (Malik et al. 2012).

Wastewater contains organic, inorganic and biological contaminants which are toxic and often carcinogenic, posing deleterious effects on human health and ecosystems (Ali 2012). Heavy metals/ metalloids (cadmium, chromium, mercury, lead, zinc, nickel, copper, arsenic and so on) are water pollutants with high toxicity showing hazardous effects. Nitrates, sulfates, phosphates, fluorides, chlorides, selenides, chromates and oxalates show hazardous effects at high concentrations; these ions also change the taste of water. Among the organic pollutants pesticides, fertilizers, hydrocarbons, phenols, plasticizers, biphenyls, detergents, oils and greases also pose toxicities to organisms (Barcelo 2004). Industrial, household and hospital wastewater contain various metabolized and non-metabolized emerging contaminants like pharmaceuticals and personal care products (PPCPs) (i.e., drugs, drugs products, cosmetics, additives, detergents, packaging materials, etc.) which are usually resistant against natural biodeterioration (Werkneh and Rene 2019).

*Corresponding Author: sushilestu18@klyuniv.ac.in

Aromatic hydrocarbons such as benzene, toluene, ethyl benzene and xylenes together labeled as BTEX are the most predominant components in treated water among other volatile contaminants. These volatile organic compounds (VOCs) can easily evaporate when water is exposed to the atmosphere (Mangun et al. 2001). Pathogenic bacteria present in wastewater are responsible for several waterborne diseases. Natural organic matter (NOM) is one of the principal characteristics of water quality which determines the strategy of purification. Sometimes NOM is represented by total organic carbon and dissolved organic carbon (Nagar and Pradeep 2020). NOM is omnipresent in animal and plant material and vitiates the effectiveness of certain techniques such as flocculation, carbon adsorption or filtration. NOM pollution in the surface and groundwater makes it unfit for drinking (Matilainen et al. 2011). Therefore, water treatment and desalination are mandatory to solve the major problems. Conventional technologies are reaching their limits in providing sufficient pure and desalinated water. Nanoparticles have great potential for wastewater treatment. Due to their high surface area, the nanoparticles can be used efficiently for removing toxic metal ions, pathogenic and non-pathogenic microbes, organic and inorganic pollutants from water. The application of nanotechnology in wastewater treatment and seawater desalination are a sustainable technique to supply fresh water in a more efficient and eco-friendly way.

7.2 IMPORTANCE OF NANOTECHNOLOGY IN WATER PURIFICATION

Nanomaterials play a key role in water purification, thereby improving and protecting water supplies. Recent advances suggest that most of the issues pertaining to water quality could be resolved or greatly ameliorated using nanoparticles, nanofiltration or other products resulting from the development of nanotechnology. Utilization of specific nanoparticles either embedded in membranes or on other structural media that can effectively, inexpensively and rapidly render unusable water into potable water is being explored at a variety of institutions. Innovative use of nanoparticles for the treatment of industrial wastewater is a potentially useful application (Kanchi 2014). Removal of contaminants and recycling of the wastewater would provide significant reductions in cost, tim. and labour to industr. and result in improved environmental stewardship. Aquifer and groundwater remediation are also critical issues, becoming more important as water supplies steadily decrease and demand continues to increase. Most of the remediation technologies available today are effective but very often costly and time-consuming, particularly pump-and-treat methods. The ability to remove toxic compounds from the subsurface and other environments that are very difficult to access *in situ.* and doing so rapidly, efficiently and within reasonable costs is the ultimate goal (Hillie and Hlophe 2007).

7.3 NANOSORBENTS

Adsorption is a surface phenomenon in which the atoms, ions or molecules from a gas, liquid or dissolved solid adsorb/attach to a surface. This can occur because of physical forces or by chemical bonds (Hu and Xu 2020). Nowadays, conventional adsorbents lose their efficiency due to small surface area and lack of selectivity. Nanoadsorbents are generally used to remove organic and inorganic pollutants from wastewater (Qu et al. 2013). These are made up of nanoparticles with a high specific surface area and connected sorption sites. The adsorbent exhibits better adsorption for organic chemicals due to its short intra-particle dispersion distance and adjustable pore size and surface chemistry (Gehrke et al. 2015). The idiosyncratic properties such as their small size, catalytic potential, high reactivity, vast surface area, ease of separation and abundant active sites-contaminants interaction embedded make them ideal adsorbents for wastewater (Ali 2012).

Different types of nanosorbents available are carbon-based (nanotubes), metal-based (metal oxides), polymeric, nanofiltration membrane, nanocatalysts, fullerene, ZnO and nano zero-valent iron (nZVI) nanosorbent (Panahi et al. 2018). Nanoparticles have two key properties that make them particularly attractive as sorbents. On a mass basis, they have much larger surface areas than bulk particles. They have great potential for novel, more efficient and faster decontamination processes aimed at the removal of organic and inorganic pollutants like heavy metals and micro-pollutants. The superior process efficacy of adsorbent material enables implementation of more compact water and wastewater treatment devices with smaller footprints, particularly for decentralized applications and point-of-use systems. Modern research activities mainly focus on the following types of nanoadsorbents: (1) carbonaceous nanomaterials (CNTs, graphene), (2) metal-based nanoadsorbents, (3) polymeric nanoadsorbent, and (4) Zeolite (Rao et al. 2007).

7.4 CARBONACEOUS NANOMATERIALS

Carbonaceous nanomaterials integrate the diacritic properties of sp^2 hybridized carbon bonds with tremendous physical, chemical and electrical properties at nanoscale. They have a wide area of applications, including environmental remediation, contaminants transformation, drug delivery, supercapacitors, catalysis, etc. Nanocellulose/nanocarbon composites are new smart hybrid materials that combine cellulose nanoparticles (nanofibrils and nanocrystals) with carbon nanoparticles (fullerenes, graphene, nanotubes and nanodiamonds) or other carbon nanostructures (carbon nanofibers, carbon quantum dots, activated carbon and carbon black) (Bacakova et al. 2020). Nanocarbon materials with nanocellulose provide the functionality of nanocarbons, using an eco-friendly, low-cost, strong, dimension-stable, non-melting, nontoxic and nonmetal matrix or carrier, which alone has versatile applications in industry, biotechnology and biomedicine (Bacakova et al. 2020, Zhang et al. 2019).

7.5 CARBON NANOTUBES

Carbon nanotubes (CNTs) have recently attracted considerable attention for the synthesis of novel membranes with attractive features for water purification. CNTs, also called buckytubes, are cylindrical carbon molecules with unique properties. CNTs, also known as tubular fullerenes, are cylindrical graphene sheets of sp^2-bonded carbon atoms. It can be divided into three categories on the basis of the number of tubes present in the CNTs. These are Single-walled CNTs (SWCNTs), Double-walled CNTs and Multi-walled CNTs (MWNTs). CNTs exhibit unique properties such as large specific areas, rich hollows and layered structures, which bring about a wide variety of remarkable applications in wastewater treatment, including the removal of organic contaminants and heavy metals (Ibrahim 2013). CNTs possess highly assessable adsorption sites and hydrophobic surfaces. Due to their hydrophobic surface, CNTs have to be stabilized in aqueous suspension in order to avoid aggregation that reduces the active surface.

7.6 GRAPHENE-BASED NANOMATERIALS

There are many types of graphene-based nanomaterials like (a) graphene, (b) multilayer graphene, (c) graphene oxide (GO), (d) reduced (GO), (e) graphene quantum dots (GQD) and (f) reduced GQD. Graphene is an infinite two-dimensional (2D) monolayer consisting of sp^2-hybridized carbon atoms. It could be viewed as the fundamental building block for other carbon allotropes such as fullerenes and carbon nanotubes. The theoretical surface area of graphene could reach to 2630 m^2/g, which is much higher than many materials. With these special structural properties, graphene has shown standout attributes and were studied in many applications (Suárez-Iglesias et al. 2017). GO is an oxidized form of graphene, a two-dimensional sheet with abundant oxygenated functional groups on the basal plane and edges. The fabrication process of GO breaks the π-conjugation network which makes it highly dispersible in water and reduces aggregation among GO sheets, displaying ample potential in wastewater treatment. (Kim et al. 2013a). The increasing electronic and conductive properties make it very effective for pollutant removal in wastewater treatment.

7.7 APPLICATIONS OF CARBONACEOUS NANOMATERIALS IN WASTEWATER TREATMENT

7.7.1 Removal of Organic Contaminants

CNTs exhibit higher efficiency than activated carbon in adsorption of various organic chemicals (Pan and Xing 2008). CNTs having functional group boost their adsorption capacity by introducing new functional groups on the surface such as— OH, –COOH and >CO (Kuzmany et al. 2004). Long and Yang (2001) showed that CNT removes dioxins more efficiently than activated carbon. Peng et al. (2003) first reported that 1,2-dichlorobenzene efficiently is removed from water by using graphitized and as-grown CNTs as adsorbents. Kuo et al. (2008) revealed the removal of direct dyes CI Direct Yellow 86 (DY86) and CI Direct Red 224 (DR224) from aqueous solutions using carbon nanotubes (CNTs). Commercial carbon nanotubes (CNTs) have been purified by acid solution and were used as adsorbents to study adsorption of trihalomethanes (THMs) from water (Lu et al. 2005). The multi-walled carbon nanotubes (PMWCNTs) prepared by catalytic decomposition of methane, with a surface area of 98 m^2/g, were valuated as an adsorbent used for tapping volatile organic compounds (VOCs). PMWCNTs were also used as a trap packing material to adsorb VOCs purged from spiked water samples. The recoveries of the tested VOCs trapped on PMWCNTs ranged from 80 to 110% which is not affected by the humidity of purge gas. (Li et al. 2004).

Graphene-based nanomaterials have shown exceptional adsorption capacity towards various organic compounds. The graphene nanosheet (GNS)/magnetite (Fe_3O_4) composite exhibited extraordinary adsorption capacity and fast adsorption rates for the removal of organic dye, methylene blue (MB), in water. The graphene/Fe_3O_4 composite could be utilized as a magnetically separable and efficient adsorbent for environmental cleaning (Ali et al. 2018).

Graphene oxide (GO) has been used as a potential adsorbent for tetracycline antibiotics removal from the aqueous solution because of its excellent hydrophilicity, special structure and high surface area. The mechanism behind the adsorption of tetracycline on graphene oxide was inferred to be via π–π interaction and cation-π bonding (Gao et al. 2012). Wang et al. (2014) investigated the adsorption of naphthalene, phenanthrene and pyrene onto graphene (GNS) and graphene oxide (GO) nanosheets. Graphene showed a high affinity to polycyclic aromatic hydrocarbons (PHAs). The hydrophobic properties and molecular sizes of the PHAs affected the adsorption of G and GO. The high affinities of the PAHs to GNS are developed due to π–π interactions with the flat surface and the sieving effect of the powerful groove areas formed by shriveling on GNS surface. The decontamination of bisphenol A (BPA) from aqueous solution by graphene adsorption was also investigated (Xu et al. 2012). The maximum adsorption capacity of graphene for BPA obtained from Langmuir isotherm was 182 mg/g at 302.15 K, which was among the highest values of BPA adsorption compared with other carbonaceous adsorbents. Maliyekkal et al. (2013) reported unpalatable adsorption of endosulfan (ES), chlorpyrifos (CP) and malathion (ML) onto reduced graphene oxide (RGO) and graphene oxide (GO) with adsorption capacities as high as ~1200, 1100 and 800 mg/g respectively from water.

7.7.2 Removal of Heavy Metals

Carbonaceous nanoparticles showed great adsorption performances and rapid rates for the removal of metals. Li et al. (2004) investigated the sorption activity of Pb(II), Cu(II) and Cd(II) onto multiwalled carbon nanotubes (MWCNTs). They reported maximum sorption capacities of 97.08 mg/g for Pb(II), 24.49 mg/g for Cu(II) and 10.86 mg/g for Cd(II) at room temperature, pH 5.0 and metal ion equilibrium concentration of 10 mg/l. Li et al. (2015) also found that the metal–ion sorption capacities of the MWCNTs were 3–4 times larger than those of powder-activated carbon and granular-activated carbon, two commonly used sorbents in water purification. Peng et al. (2005) recently developed a novel sorbent with a high surface area (189 m^2/g) comprising cerium oxide supported on carbon nanotubes (CeO_2–CNTs). They showed that the CeO_2–CNT particles are effective sorbents for As(V). Multiwall CNTs and manganese dioxide (MWCNTs/MnO_2) composite were fabricated for the separation of As(III) and As(V) from wastewater. Here, MnO_2 nanocomposite converts As(III) to As(V) which can be easily adsorbed by the MWCNTs and MnO_2 oxide itself gets reduced from Mn(IV) to Mn(II) (Saleh et al. 2011). The MWCNT–ZrO_2 with 4.85% zirconia removes arsenic from wastewater. The adsorption process is pH dependent and the adsorption capacity of the composite was 2000 µg/g and 5000 µg/g for As(III) and As(V) respectively (Ntim et al. 2012). The MWCNTs/chitosan nanocomposite used exhibited high adsorption of Cu(II) ions from the mixture of heavy metals like Cd(II), Zn(II) and Ni(II) in wastewater. Saleh et al. (2011) reported the removal of metal ions from aqueous solution in the following sequence— Cu (II) > Cd (II) ≈ Zn (III) > Ni (II).

Synthesized few-layered graphene oxide nanosheets have been used as sorbents for the removal of Cd (II) and Co (II) ions from large volumes of aqueous solutions. This adsorption of Cd (II) and Co (II) by graphene nanosheet was strongly dependent on pH and weakly dependent on ionic strength (Zhao et al. 2011). Green synthesized smart magnetic graphene (SMG) possesses an increased adsorption site with tunable and magnetic properties facilitating the adsorption and magnetic separation of aqueous Cr(VI), As(V) and Pb(II) with 99% removal efficiencies up to the concentration of 1 ppb (Gollavelli et al. 2013). The electrostatic attraction and chemical bonding such as ion exchange, dominated the adsorption process where the surface functional groups of GO and oxidized CNTs provide the major adsorption sites to heavy metals.

7.8 DISINFECTION OF WATER USING CARBON-BASED NANOMATERIALS

Carbonaceous nanomaterials exhibit strong antibacterial activity. Their lower oxidation ability could effectively inactivate pathogens under visible-light irradiation or direct contact. They have low tendency to form DBPs. Kang et al. (2007) proved that highly purified single-walled carbon nanotubes (SWNTs) exhibit strong antimicrobial activity on *Escherichia coli* (*E. coli*). They demonstrate that cell membrane damage resulting from direct contact with SWNT aggregates is the likely mechanism leading to bacterial cell death (Table 7.1). It was observed that at the same weight concentration, longer SWCNTs exhibited stronger antimicrobial activity. The longer SWCNTs aggregated with bacterial cells more effectively, whereas short SWCNTs tended to aggregate themselves without involving many bacterial cells (Yang et al. 2010).

Both graphene and graphene oxide exhibited bacterial toxicity. The extremely sharp nanowalls of graphene nanosheets directly interacting with bacteria caused RNA effluxes through the damaged cell membrane of both *E. coli* (Gram-negative) and *S. aureus* (Gram-positive) bacteria. (Akhavan and Ghaderi 2010). Azimi et al. (2014) also studied the antimicrobial effect of two functionalized GO nanostructures (graphene oxide-chlorophyllin and graphene oxide-chlorophyllin-Zn) against *E. coli*. The functionalized GO led to cell membrane damage of *E. coli*. Here, the possible toxicity mechanism of GO–Chl toward *E. coli* cells is proposed considering the surface chemistry of the new functionalized nanomaterial and the physical interaction of GO with *E. coli*.

Fullerenes showed antimicrobial activity against various bacteria, such as *E. coli*, *Salmonella* spp. and *Streptococcus* spp. This is due to inhibition of energy metabolism after internalization of the nanoparticles into the bacteria. It was evaluated that C_{60} and its derivatives showed antimicrobial activities against 6 kinds of bacteria and 2 kinds of fungi (Aoshima et al. 2009). Tegos et al. (2005) demonstrated that cationic fullerenes with one, two or three pyrrolidinium groups, incubated by illumination of white light, have a broad-spectrum antimicrobial activity and it killed more than 99.99% of bacterial and fungal cells. The SMG exhibits disinfection action (40 μg/l) toward *E. coli* bacteria with 100% killing efficacy and low toxicity (0.1 ng/nL) toward zebrafish without inducing any abnormalities (Gollavelli et al. 2013).

Table 7.1 Types of carbon-based nanomaterial and their antimicrobial mechanism.

Types of nanoparticles	Characteristic of nanoparticles as antimicrobial agent	Mechanism of antimicrobial action	Main factors that influence antimicrobial activity	References
SWCNTs & MWCNTs	High surface/volume ratio, high functionalization properties, larger core volume, highly stable.	Physical interactions with cell membrane; formation of cell-CNTs aggregates; induction the cell membrane disruption.	Surface charge, particle size, types of functional groups, types of buffers, contact time, concentration.	Vecitis et al. 2010, Arias and Yang 2009
Graphene oxide (GO)	Very sharp edge of nano wall, high surface/volume ratio, ability to functionalization, stability.	Physical interaction with cell membrane; formation of cell-GO aggregates; induction the cell membrane disruption.	Particle size; type of functional group.	Gurunathan et al. 2012, Mohammad et al. 2020
Fullerene	Stability; photodynamic therapy activity; high ability to functionalization; high surface/volume ratio; large inner volume.	Inhibit bacterial growth by impairing the respiratory chain; inhibition of energy metabolism.	Particle size; type of functional group; surface charge.	Cataldo and Da Ros 2008, Deryabin et al. 2014

7.9 METAL-BASED NANOADSORBENTS

The metal-based nanoparticles commonly known to us are- quantum dots, nanogold, nanosilver, nanocopper, iron, zinc, platinum and oxides (Ag_2O, FeO, MnO_2, CuO, Bi_2O_3, ZnO, MgO, TiO_2, CaO, Al_2O_3,), sulphides (CdSe, PbS, CdSe-CdTe, CdSe-ZnTe, CdTe-CdSe) with metal bases. Nowadays, researchers focus on the modification of nanomaterial to make it more stable during the synthesis (physical/chemical/biological) process which is non-toxic, economical and safe for the ecosystem (Salas et al. 2019). These nano-adsorbents are quite effective and cost-friendly in the process of adsorbing heavy metals and radionuclides. The key mechanism is the complexion between dissolved metals and the oxygen in metal oxides. The power here is the chemical binding or conjugated properties that metal-based nanoparticles offer. When produced at the nanoscale, metal oxides can have increased surface area and possess favorable sorption to heavy metals (Table 7.2).

Table 7.2 Types of metal-based nanomaterial in heavy metals/metalloids removal

Zero valent metal nanoparticles	Metal oxide nanoparticles	Doped metal/ metal oxide nanoparticles	Metal sulphide nanoparticles	Metal organic framework (MOF)
Ag, Cu, Au, Pt, Zn, Fe, Ni, nanoparticles, etc.	AgO_2, ZnO, MnO, CuO, TiO2, Fe_2O_3 nanoparticles.	Ag–TiO_2, Au-CuO, Pt-ZnO nanoparticles, etc.	CuS, Ag_2S, FeS, etc.	Zn-based MOF, Cu-based MOF, Mn-based MOF, etc.

7.10 APPLICATION OF METAL-BASED NANOPARTICLES IN WASTEWATER TREATMENT

Noble metal nanoparticles (Ag, Au, Pt) have been used for several biomedical applications such as anticancer, radiotherapy enhancement, drug delivery, thermal ablation, antibacterial, diagnostic assays, antifungal, gene delivery and many others. The noble metal nanoparticles have some unique properties that make it more valuable. Metal nanoparticles can be functionalized with a variety of functional groups, such as peptides, antibodies, RNA and DNA to target different cells along with potentially biocompatible polymers, for example, polyethylene glycol (Yaqoob et al. 2020).

7.11 REMOVAL OF HEAVY METALS/METALLOIDS USING METAL NANOPARTICLES

Nanosized metal oxides like iron oxides, manganese oxides, zinc oxides, titanium oxides, aluminium oxides, magnesium oxides, cerium oxides, zirconium oxides, etc. display high removal potential and specificity towards heavy metals (Table 7.3). These nanometal oxides serve as an alternative of activated carbon. Due to their high specific surface area, they feature a short intraparticle diffusion distance and are compressible without a significant reduction of surface area (Gehrke et al. 2015). Oxide nanomaterials have a large surface area, well-defined pore sizes, large pour volume and high adsorption capacity. Due to these unique characteristics, they are diversified in surface functionalization and have modified themselves easily (Borji et al. 2020). Some of the metal oxides (nanomaghemite and nanomagnetite) showed magnetic properties which facilitate separation and recovery by a low-gradient magnetic field.

Iron oxide-based nanomaterials like hematite (α-Fe_2O_3), maghemite (γ-Fe_2O_3), magnetite (Fe_3O_4), gothite (α-FeOOH) and hydrous iron oxides (HFO) have received increasing attention in removing heavy metals from wastewater over these years. It was noticed that nanoscale α-FeOOH has a higher removal capacity towards uranium than non-nanoscale goethite especially in the p^H range of 5.5 to 7.5. Khezami et al. 2016 synthesized goethite nanocrystalline powder used to treat Cd (II). There is evidence of the removal of other heavy metals such as V(V), Mn(II), Co(II), Zn(II), etc. by goethite. Nano-hematite having different morphologies like hexagonal, plate-like, nano-cubes, sub-rounded, spherical particles removed Cr(VI) in the range of 6.33–200 mg/g. Depending on different factors such as doses of adsorbents, temperature, p^H and multiple metal species, the affinity between heavy metals and nanoadsorbent fostered the following order—Pb(II) > Zn(II) > Cd(II) > Cu(II) (Zawrah et al. 2019).

Aluminum oxide or aluminium nanoparticles are commonly used as an adsorbent. They have high surface area and good thermal stability and can remove cadmium, chromium, copper, lead, mercury, etc. (Mahdavi et al. 2015). ZnO nanoparticles were effective in removing divalent ions like Cu(II), Ag(II) and Pb(II) with efficiency > 85% under UV light exposure (Le et al. 2019). Porous magnesium oxide (MnO) nanoflakes were used for the removal of As(II). These nanoflakes were reacted with As(III) which formed a magnesium-arsenite compound, contributing to the high As(III) sorption capacity of MgO nanoflakes (Liu et al. 2011).

Table 7.3 Heavy metals/metalloids removal of different metal-oxide nanoparticles

Types of metal-oxide nanoparticles	Target Heavy metals to be removed	Factors responsible for metal removal	References
Hematite (α-Fe_2O_3)	Removes Cr(VI) in the range of 6.33–200 mg/g; Order of metal removal: Pb(II) > Zn(II) > Cd(II) > Cu(II).	Doses of adsorbents, large surface area, temperature, pH, multiple metal species.	Shipley et al. 2013, Adegoke et al. 2014.
Geothite (α-FeOOH)	Possess high adsorption capacity towards Cu(II) (149.25 mg/g); removes Cd(II) at pH 7; removes uranium at pH 5.5–7.5; the order of metal removal: V(V), Mn(II), Ni(II), Co(II), Zn(II) etc.	pH, temperature, surface area.	Chen and Li 2010, Yan et al. 2011.
Maghemite (γ-Fe_2O_3)	Adsorbs 68.9 and 34.0 mg/g Pb(I) and Cu(II) respectively; Order of metal removal: Cu(II) > Cr(VI) > Mn (II) > Ni(II) > Cd(II).	Large surface area, pH, temperature, electrostatic interactions.	Akhbarizadeh et al. 2013, Rajput et al. 2017.
Magnetite (Fe_3O_4)	Maximum adsorption capacities for Cr(VI) and Ni(II) were 232.51 mg/g and 222.12 mg/g respectively; order of metal removal: Pb(II), Hg(II), Cd(II), Mn(II).	Surface-modified functional groups, electrostatic interaction, surface charge.	Schmidtke et al. 2014, Norouzian Baghani et al. 2016.
Hydrous iron oxides (HFO)	Maximum adsorption capacities for As (II) is 92 mg/g; Order of metal removal: Pb(II) > Cu(II) > Ni (II) > Cd(II).	Large surface area, ion- exchange, co-precipitation.	An et al. 2005.
Manganese oxide (MnO), Hydrous manganese oxide (HMO)	Highest adsorption capacities towards Pb (II) and Cd (II) were 318.7 and 105.1 mg/g respectively. It has also removed U, Cd(II), Cu(II), Pb(II), Zn(II), Hg(II) and so on.	The surface charge of manganese oxide $M\text{–}O^{\delta+}$ and $M\text{–}O^{\delta-}$ helps the sorption of metal ions.	Mukherjee et al. 2013, Huangfu et al. 2015.
Titanium oxide (TiO_2)	TiO_2 nanowire removes Pb(II), Cu(II), Cd(II), Zn(II). Chitosan coated TiO_2 nanoparticles showed maximum adsorption capacities towards Cd(II) and Cu(II).	Mesoporous structure, photocatalytic properties.	Anandan et al. 2009, Mahmoued et al. 2018.
Aluminium oxide (Al_2O_3)	γ-Al_2O_3 showed maximum removal efficiency towards Pb(II) and Cd(II), 97% and 87% respectively. It also removes Hg(II), As(III), Cu(II), Ni(II), Cr(VI) etc.	High surface area, mechanical strength, low temperature modification.	Tabesh et al. 2018, Shokati Poursani et al. 2015.
Zinc oxide (ZnO)	According to Langmuir model maximum adsorption capacities towards Zn(II), Cd(II) and Hg(II) were 357, 387, 714 mg/g respectively.	High surface area, p^H (4.0–8.0).	Ghiloufi et al. 2015.

Zero valent metal nanoparticles efficiently removed heavy metal(loid)s from wastewater (Table 7.4). Some zero-valent metal ions are silver, copper, iron, zinc nanoparticles and so on.

Table 7.4 Heavy metals/metalloids removal by zero-valent metal-nanoparticles

Metal nanoparticles	Target heavy metals to be removed	Factors responsible for metal removal	References
Silver nanoparticles	AgNPs coordinating with merceptosuccenic acid (MSA) removes Hg(II) (800mg/g), it also removes 24% cobalt and 77% lead from groundwater. Silvernano composite selectively removes Zn(II), Pb(II) and Fe(III) and removal efficiency was 79%, 88% and 91% respectively.	pH, temperature, optical properties, magnetic properties, large surface area, tunable surface chemistry, functionalization	Attatsi and Nsiah 2020, Sumesh et al. 2011 Darweesh et al. 2022
Iron nanoparticles	Zero valent iron nanoparticles efficiently removes Cu(II), Cd(II), Pb(II) and removal capacity was 5.56, 5.40–6.94 and 5.41–6.95 mg/g respectively. Green synthesized iron nanoparticles selectively removed Pb(II), Cd(II), Cr(VI), Cu (II); adsorption efficiency for Pb(II) and Cd(II) is 98.23% and 96.5% respectively. Iron synthesized from green tea extract removes 90.2% of Cr(VI).	High surface area, high rate of adsorption, pH	Ali 2012, Thilakan et al. 2022, Tarekegn et al. 2021
Gold (Au)	The removal capacity of Au nanoparticles using NaBH4 (reducing agent) towards Hg(0) is very high other than the ordinary adsorbents. Citrate-coated Au nanoparticles also remove Hg(II).	pH, temperature	Ojea-Jiménez et al. 2012, Qian et al. 2013

7.12 ANTIMICROBIAL ACTIVITIES OF METAL NANOPARTICLES

Metal nanoparticles such as Ag, silver oxide (Ag_2O), titanium dioxide (TiO_2), silicon (Si), copper oxide (CuO), zinc oxide (ZnO), Au, calcium oxide (CaO) and magnesium oxide (MgO) exhibit antimicrobial activity. The important factors which affect antimicrobial effects are (a) particle size, (b) surface/volume ratio, (c) surface charge, (d) shape of nanoparticles, (e) surface morphology, (f) material used in nanoparticles.

Ag and Ag_2O nanoparticles showed potential antimicrobial effects against Gram-positive, Gram-negative bacteria, food-borne bacteria, fungi, etc. Fluorescent Ag nanoparticles showed antimicrobial activity against Gram-positive (*Staphylococcus epidermis* NCIM 2493 and *Bacillus megaterium*) and Gram-negative bacteria (*Pseudomonas aeruginosa* ATCC 27853 and *Escherichia coli*) and fungal strains (*Candida albicans* and *Aspergillus niger*) (Bera et al. 2014). Sondi and Salopek (2004) showed that AgO_2 nanoparticles performed antimicrobial activity against *E. coli*. Ag nanoparticles with particle size (20–30 nm) had a greater potential of use in controlling spore-producing fungal plant pathogens (Jo et al. 2009). The antimicrobial activity of silver nanoparticles depends on the particles size and shape. AgNPs exhibit their antimicrobial potential through multifaceted mechanisms. AgNPs adhesion to microbial cells, penetration inside the cells, ROS and free radical generation, interactions with DNA and modulation of microbial signal transduction pathways have been recognized as the most prominent modes of antimicrobial action.

Cu and CuO nanoparticles are used as highly potential antimicrobial agents against various bacterial strains such as *E. coli*, *P. aeruginosa*, *K. pneumonia*, *Enterococcus faecalis*,

Shigella flexneri, *S. Typhimurium*, *S. aureus*, etc. It is noticed that copper and copper-oxide nanoparticles are highly sensitive to resistant the growth of *E. coli* and *Enterococcus faecalis* (Ahamed et al. 2014). Antibacterial activities of CuO nanoparticles depend on particle size and concentration. Azam et al. (2012a) stated that depending on the particles size, stability and concentration, CuO nanoparticles inhibit the growth of two gram-positive bacteria (*S. aureus* and *B. subtilis*) and gram-negative (*Pseudomonas aeruginosa* and *E. coli*).

Au nanoparticles are considered as antibacterial agents due to high ability to functionalization, nontoxicity, polyvalent effects, photothermal activity and ease of detection. Lima et al. (2013) studied that Au nanoparticles inhibit the growth of *E. coli* and *Salmonella typhi*, two common bacteria present in polluted water. Au nanoparticles reduced 90–95% of *E.coli* and *S. typhi* colonies. Gold nanoparticles were used to detect *Salmonella enteritidis* and *Listeria monocytogenes* within the flagella (Tiwari et al. 2011).

ZnO nanoparticles showed bacterial effects gram-positive (*S. aureus* and *B. subtilis*) and Gram-negative (*E. coli* and *P. aeruginosa)* which are highly resistant to high temperature and pressure (Azam et al. 2012b). The antibacterial activity of ZnO nanoparticles depends on concentration and larger surface area (Buzea et al. 2007). Zinc oxide nanoparticles inhibited the cell growth of *Campylobacter jejuni* (Xie et al. 2011).

$CaCO_3$/MgO composite nanoparticles showed bacterial activity against *E. coli* and *Staphylococcus aureus.* Though antibacterial action towards *Staphylococcus aureus* was greater than towards *E. coli*, the death constant was identical in both bacteria (Yamamoto et al. 2010). Vidic et al. (2013) reported that mixed nanostructure of ZnO-CaO showed high antimicrobial activity against both Gram-positive (*B. subtilis*) and Gram-negative (*E. coli*) bacterialt was noticed that nanoparticles having the same material showed different properties due to particle size. The smaller particles showed a higher surface/volume ratio than the larger particles and possessed greater antimicrobial properties. The proposed mechanisms of antimicrobial effects (Figure 7.1) are stated below.

1. Adhesion of the cell membrane
   - Alters membrane structure and permeability
   - Leakage of cellular content and ATP
   - Impair transport activity

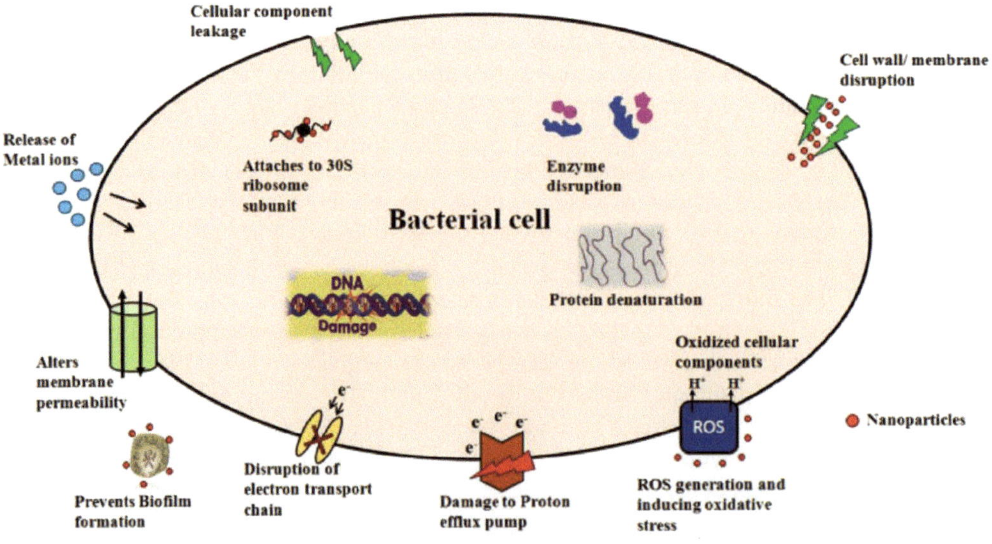

Figure 7.1 Various mechanisms of antimicrobial activity of metal/metal oxide nanoparticles (Roy et al. 2019).

2. Penetration inside the cell and nucleus
   - Mitochondrial dysfunction
   - Destabilize and denature Proteins
   - Destabilize ribosomes
   - Interact with DNA
3. Cellular toxicity and ROS generation
   - Oxidize protein and lipid
   - Oxidize DNA base
4. Modulation of cell signaling
   - Alters phosphotyrosine profile

7.13 POLYMERIC NANOADSORBENTS

Polymeric nanomaterials are used as a system in which inorganic materials are inserted to improve their chemical, mechanical and thermal sorption properties or as a bed or template to synthesized nanoparticles. Polymer hybrids possess the advantage of good adsorption capacity and good chemical stability in a wide range of p^H because of ease in functionalization through epoxide, epoxy, etc. (Khajeh et al. 2013). The polymer shell (Fe_3O_4@APS@AAco–CA MNPs) which is modified by magnetic nanoparticles (Fe_3O_4), easily removed metal ions, such as Cd^{2+}, Zn^{2+}, Pb^{2+}, Cu^{2+} with maximum adsorption capacity at p^H 5.5 (Ge et al. 2012). Bi-metal like iron (Fe) and aluminium (Al) doped micro and nanoparticles-based multifunctional polymeric adsorbents are used for the removal of fluoride and arsenic-V ions from wastewater (Kumar et al. 2011).

7.14 SILICA-BASED NANOMATERIALS

Silica-based nanomaterials are another category of nanomaterials that are widely used for heavy metals removal. They possess excellent surface area and can be functionalized with groups such as amine ($-NH_2$) and thiol (–SH) which increases the interactions with heavy metal ions. The adsorption capacity of amine-based silica nanoparticles was found in the order of Pb(II) > Cd(II) > Ni(II) (Najafi et al. 2012). Adsorption of silica nanoparticles strongly depends on pH and temperature (Kumar et al. 2011).

7.15 PHOTOCATALYSIS/NANOCATALYSTS AND REDOX-ACTIVE NANOPARTICLES

Some nanoparticles are used as photocatalysts in water purification because of their catalytic, optical and electronic properties. The large surface/volume ratio, size and shape of nanoparticles generate these properties in nanoparticles. TiO_2 nanoparticles are both oxidative and reductive catalyst that can remove organic/inorganic pollutants from polluted water (Chitose et al. 2003, Obare and Meyer 2004). TiO_2 nanoparticles reduce the toxic metal ions such as Cr(VI), Ag(I), Pt(II) and degraded organic compounds like chlorinated alkenes, benzenes, dioxin, furans, PCBs, etc. in the presence of UV light (Asahi et al. 2001).

Au nanoparticles extracted from *S. Grandiflora* extract have excellent catalytic activity on the reduction of methylene blue (Das and Velusamy 2014). Ag@CAF and Ag@TiO_2 nanomaterials exhibited strong and selective catalytic properties toward the reduction of Rodhamine B and Methyl Orange (Albukhari et al. 2019).

7.16 NANO-MEMBRANE TECHNOLOGY FOR WATER PURIFICATION AND DESALINATION

Nowadays, membrane separation associated with nanoparticles is widely used in water purification and desalination of water. There are several kinds of nanomembranes associated with it. The membranes are—(1) thin film nanocomposite (TFN) RO membrane, (2) inorganic-based membrane, (3) ceramic membrane, (4) polymeric membrane, (5) cellulose membrane and so on.

7.17 DESALINATION AND NANO-MEMBRANE

The scarcity of water resources and demand for water supplies are driving the development of technologies to use alternative water resources such as salt-water sources. Saltwater in the form of saline and brackish water represents the largest part of the global water resources and can be used as a source of drinking water through desalination processes. The desalination process through membranes plays an important role in the production of fresh water. Thin film nanocomposite membranes incorporated with interfacial polymerization techniques have resulted in tremendous achievement in producing a membrane with the right combination of water flux and salt rejection with antifouling (organic, inorganic and biological) activity (Lau et al. 2012). Nanocomposite forward osmosis (FO) membranes with MWCNT/PES substrate layer and a polyamide selective layer were used for water desalination and resulted in improved water flux and enhancement of the salt rejection from 78% to 97%, compared to FO with MWCNT-free PES substrate layer (Wang et al. 2013). The synthesized nanocomposite of reduced graphene-oxide and titanium oxide (RGO/TiO_2) embedded in polyamide layer of RO thin film membranes was used for water desalination. The RO membrane with 0.02 wt% of the RGO/TiO_2 showed 99.45% NaCl rejection and 51.3 l/m^2 h water flux (Safarpour et al. 2015). The addition of small amounts of functionalized CNTs and cellulose acetate (CA) led to an increase in the RO nanocomposite membrane permeation rates, with minimal deleterious effect on salt retention. In this CA/CNT nanocomposite membrane permeation rates were found to be 54% with a minimal decrease in NaCl (–6%) retention. The performance of nanocomposite membranes is explained by the fact that the presence of CNTs (0.0005 wt.% and 0.005 wt.%) created the connected channels between the membrane pores in the support layer and also increased the hydrophilicity of the membrane (El Badawi et al. 2014). Thin-film nanofibrous composite (TFNC)/RO membranes are found to be fabricated with two types of fibrous substrates, with one as electrospun polyacrylonitrile (PAN)/Polyethylene terephthalate (PET) and another as cellulose nanofibrous (CN/PAN/PET). Both substrates possessed sufficient mechanical strength to withstand high operation pressure for seawater desalination. CN/PAN/PET membrane had a smaller surface roughness and resulted in higher permeation than PAN/PET (Wang et al. 2017). Ghanbari et al. (2015) showed that the thin film composite (TFC) membrane exhibits higher water flux with NaCl rejection than the halloysite incorporated RO membrane. This is due to the more hydrophilic surface of the membrane. Synthesized zwitterions CNT/polyamide nanocomposite membranes with varying loadings of CNTs have been used for water desalination. The addition of zwitterions functionalized CNTs into a polyamide membrane increased both the flux and the salt rejection ratio (Chan et al. 2013). The amine-functionalized multi-walled carbon nanotubes (F-MWCNTs) membrane was used in FO membranes which exhibited high water permeability and high salt rejection properties (0.01–0.1 wt%) in comparison with TFC membrane (Amini et al. 2013).

Incorporation of aluminosilicate single-walled nanotubes (SWNT) on the surface of thin film polyamide membrane, nanocomposite membrane gives higher permeation flux, sustaining high rejection of monovalent and divalent ions (Barona et al. 2013). Surface-modified NaZ Zeolite TFN membrane in RO improves salt rejection capacity (97.8% to 98.5%) and gives higher water flux (28.5 to 40.9 l/m^2h) (Dong et al. 2011). The aPES/aTMA nanocomposite has been used in chlorine resistance. The salt rejection and water permeation values of aPES/aTMA were 98.8% and 37.8 l/m^2 h respectively as compared with non-modified membranes (Kim et al. 2013a).

Biomimetic membranes are highly selective membranes, widely used in forward osmosis and reverse osmosis application. The water permeability of aquaporin (AQP) based lipid bilayer membrane is 100 times higher than RO membrane (Hoek et al. 2014).

Nanofibrous membranes were fabricated with a combination of polyvinylidene fluoride (PVDF) and polytetrafluoroethylene (PTFE) by electrospinning. The superhydrophobic PVDF-PTFE membrane was tested for desalination by vacuum membrane distillation and a stable flux of 18.5 kg/m^2 with salt rejection higher than 99.9% was obtained (Dong et al. 2014).

Ion-exchange membranes consisting of various concentrations of graphene oxide (GO) (0.5, 1, 2, and 10 wt.%) and sulfonated polyether sulfone (SPES) have been used for water desalination. The added GO nanoparticles were successfully cross-linked with the PA and resulted in 3-times greater water permeability while maintaining the same NaCl rejection. The maximum ionic conductivity has been found in 10% GO composite which shows 3.51 $mole/m^2/h$ ionic flux and current efficiency for salt removal (Gahlot et al. 2014). Membranes fabricated with nitrogen-doped graphene oxide quantum dots (N–GOQD) by interfacial polymerization (IP) is used for RO desalination. This membrane increased water permeability and salt rejection more than the pristine polyamide membrane (Fathizadeh et al. 2019). SiO_2 and modified SiO_2 membranes are used to reject the sodium sulfate and amino salinized TiO_2 membrane rejects NaCl more efficiently (He et al. 2011).

7.18 REMOVAL OF INORGANIC CONTAMINANTS BY MEMBRANE NANOTECHNOLOGY

Different types of nanomaterials membrane have been introduced for the removal of contaminants from water/wastewater, where the membrane pore size, surface electrical charge and surface area can affect the removal efficiency of inorganic contaminants.

The nanocomposite PSF/GO membrane, fabricated using phase inversion has been evaluated for arsenic rejection from water (Rezaee et al. 2015). Incorporation of fly ash (FA) and TiO_2 NPs into nanocomposite polyurethane (PU) through a sol-gel system were analyzed for heavy metal rejections. Hg and Pb were used as heavy metal models for the evaluation of nanocomposite membrane performance (Kim et al. 2014). Organoclay embedded polysulphone (PSF) nanocomposite membranes through non-solvent induced phase separation and a loading range from 0 wt.% to 2 wt.% were used for arsenate removal from contaminated surface water. The implementation of 1.5 wt.% organoclay into PSF resulted in higher rejection, increased water flux, higher surface hydrophilicity and improved mechanical properties (Shokri et al. 2016). Organic polyethersulfone (PES) and inorganic Fe-Mn binary oxide (FEBO) particles incorporated into a mixed matrix nanocomposite membrane were used for As(III) removal from the contaminated water. An adsorption isotherm study indicated that a ratio of 1.5:1 for FMBO: PES (weight %) resulted in the best As(III) rejection performance with 94.6 L/ h and 73.5 mg/g uptake capacity (Gohari et al. 2013). Mixed matrix membranes prepared using PES and different amounts of alumina (Al_2O_3) particles by phase inversion method were used to remove copper heavy metal (Ghaemi 2016). This nanocomposite membrane exhibited 59% copper capture in comparison to only 25% capture for the pure PSF membrane. The mesoporous silica nanoparticle-embedded polyamide layers in thin film nanocomposite membranes were studied for the rejection of Na_2SO_4. The silica NPs/PA nanocomposite membrane improved the rejection of the Na_2SO_4 to more than 80%, with enhanced water flux (Wu et al. 2013).

7.19 NANO-MEMBRANE IN WATER DISINFECTION

Nanoparticles embedded into the membrane is widely used in water purification as well as wastewater treatment because of its antimicrobial properties, anti-biofouling and antiviral properties (Krishnaraj et al. 2010, Aktij et al. 2017). Generally, as microorganisms interact with filtration membranes, the permeability of normal membranes is reduced and the water flux and salt rejection efficiency are both diminished (Mohammad et al. 2015). For this reason, the fabrication/coating of membranes using various types of nanomaterials is one of the most important issues in membrane science (Ren et al. 2020). Using biocidal agents such as iron oxide (Harshiny et al. 2019), silver (Sawada et al. 2012), copper (Chen et al. 2019) or zinc (Mendes et al. 2022) on the surface of the nanocomposite membranes can improve membrane antibacterial activity and extend membrane lifetime.

Polysulfone membrane added with Ag-nanoparticles showed anti-biofouling and anti-virus properties. Incorporation of nanosilver increased membrane hydrophilicity, reducing the potential of membrane biofouling (Taurozzi et al. 2008, and Zodrow et al. 2009). Smaller Ag nanoparticles (8 nm) were most effective, while larger particle sizes result lower bactericidal activity (Pal et al. 2007). Firouzjaei et al. (2018) synthesized a thin-film nanocomposite (TFN) incorporating a graphene oxide (GO)–silver-based metal-organic framework (Ag-MOF) into TFC selective layer that showed 96% biocidal activity and antifouling properties compared to the primal membrane. Polyurethane foam coated with Ag nanofibres have shown good antibacterial properties against *E. coli*, used in water filters (Jain and Pradeep 2005). Embedded Ag nanoparticles have been reported as very effective against both Gram-positive and Gram-negative bacteria. (Savage and Diallo 2005). PSF membranes were fabricated using different sizes of Ag nanoparticles in the casting solution and studied the antibacterial activity of silver particles towards gram-negative bacteria *E. coli*. Inhibition zone and bacterial filtration indicate antibacterial activity (Mollahosseini et al. 2012). Silver-loading asymmetric hollow fiber cellulose acetate (CA) membrane incorporating in N, N-dimethylformamide (DMF) shows antibacterial activity of membranes against both Gram-negative (*E. coli*) and Gram-positive (*S. aureus*) (Chou et al. 2005). AgNPs into a PSF matrix were introduced for antibacterial and antiviral applications. Authors showed that the hydrophilicity and antibacterial activity of PSF membranes were increased by incorporating AgNPs (Zodrow et al. 2009). Antibacterial nanomaterials such as TiO_2, Cu^{2+} and chitosan-coated membrane have also been explored for their potential applications (Zhao et al. 2015, de Dicastillo et al. 2020).

Cu–NPs coated functionalized TFC–RO membranes have been used as a remarkable anti-biofouling effect with 80% to 95% anti-adhesion capacity for three model bacteria strains (*E. coli*, *P. aeruginosa*, and *S. aureus*) attributed to the toxicity of the bound Cu–NPs (Ben–Sasson et al. 2014). The adsorption of copper hydroxide on TFC–RO commercial membranes coating with copper hydroxide [$Cu(OH)_2$] produce anti-biofouling Cu^{2+} membrane. The [$Cu(OH)_2$] particles physically adsorbed on the membrane surface and bactericidal activity may be ascribed to release of Cu^{2+} ions from the membrane surface into solution. Author revealed that the bacterial killing effect depends on the p^H of the solution and this effect increases with an increase in p^H from 6.5 to 7.5 (Garcia et al. 2021).

The polymeric membrane surface grafted with polyaniline (PANI) by in-situ polymerization of aniline and in situ Cu–NPs synthesis to produce a modified PANI–CuNPs–TFN membrane which increased the water flux 17.24 kg/m^2 h and salt rejection 99.82%, as compared to unmodified membrane. The anti-biofouling properties of PANI–CuNPs–TFN modified membrane showed high inhibition zone and higher permeate fixes than the unmodified membrane (Khajouei et al. 2018). The incorporation of CuO–NPs, Cu–NPs, Cu-meta-phenylendiamine oligomers (Cu–mPD) and water-stable Cu-based MOFs into some monomer m-phenylenediamine (MDP) or trimesoyl chloride (TMC)) during IIP test has been explored as an auspicious approach to generate TFC–RO modified membrane, which enhances anti-biofouling properties (Rodriguez et al. 2018).

7.21 REMOVAL OF ORGANIC MATTER BY MEMBRANE

Nanocomposite membranes can efficiently remove all kinds of organic matter (biodegradable or non-biodegradable) such as synthetic dyes, natural organic matter, pesticides, herbicides, pharmaceuticals products, etc. The common organic dyes widely used in textile, paints, leather and paper industries produce large quantities of wastewater which are methyl blue, methyl orange, congo red, rhodamine B, acid black and direct black. Most organic dyes are highly toxic, resistant to biodegradation and even carcinogenic thus, it is urgent to remove dyes from water. Nanocomposite membranes have shown acceptable performance with optimal cost and energy consumption in removing dye contaminants from water (Zahid et al. 2021). Although naturally organic matter (NOM) has no direct impact on health, it may affect consumer satisfaction because it can contribute to the undesirable colour, tast. and odours in drinking water. In addition, the presence of NOM in water can cause difficulties in the water purification process steps (Esfahani et al. 2019).

Carbon nanotube (CNT), TiO_2 and graphene-oxide are known as promising nanomaterials that can be strung in the membrane polymer matrix such as polyacrylonitrile (PAN), polyaniline (PANI), polyethersulfone (PES), polyetherimide (PEI) and polyvinylidene (PVDF) membrane. The main mechanism to remove dye materials is the electrostatic interaction between dyes and the membrane surface (Yin and Deng, 2015, Wen et al. 2019). Using graphene oxide and organo-modified graphene oxide, Nafion/organo-modified mixed matrix membranes are prepared which remove methyl orange (MO) up to 90% from the aqueous solution (Filice et al. 2015). Totally biobased cellulose nanocomposite membrane fabricated with chitosan matrix (10 wt%) successfully remove 98%, 84% and 70% respectively of positively charged dyes like Victoria Blue 2B, Methyl Violet 2B and Rhodamine 6G (Karim et al. 2014). The fly ash-doped TiO_2 nanofibers were prepared by electrospinning process for the removal of organic pollutants and microorganism from water. Using a mixture of fly ash-TiO_2 nanoparticles in the PVP matrix, the complete degradation of MB dye occurs within 120 min while only 70% dye was degraded by TiO_2 nanofiber within the same time (Saud et al. 2015). The drastic change in the degradation rate of methylene blue was attributed to the synergistic effect of adsorption properties of fly ash and degradation properties of TiO_2 nanofiber.

A thin-film nanocomposite (TFN) membrane incorporated with Amine-functionalized boron nitride $BN(NH_2)$ nanosheets is used to remove natural organic matter (NOM). Also, this modified membrane has the capability of removing humic acids and lower molecular-sized compounds from water which are detected by liquid chromatography and fluorescence excitation-emission matrices (FEEM) (Abdikheibari et al. 2020).

Multi-wall carbon nanotubes (MWCNTs)/polyaniline (PANI)/polyethersulfone (PES) membranes have been fabricated for the removal of humic acid. This membrane removed 80% of humic acid and showed higher permeability than the primitive PES membrane. The mechanism involved in the natural organic matter rejections is the positively charged surface of MWCNTs/PANI complex membrane which increased adsorption capacity by electrostatic interaction (Lee et al. 2016). Cheshomi et al. (2018) investigated the new thin film nanocomposite (TFN) Pebax/(Psf-PEs) membrane fabricated with hydrophilic TiO_2 nanoparticles via dip coating techniques used in humic acid removal from aqueous solution. The maximum permeate flux and humic acid rejection of the modified membrane were obtained at 0.03 and 0.01 wt.% TiO_2 loadings, respectively. The main mechanisms for HA removal are thought to be the electrostatic interaction between and carboxylate groups of HA and the hydrogen bonding between the OH groups of and the carboxylate. The zinc oxide nanoparticles (ZnO–NPs) fabricated with polyethersulfone (PES) matrix to prepare mixed matrix membrane which showed the highest HA rejection (97.98) at the 3.75 wt.% of ZnO-NPs, while the HA rejection of the PES membrane was 94.45% (Ahmad et al. 2015). PSF mixed matrix membrane incorporated with graphene oxide nanoparticles (GO) showed high rejection performance for bovine serum (around 95%) but low rejection for ovalbumin

(Zhao et al. 2013). The removal of organic fraction and ionic species (cations: lithium, sodium, potassium, magnesium, calcium and anions: fluoride, chloride, nitrite, nitrate, sulfate) from oil sands process- affected water and those are removed by MWCNT/PSF and MWCNT/PA thin film nanocomposite membranes (Kim et al. 2013b).

Pharmaceuticals are present in sewage as a result of the direct consumption and excretion of the parent compound or metabolites in human waste products and to a lesser extent, by direct disposal of unused medications (Galus et al. 2013). Huang et al. (2016) revealed that chitosan/silver-coated composite NF membrane enhanced antibacterial activity and up to 84% rejection of PPCPs which was mainly influenced by electrostatic repulsion with some adsorptive interactions during the initial stage. The nanocomposite membrane was positively charged and the selected PPCPs: atenolol > carbamazepine > ibuprofen, which were positive, neutral and negatively charged respectively. Photo-regenerable multi-walled carbon nanotube membranes embedded with TiO_2 nanoparticles (having porous and spherical morphology) were used in adsorptive removal of carbamazepine, ibuprofen and acetaminophen from wastewater. The initial removal percentages of the pharmaceuticals by the MWNT–TiO_2 membranes were 80%, 45% and 24% for carbamazepine, ibuprofen and acetaminophen, respectively. However, following photocatalytic regeneration membrane removal declined from 55% to 19% (Zaib et al. 2013). Nanocomposite membranes containing single-walled and multi-walled carbon nanotubes (SWC NT and MWCNT) were fabricated and used in the filtration of triclosan (TCS), acetaminophen (AAP) and ibuprofen (IBU) to determine the capabilities and mechanisms of PPCP removal. The removals ranged from approximately 10% to 90% which increased with increasing number of aromatic rings (AAP ≈ IBU < TCS), decreasing surface oxygen content (oxidized MWCNT < MWCNT) and increasing specific surface area (MWCNT < SWCNT). The variation of pH from 4 to 10 influenced PPCP removal by upto 70% (Wang et al. 2015).

Niedergall et al. (2014) synthesized nanoscale spherical polymer adsorbers in polyethersulfone (PES) matrices and demonstrated that trimethylolpropane trimethacrylate (TRIM) and 2-hydroxy-3-methacryloyloxypropyltrimethyl-ammonium chloride (HYMOPTA) particle-loaded (50–500 nm size range) membranes had higher BPA adsorption capacity when compared to the unmodified 'reference' membrane. The concentrations of BPA in the permeate samples were 115 ppb and 10 ppb when filtering 50 ml of 120 ppb aqueous BPA solution by the reference membrane and the particle-loaded 'anti-BPA membrane', respectively. Bellona and Drewes (2005) observed that using NF-200 membrane, 75% ibuprofen was removed at pH 7 at a recovery rate of 3% which coincides with current finding. On the other hand, they also reported about 85% rejection of the compound at pH 7 by NF-90 at a recovery rate of 9% while almost complete rejection was found in our study. The relatively lower rejection of ibuprofen in the former case could be due to the presence of divalent cations which resulted in a reduction of effective surface charge of NF-90 (Bellona et al. 2004).

7.22 CONCLUSIONS

It can be concluded that nanotechnology as well as nanomaterials play a unique role in water purification and desalination by removing micropollutants and intensifying the filtration process, hence, rendering potable water. Nanoadsorbents, nanometals, nanomembranes and photocatalysts such as nanoengineered materials offer novel technologies in water treatment processes that can be easily adapted to customer–specific applications. The ease of synthesis, economic feasibilit. and easy surface modification are vital features that have led to the development of this novel technology. However, there are certain drawbacks of this technology like the release of nanoparticles from the nanomaterial surfaces exhibiting metal toxicity posing a high risk on humans and environment. Thereby, elaborate research is needed to overcome the disadvantages of nanoparticles/nanomaterials.

References

Abdikheibari, S., Dumée, L.F., Jegatheesan, V., Mustafa, Z., Le-Clech, P., Lei, W. and Baskaran, K. 2020. Natural organic matter removal and fouling resistance properties of a boron nitride nanosheet-functionalized thin film nanocomposite membrane and its impact on permeate chlorine demand. J. Water Process. Eng. 34: 101160.

Adegoke, H.I., AmooAdekola, F., Fatoki, O.S. and Ximba, B.J. 2014. Adsorption of Cr (VI) on synthetic hematite (α-Fe_2O_3) nanoparticles of different morphologies. Korean J. Chem Eng. 31(1): 142–154.

Ahamed, M., Alhadlaq, H.A., Khan, M.A., Karuppiah, P. and Al-Dhabi, N.A. 2014. Synthesis, characterizatio. and antimicrobial activity of copper oxide nanoparticles. J. Nanomater. 2014: 637858.

Ahmad, A.L., Abdulkarim, A.A., Ismail, S. and Ooi, B.S. 2015. Preparation and characterisation of PES-ZnO mixed matrix membranes for humic acid removal. Desalin. Water Treat. 54(12): 3257–3268.

Akhavan, O. and Ghaderi, E. 2010. Toxicity of graphene and graphene oxide nanowalls against bacteria. ACS Nano. 4(10): 5731–5736.

Akhbarizadeh, R., Shayestefar, M.R. and Darezereshki, E. 2014. Competitive removal of metals from wastewater by maghemite nanoparticles: a comparison between simulated wastewater and AMD. Mine Water Environ. 33(1): 89–96.

Aktij, S.A., Rahimpour, A. and Figoli, A. 2017. Low content nano-polyrhodanine modified polysulfone membranes with superior properties and their performance for wastewater treatment. Environ. Sci. Nano. 4(10): 2043–2054.

Albukhari, S.M., Ismail, M., Akhtar, K. and Danish, E.Y. 2019. Catalytic reduction of nitrophenols and dyes using silver nanoparticles@ cellulose polymer paper for the resolution of waste water treatment challenges. Colloids Surf. A Physicochem. Eng. Asp. 577: 548–561.

Ali, I. 2012. New generation adsorbents for water treatment. Chem. Rev. 112(10): 5073–5091.

Ali, I., Alharbi, O.M., Tkachev, A., Galunin, E., Burakov, A. and Grachev, V.A. 2018. Water treatment by new-generation graphene materials: hope for bright future. Environ. Sci. Pollut. Res. 25(8): 7315–7329.

Amini, M., Jahanshahi, M. and Rahimpour, A. 2013. Synthesis of novel thin film nanocomposite (TFN) forward osmosis membranes using functionalized multi-walled carbon nanotubes. J. Membr. Sci. 435: 233–241.

An, B., Steinwinder, T.R. and Zhao, D. 2005. Selective removal of arsenate from drinking water using a polymeric ligand exchanger. Water Res. 39(20): 4993–5004.

Anandan, S., Kathiravan, K., Murugesan, V. and Ikuma, Y. 2009. Anionic (IO^{3-}) non-metal doped TiO_2 nanoparticles for the photocatalytic degradation of hazardous pollutant in water. Catal. Commun. 10(6): 1014–1019.

Aoshima, H., Kokubo, K.E.N., Shirakawa, S., Ito, M., Yamana, S. and Oshima, T. 2009. Antimicrobial activity of fullerenes and their hydroxylated derivatives. Biocontrol Sci. 14(2): 69–72.

Arias, L.R. and Yang, L. 2009. Inactivation of bacterial pathogens by carbon nanotubes in suspensions. Langmuir. 25(5): 3003–3012.

Asahi, R.Y.O.J.I., Morikawa, T.A.K.E.S.H.I., Ohwaki, T., Aoki, K. and Taga, Y. 2001. Visible-light photocatalysis in nitrogen-doped titanium oxides. Sci. 293(5528): 269–271.

Attatsi, I.K. and Nsiah, F. 2020. Application of silver nanoparticles toward Co (II) and Pb (II) ions contaminant removal in groundwater. Appl. Water Sci. 10(6): 1–13.

Azam, A., Ahmed, A.S., Oves, M., Khan, M.S. and Memic, A. 2012a. Size-dependent antimicrobial properties of CuO nanoparticles against Gram-positive and-negative bacterial strains. Int. J. Nanomed. 7: 3527–3535.

Azam, A., Ahmed, A.S., Oves, M., Khan, M.S., Habib, S.S. and Memic, A. 2012b. Antimicrobial activity of metal oxide nanoparticles against Gram-positive and Gram-negative bacteria: a comparative study. Int. J. Nanomed. 7: 6003– 6009.

Azimi, S., Behin, J., Abiri, R., Rajabi, L., Derakhshan, A.A. and Karimnezhad, H. 2014. Synthesis, characterization and antibacterial activity of chlorophyllin functionalized graphene oxide nanostructures. Sci. Adv. Mater. 6(4): 771–781.

Bacakova, L., Pajorova, J., Tomkova, M., Matejka, R., Broz, A., Stepanovska, J. and Kallio, P. 2020. Applications of nanocellulose/nanocarbon composites: Focus on biotechnology and medicine. Nanomater. 10(2): 196.

Barceló, D. (ed.). 2004. Emerging Organic Pollutants in Waste Waters and Sludge. Springer Berlin, Heidelberg.

Baroña, G.N.B., Lim, J., Choi, M. and Jung, B. 2013. Interfacial polymerization of polyamide-aluminosilicate SWNT nanocomposite membranes for reverse osmosis. Desalination. 325: 138–147.

Bellona, C., Drewes, J.E., Xu, P. and Amy, G. 2004. Factors affecting the rejection of organic solutes during NF/RO treatment—a literature review. Water Res. 38(12): 2795–2809.

Bellona, C. and Drewes, J.E. 2005. The role of membrane surface charge and solute physico-chemical properties in the rejection of organic acids by NF membranes. J. Membr. Sci. 249(1–2): 227–234.

Ben-Sasson, M., Zodrow, K.R., Genggeng, Q., Kang, Y., Giannelis, E.P. and Elimelech, M. 2014. Surface functionalization of thin-film composite membranes with copper nanoparticles for antimicrobial surface properties. Environ. Sci. Technol. 48(1): 384–393.

Bera, R.K., Mandal, S.M. and Raj, C.R. 2014. Antimicrobial activity of fluorescent Ag nanoparticles. Lett. Appl. Microbiol. 58(6): 520–526.

Borji, H., Ayoub, G.M., Bilbeisi, R., Nassar, N. and Malaeb, L. 2020. How effective are nanomaterials for the removal of heavy metals from water and wastewater? Water Air Soil Pollut. 231(7): 1–35.

Buzea, C., Pacheco, I.I. and Robbie, K. 2007. Nanomaterials and nanoparticles: sources and toxicity. Biointerphases. 2(4): MR17–MR71.

Cataldo, F. and Da Ros, T. (eds). 2008. Medicinal Chemistry and Pharmacological Potential of Fullerenes and Carbon Nanotubes (Vol. 1). Springer Science & Business Media.

Chan, W.F., Chen, H.Y., Surapathi, A., Taylor, M.G., Shao, X., Marand, E. and Johnson, J.K. 2013. Zwitterion functionalized carbon nanotube/polyamide nanocomposite membranes for water desalination. Acs Nano. 7(6): 5308–5319.

Chen, Y.H. and Li, F.A. 2010. Kinetic study on removal of copper (II) using goethite and hematite nano-photocatalysts. J. Colloid Interface Sci. 347(2): 277–281.

Chen, H.F., Wu, J.J., Wu, M.Y. and Jia, H. 2019. Preparation and antibacterial activities of copper nanoparticles encapsulated by carbon. New Carbon Mater. 34(4): 382–389.

Cheshomi, N., Pakizeh, M. and Namvar-Mahboub, M. 2018. Preparation and characterization of TiO_2/Pebax/(PSf-PES) thin film nanocomposite membrane for humic acid removal from water. Polym. Adv. Technol. 29(4): 1303–1312.

Chitose, N., Ueta, S., Seino, S. and Yamamoto, T.A. 2003. Radiolysis of aqueous phenol solutions with nanoparticles. 1. Phenol degradation and TOC removal in solutions containing TiO_2 induced by UV, γ-ray and electron beams. Chemosphere. 50(8): 1007–1013.

Chou, W.L., Yu, D.G. and Yang, M.C. 2005. The preparation and characterization of silver-loading cellulose acetate hollow fiber membrane for water treatment. Polym. Adv. Technol. 16(8): 600–607.

Darweesh, M.A., Elgendy, M.Y., Ayad, M I., Ahmed, A.M.M., Elsayed, N.K. and Hammad, W.A. 2022. A unique, inexpensiv. and abundantly available adsorbent: composite of synthesized silver nanoparticles (AgNPs) and banana leaves powder (BLP). Heliyon. 8(4): e09279.

Das, J. and Velusamy, P. 2014. Catalytic reduction of methylene blue using biogenic gold nanoparticles from Sesbania grandiflora L.J. Taiwan Inst. Chem. Eng. 45(5): 2280–2285.

de Dicastillo, C.L., Correa, M.G., Martínez, F.B., Streitt, C. and Galotto, M.J. 2020. Antimicrobial effect of titanium dioxide nanoparticles. Antimicrobial Resistance—A One Health Perspective. IntechOpen. Available at: http://dx.doi.org/10.5772/intechopen.90891.

Deryabin, D.G., Davydova, O.K., Yankina, Z.Z., Vasilchenko, A.S., Miroshnikov, S.A., Kornev, A.B. and Troshin, P.A. 2014. The activity of [60]fullerene derivatives bearing amine and carboxylic solubilizing groups against *Escherichia coli*: a comparative study. J. Nanomater. 2014: 907435.

Dong, H., Qu, X.Y., Zhang, L., Cheng, L.H., Chen, H.L. and Gao, C.J. 2011. Preparation and characterization of surface-modified zeolite-polyamide thin film nanocomposite membranes for desalination. Desalin. Water Treat. 34(1–3): 6–12.

Dong, Z.Q., Ma, X.H., Xu, Z.L., You, W.T. and Li, F.B. 2014. Superhydrophobic PVDF–PTFE electrospun nanofibrous membranes for desalination by vacuum membrane distillation. Desalination. 347: 175–183.

El Badawi, N., Ramadan, A.R., Esawi, A.M. and El-Morsi, M. 2014. Novel carbon nanotube–cellulose acetate nanocomposite membranes for water filtration applications. Desalination. 344: 79–85.

Esfahani, M.R., Aktij, S.A., Dabaghian, Z., Firouzjaei, M.D., Rahimpour, A., Eke, J. and Koutahzadeh, N. 2019. Nanocomposite membranes for water separation and purification: Fabrication, modificatio. and applications. Sep. Purif. Technol. 213: 465–499.

Fathizadeh, M., Tien, H.N., Khivantsev, K., Song, Z., Zhou, F. and Yu, M. 2019. Polyamide/nitrogen-doped graphene oxide quantum dots (N-GOQD) thin film nanocomposite reverse osmosis membranes for high flux desalination. Desalination. 451: 125–132.

Filice, S., D'Angelo, D., Libertino, S., Nicotera, I., Kosma, V., Privitera, V. and Scalese, S. 2015. Graphene oxide and titania hybrid Nafion membranes for efficient removal of methyl orange dye from water. Carbon. 82: 489–499.

Firouzjaei, M.D., Shamsabadi, A.A., Aktij, S.A., Seyedpour, S.F., Sharifian Gh, M., Rahimpour, A. and Soroush, M. 2018. Exploiting synergetic effects of graphene oxide and a silver-based metal–organic framework to enhance antifouling and anti-biofouling properties of thin-film nanocomposite membranes. ACS Appl. Mater. Interfaces. 10(49): 42967–42978.

Gahlot, S., Sharma, P.P., Gupta, H., Kulshrestha, V. and Jha, P.K. 2014. Preparation of graphene oxide nano-composite ion-exchange membranes for desalination application. RSC Adv. 4(47): 24662–24670.

Galus, M., Jeyaranjaan, J., Smith, E., Li, H., Metcalfe, C. and Wilson, J.Y. 2013. Chronic effects of exposure to a pharmaceutical mixture and municipal wastewater in zebrafish. Aquat. Toxicol. 132: 212–222.

Gao, Y., Li, Y., Zhang, L., Huang, H., Hu, J., Shah, S.M. and Su, X. 2012. Adsorption and removal of tetracycline antibiotics from aqueous solution by graphene oxide. J. Colloid Interface Sci. 368(1): 540–546.

García, A., Rodríguez, B., Giraldo, H., Quintero, Y., Quezada, R., Hassan, N. and Estay, H. 2021. Copper-modified polymeric membranes for water treatment: a comprehensive review. Membranes. 11(2): 93.

Ge, F., Li, M.M., Ye, H. and Zhao, B.X. 2012. Effective removal of heavy metal ions Cd^{2+}, Zn^{2+}, Pb^{2+}, Cu^{2+} from aqueous solution by polymer-modified magnetic nanoparticles. J. Hazard. Mater. 211: 366–372.

Gehrke, I., Geiser, A. and Somborn-Schulz, A. 2015. Innovations in nanotechnology for water treatment. Nanotechnol. Sci. Appl. 8: 1.

Ghaemi, N. 2016. A new approach to copper ion removal from water by polymeric nanocomposite membrane embedded with γ-alumina nanoparticles. Appl. Surf. Sci. 364: 221–228.

Ghanbari, M., Emadzadeh, D., Lau, W.J., Matsuura, T. and Ismail, A.F. 2015. Synthesis and characterization of novel thin film nanocomposite reverse osmosis membranes with improved organic fouling properties for water desalination. Rsc Adv. 5(27): 21268–21276.

Ghiloufi, I., Khezami, L. and El Mir, L. 2015. Preparation and characterization of nanoporous resin for heavy metal removal from aqueous solution. J. Water Supply: Res. Technol.—AQUA. 64(3): 316–325.

Gohari, R.J., Lau, W.J., Matsuura, T. and Ismail, A.F. 2013. Fabrication and characterization of novel PES/Fe–Mn binary oxide UF mixed matrix membrane for adsorptive removal of As(III) from contaminated water solution. Sep. Purif. Technol. 118: 64–72.

Gollavelli, G., Chang, C.C. and Ling, Y.C. 2013. Facile synthesis of smart magnetic graphene for safe drinking water: heavy metal removal and disinfection control. ACS Sustain. Chem. Eng. 1(5): 462–472.

Gurunathan, S., Han, J.W., Dayem, A.A., Eppakayala, V. and Kim, J.H. 2012. Oxidative stress-mediated antibacterial activity of graphene oxide and reduced graphene oxide in *Pseudomonas aeruginosa*. Int. J. Nanomedicine. 7: 5901.

Harshiny, M., Aiswarya Devi, S. and Matheswaran, M. 2019. Spiny amaranth leaf extract mediated iron oxide nanoparticles: biocidal photocatalytic propensity, stability, dissolubility and reusability. Biocatal. Agric. Biotechnol. 21: 101296.

He, J., Lin, X.M., Chan, H., Vukovic, L., Král, P. and Jaeger, H.M. 2011. Diffusion and filtration properties of self-assembled gold nanocrystal membranes. Nano Lett. 11(6): 2430–2435.

Hillie, T. and Hlophe, M. 2007. Nanotechnology and the challenge of clean water. Nat. Nano Technolol. 2(11): 663–664.

Hoek, E.M., Pendergast, M.T.M. and Ghosh, A.K. 2014. Nanotechnology-based membranes for water purification. pp. 133–154. *In*: Street, A., Sustich, R., Duncan, J. and Savage, N. (eds). Nanotechnology Applications for Clean Water: Solutions for Improving Water Quality. 2nd Ed. Elsevier.

Hu, H. and Xu, K. 2020. Physicochemical technologies for HRPs and risk control. pp. 169-207. *In*: Ren, H. and Zhang, X. (eds). High-risk Pollutants in Wastewater. Elsevier.

Huang, Z.H., Yin, Y.N., Aikebaier, G.L.M.L. and Zhang, Y. 2016. Preparation of a novel positively charged nanofiltration composite membrane incorporated with silver nanoparticles for pharmaceuticals and personal care product rejection and antibacterial properties. Water Sci. Technol. 73(8): 1910–1919.

Huangfu, X., Jiang, J., Lu, X., Wang, Y., Liu, Y., Pang, S.Y. and Ma, J. 2015. Adsorption and oxidation of thallium (I) by a nanosized manganese dioxide. Water Air Soil Pollut. 226(1): 1–9.

Ibrahim, K.S. 2013. Carbon nanotubes-properties and applications: a review. Carbon Lett. 14(3): 131–144.

Jain, P. and Pradeep, T. 2005. Potential of silver nanoparticle-coated polyurethane foam as an antibacterial water filter. Biotechnol. Bioeng. 90(1): 59–63.

Jo, Y.K., Kim, B.H. and Jung, G. 2009. Antifungal activity of silver ions and nanoparticles on phytopathogenic fungi. Plant Dis. 93(10): 1037–1043.

Kanchi, S. 2014. Nanotechnology for water treatment. J. Environ. Anal. Chem. 1(2): 1–3.

Kang, S., Pinault, M., Pfefferle, L.D. and Elimelech, M. 2007. Single-walled carbon nanotubes exhibit strong antimicrobial activity. Langmuir. 23(17): 8670–8673.

Karim, Z., Mathew, A.P., Grahn, M., Mouzon, J. and Oksman, K. 2014. Nanoporous membranes with cellulose nanocrystals as functional entity in chitosan: removal of dyes from water. Carbohydr. Polym. 112: 668–676.

Khajeh, M., Laurent, S. and Dastafkan, K. 2013. Nanoadsorbents: classification, preparatio. and applications (with emphasis on aqueous media). Chem. Rev. 113(10): 7728–7768.

Khajouei, M., Jahanshahi, M. and Peyravi, M. (2018). Biofouling mitigation of TFC membrane by in-situ grafting of PANI/Cu couple nanoparticle. J. Taiwan Inst. Chem. Eng. 85: 237–247.

Khezami, L., Ould M'hamed, M., Lemine, O.M., Bououdina, M. and Bessadok-Jemai, A. 2016. Milled goethite nanocrystalline for selective and fast uptake of cadmium ions from aqueous solution. Desalin. Water Treat. 57(14): 6531–6539.

Kim, E.S., Liu, Y. and El-Din, M.G. 2013a. An in-situ integrated system of carbon nanotubes nanocomposite membrane for oil sands process-affected water treatment. J. Membr. Sci. 429: 418–427.

Kim, S.G., Hyeon, D.H., Chun, J.H., Chun, B.H. and Kim, S.H. 2013b. Nanocomposite poly (arylene ether sulfone) reverse osmosis membrane containing functional zeolite nanoparticles for seawater desalination. J. Membr. Sci. 443, 10–18.

Kim, H.J., Pant, H.R., Kim, J.H., Choi, N.J. and Kim, C.S. 2014. Fabrication of multifunctional TiO_2–fly ash/polyurethane nanocomposite membrane via electrospinning. Ceram. Int. 40(2): 3023–3029.

Krishnaraj, C., Jagan, E.G., Rajasekar, S., Selvakumar, P., Kalaichelvan, P.T. and Mohan, N. 2010. Synthesis of silver nanoparticles using *Acalypha indica* leaf extracts and its antibacterial activity against water borne pathogens. Colloids Surf. B: Biointerfaces. 76(1): 50–56.

Kumar, V., Talreja, N., Deva, D., Sankararamakrishnan, N., Sharma, A. and Verma, N. 2011. Development of bi-metal doped micro-and nano multi-functional polymeric adsorbents for the removal of fluoride and arsenic (V) from wastewater. Desalination. 282: 27–38.

Kunduru, K.R., Nazarkovsky, M., Farah, S., Pawar, R.P., Basu, A. and Domb, A.J. 2017. Nanotechnology for water purification: applications of nanotechnology methods in wastewater treatment. Water Purif. 33–74.

Kuo, C.Y., Wu, C.H. and Wu, J.Y. 2008. Adsorption of direct dyes from aqueous solutions by carbon nanotubes: Determination of equilibrium, kinetics and thermodynamics parameters. J. Colloid Interface Sci. 327(2): 308–315.

Kuzmany, H., Kukovecz, A., Simon, F., Holzweber, M., Kramberger, C. and Pichler, T. 2004. Functionalization of carbon nanotubes. Synth. Met. 141(1–2): 113–122.

Lau, W.J., Ismail, A.F., Misdan, N. and Kassim, M.A. 2012. A recent progress in thin film composite membrane: A review. Desalination. 287: 190–199.

Le, A.T., Pung, S.Y., Sreekantan, S. and Matsuda, A. 2019. Mechanisms of removal of heavy metal ions by ZnO particles. Heliyon. 5(4): e01440.

Lee, A., Elam, J.W. and Darling, S.B. 2016. Membrane materials for water purification: design, developmen. and application. Environ. Sci. Water Res. Technol. 2(1): 17–42.

Li, Q.L., Yuan, D.X. and Lin, Q.M. 2004. Evaluation of multi-walled carbon nanotubes as an adsorbent for trapping volatile organic compounds from environmental samples. J. Chromatogr. A. 1026(1–2): 283–288.

Li, R., Zhang, L. and Wang, P. 2015. Rational design of nanomaterials for water treatment. Nanoscale. 7(41): 17167–17194.

Lima, E., Guerra, R., Lara, V. and Guzmán, A. 2013. Gold nanoparticles as efficient antimicrobial agents for Escherichia coli and Salmonella typhi. Chem. Cent. J. 7(1): 1–7.

Liu, Y., Li, Q., Gao, S. and Shang, J.K. 2011. Exceptional As(III) sorption capacity by highly porous magnesium oxide nanoflakes made from hydrothermal synthesis. J. Am. C. 94(1): 217–223.

Long, R.Q. and Yang, R.T. 2001. Carbon nanotubes as superior sorbent for dioxin removal. J. Am. Chem. Soc. 123(9): 2058–2059.

Lu, C., Chung, Y.L. and Chang, K.F. 2005. Adsorption of trihalomethanes from water with carbon nanotubes. Water Res. 39(6): 1183–1189.

Mahdavi, S., Jalali, M. and Afkhami, A. 2015. Heavy metals removal from aqueous solutions by Al_2O_3 nanoparticles modified with natural and chemical modifiers. Clean Technol. Environ. Policy. 17(1): 85–102.

Mahmoud, M.E., Abou Ali, S.A. and Elweshahy, S.M. 2018. Microwave functionalization of titanium oxide nanoparticles with chitosan nanolayer for instantaneous microwave sorption of Cu (II) and Cd (II) from water. Int. J. Biol. Macromol. 111: 393–399.

Malik, A., Yasar, A., Tabinda, A.B. and Abubakar, M. 2012. Water-borne diseases, cost of illness and willingness to pay for diseases interventions in rural communities of developing countries. Iran. J. Public Health. 41(6): 39.

Maliyekkal, S.M., Sreeprasad, T.S., Krishnan, D., Kouser, S., Mishra, A.K., Waghmare, U.V. and Pradeep, T. 2013. Graphene: a reusable substrate for unprecedented adsorption of pesticides. Small. 9(2): 273–283.

Mangun, C.L., Yue, Z., Economy, J., Maloney, S., Kemme, P. and Cropek, D. 2001. Adsorption of organic contaminants from water using tailored ACFs. Chem. Mater. 13(7): 2356–2360.

Matilainen, A., Gjessing, E.T., Lahtinen, T., Hed, L., Bhatnagar, A. and Sillanpää, M. 2011. An overview of the methods used in the characterisation of natural organic matter (NOM) in relation to drinking water treatment. Chemosphere. 83(11): 1431–1442.

Mendes, C.R., Dilarri, G., Forsan, C.F., Sapata, V.de.M.R., Lopes, P.R.M., de Moraes, P.B. and Bidoia, E.D. 2022. Antibacterial action and target mechanisms of zinc oxide nanoparticles against bacterial pathogens. Sci. Rep. 12(1): 1–10.

Mohammad, A.W., Teow, Y.H., Ang, W.L., Chung, Y.T., Oatley-Radcliffe, D.L. and Hilal, N. 2015. Nanofiltration membranes review: recent advances and future prospects. Desalination. 356: 226–254.

Mohammed, H., Kumar, A., Bekyarova, E., Al-Hadeethi, Y., Zhang, X., Chen, M. and Rimondini, L. 2020. Antimicrobial mechanisms and effectiveness of graphene and graphene-functionalized biomaterials. a scope review. Front. Bioeng. Biotechnol. 8: 465.

Mollahosseini, A., Rahimpour, A., Jahamshahi, M., Peyravi, M. and Khavarpour, M. 2012. The effect of silver nanoparticle size on performance and antibacteriality of polysulfone ultrafiltration membrane. Desalination. 306: 41–50.

Mukherjee, J., Ramkumar, J., Chandramouleeswaran, S., Shukla, R. and Tyagi, A.K. 2013. Sorption characteristics of nano manganese oxide: efficient sorbent for removal of metal ions from aqueous streams. J. Radioanal. Nucl. Chem. 297(1): 49–57.

Nagar, A. and Pradeep, T. 2020. Clean water through nanotechnology: needs, gap. and fulfillment. ACS Nano. 14(6): 6420–6435.

Najafi, M., Yousefi, Y. and Rafati, A.A. 2012. Synthesis, characterization and adsorption studies of several heavy metal ions on amino-functionalized silica nano hollow sphere and silica gel. Sep. Purif. Technol. 85: 193–205.

Niedergall, K., Bach, M., Hirth, T., Tovar, G.E. and Schiestel, T. 2014. Removal of micropollutants from water by nanocomposite membrane adsorbers. Sep. Purif. Technol. 131: 60–68.

Norouzian-Baghani, A., Mahvi, A.H., Gholami, M., Rastkari, N. and Delikhoon, M. 2016. One-Pot synthesis, characterization and adsorption studies of amine-functionalized magnetite nanoparticles for removal of Cr (VI) and Ni (II) ions from aqueous solution: Kinetic, isotherm and thermodynamic studies. J. Environ. Health Sci. Eng. 14(1): 1–12.

Ntim, S.A. and Mitra, S. 2012. Adsorption of arsenic on multiwall carbon nanotube–zirconia nanohybrid for potential drinking water purification. J. Colloid Interface Sci. 375(1): 154–159.

Obare, S.O. and Meyer, G.J. 2004. Nanostructured materials for environmental remediation of organic contaminants in water. J. Environ. Sci. Health A. 39(10): 2549–2582.

Ojea-Jiménez, I., López, X., Arbiol, J. and Puntes, V. 2012. Citrate-coated gold nanoparticles as smart scavengers for mercury (II) removal from polluted waters. ACS Nano. 6(3): 2253–2260.

Pal, S., Tak, Y.K. and Song, J.M. 2007. Does the antibacterial activity of silver nanoparticles depend on the shape of the nanoparticle? A study of the gram-negative bacterium *Escherichia coli*. Appl. Environ. Microbiol. 73(6): 1712–1720.

Pan, B. and Xing, B. 2008. Adsorption mechanisms of organic chemicals on carbon nanotubes. Environ. Sci. Technol. 42(24): 9005–9013.

Panahi, Y., Mellatyar, H., Farshbaf, M., Sabet, Z., Fattahi, T. and Akbarzadehe, A. 2018. Biotechnological applications of nanomaterials for air pollution and water/wastewater treatment. Mater.Today Proc. 5(7): 15550–15558.

Peng, X., Li, Y., Luan, Z., Di, Z., Wang, H., Tian, B. and Jia, Z. 2003. Adsorption of 1, 2-dichlorobenzene from water to carbon nanotubes. Chem. Phys. Lett. 376(1–2): 154–158.

Peng, X., Luan, Z., Ding, J., Di, Z., Li, Y. and Tian, B. 2005. Ceria nanoparticles supported on carbon nanotubes for the removal of arsenate from water. Mater. Lett. 59(4): 399–403.

Poursani, A.S., Nilchi, A., Hassani, A.H., Shariat, M. and Nouri, J. 2015. A novel method for synthesis of nano-γ-Al_2O_3: study of adsorption behavior of chromium, nickel, cadmium and lead ions. Int. J. Environ. Sci. Technol. 12(6): 2003–2014.

Qian, H., Pretzer, L.A., Velazquez, J.C., Zhao, Z. and Wong, M.S. 2013. Gold nanoparticles for cleaning contaminated water. J. Chem. Technol. Biotechnol. 88(5): 735–741.

Qu, X., Brame, J., Li, Q. and Alvarez, P.J. 2013. Nanotechnology for a safe and sustainable water supply: enabling integrated water treatment and reuse. Acc. Chem. Res. 46(3): 834–843.

Rajput, S., Singh, L.P., Pittman Jr, C.U. and Mohan, D. 2017. Lead (Pb^{2+}) and copper (Cu^{2+}) remediation from water using superparamagnetic maghemite (γ-Fe_2O_3) nanoparticles synthesized by Flame Spray Pyrolysis (FSP). J. Colloid Interface Sci. 492: 176–190.

Rao, G.P., Lu, C. and Su, F. 2007. Sorption of divalent metal ions from aqueous solution by carbon nanotubes: a review. Sep. Purif. Technol. 58(1): 224–231.

Ren, L., Ping, M. and Zhang, X. 2020. Membrane biofouling control by surface modification of quaternary ammonium compound using atom-transfer radical-polymerization method with silica nanoparticle as interlayer. Membr. 10(12): 417.

Rezaee, R., Nasseri, S., Mahvi, A.H., Nabizadeh, R., Mousavi, S.A., Rashidi, A. and Nazmara, S. 2015. Fabrication and characterization of a polysulfone-graphene oxide nanocomposite membrane for arsenate rejection from water. J. Environ. Health Sci. Eng. 13(1): 1–11.

Rodríguez, B., Oztürk, D., Rosales, M., Flores, M. and García, A. 2018. Antibiofouling thin-film composite membranes (TFC) by in situ formation of Cu-(m-phenylenediamine) oligomer complex. J. Mater. Sci. 53(9): 6325–6338.

Roy, A., Bulut, O., Some, S., Mandal, A.K. and Yilmaz, M.D. 2019. Green synthesis of silver nanoparticles: biomolecule-nanoparticle organizations targeting antimicrobial activity. RSC Adv. 9(5): 2673–2702.

Safarpour, M., Khataee, A. and Vatanpour, V. 2015. Thin film nanocomposite reverse osmosis membrane modified by reduced graphene oxide/TiO_2 with improved desalination performance. J. Membr. Sci. 489, 43–54.

Salas, P., Odzak, N., Echegoyen, Y., Kägi, R., Sancho, M.C. and Navarro, E. 2019. The role of size and protein shells in the toxicity to algal photosynthesis induced by ionic silver delivered from silver nanoparticles. Sci. Total Environ. 692: 233–239.

Saleh, T.A., Agarwal, S. and Gupta, V.K. 2011. Synthesis of MWCNT/MnO_2 and their application for simultaneous oxidation of arsenite and sorption of arsenate. Appl. Catal. B: Environ. 106(1–2): 46–53.

Saud, P.S., Pant, B., Park, M., Chae, S.H., Park, S.J., Mohamed, E.N. and Kim, H.Y. 2015. Preparation and photocatalytic activity of fly ash incorporated TiO_2 nanofibers for effective removal of organic pollutants. Ceram. Int. 41(1): 1771–1777.

Savage, N. and Diallo, M.S. 2005. Nanomaterials and water purification: opportunities and challenges. J. Nanoparticle Res. 7: 331–342.

Sawada, I., Fachrul, R., Ito, T., Ohmukai, Y., Maruyama, T. and Matsuyama, H. 2012. Development of a hydrophilic polymer membrane containing silver nanoparticles with both organic antifouling and antibacterial properties. J. Membr. Sci. 387: 1–6.

Schmidtke, C., Eggers, R., Zierold, R., Feld, A., Kloust, H., Wolter, C. and Weller, H. 2014. Polymer-assisted self-assembly of superparamagnetic iron oxide nanoparticles into well-defined clusters: controlling the collective magnetic properties. Langmuir. 30(37): 11190–11196.

Shipley, H.J., Engates, K.E. and Grover, V.A. 2013. Removal of Pb (II), Cd (II), Cu (II. and Zn (II) by hematite nanoparticles: effect of sorbent concentration, pH, temperatur. and exhaustion. Environ. Sci. Pollut. Res. 20(3): 1727–1736.

Shokri, E., Yegani, R., Pourabbas, B. and Kazemian, N. 2016. Preparation and characterization of polysulfone/organoclay adsorptive nanocomposite membrane for arsenic removal from contaminated water. Appl. Clay Sci. 132: 611–620.

Sondi, I. and Salopek-Sondi, B. 2004. Silver nanoparticles as antimicrobial agent: a case study on *E. coli* as a model for Gram-negative bacteria. J. Colloid Interface Sci. 275(1): 177–182.

Suárez-Iglesias, O., Collado, S., Oulego, P. and Díaz, M. 2017. Graphene-family nanomaterials in wastewater treatment plants. Chem. Eng. J. 313: 121–135.

Sumesh, E., Bootharaju, M.S. and Pradeep, T. 2011. A practical silver nanoparticle-based adsorbent for the removal of Hg^{2+} from water. J. Hazard. Mater. 189(1–2): 450–457.

Tabesh, S., Davar, F. and Loghman-Estarki, M.R. 2018. Preparation of γ-Al_2O_3 nanoparticles using modified sol-gel method and its use for the adsorption of lead and cadmium ions. J. Alloys Compd. 730: 441–449.

Tarekegn, M.M., Hiruy, A.M. and Dekebo, A.H. 2021. Correction: Nano zero valent iron (nZVI) particles for the removal of heavy metals (Cd^{2+}, Cu^{2+} and Pb^{2+}) from aqueous solutions. RSC Adv. 11(43): 27084–27084.

Taurozzi, J.S., Arul, H., Bosak, V.Z., Burban, A.F., Voice, T.C., Bruening, M.L. and Tarabara, V.V. 2008. Effect of filler incorporation route on the properties of polysulfone–silver nanocomposite membranes of different porosities. J. Membr. Sci. 325(1): 58–68.

Tegos, G.P., Demidova, T.N., Arcila-Lopez, D., Lee, H., Wharton, T., Gali, H. and Hamblin, M.R. 2005. Cationic fullerenes are effective and selective antimicrobial photosensitizers. Chem. Biol. 12(10): 1127–1135.

Thilakan, D., Patankar, J., Khadtare, S., Wagh, N.S., Lakkakula, J., El-Hady, K.M. and Tarique, M. 2022. Plant-derived iron nanoparticles for removal of heavy metals. Int. J. Chem. Eng. 2022.

Tiwari, P.M., Vig, K., Dennis, V.A. and Singh, S.R. 2011. Functionalized gold nanoparticles and their biomedical applications. Nanomaterials. 1(1): 31–63.

Vecitis, C.D., Zodrow, K.R., Kang, S. and Elimelech, M. 2010. Electronic-structure-dependent bacterial cytotoxicity of single-walled carbon nanotubes. ACS Nano. 4(9): 5471–5479.

Vidic, J., Stankic, S., Haque, F., Ciric, D., Le Goffic, R., Vidy, A. and Delmas, B. 2013. Selective antibacterial effects of mixed ZnMgO nanoparticles. J. Nanopart. Res. 15(5): 1–10.

Wang, Y., Ou, R., Ge, Q., Wang, H. and Xu, T. 2013. Preparation of polyethersulfone/carbon nanotube substrate for high-performance forward osmosis membrane. Desalination. 330: 70–78.

Wang, J., Chen, Z. and Chen, B. 2014. Adsorption of polycyclic aromatic hydrocarbons by graphene and graphene oxide nanosheets. Environ. Sci. Technol. 48(9): 4817–4825.

Wang, Y., Zhu, J., Huang, H. and Cho, H.H. 2015. Carbon nanotube composite membranes for microfiltration of pharmaceuticals and personal care products: capabilities and potential mechanisms. J. Membr. Sci. 479: 165–174.

Wang, X., Ma, H., Chu, B. and Hsiao, B.S. 2017. Thin-film nanofibrous composite reverse osmosis membranes for desalination. Desalination. 420: 91–98.

Wen, Y., Yuan, J., Ma, X., Wang, S. and Liu, Y. 2019. Polymeric nanocomposite membranes for water treatment: a review. Environ. Chem. Lett. 17(4): 1539–1551.

Werkneh, A.A. and Rene, E.R. 2019. Applications of nanotechnology and biotechnology for sustainable water and wastewater treatment. pp. 405–430. *In*: Bui, X.T., Chiemchaisri, C., Fujioka, T. and Varjani, S. (eds). Water and Wastewater Treatment Technologies. Energy, Environment, and Sustainability. Springer, Singapore.

Wu, H., Tang, B. and Wu, P. 2013. Optimizing polyamide thin film composite membrane covalently bonded with modified mesoporous silica nanoparticles. J. Membr. Sci. 428: 341–348.

Xie, Y., He, Y., Irwin, P.L., Jin, T. and Shi, X. 2011. Antibacterial activity and mechanism of action of zinc oxide nanoparticles against *Campylobacter jejuni*. Appl. Environ. Microbiol. 77(7): 2325–2331.

Xu, J., Wang, L. and Zhu, Y. 2012. Decontamination of bisphenol a from aqueous solution by graphene adsorption. Langmuir. 28(22): 8418–8425.

Yamamoto, O., Ohira, T., Alvarez, K. and Fukuda, M. 2010. Antibacterial characteristics of $CaCO_3$–MgO composites. Mater. Sci. Eng. B. 173(1–3): 208–212.

Yan, L., Qiaohui, F. and Wangsuo, W. 2011. Sorption of Th (IV) on goethite: effects of pH, ionic strength, FA and phosphate. J. Radioanal. Nucl. Chem. 289(3): 865–871.

Yang, C., Mamouni, J., Tang, Y. and Yang, L. 2010. Antimicrobial activity of single-walled carbon nanotubes: length effect. Langmuir. 26(20): 16013–16019.

Yaqoob, A.A., Ahmad, H., Parveen, T., Ahmad, A., Oves, M., Ismail, I.M. and Mohamad Ibrahim, M.N. 2020. Recent advances in metal decorated nanomaterials and their various biological applications: a review. Front. Chem. 8: 341.

Yin, J. and Deng, B. 2015. Polymer-matrix nanocomposite membranes for water treatment. J. Membr. Sci. 479: 256–275.

Zahid, M., Ahmad, H., Drioli, E., Rehan, Z. A., Rashid, A., Akram, S. and Khalid, T. 2021. Role of polymeric nanocomposite membranes for the removal of textile dyes from wastewater. pp. 91–103. *In*: Kamal, A. Abd-Elsalam and Muhammad Zahid (eds). Aquananotechnology: Applications of Nanomaterials for Water Purification. Elsevier.

Zaib, Q., Mansoor, B. and Ahmad, F. 2013. Photo-regenerable multi-walled carbon nanotube membranes for the removal of pharmaceutical micropollutants from water. Environ. Sci.: Process. Impacts. 15(8): 1582–1589.

Zawrah, M.F., El Shereefy, E.S.E. and Khudir, A.Y. 2019. Reverse precipitation synthesis of $\leq$ 10 nm magnetite nanoparticles and their application for removal of heavy metals from water. Silicon. 11(1): 85–104.

Zhang, C., Wang, W., Duan, A., Zeng, G., Huang, D., Lai, C. and Yang, Y. 2019. Adsorption behavior of engineered carbons and carbon nanomaterials for metal endocrine disruptors: experiments and theoretical calculation. Chemosphere. 222: 184–194.

Zhao, G., Li, J., Ren, X., Chen, C. and Wang, X. 2011. Few-layered graphene oxide nanosheets as superior sorbents for heavy metal ion pollution management. Environ. Sci. Technol. 45(24): 10454–10462.

Zhao, C., Xue, J., Ran, F. and Sun, S. 2013. Modification of polyethersulfone membranes–a review of methods. Prog. Mater. Sci. 58(1): 76–150.

Zhao, Q., Hou, J., Shen, J., Liu, . and Zhang, Y. 2015. Long-lasting antibacterial behavior of a novel mixed matrix water purification membrane. J. Mater. Chem. A. 3(36): 18696–18705.

Zodrow, K., Brunet, L., Mahendra, S., Li, D., Zhang, A., Li, Q. and Alvarez, P.J. 2009. Polysulfone ultrafiltration membranes impregnated with silver nanoparticles show improved biofouling resistance and virus removal. Water Res. 43(3): 715–723.

Chapter 8

Functionalization of 3D Graphene-based Materials for Removal of Impurities from Water

Malgorzata Skorupska[1] **and Anna Ilnicka***[1]

[1]Faculty of Chemistry, Nicolaus Copernicus University in Torun, Gagarina 7, 87-100 Torun, Poland

Jerzy P. Lukaszewicz[1,2]

[2]Centre for Modern Interdisciplinary Technologies, Nicolaus Copernicus University in Torun, Wilenska 4, 87-100 Torun, Poland

8.1 INTRODUCTION

The year 2004, when Novoselov and Geim discovered graphene, marked the beginning of a carbon revolution (Huang et al. 2009). Graphene, possesses unique properties such as high specific surface area, excellent thermal and electrical conductivity, unique conductive properties, high mechanical strength, and the ability to create not only two-dimensional structures (graphene nanosheets, graphene oxide (GO)) but also one-dimensional structures (graphene quantum dots) and three-dimensional structures (graphene foams, three-dimensional structured graphene). Therefore, graphene has attracted the attention of researchers in various fields, e.g., engineering, energy, environmental protection, and medicine (Allen et al. 2010, Choi et al. 2010, Yu et al. 2020). Graphene is used in biosensors, electronic devices, energy storage devices, desalination, and water purification (Nguyen and Nguyen, 2016, Han et al. 2017, Chen et al. 2021). The latter application has a significant environmental impact due to the vital importance of water in the life of all living organisms. Today the world is facing the problem of the quality and quantity of available drinking water. The problem is caused by many factors, including climate change, affecting the 21st century. The increasing average temperature means glaciers are beginning to melt, thus increasing the amount of water in nature's water cycle, which directly impacts the

*Corresponding Author: ailnicka@umk.pl

intensity of rainfall and droughts. The low quality of drinking water caused by chemical pollutants such as heavy metals or pesticides, pathogens, industrial chemicals, dyes, and toxins accumulated by aquatic living organisms in surface water bodies is a global problem (Schwarzenbach et al. 2010). One of the possible solutions for water purification is the use of porous carbon nanomaterials (Shi et al. 2014, Thines et al. 2017). However, the question to be addressed is: can graphene-based filtering materials effectively purify water of all kinds of contaminants by means of surface functionalization and an extended structure?

This chapter will focus on recent advances in fine-tuning the structure of three-dimensional (3D) graphene materials and the mechanism of action of functional groups found on the surface of nanosheets, which assist in the removal of various impurities found in water. The main issues addressed in the monograph are the relationship between structural and morphological factors, the existence of surface functional groups, and the efficiency of water purification or desalination.

8.2 TRENDS IN THE FUNCTIONALIZATION OF 3D GRAPHENE-BASED MATERIALS

Due to the defective structure at the edges of graphene sheets, the material is susceptible to so-called functionalization, which in turn enables diversified application (Hiew et al. 2018). The hexagonal arrangement of carbon atoms in two-dimensional graphene planes resulting from hypothetical sp^2-hybridization of the atoms results in a large surface area and a delocalized π-electron system, which may act on all kinds of impurities (Dieu Le and Vinh Tran 2022). Two-dimensional graphene materials can agglomerate, which is unfavorable, leading to relatively large 3D structures with radically reduced surface area. Scientists are trying to resolve this problem with several technological measures to prevent agglomeration, increase the number of adsorption centers (proportional to the surface area), and facile the accessibility of impurities to active sites. To produce a material with good hydrophilic properties, a well-developed surface area and pore structure combined with excellent adsorption capacity have become a major challenge for researchers. The hydrothermal method is often used to achieve this goal in combination with the appropriate application of lyophilization. Weak hydrogen bonds, van der Waals forces, and π-π interactions are responsible for the formation of cross bonds that make the three-dimensional structure possible. Thin graphene sheets' walls are stacked together and linked through cross-linking bonds, enabling them to merge into flexible layers, ultimately forming a stable 3D structural network. Therefore, these inherently flexible graphene layers become an important factor in the construction of 3D structures (Xu et al. 2010). On the other hand, improper stacking of graphene sheets can reduce the surface area and thus the porosity; this results in a higher density of graphene materials and ultimately weakens the adsorption capacity (Ren et al. 2016). An ultrathin 3D material heteroatom-doped graphene was obtained using a hydrothermal method and a freeze-drying process. What and how many reagents and precursors we use affect the final adsorption mechanism and the hydrophilic properties of the final material. For example, while using poly(vinyl alcohol) (PVA) as a GO crosslinking stabilizer, it can be seen that the backbone of graphene flakes is effectively strengthened, and the addition of PVA does not determine the final adsorption capacity of the resulting composite for the corresponding organic dyes. Nevertheless, it is worth noting that the adsorption efficiency also depends on the type of impurity and so Methylene Blue molecules effectively interact with GO layers through hydrogen bonds and π - π ring coupling and weakly interact with PVA chains. On the other hand, Congo Red is effectively adsorbed on graphene layers and through polymer chains (Dai et al. 2016).

3D graphene materials have been used for the purification of water from toxins because of their ability to be doped with heteroatoms, which can become potential active sites for the adsorption of organic or inorganic pollutants (Shi et al. 2016a, Singh et al. 2018, Yao et al. 2018).

The large specific surface area, and extended 3D structure, allow access to these sites, making graphene an excellent adsorbent. Researchers are constantly improving the properties of graphene by introducing heteroatoms into the structure, making the material even more attractive in terms of its applicability. In the main, oxygen (Zhao et al. 2019), nitrogen (Sun et al. 2013), or sulfur (Sivakumar et al. 2022) is introduced into the structure, including boron (Sun et al. 2013, Singh et al. 2018), phosphorus (Shi et al. 2019), or doped with more than one element like N, S (Shi et al. 2016a, Ren et al. 2018, Jiang et al. 2019b, Sivakumar et al. 2022) and O, N (Henriques et al. 2016) or O, N, P (Chen et al. 2018a). These types of doping are important to remove contaminants or heavy metals from water. The modified surface through the application of templates, stabilizers, and polymers, and the content of numerous functional groups and transition metal oxides, makes the adsorption and interaction between the contaminants and active sites very fast, and the selectivity for organic and inorganic contaminants very high (Chowdhury and Balasubramanian, 2014, Kim et al. 2018, Hossain et al. 2020). An additional advantage of graphene as an adsorbent is its price when comparing the cost of pure carbon nanotubes (Hossain et al. 2020). Heteroatom doping of graphene structure is useful for applications such as removing dyes, organic compounds, heavy metals, and water desalination (Wang et al. 2020).

These are the main applications of doped graphene in environmental protection. The following will discuss the preparation of doped materials and the mechanisms of functional groups that eliminate toxins and contaminants from the aquatic environment. There are many methods and techniques for removing contaminants such as heavy metals or organic dyes through chemical, physical or even biological methods. Well-known techniques include membrane filtration, adsorption of pollutants onto carbon adsorbents, electrochemical treatment, photocatalysis, ion exchange, and coagulation. All water purification techniques have many advantages but are also not without disadvantages and limitations (Carolin et al. 2017). The most effective and safe but also very scot-free method of purifying water from contaminants, among other techniques, is the adsorption process. Rapid industrial development is intensifying research into various wastewater and water purification techniques, so there is a great chance that the existing ideas applied at the laboratory scale will form the basis for the future commercialization of 3D graphene materials for water purification.

8.2.1 Oxygen Functional Groups and Pi-Pi Stacking

The most common and effective method for purifying water from various contaminants is introducing oxygen functional groups into the graphene structure, which significantly improves the properties of pure graphene. Due to their polarity, the oxygen functional groups interact with the impurities, leading to their partial or complete removal from the aqueous environment. The oxygen functional groups are characterized by a negative charge accumulated on O atoms, which increases the hydrophilic properties in aqueous solutions and promotes adsorption interactions between positively charged impurities, metal ions, or organic compounds (Hossain et al. 2020). The oxygen groups like –OH, –C–O, –C=O, C–O–C, –O–C–O– or –COOH (Figure 8.1) can be introduced by surface oxidation (Lobato-Peralta et al. 2021).

Oxygen groups on the surface of the graphene adsorbent are introduced using strong oxidants $K_2Cr_2O_7$, $KMnO_4$, and HNO_3, H_3PO_4, or H_2SO_4 or a mixture of these acids (Marcano et al. 2010, Liu et al. 2018, Panicker et al. 2021, Shiyanova et al. 2021, Choi et al. 2022). The use of oxidizing/acidic reagents increases the adsorption of heavy metals (Zhao et al. 2011). The amount and type of oxygen functional groups introduced into the structure depend largely on the type and amount of the oxidizing agent and the synthesis conditions. Well-optimized synthesis conditions and the number of suitable substrates favor the introduction of oxygen functional groups. A freeze-drying process permits the formation of a 3D structure (Figures 8.2A and 8.2B) (Wu et al. 2016, Liu et al. 2017a). The freeze-drying method, with or without a template, is relatively facile and uncomplicated. The use of specific additives, i.e., precursors of oxygen groups, such as chitosan

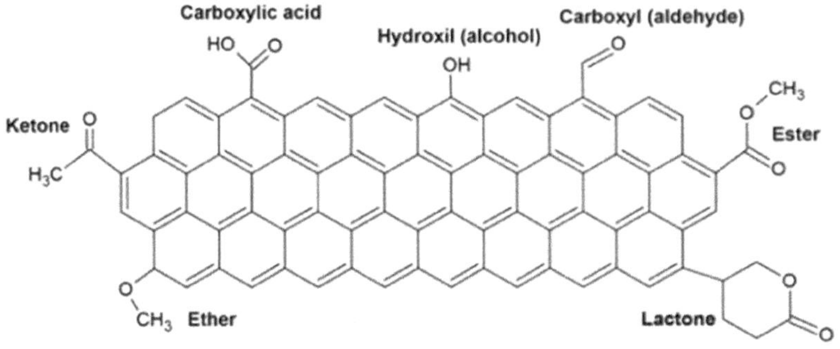

Figure 8.1 Examples of oxygen-based functional groups on the surface of activated carbon (based on FTIR analysis of AC derived from potato peel) (Lobato-Peralta et al. 2021), with permission from the Journal of Environmental Chemical Engineering, 2021.

or xanthan gum, facilitates the formation of aerogels or hybrid hydrogels with graphene oxide (Yu et al. 2013, Sahraei and Ghaemy 2017). These materials can be obtained by mixing GO along with specific additives. At first, the whole set of materials is frozen. Then, under reduced pressure, water sublimates and leaves voids after the removed water crystallizes. This way, a 3D network of interconnected GO flakes is created, exhibiting a specific porosity (Shehzad et al. 2016). Alternatively, a 3D graphene-based structure may be formatted in other ways, including chemical vapor deposition (CVD), where a hard nickel or copper template is used, on which the growth of graphene layers proceeds. After removing the hard template matrix (etching), one obtains a 3D porous graphene matrix (Bhaviripudi et al. 2010, Shehzad et al. 2016). Metal oxides (such as ZnO or MgO) and salts may be successively used, too. (Kang et al. 2011, Li et al. 2013, Ning et al. 2014, Shi et al. 2016b). Shi et al. obtained graphene foams using the CVD method, where methane served as a carbon phase precursor, and hydrochloric acid was used to etch the remains of the template. Such foams had a macropore structure, in which pores of 300 nm diameter dominated, allowing effective adsorption of organic pollutants such as ethanol, n-hexane, gasoline, or various oils (Shi et al. 2016b). It can often be seen that the application of strong acids may have a negative impact on the structure and final adsorption properties by changing the specific surface area and reducing the total pore size (Yang et al. 2019). The oxidized graphene structure mainly interacts through carboxyl groups (–COOH) with metal ions such as Cr(VI), Cu(II), Zn(II), and U(VI) because of its chelating properties and strong electrostatic attraction (Cho et al. 2010, Park et al. 2011, Yan et al. 2016, Zhao et al. 2019). Zhao et al. (Zhao et al. 2019) obtained graphene oxide sponges by a modified Hummers method and then treated them with chloroacetic acid to introduce carboxyl groups to the surface. The freeze-drying process was used to obtain a 3D structure. The resulting material was tested for soil and water remediation. It was shown that the epoxy and the hydroxyl groups did not interact and showed weaker binding to copper ions relative to the carboxyl methyl group. The investigated Cu(II) adsorption capacity was 93.8 mg g^{-1}. Other ions to be reduced from aqueous media are Co(II) or Cd(II) ions. The first report on the adsorption of these metal ions with graphene oxide was demonstrated by Zhao et al. (Zhao et al. 2011). At a temperature of 30°C and a pH of 6.0, the sorption capacities for Co(II) and Cd(II) were 68.2 mg g^{-1} and 106.3 mg g^{-1}, respectively. The good sorption capacities of the graphene materials are not only obtained by numerous oxygen functional groups on the carbon surface but also by the separated graphene sheets with a gap of 0.87 nm or even 0.96 nm, thanks to which metal particles can perfectly penetrate the graphene spaces (McAllister et al. 2007, Fan et al. 2008, Zhao et al. 2011, Qi et al. 2018). Wang et al. (Wang et al. 2018) showed that the oxygen-doped material, compared to the nitrogen-doped material, exhibited better sorption properties towards the Cu(II) ion, for GO, the value was 43.6 mg g^{-1}. In comparison, for nitrogen-doped graphene oxide, the value was 27.3 mg g^{-1}. In a research paper, Fang et al. investigated the effect of the type of oxygen groups on the surface of porous materials on the adsorption of heavy metals, mainly Pb(II) (Fang et al. 2017a).

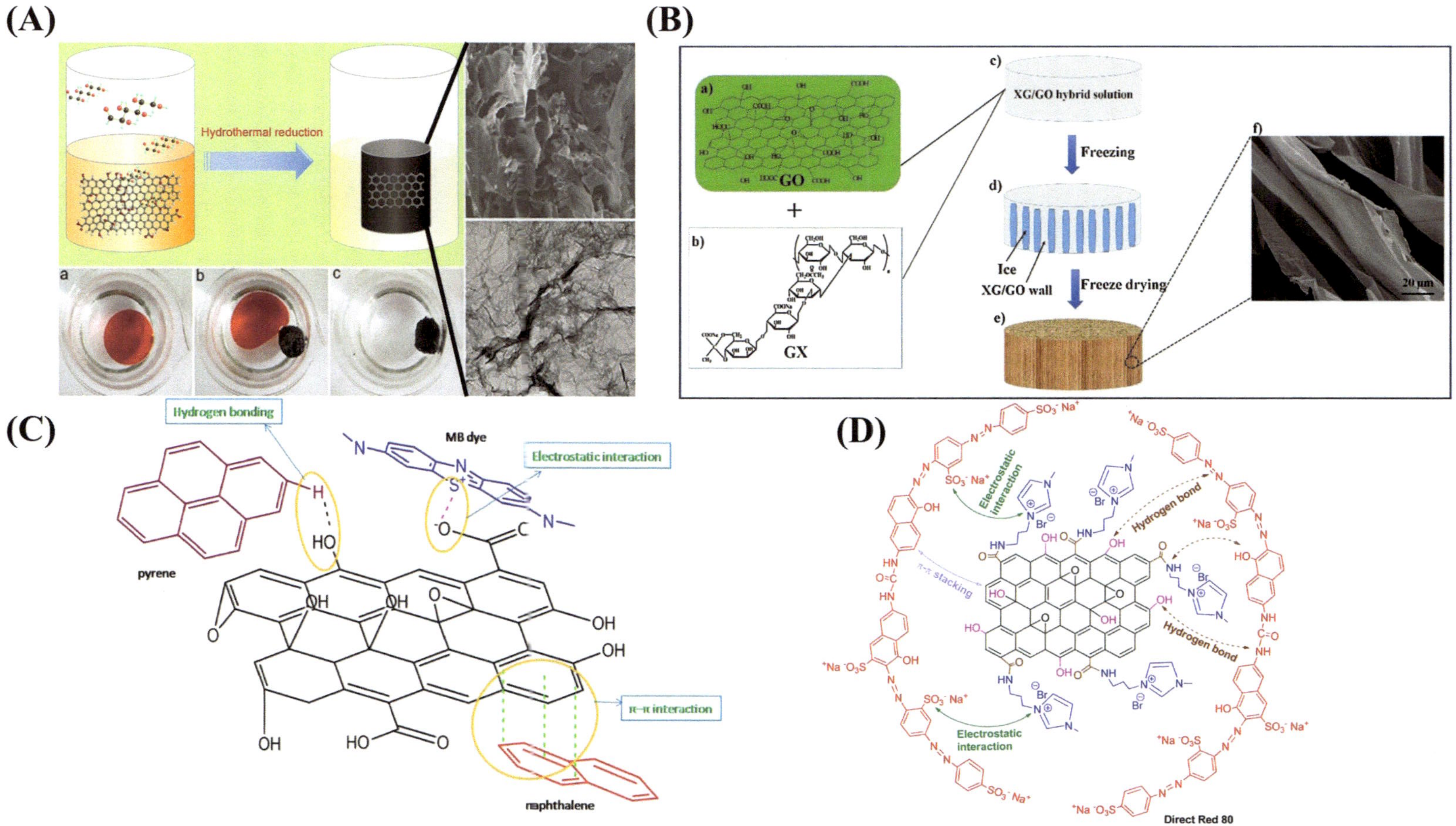

Figure 8.2 **(A)** Schematic of one-step hydrothermal synthesis of obtaining graphene foams, along with SEM images and the process of adsorption of dodecane over time (Wu et al. 2016), with permission from Applied Surface Science, 2016. **(B)** Steps of green synthesis of graphene oxide-based hybrid aerogels along with the structure and SEM image of the obtained graphene materials (Liu et al. 2017a), with permission from Carbohydrate Polymers, 2017. **(C)** Interactions occurring between graphene oxide and impurities (Thakur and Kandasubramanian, 2019), with permission from American Chemical Society, 2019. **(D)** The adsorption process of DR80 on a sponge made of modified graphene oxide (Zambare et al. 2017), with permission from American Chemical Society, 2017.

They investigated two types of oxygen groups, hydroxyl and carboxyl, on the adsorption process of Pb^{2+} ions. They showed that the specific surface area does not have a decisive effect on the adsorption of heavy metals; however, pore size favors better penetration to active sites where metal ions can be anchored. The main mechanism of adsorption of metal ions on graphene surfaces is due to electrostatic attraction, ion exchange, and complexation of heavy metals where the oxidized graphene surface can occur (Xu and Wang 2017, Duan et al. 2020). The amount of oxygen functional groups and their type determines the effective mechanism and formation of active adsorption sites. The appropriate oxygen content on the surface of the graphene material and the electrostatic attraction or chelating properties can influence the final removal of heavy metals from water. Complexation on the graphene oxide surface or electrostatic attraction and a large surface area result in an efficient adsorption process of metal ions (Wu et al. 2013). The adsorption mechanism may be due to several factors; it can be slightly different for heavy metals or organic dyes. The mechanism can also be determined by the extensive porous structure of graphene materials exposing the active sites of oxygen functional groups, providing access and the possibility to attach contaminants, thus removing them from the aqueous environment (Wei et al. 2017). The oxygen functional groups result in negatively charged active sites, which trap positively charged domains in dye molecules or metal cations, connecting through π-π interactions between the dye rings and the graphene structure, electrostatic interactions, and hydrogen bonds (Figures 8.2C and 8.2D) (Zambare et al. 2017).

As shown in Figure 8.2D the adsorption mechanism is complex, and several interactions may occur on the surface of the adsorbent and adsorbate. In low pH environments, hydrogen bonds can play a visible role, which has a beneficial effect on sorption processes. At high pH values, on the other hand, the oxygen functional groups are deprotonated and then, the π-π interactions between graphene planes and the aromatic rings of the impurities, may have a dominant effect on adsorption. The π-π aromatic ring stacking of the adsorbent and the adsorbate does not depend on the surface charge of the graphene materials, so the graphene ring system couples to the aromatic ring of the dyes in different pH environments (Lai et al. 2019). Given the multiple possibilities and the synergistic effect of interactions at the graphene-pollutant interface, graphene materials are favorable during water treatment.

The acidic or alkaline environment in which dyes and contaminants can be absorbed is a key factor in effective sorption on graphene materials (Thakur and Kandasubramanian 2019). Organic dyes can be divided into two categories: acidic and basic, otherwise known as cationic and anionic dyes, respectively. They are usually water-soluble, which does not necessarily pose much of a threat, but they also contain toxic, non-biodegradable, water-insoluble dyes. Therefore, the adsorbents used are targeted at the specific type of adsorbent and, more precisely, their charge properties. At high pH, the highest adsorption is observed for anionic dyes, while at low pH, cationic dyes have the highest adsorption value. At high pH deprotonation process of oxygen functional groups occurs, while at low pH the formation of the $-OH^{2+}$ group is possible, thus increasing the electrostatic attraction of anionic dyes to the surface of graphene materials (Lai et al. 2019). Yan et al. synthesized a material comprising GO and chitosan, which showed effective adsorption by removing Cu(II) and Cr(VI) from water and the cationic dyes of methylene blue (MB), and the anionic dyes of methyl orange (MO) (Yan et al. 2016). They proved that all impurities in the MB and Cu(II) configuration, as well as MO and Cr(VI) under appropriate pH conditions can be removed by the synthesized graphene-based material due to its amphoteric properties.

As expected, pH value above 6.0 causes the precipitation of Cu(II) ions. Below this value, copper ions are more effectively complexed on the surface of graphene oxide containing carboxyl and hydroxyl functional groups. During copper complexation on the surface, deprotonation of the –COOH and –OH groups may occur, which promotes the adsorption process of metal ions. Thus, pH change is the reason for oxygen functional group charging (Wu et al. 2013). Summary of pollutant kind and removal conditions from water by oxygen-doped graphene is presented in Table 8.1.

Table 8.1 Summary of oxygen-doped graphene parameters for water pollutants

Name of graphene materials	Type of oxygen functional group	Hetero-atom content	S_{BET}[1] ($m^2 g^{-1}$)	Kind of pollutants	Q_{max} ($mg g^{-1}$)	pH adsor-ption	Ref.
GO	OCH_2COOH	31,0% at.	—	Cu(II)	93.8	4–6	Zhao et al. 2019.
GO	C–O, C=O, O–C=O,	18.6% at. 9.8% at. 0.2% at	—	Cd(II) Co(II)	106.3 68.2	6	Zhao et al. 2011
GS	–OH –COOH	14.6% at.	—	dodecane crude oil machine oil	23–35g/g	—	Wu et al. 2016
XG/GO	–OH –COOH	—	—	Rh B[1] MB[2] Cu(II)	244.36 290.57 53.2	—	Liu et al. 2017a
CS-GO	–OH –COOH	—	—	MB, MO[3], Cu(II), Cr(IV)	1.60mmol/g 0.80mmol/g 1.34mmol/g 0.55mmol/g	5	Yan et al. 2016
GO MGO[5]	–OH, C=O, –COOH	—	—	Co(II	43.6 23.2	5	Wang et al. 2018
Gl-RGO, La-RGO, Su-RGO	Hydroxyl Lactones Carboxyl Hydrogen bonded carboxyl	—	22.18–28.52	Pb(II)	101.1 86.81 42.00	2–7	Fang et al. 2017a
GO	–OH –COOH	—	—	Cu(II)	10.75- 117.5	1–5.3	Wu et al. 2013
GO/MCC[6]	–OH	34.59% wt. in GO 0.3 do 14% wt. GO	30.5–43.7	MB	2630	—	Wei et al. 2017
GO sponge	C–O C=O	—	—	DR80[7]	97.9	2	Zambare et al. 2017
GCAMs[8]	C–O C=O (C–N)	—	—	Pb(II), Cu(II), Cr(VI), MB, Rh B, MO, EY[9], phenol	747.5 457.5 292.8 584.6 492.8 189.4 124.8 9.5–73.1	5.0 4.9 7.0 6.0 6.5 7.5 7 6	Yu et al. 2017
GO/PVA	–OH –COOH	—	—	CR[10] MB MG[11],NR[12], ABK[13], ES[14] ethanol, cyclohexane, DMF[15] dichloroethane	CR: 11–33 MB: 4–24	—	Dai et al. 2016

[1]Rh B – rhodamine B, [2]MB – methylene blue, [3]MO – methyl orange, [5]MGO – magnetic graphene oxide, [6]MCC – Microcrystalline cellulose, [7] DR80 – direct red, [8]GCAMs – graphene oxide aerogel microsphere, [9]EY – Eosin Y, [10]CR – Congo red, [11]MG – Malachite green, [12]NR – Neutral red, [13]ABK – Acid chrome blue k, [14]ES – Eosin, [15]DMF – N, N-dimethylformamide.

8.2.2 Nitrogen Functional Groups

Given the favorable adsorption properties of nitrogen, the researchers paid particular attention to introducing nitrogen atoms to the graphene structure. Nitrogen functional groups bonded to graphene sheets (potentially interacting with various water contaminants) are presented in Figure 8.3 (Lobato-Peralta et al. 2021). Setshedi et al. demonstrated how graphene oxide might be converted into nitrogen-doped graphene (Setshedi et al. 2015). Polypyrrole monomer, as a nitrogen precursor was used along with sonication to mix it with water-dispersed graphene oxide. Then, polymerization was run for 3 h with continuous stirring. The final product was washed, filtered, and dried under a vacuum. The resulting material showed Cr(VI) adsorption capacity of 99.98–53.27% in the pH range of 2–9. Below pH of 6.2, the surface of Py-GO NC was charged positively due to protonation of the embedded nitrogen atoms in the GO structure. The use of polyethylenimine (PEI) may have a similar effect, which can also effectively introduce $-H_2$ groups (Wang et al. 2017). Another method of introducing nitrogen functional groups is the treatment of carbon materials with nitrogen plasma, as Soudani et al. (Soudani et al. 2017), incorporating 6.1% at. into the carbon structure in the form of pyridine structure, amides, amines, and quaternary nitrogen. It is also possible to incorporate amino groups into the carbon structure by acid treatment, using mixtures of HNO_3 and H_2SO_4 acids (Ohashi et al. 2008). It is also worth noting that amino groups and triazoles can interact easily with impurities to form binding sites (Ding et al. 2019, Masís-Mora et al. 2019, Ouyang et al. 2020).

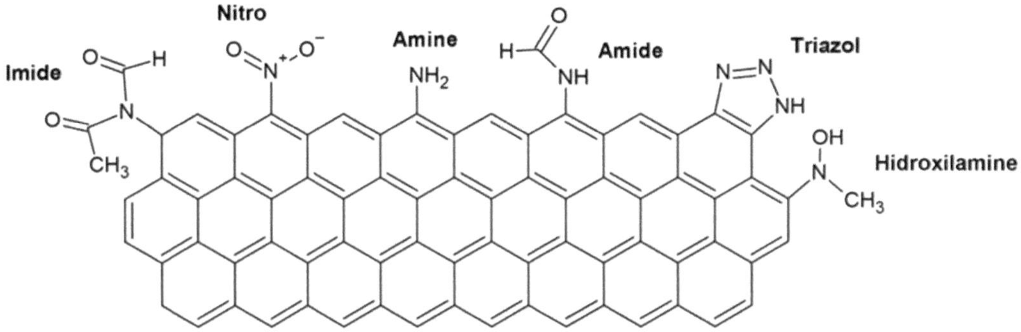

Figure 8.3 Examples of embedded nitrogen functional groups in the structure of activated carbon (Lobato-Peralta et al. 2021), with permission from the Journal of Environmental Chemical Engineering, 2021.

Furthermore, the introduction of pyridine-N, pyrrole-N, and quaternary-N functional groups changes the properties of graphene (Seredych et al. 2008, Shen et al. 2021). These groups can be formed by the interaction of appropriate nitrogen precursors on the graphene surface, e.g., melamine (Yokwana et al. 2018a, b, Wang et al. 2019), urea (Tabani et al. 2021), ammonia (Jiang et al. 2017), or aliphatic amines (Jiang et al. 2019a), pyrrole (Rahmani et al. 2018), hydrazine (Yang et al. 2016) or at least chitosan (Yuan et al. 2016, Qi et al. 2018). Jiang et al. (Jiang et al. 2019a) applied a primary-substituted amine to form pyridine groups on the graphene surface, while pyrrole groups were obtained by using a secondary amine. These pyridine and pyrrole functional groups are considered catalytically active sites, particularly when located at graphene plane edges. Moreover, due to free electron pairs nitrogen atoms can coordinatively bind some contaminants (Rahmani et al. 2018). Well-developed 3D structure of graphene-based materials facilitates the diffusion of impurities (Lai et al. 2019). Fang et al. (Fang et al. 2017b) carefully analyzed the effect of a graphene structure on the adsorption of organic contaminants with particular focus on methylene blue and rhodamine B dyes. The synthesized material showed a high adsorption affinity towards the organic impurities better than a nitrogen-free material. The presence of nitrogen in the graphene-based structures causes intensive π–π interactions between aromatic carbon rings and specific organic molecules (Fang et al. 2017b). Using phenol as an

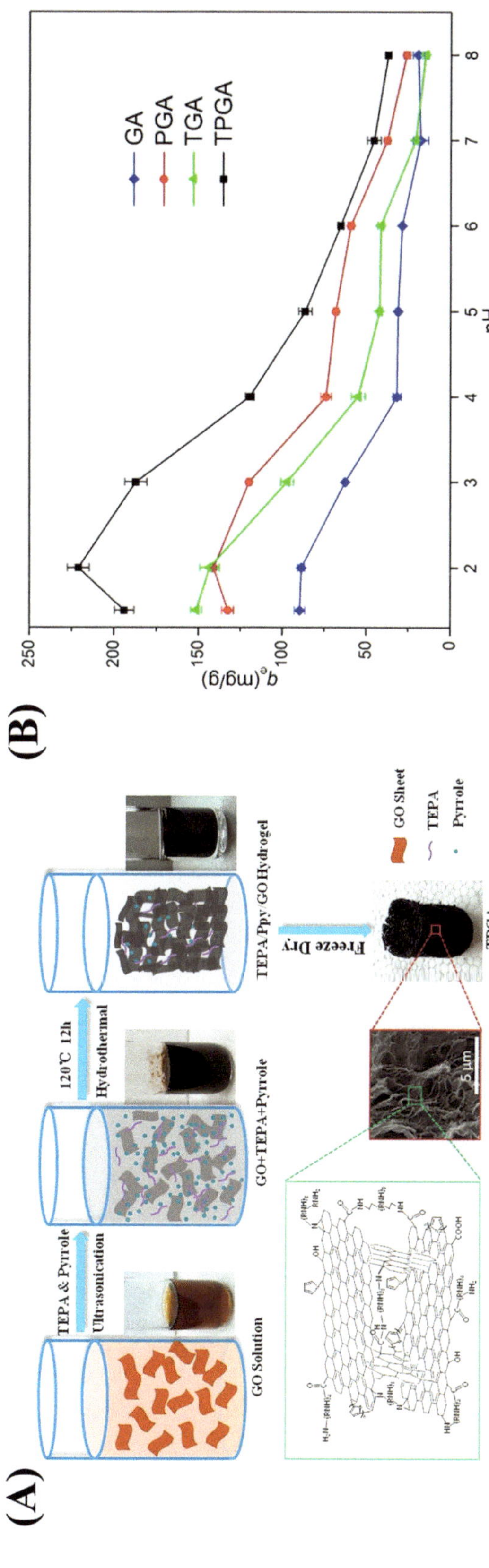

Figure 8.4 (A) Synthesis steps for synthesis of nitrogen-doped graphene foams, (B) dependence of pH value on adsorption capacity. (Liang et al. 2018) with permission from Chemical Engineering Journal, 2018.

Table 8.2 Different contaminants adsorption on nitrogen-doped graphene.

Name of graphene materials	Type of nitrogen functional group	Heteroatom content	S_{BET}[1] ($m^2\ g^{-1}$)	Kind of pollutants	Q_{max} ($mg\ g^{-1}$)	pH adsorption	Ref.
NSGAs[1]	N pyridinic N pyrrolic N substitutional	0.71–1.66% at.	107–132	AO[2]	87%	4-9	Jiang et al. 2019b
3DGON[3] 3DGOSN[4]	N pyridinic N amino N pyrrolitic N graphitic oxidized N	31.9% at. 42.1% at. 16.2% at. 4.1% at. 5.7% at.	—	Hg(II)	4515 $\mu g\ g^{-1}$	3–9	Henriques et al. 2016
NGO[5] NMG[6]	C=N C–N	—	—	Co(II)	27.3 14.6	5	Wang et al. 2018
mim GO sponge	C–N, N–C=O	3.44% at.	31.71	DR80	501.3	2	Zambare et al. 2017
NGA/PMS[12]	N Pyridinic N pyrrolic N graphitic	3.23–4.76% at.	393–420	IBP[13]	4.7–12.5%	—	Wang et al. 2019 Jiang et al. 2017
NGA–EDA[7] NGA–DETA[8] NGA–TEPA[9]	Pyridinic N Pyrrolic N Quaternary N	8.22–13.75% at.	—	AO	95%	4–9	Jiang et al. 2019a
NGA	N pyridinic N pyrrolic/pyridine amine or N graphitic	4.16% at.	340	Crude oil	210	—	Rahmani et al. 2018
N–GNSs–F[10]	N Pyridinic N Pyrrolic N Formamide	3–3.17% at.	92.1	MB, RhB	439.9 317.6	—	Fang et al. 2017b
3DNG[11]	N pyridinic N pyrrolitic N graphitic oxidized N	29.05% at. 22.76% at. 33.57% at. 14.62% at.	536	MB	20–53	—	Maouche et al. 2020
N-rGO/PMS	N pyridinic N pyrrolic N graphitic oxidized N	6.2% at.		Phenol SCP[14]	—	3–12	Kang et al. 2020

Name of graphene materials	Type of nitrogen functional group	Heteroatom content	S_{BET}[1] ($m^2 g^{-1}$)	Kind of pollutants	Q_{max} ($mg g^{-1}$)	pH adsorption	Ref.
NG	N pyridinic N pyrrolitic N graphitic	38.4% at. 31.8% at. 25.9% at.	8–515	SAM[15]	—		Chen et al. 2018b
N–GO–x	N pyridinic N pyrrolic N graphitic	N/C= 0.06–0.08	—	Boron	58.70	8.5	Chen et al. 2017
NGOA NGA	N pyridinic N pyrrolic Pyridine Noxide	4.3–4.8	651–852	n-hexane, THF[16], chlorobenzenexylene, tetrachloromethane, NMP[17]	41.6–68.6 49.5–71.4 62.1–83.9 46.2–52.6 82.4–111.6 42.7–60.2	—	Ren et al. 2016
TPGA[18]	–NH(R) N pyrrolic O=C–N $-NH_2/NH_3^+$	—	—	Cr(VI)	408.48	1.5–3	Liang et al. 2018
GOPEI[19]	C–N–C N–C=O $C-NH_2$	41.64% at	357.67	As(V) As(III)	4.80 4.26	4 7	Singh et al. 2016
$GO-NH_2$	C–N	3.7% at.	308–320	Co(II)	116.35	6–7	Fang et al. 2014
CNG[20]	–OH, –COOH	21.8% at.	695	Pb(II) Cd(II)	521 498	-	Liu et al. 2017a
PAH–GO[21]	N–H C–N	9.99% at.	172.24	Cu(II)	349.03	6	Xing et al. 2015

[1]NSGAs–N/S codoped graphene aerogels, [2]AO–acridine orange, 3DGO–3-dimensional graphene oxide [3]N-doped, [4]NS-doped, [5]NGO–N-doped GO, [6]NMGO–Ndoped MGO, NGAs–N-doped graphene aerogels from [7]EDTA–ethylenediamine, [8]DETA–diethylenetriamine, [9]TEPA–tetraethylenepentamine, [10]NGNSsF–Ndoped 3D Graphene nanoscrollbased foam, [11]3DNG–3D nitrogen-doped graphene aerogel, [12]PMS–peroxymonosulfate, [13]IBP–ibuprofen, [14]SCP–sulfachlorpyridazine, [15]SAM–sulfacetamide,[16]THF–tetrahydrofuran,[17]NMP–N-methyl-2-pyrrolidone,[18]TPGA–Tetraethylenetetramine/polypyrrole/GO aerogel, [19]GOPEI–GO/polyethyleneimine, [20]CNG–crumpled N-doped graphene, [21]PAH-GO–composite poly(allylamine hydrochloride) and GO.

example, it can be deduced that because of the donor heteroatoms, aromatic compounds' adsorption increases, resulting in an increase in π–π interactions (Soudani et al. 2017). Kim et al. (Kim et al. 2015) compared the behavior of two dyes: a cationic dye—Methylene blue, and an anionic dye-Acid Red 1. It was proven that hydrogen bonds were also involved during adsorption besides π–π stacking (dye molecule—reduced graphene oxide). Better accessibility to active sites and defects (like heteroatoms) resulting from the presence of a properly tailored structure is the source of a synergistic effect (Gogoi et al. 2019, Maouche et al. 2020). Nitrogen atom doping of graphene can promote the adsorption processes of pollutants and become a catalyst for activating materials such as peroxymonosulfate, which can effectively degrade pharmaceuticals in water (Chen et al. 2018b, Wang et al. 2019, Kang et al. 2020, Li et al. 2022). Nitrogen-doped materials can also effectively remove boron from seawater through the increased surface charge of graphene-based structures (Chen et al. 2017). In such a case, the sorption capacity depends, as mentioned in the previous section, on pH, pore volume, and hydrophobic properties. The expanded structure and large pore volume, which contribute to the increase of active sites available for organic solvents, favored the removal of organic solvents such as n-hexane xylene, chlorobenzene, or tetrahydrofuran from water. Liang et al. (Liang et al. 2018) synthesized a nitrogen-doped 3D graphene material for Cr(VI) adsorption in aqueous media using a one-step method. Pyrrole and tetraethylenetetramine were used as nitrogen precursors to introduce amine functional groups. A diagram of this synthesis is shown in Figure 8.4, along with the effect of pH on the adsorption capacity of the obtained materials. Additionally, pyrrole and tetraethylenetetramine perform the function of crosslinking and stabilizing the distribution of graphene oxide sheets in three-dimensional space.

The use of polyethylenimine, on the other hand, changes the charge of the graphene surface, making the obtained material an effective adsorbent for the removal of As(V) and As(III) ions from the aqueous environment (Singh et al. 2016). The resulting nitrogen and oxygen co-doped materials are also used as membranes for heavy metal removal (Fang et al. 2014). The combination of a nitrogen atom and a large surface area of the material's proprietary surface with a sufficiently large pore volume can promote excellent diffusion of heavy metal ions and create an accessible space to the active site (Liu et al. 2017b). Nitrogen functional groups, such as $-NH_2$ and $-NH$ and embedded oxygen atoms, can effectively contribute to the adsorption of heavy metal ions due to a synergic effect (Xing et al. 2015). At low pH values, the concentration of adsorbed hydrogen ions increases and charges the surface positively, which may be the reason for hindered adsorption of metal cations. Hydrogen cations compete with metal ions during adsorption on nitrogen surface species by blocking the electron pairs in $-NH_2$ and $-NH$ (nitrogen groups protonation). Therefore, at low pH, copper (II) metal ions are rather repelled from the positively charged surface (Xing et al. 2015). The electron pair at the nitrogen atoms in functional groups on graphene surface is responsible for the complexation of metal ions but also causes the increase of hydrophilic properties, which can facilitate contact with metal ions (Yu et al. 2018). The influence of different factors on metal ion adsorption on nitrogen-modified graphene is presented in Table 8.2.

8.2.3 Sulfuration

By introducing the sulfur atoms into the three-dimensional graphene, it is possible to increase the molecular defects of the structure and contribute to the formation of additional adsorption sites. Sulfurization causes a change in the distribution of electrons, which directly affects the reactivity of the final material (Guo et al. 2020). Through the insertion of sulfur atoms, also hydrophilic properties are changed, which additionally affects the sorption properties of such obtained materials and the removal of contaminants from water. The sulfur atoms can be introduced in the form of several functional groups, as presented in Figure 8.5. Insertion of sulfonic groups ($-SO_3H$) makes the graphene surface negatively charged over a wide pH range from 2 to 10, which favorably affects the adsorption capacity of positively charged metal ions (Wu et al. 2015). The introduction of sulfur directly into the carbon rings is likely to distort the graphene structure due to the large size of the sulfur atom comparing with the carbon atom size. The long bond between sulfur and carbon can change the hybridization of neighboring carbon atoms from sp^2 to sp^3 (Wang et al. 2014). A wide range of silica templates and stabilizers are used to prevent the re-layering of graphene layers, which promotes the introduction of sulfur atoms throughout the carbon structure (Manna and Raj 2018) while creating an active site and a porous material that promotes electrolyte diffusion.

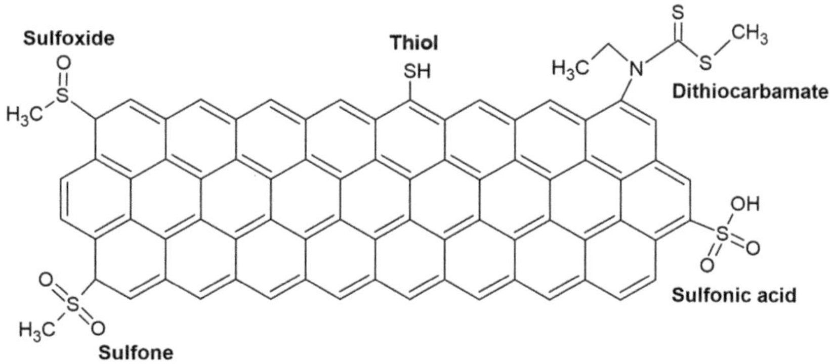

Figure 8.5 Examples of sulfur functional groups introduced in the structure of graphene (Lobato-Peralta et al. 2021), with permission from the Journal of Environmental Chemical Engineering, 2021.

It is worth noting that sulfur functional groups effectively remove several heavy metal cations from aqueous solutions: copper, lead, and cadmium but also mercury, zinc, or nickel (Pirveysian and Ghiaci 2018). Thiol groups have a higher affinity for the aforementioned metal ions because of their ability to form stable complexes together with heavy metals (Yang and Shao 2000, Kumar et al. 2016, Wei et al. 2018, Ahmad et al. 2020). These distinct functional groups have strong binding possibilities because, according to the Hard-Soft-Acid-Base (HSAB) theory, they are considered soft bases, while metal ions are classified as soft acids (Mohammadnia et al. 2019). Introducing sulfonyl groups into graphene is not a simple synthesis. Nevertheless, Zhang et al. (Zhang et al. 2016) carried out the synthesis in a single vessel using sulfonic acid (50% SO_3/H_2SO_4) and graphene oxide, into which sulfur functional groups were incorporated. They demonstrated excellent adsorption properties towards U(VI) ions from solutions and proved that the process was spontaneous and endothermic. Uranium removal from water compounds using non-sulfurized graphene oxide (obtained by modified Hummer's method) was less effective (adsorption capacity of 97.5 $mg\,g^{-1}$ (Zhao et al. 2012)) than for the analogous but sulfurized material (adsorption capacity of 309.09 $mg\,g^{-1}$). Heavy metal ions are not the only ions an S-modified graphene surface can capture in aqueous media. Typically, an organic cationic dye, methylene blue (MB) was used to determine the adsorption capacity. Shen et al. (Shen and Chen 2015) tested a series of graphene-originated materials: "pure" graphene oxide, reduced graphene oxide, and sulfonated graphene. They proved that S-heteroatoms embedded in the graphene sheets contribute to a higher sorption capacity than non-sulfurized materials. Manna et al. (Manna and Raj 2018) obtained a highly porous material doped with sulfur using BDS. They investigated its selective adsorption capacity towards Hg(II) ions in the presence of other heavy metal ions. They observed that as the pH increases, mercury ion adsorption increases from 4 to 7. Although the specific surface area of the undoped sulfur material (*p* rGO) was the same as for S-doped one (*p* S-rGO), the latter material showed a double absorption capacity. Then, they compared an S-doped porous material with its non-porous counterpart and concluded that the porous doped material showed one and a half times higher removal capacity of Hg(II) ions. Therefore, the porosity of graphene materials and the sulfur introduced into the carbon structure play a crucial role in removing heavy metal ions by avoiding ion migration problems. The kinetics of the process was studied, too. Thakur et al.

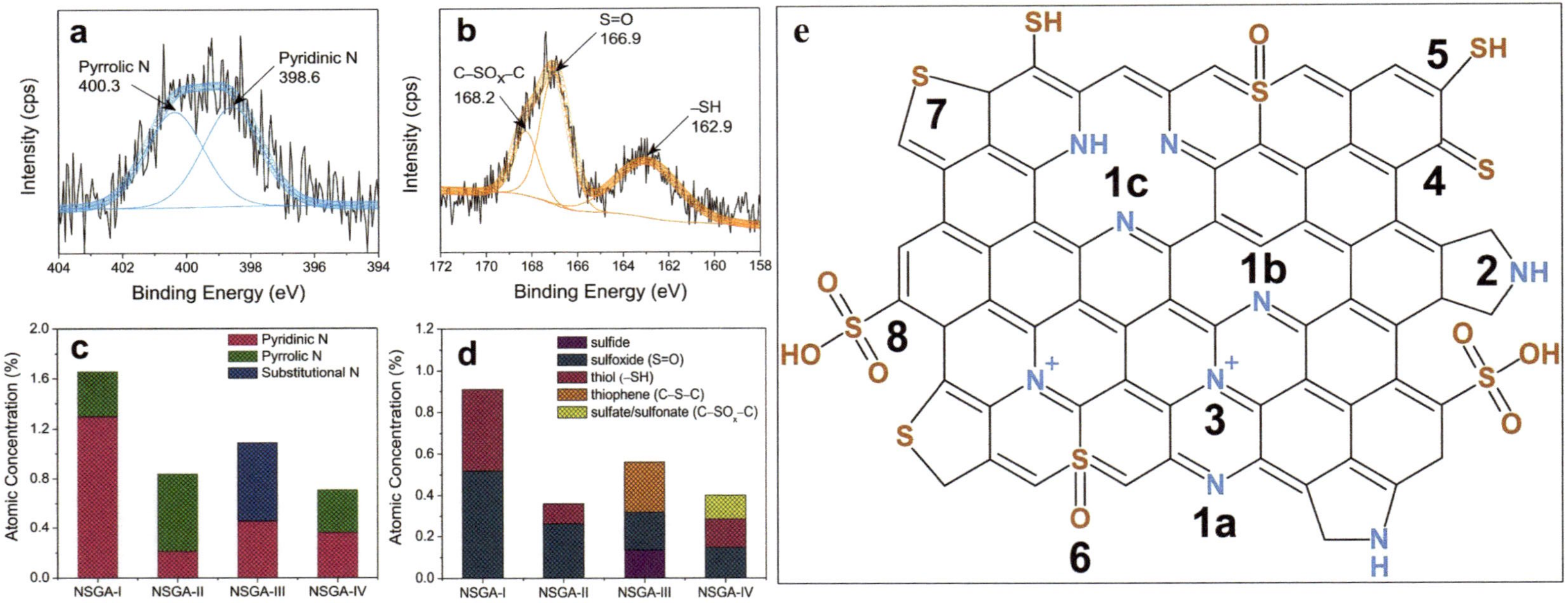

Figure 8.6 High-resolution XPS spectrum for **(a)** N1s, **(b)** S2p, distribution of individual, **(c)** nitrogen, **(d)** sulfur functional groups, **(e)** types of embedded nitrogen heteroatom functional groups: 1(a-c) and 2–N pyridine, 3–N substitution, and sulfur functional groups: 4–sulfide group, 5–thiol group, 6–sulfoxide, 7–thiophene groups, 8–sulfone groups (Jiang et al. 2019b) with permission from Environmental Pollution, 2019.

(Thakur et al. 2013) obtained sulfur-doped graphene using thiosulfation and graphene oxide with lime juice as the reducing agent. They studied the adsorption kinetics under standard conditions at neutral pH and ambient temperature. At first, the adsorption of Hg(II) ions, on the sulfur-doped material was at 90% of the final value within 15 minutes from the beginning of the conducted test. After that, the adsorption rate decreased gradually to reach an equilibrium state. After 30 minutes, Hg(II) ions were almost completely removed, proving rapid adsorption. The adsorption was fast not because of the interaction between Hg(II) ions and sulfur atoms embedded on the surfaces of the graphene layers but also by the electrostatic attraction of the metal ions to the entire surface of the graphene sheets.

Besides the positive influence of S only insertion, similarly positive results were achieved in the case of co-doping with two heteroatoms (Zheng et al. 2020). Typically, a synergistic effect may be expected in the case of dual doping. In most dual doping, sulfur atoms are inserted along with nitrogen atoms, as described by Jiang et al. (Jiang et al. 2019b). Nitrogen and sulfur functional groups were introduced via the hydrothermal method. Ammonium thiocyanate (NH_4SCN) was used as a precursor of sulfur and nitrogen. The groups incorporated into the graphene structure were characterized by XPS (Figures 8.6a and 8.6b). The XPS results indicated that the use of graphene oxide and NH4SCN resulted in the incorporation of nitrogen functional groups (Figures 8.6c and 8.6e) such as pyridine and pyrrole, and sulfur functional groups (Figures 8.6d and 8.6e) like sulfonate ($–SO_3^-$), oxidized sulfur bonds ($C–SO_x–C$), thiol (–SH), thiophene (C–S–C), and sulfoxide, (S=O). The materials have been used as effective oil (petroleum) adsorbents and organic solvents such as toluene-benzene and cyclohexane. Moreover, the materials were also tested as active compounds for photocatalytic degradation of AO dye and exhibited a high photocatalytic activity leading to the decolorizing of 95.29–99.23% of the solution within four hours (Jiang et al. 2019b).

Co-doping (nitrogen and sulfur) contributes to improved sorption and catalytic properties, and decrease the original hydrophobicity. Shi et al. (Shi et al. 2016a) performed the successful synthesis of porous graphene materials that contain both sulfur and nitrogen in their structure. The applied method relied on glutathione, which additionally served as a binding and reducing agent. Glutathione having amine groups ($–NH_2$) and sulfhydryl groups (–SH) can form hydrogen and covalent bonds between graphene sheets, and therefore individual graphene layers get separated. Re-agglomeration caused by π-π interactions is hindered (Chen et al. 2013). The nitrogen functional group of glutathione reacts with oxygen functional groups of graphene oxide, forming amide groups. In parallel SH groups of glutathione react with epoxide groups of graphene oxide, yielding covalent bonds interconnecting the latter material (Shi et al. 2016a). The contents of nitrogen and sulfur significantly affect the adsorption properties. Again, a high surface area exposing numerous active sites in the graphene sheets favors the adsorption of contaminants. The process of dual doping (nitrogen and sulfur) can be considered theoretically using density functional theory (DFT). Nitrogen atoms have an electronegativity (3.04) greater than carbon atoms (2.55), so the pyridine groups formed and graphitized N forms a positively charged adjacent carbon atom, promoting electron transfer (Duan et al. 2015, Tian et al. 2016). On the other hand, sulfur has a similar electronegativity (2.58) to carbon, effectively reducing the energy value between the molecular orbitals HOMO and LUMO and efficiently transforming the electron configuration. To improve the catalytic activity, adsorption capacity, and chemically inert graphene properties, the most important thing is the appropriate ratio of nitrogen, sulfur, and carbon. Too much doping and high functional group content can weaken the synergistic effect during the adsorption of impurities (Tian et al. 2016). The influence of different factors on the removal of contaminants is presented in Table 8.3.

Table 8.3 The influence of different factors on the contaminants capture by S-doped graphene.

Name of graphene materials	Type of oxygen functional group	Heteroatom content	S_{BET}[1] ($m^2\ g^{-1}$)	Kind of pollutants	Parameters of adsorption	pH adsorption	References
NSGAs[1]	S=O –SH C–S–C sulfide	0.40–0.91	107–132	AO	87%	4–9	Jiang et al. 2019b
3D-SRGO[2]	$-SO_3H$	3.00% at.	330.29	Cd(II)	234.8	6	Wu et al. 2015
*p*S-rGO[3]	–C–S–C– $-SO_x-$	9.96% at.	449.43	Hg(II)	829.27	7	Manna and Raj 2018
S–Chitin	–SH	1.6–2.1% iodimetry	—	Cu(II) Cd(II) Cr(III) Ni(II) Pb(II)	94.7 57.1 0.90 1.9 108.3	6–8	Yang and Shao 2000
MBT–GO[4]	C=N	5.48% at.	—	Hg(II)	107.52	5.4–6.9	Kumar et al. 2016
SGO[5]	$-SO_3H$	2.763% at.	—	MB Pb(II)	2530 415	—	Wei et al. 2018
GO/Fe_3O_4–Si–Pr–SH	–S–H	—	—	Hg(II)	129.7	2–8	Mohammadnia et al. 2019
GOS[6]	$-SO_3H$	4.33% at.	—	U(VI)	211.10	6	Zhang et al. 2016
GS[7]	$-SO_3H$	2.57% at.	616	PHE[8] MB Cd(II)	400 906 58	—	Shen and Chen 2015
PN S–CN/rGO	Ph-SO_3^-	0.28% at. S	188.5	RhB Cr(VI)	95% 73.9%	3	Zheng et al. 2020
GO–SH	–SH	—	—	PAT[9]	24.75 μg/mg	5–8	Liu et al. 2021

[1]NSGAs– N/S codoped graphene aerogels, [2]3D-SRGO– 3D sulfonated rGO, [3]pS-rGO–porous S–doped rGO, [4] MBT-GO– 2-Mercaptobenzothiazole, [5]SGO– sulfonated graphene oxide, [6]GOS–sulfonated GO, [7]GS– sulfonated graphene, [8]PHE–phenanthren, [9]PAT– patulin.

8.3 Conclusions and Perspectives

In conclusion, much research effort has been made towards obtaining effective adsorbents by heteroatom-doping of graphene and its derivatives in recent years. This is due to the enormous societal and environmental impact of the water purification issue. The quality and quantity of available drinking water has become a global problem, and many solutions are elaborated to provide adequate drinking water resources. Recently, the use of nanomaterials, among them graphene, has become attractive due to the possible surface modifications, which are changing the surface chemistry of graphene-based materials via embedded heteroatoms. However, problems are still encountered in modifying nanomaterials and producing 3D structures. Expectations for suitable materials for the degradation or removal of a broad range of contaminants, are still a challenge. The hydrothermal method combined with exploiting graphene oxide and the freeze-drying process has become an attractive alternative for 3D structuring of graphene. Such sheet assembly results in an extensive pore system, which can be penetrated by contaminants and finally adsorbed. The use of hard templates is also an effective method for obtaining 3D structures; however, it should be noted that residual templates must be removed by additional treatment like etching. It affects whole manufacturing, which may be unattractive economically. The problem may be solved by applying so-called soft templates, which are also effective for improving the volume and size of pores. Simultaneously, heteroatoms may be inserted. Thus, the whole structure is tuned to the needs of contaminants removal from water. So far, nitrogen, oxygen, and sulfur functional groups are recognized as adsorption centers of heavy metal ions, organic oils, inorganic oils, and antibiotics.

Heavy metal cations are effectively removed through graphene-based materials incorporating sulfur, nitrogen, or oxygen into their structure. In contrast, materials based on oxygen functional groups are tuned to purify water from heavy metals in an anionic form. Multielement doping of 3D structured graphene is possible, which positively affects hydrophilic properties. Some synergistic effects may occur like improved adsorption kinetics and increased potential adsorption site. The appropriate concentration and functional group types may improve some contaminants' adsorption capacity. However, an inadequate heteroatom ratio may block access to active sites and cause the agglomeration of graphene sheets. The environment in which adsorption occurs is also important. pH is one of the factors that determine the rate and adsorption capacity of 3D graphene materials. It is related to the charge of the pollutant and the charge of the surface on which the target pollutant is to be adsorbed. Anionic heavy metals, organic dyes, and oils can be effectively adsorbed in aqueous solutions of slightly acidic pH, while cationic contaminants are better adsorbed in neutral or slightly alkaline environments.

The problem to be solved is the selectivity of the adsorbents. Wastewater or processing water is contaminated with various pollutants: bacteria, nutrients. It is crucial for future 3D graphene materials to have the ability to capture specific contaminants or several contaminants at the same time.

References

Ahmad, S.Z.N., Salleh, W.N.W., Ismail, A.F., Yusof, N., Yusop, M.Z.M. and Aziz, F. 2020. Adsorptive removal of heavy metal ions using graphene-based nanomaterials: toxicity, roles of functional groups and mechanisms. Chemosphere. 248: 126008.

Allen, M.J., Tung, V.C. and Kaner, R.B. 2010. Honeycomb carbon: a review of graphene. Chem. Rev. 110(1): 132–145.

Bhaviripudi, S., Jia, X., Dresselhaus, M.S. and Kong, J. 2010. Role of kinetic factors in chemical vapor deposition synthesis of uniform large area graphene using copper catalyst. Nano Lett. 10(10): 4128–4133.

Carolin, C.F., Kumar, P.S., Saravanan, A., Joshiba, G.J., and Naushad, M. 2017. Efficient techniques for the removal of toxic heavy metals from aquatic environment: a review. J. Environ. Chem. Eng. 5(3): 2782–2799.

Chen, P., Yang, J.-J., Li, S.-S., Wang, Z., Xiao, T.-Y., Qian, Y.-H., and Yu, S.-H. 2013. Hydrothermal synthesis of macroscopic nitrogen-doped graphene hydrogels for ultrafast supercapacitor. Nano Energy. 2(2): 249–256.

Chen, F., Guo, L., Zhang, X., Leong, Z.Y., Yang, S. and Yang, H.Y. 2017. Nitrogen-doped graphene oxide for effectively removing boron ions from seawater. Nanoscale. 9(1): 326–333.

Chen, Y., Song, X., Zhao, T., Xiao, Y., Wang, Y. and Chen, X. 2018a. A phosphorylethanolamine-functionalized super-hydrophilic 3D graphene-based foam filter for water purification. J. Hazard. Mater. 343: 298–303.

Chen, X., Oh, W.-D., Hu, Z.-T., Sun, Y.-M., Webster, R.D., Li, S.-Z. and Lim, T.-T. 2018b. Enhancing sulfacetamide degradation by peroxymonosulfate activation with N-doped graphene produced through delicately-controlled nitrogen functionalization via tweaking thermal annealing processes. Appl. Catal., B: Environ. 225: 243–257.

Chen, B., Johnson, Z.T., Sanborn, D., Hjort, R.G., Garland, N.T., Soares, R.R. and Jing, D. 2021. Tuning the structure, conductivity, and wettability of laser-induced graphene for multiplexed open microfluidic environmental biosensing and energy storage devices. ACS Nano. 16(1): 15–28.

Cho, H.-H., Wepasnick, K., Smith, B.A., Bangash, F.K., Fairbrother, D.H. and Ball, W.P. 2010. Sorption of aqueous Zn[II] and Cd[II] by multiwall carbon nanotubes: the relative roles of oxygen-containing functional groups and graphenic carbon. Langmuir. 26(2): 967–981.

Choi, W., Lahiri, I., Seelaboyina, R. and Kang, Y.S. 2010. Synthesis of graphene and its applications: a review. Crit. Rev. Solid State Mater. Sci. 35(1): 52–71.

Choi, J.-S., Im, K.-S., Lee, T.-K., Choi, Y.-J. and An, S.-J. 2022. Electrical and structural properties of CVD-graphene oxidized using $KMnO_4/H_2SO_4$ Solution. Crystals. 12(4): 439.

Chowdhury, S. and Balasubramanian, R. 2014. Recent advances in the use of graphene-family nanoadsorbents for removal of toxic pollutants from wastewater. Adv. Colloid Interface Sci. 204: 35–56.

Dai, J., Huang, T., Tian, S.-Q., Xiao, Y.-J., Yang, J.-H., Zhang, N. and Zhou, Z.-W. 2016. High structure stability and outstanding adsorption performance of graphene oxide aerogel supported by polyvinyl alcohol for waste water treatment. Mater. Des. 107: 187–197.

Dieu Le, T. and Vinh Tran, H. 2022. Graphene oxide-based adsorbents for organic-dyes removal from contaminated water: a review. Zeitschrift für Anorganische und Allgemeine Chemie. 648(18): e202200140.

Ding, J., Yan, Z., Feng, L., Zhai, F., Chen, X., Xu, Y., and Pan, N. 2019. Benzotriazole decorated graphene oxide for efficient removal of U(VI). Environ. Pollut. 253: 221–230.

Duan, X., O'Donnell, K., Sun, H., Wang, Y. and Wang, S. 2015. Sulfur and nitrogen co-doped graphene for metal-free catalytic oxidation reactions. Small. 11(25): 3036–3044.

Duan, C., Ma, T., Wang, J. and Zhou, Y. 2020. Removal of heavy metals from aqueous solution using carbon-based adsorbents: a review. J. Water Process Eng. 37: 101339.

Fan, X., Peng, W., Li, Y., Li, X., Wang, S., Zhang, G. and Zhang, F. 2008. Deoxygenation of exfoliated graphite oxide under alkaline conditions: a green route to graphene preparation. Adv. Mater. 20(23): 4490–4493.

Fang, F., Kong, L., Huang, J., Wu, S., Zhang, K., Wang, X., and Huang, X.-J. 2014. Removal of cobalt ions from aqueous solution by an amination graphene oxide nanocomposite. J. Hazard. Mater. 270: 1–10.

Fang, Q., Zhou, X., Deng, W. and Liu, Z. 2017a. Hydroxyl-containing organic molecule induced self-assembly of porous graphene monoliths with high structural stability and recycle performance for heavy metal removal. Chem. Eng. J. 308: 1001–1009.

Fang, Q., Zhou, X., Deng, W., Liu, Y., Zheng, Z. and Liu, Z. 2017b. Nitrogen-doped graphene nanoscroll foam with high diffusion rate and binding affinity for removal of organic pollutants. Small. 13(14): 1603779.

Gogoi, J., Choudhury, A.D. and Chowdhury, D. 2019. Graphene oxide clay nanocomposite as an efficient photo-catalyst for degradation of cationic dye. Mater. Chem. Phys. 232: 438–445.

Guo, Z., Feng, Y., Zhang, C., Huang, G., Chi, J., Yao, Q., and Chen, X. 2020. Three dimensional graphene materials doped with heteroatoms for extraction and adsorption of environmental pollutants in wastewater. J. Environ. Sci. Health. Part C. 39(1): 17–43.

Han, T.-H., Kim, H., Kwon, S.-J. and Lee, T.-W. 2017. Graphene-based flexible electronic devices. Mater. Sci. Eng.: R: Reports. 118: 1–43.

Henriques, B., Gonçalves, G., Emami, N., Pereira, E., Vila, M. and Marques, P.A. 2016. Optimized graphene oxide foam with enhanced performance and high selectivity for mercury removal from water. J. Hazard. Mater. 301: 453–461.

Hiew, B.Y.Z., Lee, L.Y., Lee, X.J., Thangalazhy-Gopakumar, S., Gan, S., Lim, S.S. and Khiew, P.S. 2018). Review on synthesis of 3D graphene-based configurations and their adsorption performance for hazardous water pollutants. Process Saf. Environ. Prot. 116: 262–286.

Hossain, M.F., Akther, N. and Zhou, Y. 2020. Recent advancements in graphene adsorbents for wastewater treatment: current status and challenges. Chin. Chem. Lett. 31(10): 2525–2538.

Huang, J.Y., Ding, F., Yakobson, B.I., Lu, P., Qi, L. and Li, J. 2009. *In situ* observation of graphene sublimation and multi-layer edge reconstructions. PNAS. 106(25): 10103–10108.

Jiang, Y., Chowdhury, S. and Balasubramanian, R. 2017. Nitrogen-doped graphene hydrogels as potential adsorbents and photocatalysts for environmental remediation. Chem. Eng. J. 327: 751–763.

Jiang, Y., Chowdhury, S. and Balasubramanian, R. 2019a. New insights into the role of nitrogen-bonding configurations in enhancing the photocatalytic activity of nitrogen-doped graphene aerogels. J. Colloid Interface Sci. 534: 574–585.

Jiang, Y., Chowdhury, S. and Balasubramanian, R. 2019b. Nitrogen and sulfur codoped graphene aerogels as absorbents and visible light-active photocatalysts for environmental remediation applications. Environ. Pollut. 251: 344–353.

Kang, C.G., Kang, J.W., Lee, S.K., Lee, S.Y., Cho, C.H., Hwang, H.J. and Yang, H. 2011. Characteristics of CVD graphene nanoribbon formed by a ZnO nanowire hardmask. Nanotechnology. 22(29): 295201.

Kang, J., Zhou, L., Duan, X., Sun, H. and Wang, S. 2020. Catalytic degradation of antibiotics by metal-free catalysis over nitrogen-doped graphene. Catalysis Today. 357: 341–349.

Kim, H., Kang, S.-O., Park, S. and Park, H.S. 2015. Adsorption isotherms and kinetics of cationic and anionic dyes on three-dimensional reduced graphene oxide macrostructure. J. Ind. Eng. Chem. 21: 1191–1196.

Kim, S., Park, C.M., Jang, M., Son, A., Her, N., Yu, M. and Yoon, Y. 2018. Aqueous removal of inorganic and organic contaminants by graphene-based nanoadsorbents: a review. Chemosphere. 212: 1104–1124.

Kumar, A.S.K., Jiang, S.-J. and Tseng, W.-L. 2016. Facile synthesis and characterization of thiol-functionalized graphene oxide as effective adsorbent for Hg(II). J. Environ. Chem. Eng. 4(2): 2052–2065.

Lai, K.C., Lee, L.Y., Hiew, B.Y.Z., Thangalazhy-Gopakumar, S. and Gan, S. 2019. Environmental application of three-dimensional graphene materials as adsorbents for dyes and heavy metals: review on ice-templating method and adsorption mechanisms. J. Environ. Sci. 79: 174–199.

Li, W., Gao, S., Wu, L., Qiu, S., Guo, Y., Geng, X. and Gong, Y. 2013. High-density three-dimension graphene macroscopic objects for high-capacity removal of heavy metal ions. Sci. Rep. 3(1): 1–6.

Li, Y., Xie, M., Zhang, S., Zhao, L., Kong, L., Zhan, J. and Zhao, R.-S. 2022. Porous 3D superstructure of nitrogen doped carbon decorated with ultrafine cobalt nanodots as peroxymonosulfate activator for the degradation of sulfonamides. Chem. Eng. J. 428: 131329.

Liang, Q., Luo, H., Geng, J. and Chen, J. 2018. Facile one-pot preparation of nitrogen-doped ultra-light graphene oxide aerogel and its prominent adsorption performance of Cr(VI). AEJ. 338: 62–71.

Liu, S., Yao, F., Oderinde, O., Zhang, Z. and Fu, G. 2017a. Green synthesis of oriented xanthan gum–graphene oxide hybrid aerogels for water purification. Carbohydr. Polym. 174: 392–399.

Liu, L., Guo, X., Tallon, R., Huang, X. and Chen, J. 2017b. Highly porous N-doped graphene nanosheets for rapid removal of heavy metals from water by capacitive deionization. Chem. Commun. 53(5): 881–884.

Liu, X., Shi, L., Jiang, W., Zhang, J. and Huang, L. 2018. Taking full advantage of $KMnO_4$ in simplified Hummers method: a green and one pot process for the fabrication of alpha MnO_2 nanorods on graphene oxide. Chem. Eng. Sci. 192: 414–421.

Lobato-Peralta, D.R., Duque-Brito, E., Ayala-Cortes, A., Arias, D., Longoria, A., Cuentas-Gallegos, A.K. and Okoye, P.U. 2021. Advances in activated carbon modification, surface heteroatom configuration, reactor strategies, and regeneration methods for enhanced wastewater treatment. J. Environ. Chem. Eng. 9(4): 105626.

Manna, B. and Raj, C.R. 2018. Nanostructured sulfur-doped porous reduced graphene oxide for the ultrasensitive electrochemical detection and efficient removal of Hg(II). ACS Sustainable Chem. Eng. 6(5): 6175–6182.

Maouche, C., Zhou, Y., Peng, J., Wang, S., Sun, X., Rahman, N. and Yang, J. 2020. A 3D nitrogen-doped graphene aerogel for enhanced visible-light photocatalytic pollutant degradation and hydrogen evolution. RSC Advances. 10(21): 12423–12431.

Marcano, D.C., Kosynkin, D.V., Berlin, J.M., Sinitskii, A., Sun, Z., Slesarev, A. and Tour, J.M. 2010. Improved synthesis of graphene oxide. ACS Nano. 4(8): 4806–4814.

Masís-Mora, M., Lizano-Fallas, V., Tortella, G., Beita-Sandí, W. and Rodríguez-Rodríguez, C.E. 2019. Removal of triazines, triazoles and organophophates in biomixtures and application of a biopurification system for the treatment of laboratory wastewaters. Chemosphere. 233: 733–743.

McAllister, M.J., Li, J.-L., Adamson, D.H., Schniepp, H.C., Abdala, A.A., Liu, J. and Prud'homme, R.K. 2007. Single sheet functionalized graphene by oxidation and thermal expansion of graphite. Chem. Mater. 19(18): 4396–4404.

Mohammadnia, E., Hadavifar, M. and Veisi, H. 2019. Kinetics and thermodynamics of mercury adsorption onto thiolated graphene oxide nanoparticles. Polyhedron. 173: 114139.

Nguyen, B.H. and Nguyen, V.H. 2016. Promising applications of graphene and graphene-based nanostructures. Adv. Nat. Sci.: Nanosci. Nanotechnol. 7(2): 023002.

Ning, J., Xu, X., Liu, C. and Fan, D. 2014. Three-dimensional multilevel porous thin graphite nano-superstructures for Ni(OH) 2-based energy storage devices. J. Mater. Chem. A. 2(38): 15768–15773.

Ohashi, M., Kapoor, M.P. and Inagaki, S. 2008. Chemical modification of crystal-like mesoporous phenylene-silica with amino group. Chem. Commun. (7): 841–843.

Ouyang, J., Zhou, L., Liu, Z., Heng, J.Y. and Chen, W. 2020. Biomass-derived activated carbons for the removal of pharmaceutical mircopollutants from wastewater: a review. Sep. Purif. Technol. 253: 117536.

Panicker, N.J., Das, J. and Sahu, P. 2021. Synthesis of highly oxidized graphene (HOG) by using HNO_3 and $KMnO_4$ as oxidizing agents. Mater. Today: Proc. 46: 6270–6274.

Park, M., Kim, B.-H., Kim, S., Han, D.-S., Kim, G. and Lee, K.-R. 2011. Improved binding between copper and carbon nanotubes in a composite using oxygen-containing functional groups. Carbon. 49(3): 811–818.

Pirveysian, M. and Ghiaci, M. 2018. Synthesis and characterization of sulfur functionalized graphene oxide nanosheets as efficient sorbent for removal of Pb^{2+}, Cd^{2+}, Ni^{2+} and Zn^{2+} ions from aqueous solution: a combined thermodynamic and kinetic studies. Appl. Surf. Sci. 428: 98–109.

Qi, C., Zhao, L., Lin, Y. and Wu, D. 2018. Graphene oxide/chitosan sponge as a novel filtering material for the removal of dye from water. J. Colloid Interface Sci. 517: 18–27.

Rahmani, Z., Rashidi, A.M., Samadi, M.T. and Rahmani, A.R. 2018. N-doped reduced graphene oxide aerogel for the selective adsorption of oil pollutants from water: Isotherm and kinetic study. J. Ind. Eng. Chem. 61: 416–426.

Ren, H., Shi, X., Zhu, J., Zhang, Y., Bi, Y. and Zhang, L. 2016. Facile synthesis of N-doped graphene aerogel and its application for organic solvent adsorption. J. Mater. Sci. 51(13): 6419–6427.

Ren, X., Feng, J., Si, P., Zhang, L., Lou, J. and Ci, L. 2018. Enhanced heterogeneous activation of peroxydisulfate by S, N co-doped graphene via controlling S, N functionalization for the catalytic decolorization of dyes in water. Chemosphere. 210: 120–128.

Sahraei, R. and Ghaemy, M. 2017. Synthesis of modified gum tragacanth/graphene oxide composite hydrogel for heavy metal ions removal and preparation of silver nanocomposite for antibacterial activity. Carbohydr. Polym. 157: 823–833.

Schwarzenbach, R.P., Egli, T., Hofstetter, T.B., Von Gunten, U. and Wehrli, B. 2010. Global water pollution and human health. Annu. Rev. Environ. Resour. 35(1): 109–136.

Seredych, M., Hulicova-Jurcakova, D., Lu, G.Q. and Bandosz, T.J. 2008. Surface functional groups of carbons and the effects of their chemical character, density and accessibility to ions on electrochemical performance. Carbon. 46(11): 1475–1488.

Setshedi, K.Z., Bhaumik, M., Onyango, M.S. and Maity, A. 2015. High-performance towards Cr(VI) removal using multi-active sites of polypyrrole–graphene oxide nanocomposites: batch and column studies. Chem. Eng. J. 262: 921–931.

Shehzad, K., Xu, Y., Gao, C. and Duan, X. 2016. Three-dimensional macro-structures of two-dimensional nanomaterials. Chem. Soc. Rev. 45(20): 5541–5588.

Shen, Y. and Chen, B. 2015. Sulfonated graphene nanosheets as a superb adsorbent for various environmental pollutants in water. Environ. Sci. Technol. 49(12): 7364–7372.

Shen, Q., Jiang, Z., Tian, X., Zhu, H. and Jiang, Z.-J. 2021. Effects of nitrogen-doping structural changes of spherical hollow graphene on the growth of MoS_{2+x} nanosheets and the enhanced hydrogen evolution reaction. J. Alloys Compd. 884: 161073.

Shi, K., Ren, M. and Zhitomirsky, I. 2014. Activated carbon-coated carbon nanotubes for energy storage in supercapacitors and capacitive water purification. ACS Sustainable Chem. Eng. 2(5): 1289–1298.

Shi, Y.-C., Wang, A.-J., Wu, X.-L., Chen, J.-R. and Feng, J.-J. 2016a. Green-assembly of three-dimensional porous graphene hydrogels for efficient removal of organic dyes. J. Colloid Interface Sci. 484: 254–262.

Shi, L., Chen, K., Du, R., Bachmatiuk, A., Rümmeli, M.H., Xie, K. and Liu, Z. 2016b. Scalable seashell-based chemical vapor deposition growth of three-dimensional graphene foams for oil–water separation. J. Am. Chem. Soc. 138(20): 6360–6363.

Shi, Y., Liu, G., Wang, L. and Zhang, H. 2019. Heteroatom-doped porous carbons from sucrose and phytic acid for adsorptive desulfurization and sulfamethoxazole removal: a comparison between aqueous and non-aqueous adsorption. J. Colloid Interface Sci. 557: 336–348.

Shiyanova, K.A., Gudkov, M.V., Rabchinskii, M.K., Sokura, L.A., Stolyarova, D.Y., Baidakova, M.V. and Komarov, I.A. 2021. Graphene oxide chemistry management via the use of $KMnO_4/K_2Cr_2O_7$ oxidizing agents. Nanomaterials. 11(4): 915.

Singh, D.K., Kumar, V., Singh, V.K. and Hasan, S.H. 2016. Modeling of adsorption behavior of the amine-rich GOPEI aerogel for the removal of As(III) and As(V) from aqueous media. RSC Ad. 6(61): 56684–56697.

Singh, M., Kaushal, S., Singh, P. and Sharma, J. 2018. Boron doped graphene oxide with enhanced photocatalytic activity for organic pollutants. J. Photochem. Photobiol., A: Chemistry. 364: 130–139.

Sivakumar, M., Widakdo, J., Hung, W.-S., Wang, C.-F., Hu, C.-C., Lee, K.-R. and Lai, J.-Y. 2022. Porous graphene nanoplatelets encompassed with nitrogen and sulfur group for heavy metal ions removal of adsorption and desorption from single or mixed aqueous solution. Sep. Purif. Technol. 288: 120485.

Soudani, N., Najar-Souissi, S., Abderkader-Fernandez, V.K. and Ouederni, A. 2017. Effects of nitrogen plasma treatment on the surface characteristics of olive stone-based activated carbon. Environ. Technol. 38(8): 956–966.

Sun, H., Wang, Y., Liu, S., Ge, L., Wang, L., Zhu, Z. and Wang, S. 2013. Facile synthesis of nitrogen doped reduced graphene oxide as a superior metal-free catalyst for oxidation. Chem. Commun. 49(85): 9914–9916.

Tabani, H., Bameri, A.E., Abedi, H., Hatefi, R., Gorjizadeh, A. and Moghaddam, A.Z. 2021. Introduction of nitrogen doped graphene nanosheets as efficient adsorbents for nitrate removal from aqueous samples. J. Environ. Health Sci. Eng. 19(2): 1875–1886.

Thakur, S., Das, G., Raul, P.K. and Karak, N. 2013. Green one-step approach to prepare sulfur/reduced graphene oxide nanohybrid for effective mercury ions removal. J. Phys. Chem. C. 117(15): 7636–7642.

Thakur, K. and Kandasubramanian, B. 2019. Graphene and graphene oxide-based composites for removal of organic pollutants: a review. J. Chem. Eng. Data. 64(3): 833–867.

Thines, R., Mubarak, N., Nizamuddin, S., Sahu, J., Abdullah, E. and Ganesan, P. 2017. Application potential of carbon nanomaterials in water and wastewater treatment: a review. J. Taiwan Inst. Chem. Eng. 72: 116–133.

Tian, W., Zhang, H., Duan, X., Sun, H., Tade, M.O., Ang, H.M. and Wang, S. 2016. Nitrogen-and sulfur-codoped hierarchically porous carbon for adsorptive and oxidative removal of pharmaceutical contaminants. ACS Appl. Mater. Interfaces. 8(11): 7184–7193.

Wang, X., Sun, G., Routh, P., Kim, D.-H., Huang, W. and Chen, P. 2014. Heteroatom-doped graphene materials: syntheses, properties and applications. Chem. Soc. Rev. 43(20): 7067–7098.

Wang, D., Zhang, G., Zhou, L., Wang, M., Cai, D. and Wu, Z. 2017. Synthesis of a multifunctional graphene oxide-based magnetic nanocomposite for efficient removal of Cr(VI). Langmuir. 33(28): 7007–7014.

Wang, X., Liu, Y., Pang, H., Yu, S., Ai, Y., Ma, X. and Wang, X. 2018. Effect of graphene oxide surface modification on the elimination of Co(II) from aqueous solutions. Chem. Eng. J. 344: 380–390.

Wang, J., Duan, X., Dong, Q., Meng, F., Tan, X., Liu, S. and Wang, S. 2019. Facile synthesis of N-doped 3D graphene aerogel and its excellent performance in catalytic degradation of antibiotic contaminants in water. Carbon. 144: 781–790.

Wang, H., Mi, X., Li, Y. and Zhan, S. 2020. 3D graphene-based macrostructures for water treatment. Adv. Mater. 32(3): 1806843.

Wei, X., Huang, T., Yang, J.-H., Zhang, N., Wang, Y. and Zhou, Z.-W. 2017. Green synthesis of hybrid graphene oxide/microcrystalline cellulose aerogels and their use as superabsorbents. J. Hazard. Mater. 335: 28–38.

Wei, M.-P., Chai, H., Cao, Y.-L. and Jia, D.-Z. 2018. Sulfonated graphene oxide as an adsorbent for removal of Pb^{2+} and methylene blue. J. Colloid Interface Sci. 524: 297–305.

Wu, W., Yang, Y., Zhou, H., Ye, T., Huang, Z., Liu, R. and Kuang, Y. 2013. Highly efficient removal of Cu(II) from aqueous solution by using graphene oxide. Water Air Soil Pollut. 224(1): 1–8.

Wu, S., Zhang, K., Wang, X., Jia, Y., Sun, B., Luo, T. and Shen, W. 2015. Enhanced adsorption of cadmium ions by 3D sulfonated reduced graphene oxide. Chem. Eng. J. 262: 1292–1302.

Wu, R., Yu, B., Liu, X., Li, H., Wang, W., Chen, L. and Yang, S.-T. 2016. One-pot hydrothermal preparation of graphene sponge for the removal of oils and organic solvents. Appl. Surf. Sci. 362: 56–62.

Xing, H.T., Chen, J.H., Sun, X., Huang, Y.H., Su, Z.B., Hu, S.R. and Wu, W.B. 2015. NH_2-rich polymer/graphene oxide use as a novel adsorbent for removal of Cu(II) from aqueous solution. Chem. Eng. J. 263: 280–289.

Xu, Y., Sheng, K., Li, C. and Shi, G. 2010. Self-assembled graphene hydrogel via a one-step hydrothermal process. ACS Nano. 4(7): 4324–4330.

Xu, L. and Wang, J. 2017. The application of graphene-based materials for the removal of heavy metals and radionuclides from water and wastewater. Crit. Rev. Env. Sci. Technol. 47(12): 1042–1105.

Yan, H., Yang, H., Li, A. and Cheng, R. 2016. pH-tunable surface charge of chitosan/graphene oxide composite adsorbent for efficient removal of multiple pollutants from water. Chem. Eng. J. 84: 1397–1405.

Yang, W., Gao, H., Zhao, Y., Bi, K. and Li, X. 2016. Facile preparation of nitrogen-doped graphene sponge as a highly efficient oil absorption material. Mater. Lett. 178: 95–99.

Yang, X., Wan, Y., Zheng, Y., He, F., Yu, Z., Huang, J. and Gao, B. 2019. Surface functional groups of carbon-based adsorbents and their roles in the removal of heavy metals from aqueous solutions: a critical review. Chem. Eng. J. 66: 608–621.

Yang, Y. and Shao, J. 2000. Synthesis of sulfhydryl chitin and its adsorption properties for heavy metal ions. J. Appl. Polym. Sci. 77(1): 151–155.

Yao, Y., Zhang, J., Gao, M., Yu, M., Hu, Y., Cheng, Z. and Wang, S. 2018. Activation of persulfates by catalytic nickel nanoparticles supported on N-doped carbon nanofibers for degradation of organic pollutants in water. J. Colloid Interface Sci.. 529: 100–110.

Yokwana, K., Kuvarega, A.T., Mhlanga, S.D. and Nxumalo, E.N. 2018. Mechanistic aspects for the removal of Congo red dye from aqueous media through adsorption over N-doped graphene oxide nanoadsorbents prepared from graphite flakes and powders. Phys. Chem. Earth, Parts A/B/C. 107: 58–70.

Yokwana, K., Ray, S.C., Khenfouch, M., Kuvarega, A.T., Mamba, B.B., Mhlanga, S.D. and Nxumalo, E.N. 2018. Facile synthesis of nitrogen doped graphene oxide from graphite flakes and powders: a comparison of their surface chemistry. J. Nanosci. Nanotechnol. 18(8): 5470–5484.

Yu, B., Xu, J., Liu, J.-H., Yang, S.-T., Luo, J., Zhou, Q. and Liu, Y. 2013. Adsorption behavior of copper ions on graphene oxide–chitosan aerogel. J. Environ. Chem. Eng. 1(4): 1044–1050.

Yu, W., Lian, F., Cui, G. and Liu, Z. 2018. N-doping effectively enhances the adsorption capacity of biochar for heavy metal ions from aqueous solution. Chemosphere. 193: 8–16.

Yu, W., Sisi, L., Haiyan, Y. and Jie, L. 2020. Progress in the functional modification of graphene/graphene oxide: A review. RSC Adv. 10(26): 15328–15345.

Yuan, H., Meng, L.-Y. and Park, S.-J. 2016. A review: synthesis and applications of graphene/chitosan nanocomposites. Carbon Lett. 17(1): 11–17.

Zambare, R., Song, X., Bhuvana, S., Antony Prince, J.S. and Nemade, P. 2017. Ultrafast dye removal using ionic liquid–graphene oxide sponge. ACS Sustainable Chem. Eng. 5(7): 6026–6035.

Zhang, Z.-B., Qiu, Y.-F., Dai, Y., Wang, P.-F., Gao, B., Dong, Z.-M. and Le, Z.-G. 2016. Synthesis and application of sulfonated graphene oxide for the adsorption of uranium (VI) from aqueous solutions. J. Radioanal. Nucl. Chem. 310(2): 547–557.

Zhao, G., Li, J., Ren, X., Chen, C. and Wang, X. 2011. Few-layered graphene oxide nanosheets as superior sorbents for heavy metal ion pollution management. Environ. Sci. Technol. 45(24): 10454–10462.

Zhao, G., Wen, T., Yang, X., Yang, S., Liao, J., Hu, J. and Wang, X. 2012. Preconcentration of U(vi) ions on few-layered graphene oxide nanosheets from aqueous solutions. Dalton Trans. 41(20): 6182–6188.

Zhao, L., Chen, J., Xiong, N., Bai, Y., Yilihamu, A., Ma, Q. and Yang, S.-T. 2019. Carboxylation as an effective approach to improve the adsorption performance of graphene materials for Cu^{2+} removal. Sci. Total Environ. 682: 591–600.

Zheng, Y., Liu, Y., Guo, X., Chen, Z., Zhang, W., Wang, Y. and Zhao, Y. 2020. Sulfur-doped g-C_3N_4/rGO porous nanosheets for highly efficient photocatalytic degradation of refractory contaminants. J. Mater. Sci. Technol. 41: 117–126.

Chapter 9

Biochar-based Nanocomposites: A Novel and Sustainable Solution to (Waste)Water Contamination

Abhishek Kumar[1]

[1] Department of Civil and Environmental Engineering, Birla Institute of Technology, Mesra, Ranchi-835215, Jharkhand, India

Wasim Akram Shaikh[*1,2] **and Chandra Mukherjee**[2]

[2] Department of Basic Sciences, School of Science and Technology, The Neotia University, Diamond Harbour Road, Sarisha-743368, 24 Parganas (South), West Bengal, India

Jayanta Kumar Biswas[3,4]

[3]Enviromicrobiology, Ecotoxicology and Ecotechnology Research Laboratory, Department of Ecological Studies, University of Kalyani, Kalyani-741235, Nadia, West Bengal, India

[4]International Centre for Ecological Engineering, University of Kalyani, Kalyani-741235, West Bengal, India

9.1 INTRODUCTION

Biochar is a carbon-rich, efficient adsorbent with abundant mesopores and/or micropores (Kumar et al. 2021a, 2022a, b). Its role as an adsorbent can further be increased manifold by doping of nanoparticles (NPs) on its surface, resulting in the formation of the biochar-based nanocomposite. NPs doping is one of the most efficient and effective methods of advanced hybrid modification of biochar, applied to increase the removal efficiency of biochar (Kumar et al. 2022c). Doping of nanoparticles on biochar surface could also increase the electrical conductivity of biochar thereby enhancing the transfer of electrons and capacity for increasing specific chemical bonding and

*Corresponding Author: wasim.bu12@gmail.com; wasimakram.shaikh@tnu.in

simpler electrostatic attraction interaction of dye molecules with nanoparticles on nanocomposite (Amen et al. 2020).

Metal nanoparticle-based nanocomposites are the most efficient techno-economic, eco-friendly, non-toxic, and low-cost adsorbent used for a wide range of environmental applications. These metal nanoparticle-based nanocomposites are highly carbon-rich with little or no oxygen derived from pyrolyzed biomass (like animal litters, crop residues, solid wastes, and wood biomass) as a functional matrix for metal nanoparticles. Metal nanoparticles were then fabricated through a modified chemical co-precipitation method (Faraji et al. 2023). Silver (AgNP) and Iron nanoparticles (nZVI) are the most commonly used metal nanoparticles to fabricate biochar-based metal nanocomposite for the remediation of environmental contaminants (Lenka et al. 2021, Shaikh et al. 2022a). Several feedstocks have been used as the supporting matrices of the biochar-based metal nanocomposite, including bamboo biomass, chestnut biochar, waste biomass biochar, leaf litter biochar, activated sludge biochar, and pinewood biochar (Yoon et al. 2023).

NPs add additional unique properties, including high surface area, active functional groups with a decreased tendency of agglomeration and increase the stability of NPs (Amen et al. 2020, Shaheen et al. 2023). Combined influence of surface functional groups of biochar and doped nanomaterials enable H-bonding, electrostatic attraction, chemisorption, and surface complex formation with contaminants (dye, heavy metal, antibiotic etc.) (Yan et al. 2022). The negatively charged functional groups (including SO_3^- groups, COO^-) get attracted towards positively charged active surface functional groups (including =NH, $-NH_2$, ≡CH, $=CH_2$ etc.) of nanocomposite matrix (Wu et al. 2017). Additionally, O-rich surface functional groups (NO SO, C=O, and –OH) of nanocomposite create van der Waals interaction, H-bonding, and surface complexation with surface functional groups of pollutants (Shaikh et al. 2023, 2022c).

Biochar-based nanoparticles are the most advanced and widely utilized biogenic adsorbent used in different environmental applications. The multiple functional composites can exert synergistic effects of biochar and coating functional nanoparticles on heavy metal removal simultaneously. Heavy metal removal is an important research area of biochar-based nano-composites for remediation of wastewater pollution. Different metal(loid)s were successfully removed by using biochar-based nano-composites including As(III and V) (Shaikh et al. 2020a, Yoon et al. 2023), Cd(II) (Ma et al. 2023), Cr(VI) (Abad et al. 2022), Hg(II) (Salcedo et al. 2021), and Pb(II) (Zeng et al. 2022).

9.2 DIFFERENT FUNCTIONALIZING STRATEGIES FOR BIOCHAR-BASED NANO-COMPOSITE

Incorporating biochar in different nano-structured framework enables fabrication of biochar-supported hybrid systems, which augments the efficiency and performance of the composite several times. Such an enhancement could be achieved through:

9.3 SURFACE DOPING AND MODIFICATION

The pores and functional groups of biochar, support numerous nanoparticles, which eventually enhances its functionality and contaminant removal efficiency through adsorption or photo-catalytic degradation (Lyu et al. 2020). Heteroatoms like B, S, O, N, and P could be doped on active sites of biochar to effectively enhance electrical properties like conductivity, durability, charge-discharge rate, charge transfer, wettability, and interaction of π-electrons of carbon with lone pair of heteroatoms, which helps in fabricating materials efficient in storing energy (Figure 9.2). Nitrogen-based biomasses (e.g., Azolla ferns and marine micro and macro-algae) could be valorized into

N-doped biochar through pyrolysis and hydrothermal carbonisation (Hu et al. 2021). N doping could also be achieved by treating carbon-based biomass with $(NH_4)_2$ HPO_4, $(NH_4)_2SO_4$, NH_3, thiourea, melamine, NH_4Cl, and urea followed by thermo-chemical conversion (Norouzi et al. 2017). Doping of P, B, O, and S onto biochar helps increase gravimetric capacitance as well as electro-chemical storage of electricity in super-capacitors. Pyridine and sp^3-N were used for increasing specific capacitance by Lian et al. (2019). One-step carbonization involving NH_3 as N-dopant enabled fabrication of ultra-nitrogen-doped biochar to be successfully used as working electrode in super-capacitors. In another study, N-doped kelp-derived biochars were observed to demonstrate impressive specific capacitance results (Li et al. 2015). P-doping shows superior electro-chemical behaviour, compared to nitrogen, because of a larger structural diversion in hexagonal C framework (and rolls out better from graphene plane) and a larger diameter, which eventually improves conductivity. Co-doping of biochar with N and P (through $NH_4H_2PO_4$ and $KHCO_3$ dual activation method) helps in the fabrication of highly efficient electrodes for deionization capacitors. In a study, co-doped biochar showed presence of abundant mesopores and very high surface area (2726 m^2 g^{-1}) (Xu et al. 2017). P and O codoped carbon spheres showed enhanced specific capacity (186.8 F g^{-1} at 0.1 A g^{-1} current density) in a study (Wang et al. 2012). The performance of super-capacitors is highly affected by graphitization level, surface area, and surface morphology (Wang et al. 2012). Non-activated biochar with unique porosity was modified to enhance surface area and obtain improved 2D porous structures with graphitic layers, which demonstrate minimized ion-transport resistance, reduced diffusion distance, and increased electrolyte penetration (Norouzi et al. 2020). Biochar could be doped with transition metal oxide nanoparticles (e.g., Fe_3O_4) to efficiently increase pseudo-capacitance and electrical double-layer capacitance.

9.4 SURFACE RECOMBINATION

Surface recombination occurs because of the presence of defects or charge layers on the surface where carriers recombine. Such a type of functionalization involves techniques like metal loading or chemical reduction/oxidation to fabricate modified biochars (Figure 9.2). Chemical oxidation is used for preparing hydrophilic biochar. Hydrophilic properties of biochar-based electrodes could be enhanced by facilitating oxidants (e.g., H_3PO_4, H_2O_2, HNO_3, and HCl), which enhances specific capacitance. Biochars modified with oxidants show high resistance to oxidation and demonstrate improved stability of rings (Yang et al. 2019). Chemical reduction is used for fabricating biochar-based electrodes by using reducing agents (like NH_4OH, KOH, and NaOH) to enhance surface area and porous properties. KOH activation improves 3D inter-connected structure (macro-, meso-, and micro-pore architecture) of biochars, where macro-pores enable electrolyte transportation and ion diffusion, meso-pores facilitate reservoirs as ion-buffers (sites for ion inclusion and exclusion), and micro-pores function in Li-ion storage (Norouzi et al. 2020). Biochar hybridization could increase active sites and surface area of photo-catalyst nano-materials, consequently enhancing binding sites, photo-catalytic active sites, and light scattering, thereby making the photo-degradation of organic pollutants more efficient (Shaikh et al. 2018, Shaikh et al. 2020b). Organic pollutants are adsorbed by biochars through electrostatic interactions, H-bonding, and π–π assembling (Shaikh et al. 2021, 2022a, b, c).

9.5 TRANSPORTING ELECTRONS

The reactivity of biochar-based nano-materials is enhanced by the semiconductor nature of biochar, which transfers electrons from bulk chemical electron donors to electron receptors via condensed aromatic ring and redox active networks. The hybridisation of Fe_3O_4 nano-materials on carbon-rich gel networks makes continuous and stable transport of electrons and ions to electrode surface, as suggested by Wu et al. (2013). Electron transfer minimizes electron-hole recombination, correspondingly increasing the efficacy of photo-degradation of contaminants (Shaikh et al. 2020b). Delocalization of electrons from binding sites to receptors on organic contaminants is enabled by condensed aromatic ring structures. Nano functionalized biochar showed improved electrochemical property in a study by Irshad et al. (2022).

9.6 ELECTRON RESERVOIR

Biochar acts as an excellent reservoir of electrons by aiding in electron storage. While the redox activity of quinone molecules of biochar enhances electron storage, biochar's condensed aromatic ring structure facilitates electron transport. The variation in feedstock and production conditions influences electron storage potential of biochar. Plant-based biochars could store ~2 mmol of electrons per gram, as suggested by Kluepfel et al. (2014).

9.7 ENHANCED CHARGE SEPARATION

The activation and deactivation of photo-catalysts are dictated by electron-hole separation and fast recombination times respectively (Figure 9.1). Consequently, biochar is used to support photo-catalysts to maximize the recombination time. In an investigation, Shi (2013) suggested that light irradiation excites photo-catalysts (Eq. 1), but with time (short period) get deactivated (Eq. 2).

$$\text{Photo-catalyst} \rightarrow \text{Photo-catalyst*} \quad (1)$$

$$\text{Photo-catalyst*} \rightarrow \text{Photo-catalyst} \quad (2)$$

Biochar is capable of activating photo-catalysts without causing any damage to the active sites (Eq. 3). Since the deactivation is not easy, kinetic reactions get accelerated, thereby enhancing the photo-degradation of contaminants (Eq. 4).

$$\text{Biochar-Photo-catalyst} \rightarrow \text{Biochar-Photo-catalyst*} \quad (3)$$

$$\text{Biochar} + \text{Photo-catalyst} \rightarrow \text{Biochar*/Photo-catalyst*} \quad (4)$$

9.7 REDUCED BAND GAP ENERGY

The hybridisation of photo-catalysts with biochar reduces the band gap energy. In a study, formulation of TiO_2 using carbon doping caused a shift in light sensitivity from UV to visible region (Colmenares et al. 2013). The conduction band of carbon is placed above that of TiO_2 to enhance the sensitization (Colmenares et al. 2013). Ti–O–C bonds enable creation of a delocalized state in band gap, inhibiting charge carrier recombination (Colmenares et al. 2013).

9.8 PORE STRUCTURE TAILORING

Pore properties of biochars depend on the variation in feedstock material and pyrolysis conditions. The surface area, total pore volume, and porosity of biochars showed variation from 1.06 to 70.22 m^2/g, 1.28 to 3.68 cm^3/g and 58 to 80%, respectively. Pore size distribution showed bimodal peaks of 5–15 and 1.5–5 μm for broad-leaf forest and herbaceous plant biochars and 6–25 and 1.5–3 μm for coniferous forest biochars. The herbaceous plant biochars possessed larger transmission pore volume in comparison to coniferous forest and broad-leaf biochars. The better is the detailing of pore structure characterization in biochars, the improved is the efficiency of prediction of biochars as sorbents of pollutants.

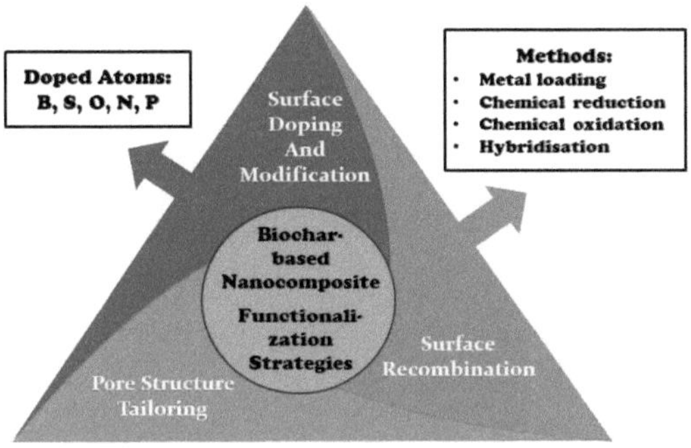

Figure 9.1 Different functionalizing strategies for biochar-based nanocomposite

9.9 TYPES OF BIOCHAR-BASED NANOCOMPOSITE

The fabrication of biochar composites leads to the three variety, namely, metal oxide/hydroxide biochar-based nano-composites, biochar composites with varied magnetic behavioural, and biochar composites coated with nanoparticles (Figure 9.2). The above categorisation is done on the basis of different nano-materials that are used to functionalise biochar either through pre-treatment of biomass or post-treatment of biochar. Biochar composites thus fabricated enhanced the contaminants removal efficiency, catalytic degradation ability, and easy separation through the modification of surface morphology which includes surface-active sites, specific surface area, pore volume and size distribution.

9.10 NANO METAL OXIDE/HYDROXIDE BIOCHAR NANOCOMPOSITES

Metal oxide/hydroxide biochar nanocomposites can be prepared through pyrolysis followed by the incorporation of specific element by the process of biomass pre-treating and pre-accumulation using metal salt and incorporation of nano metal oxide (Figure 9.2).

The activity of the specific elements can be improved furthermore by transforming it into oxides and hydroxides nano particles after thermal treatment. Mg enriched tomato tissues was engineered to produce an innovative biochar through slow pyrolysis (Yao et al. 2013). The surface of Mg-biochar has been modified by incorporation of $Mg(OH)_2$ and MgO nanoparticles which in turn act as effective phosphate adsorption sites. The bioaccumulation of Al, Zn, Ca, Fe, Mn on biomass and introduction of nano-metal oxide/hydroxide into resulted biochar have noticeable impact on the physiochemical properties. Some commonly used metal salts like $CaCl_2$ Liu et al. 2016a), $MgCl_2$ (Liu et al. 2016a), $KMnO_4$ (Yao et al. 2013), $MnCl_2$ (Gan et al. 2015), and $ZnCl_2$ (Zhang et al. 2012) are used for the biomass pre-treatment. This result in the formation of respective metal oxides and hydroxides nanoparticles on the surface of biochar by slow pyrolysis method. For instance, the removal efficiencies of phosphate and nitrate has highly been enhanced by using porous MgO-biochar nanocomposites. The biochar surfaces are greatly been modified by nanocrystalline AlOOH flakes which shows a drastic increase in the area and reactive sites to attract various water contaminants including heavy metals and other organic and inorganic components. Cr(VI) been effectively removed by $ZnCl_2$ induced ZnO–biochar nanocomposites.

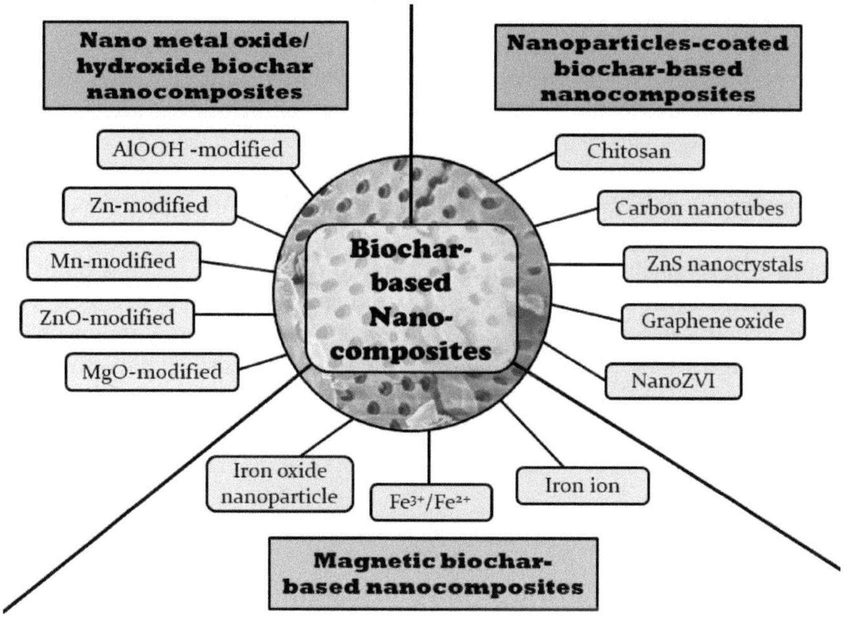

Figure 9.2 Types of biochar-based nanocomposite

Methods like evaporation and heat treatment as well as direct hydrolysis and conventional wet impregnation technique has been utilised for incorporation of metal oxide nanoparticles after pyrolysis (Tan et al. 2016). Surface area of $Fe(NO_3).39H_2O$ modified biochar prepared via an evaporative route has dramatically been increased by the order of 2.5 approximately. The removal capacity of Cu^{2+} has been increased manifold due to the presence of MnOx complexes on the surface of biochar. Hydrated MnOx impinched biochar obtained by a conventional wet impregnation method, showed an enhanced adsorption capacity towards Pb^{2+}. Fe salt onto hickory biochar is directly hydrolysed to prepare Fe-loaded biochar which act as a potential As adsorbent.

9.11 MAGNETIC BIOCHAR-BASED NANOCOMPOSITES

Iron oxides and iron ion onto biochar are subjected to chemical co-precipitation and pre-treatment method respectively in order to incorporate magnetic behaviour that results in to separate aqueous biochar easily. Active sites on the surface of biochar, coated with magnetic iron oxide nanoparticles including Fe_3O_4, and $CoFe_2O_4$, are utilized for the removal of contaminants. Adsorption capacity of the biochar is increased (Amen et al. 2020) in substantial amount with the introduction of magnetic property within it and can be subjected to magnetic separation as well.

Orange peel powder are transformed to magnetic biochar by chemical co-precipitation of Fe^{3+}/Fe^{2+} prior to pyrolysis. Biochar surface embedded with spinel cobalt ferrite magnify the adsorption capacity of Pb^{2+} and Cd^{2+}. Furthermore, magnetic biochar can also be prepared from natural hematite mineral upon treatment with biomass (Amen et al, 2020). Carbon surface loaded with c-Fe_2O_3 particles shows active sites for As adsorption.

9.12 NANOPARTICLES-COATED BIOCHAR-BASED NANOCOMPOSITES

Biochar composite with surface coated functional nanoparticles (chitosan, carbon nanotubes, ZnS nanocrystals, graphene, graphene oxide, nanoscale zero-valent Fe, layered double hydroxides, graphitic C_3N_4) are capable of removing various contaminants (Figure 9.2). The key factor for better performance of biochar in contaminants removal is the modification in surface functional nanoparticles viz. porosity, surface area and thermal stability. Graphene coated biochar based nanoparticles shows organic contaminants degradation capability in addition to the adsorption ability.

Surface morphology modification of biochar-based functional composites is carried out by a dip-coating technique preceding sonication. Ultrasonicator is used for sonication results in the enhanced thermal stability, porosity and surface area of biochars. Chitosan are coated onto biochar surfaces to develop chitosan-modified biochar (Tan et al. 2016). This kind of modification serves dual advantages to chitosan viz, increased porous network and surface area with enhanced chemical affinity. On the contrary, decrease in the surface area by chitosan has also been exhibited due to partial blockage of pores after synthesis. Hence, the synthesis techniques demand rigorous research work during the loading of specific nanomaterials onto biochar in order to minimize the undesirable effect on the expected properties.

9.12 APPLICATIONS OF BIOCHAR-BASED NANOCOMPOSITE

9.12.1 Degradation of Environmental Contaminants

Pristine biochar could be associated with limitations such as low adsorption capacity, limited surface functionality, hydrophilicity, surface area, and active sites, which prevent them from meeting industrial requirements for large-scale commercial use (Wang et al. 2018a). Correspondingly, biochar-supported photo-catalysts are used as an alternative. They are cost-effective and with the addition of hybrid nanoparticles enhance photo-degradation and adsorption capacity (Chen et al. 2018). The adsorption of methylene blue by Al_2O_3/biochar was investigated by Wang et al. (2017a), where a high adsorption capacity was observed for the hybrid photo-catalyst (35 mg g^{-1} more than the pristine biochar). Kulaksiz et al. (2017) used Fe modified/biochar for photo-catalytically degrading malachite green. The photo-catalyst possessed higher surface area and pore volume than pure biochar, which facilitated photo-catalytic degradation. The impregnation of magnetic particles improves the removal efficacy by enhancing magnetic properties which minimises secondary pollution, improves particle dispersion, porosity, surface area, and recovery.

9.12.2 Catalytic Degradation

Biochar behaves both as an adsorbent and a catalyst, although it shows lesser degradation potential compared to biochar assisted-photo-catalysts. Thines et al. (2017) studied a photo-degradation study of phenol using various hybrid biochar supported nanocomposite. Lisowski et al. (2017) fabricated TiO_2/biochar to photo-degrade phenol by UV/visible light irradiation. A number of adsorbents were prepared from different raw materials such as softwood pellets and *Miscanthus* straw at various pyrolysis temperatures (550–700°C). Hybrid biochar was more efficient than pure biochar for phenol photo-degradation. A metal-free g-C_3N_4–C/biochar showed better photo-degradation efficiency for the catalytic decomposition of MBT than pure g-C_3N_4 (Zhu et al. 2018). Generally, the doped nano material (DnM) acts as a catalyst in a system by enabling "electron-shuttling" process more effectively by passing electrons to environmental pollutants (EP) (acceptor) from $NaBH_4$ (donor) (Varadavenkatesan et al. 2019). Electron relay alters surface of DnM, which appears as blue spectral shift visually in plasmon resonance bands in DnM's surfaces (Pradhan et al. 2002). The spectral blue shifts of DnM overlapped the absorption peaks of EP (Varadavenkatesan et al. 2019). The catalyst makes electron transfer smoother and easier, when its redox potential lies between acceptor and donor (Tripathi et al. 2013).

9.12.3 Photocatalytic degradation

Biochar-supported photo-catalysts also cost-efficiently helps in photo-catalytic degradation of EP, enabling treatment of wastewater (Shaikh et al. 2022c). Photo-catalysts like Ag/biochar, CdS/biochar, ZnO/biochar, and TiO_2/biochar have been utilised previously for photo-degradation of EP (Shaikh et al. 2020b). These catalytic nanoparticles are used for doping biochars to enhance their optical property, stability, conductivity, and porous structure (Shaikh et al. 2022b). Further, the catalytic property of nanoparticles helps increase the efficiency of photo-degradation of EP (Shaikh et al. 2018). Zhang and Lu (2018) fabricated TiO_2–doped coconut shell-based biochar to photo-catalytically decolourise dye reactive brilliant blue, who observed evenly distributed TiO_2 nanoparticles on biochar surface. Upon absorbing photons of energy equal or greater than band gap, produce electrons and holes, which by reacting with water help in producing OH and superoxide radicals (reactive oxygen species), which help in photo-degradation of dye molecules and organic contaminants Zhang and Lu (2018).

9.13 ADSORPTIVE REMOVAL OF ENVIRONMENTAL CONTAMINANTS

Biochar-based nano-composites possess high removal efficiency for organic/inorganic contaminants like heavy metals, dyes, antibiotics, and anions like phosphate, nitrate, and fluoride (Amen et al. 2020, Jang and Kan, 2019, Shaikh et al. 2022c). Biochar-based nano-composites showed phosphate adsorption capacity ranging from 17.41–886.2 mg g^{-1}, where Mg–Al based nano-composites demonstrated best adsorption efficiency (Jung and Ahn 2016). The removal is governed by multiple mechanisms such as the sorption on an enhanced surface area of composite or impregnated nano-composites which dramatically augments contaminant removal from water (Creamer et al. 2014). The colloids/nano-materials on biochar surface act as reactive sites involved in removal of inorganic contaminant (Amen et al. 2020). The interactions between contaminants and nanoparticles in biochar matrix include ion exchange, hydrogen bonding, precipitation, and electrostatic attraction (Alam et al. 2018, Shaikh et al. 2022b, c).

9.14 REMOVAL OF INORGANIC CONTAMINANTS

9.14.1 Nutrients

Nutrients like phosphate, ammonia, and nitrate promote growth of photosynthetic organisms, eventually leading to eutrophication in aquatic environments. Biochar has been reported to remove phosphate, ammonia, and nitrate from aqueous media (Yao 2013), despite negative results in certain studies (Gai et al. 2014). At higher pH, anion adsorption reduces because of the negatively charged surface of biochar. Nevertheless, modified biochar shows high nutrient removal efficiency from aquatic systems compared to raw biochar. Modification of biochar could influence adsorption differently for the same contaminant, possibly due to changes in feedstock (Kumar et al. 2022c). MgO-modified biochar showed 66.7% phosphate removal enabled by strong affinity of MgO for aqueous anions via mono-, bi-, and tri-nuclear complex formation, with the amount and size of MgO particles influencing the mechanism of complexation (Yao 2013). Interestingly, negligible phosphate removal (0.5%) and low nitrate removal (3.6%) was also observed by MgO-modified biochar, with electrostatic repulsion between anion and biochar surface being suggested as the reason for low removal (Yao 2013). Chitosan-modified biochar showed low phosphate removal, because of net negative surface charge of modified biochar (Zhou et al. 2014). Zero-valent iron-modified biochar showed high phosphate removal, and rose above 56% with an additional increase in Fe content (Almeelbi and Bezbaruah 2014). Moreover, low pH of solution (5.7) compared to pHpzc of the adsorbent (7.7) supported the existence of a predominant cationic form, which could have enhanced removal of phosphate. Clay modification has also been shown to increase adsorption capacities with respect to phosphate and ammonium, enabled by an augmented number of binding sites and the elevated surface area (Alam et al. 2018). While monolayer chemisorption controls phosphate and ammonium removal at low concentrations, multilayer chemical and physical adsorption participated in removal at higher concentrations (Chen et al. 2017).

9.14.2 Trace Metals

The existence and increase of metal concentrations in aquatic ecosystems lead to threatening ecological concerns (Inyang and Dickenson 2015) (Figure 9.3). These metals are inclusive of Cd, Cr, Cu, Zn, As, Hg and Pb, with United States Environmental Protection Agency fixing the maximum permissible limits of these metal(loid)s in potable waters and wastewaters. The threatening rise in metal concentrations has led to development of methods with an aim of minimising the concentrations of metals below toxic levels. These methods include electrodialysis, reverse osmosis, solvent extraction, ion exchange, chemical precipitation, and adsorption. Adsorption is considered relatively less expensive in comparison to other methods for removing metals from aqueous systems. In previous studies, As (Vithanage et al. 2017), Cd (Wang et al. 2015a), Cr (Mondal et al. 2017), Cu (Wang et al. 2015a), Hg (Wang et al. 2020), and Pb (Liu et al. 2016b) were removed using biochar-based composites. Nano-materials such as metal oxide/hydroxide (AlOOH, Al_2O_3, MgO, CaO, $Mg(OH)_2$, ZnO, and MnOx), iron oxides ($CoFe_2O_4$, Fe_2O_3, and Fe_3O_4) and other nano-particles (LDHs, ZnS nano-crystals, graphene, GO, CNTs, and chitosan) have been used for improving adsorption capacity of composites (Amen et al. 2020).

In a study, biochar composites modified with different levels of Mn were assessed for removing lead (Wang et al. 2015b). The study reported a maximum Pb(II) removal of 98.9% for composite containing 3.65% Mn, while elevated loading of Mn minimized surface area (due to pore blockage) and consequently decreased the Pb removal efficiency. Further, solution pH was considered critical in influencing removal efficiency, primarily because of the changes in net composite surface charges with changes in solution pH (Wang et al. 2015b). Stillage residue-derived biochars (biomass obtained from biofuel plant manufacturing ethanol) removed 97.4% of Ag(I) from aqueous systems with multiple mechanisms governing the process of adsorption which occurred on a multi-layered surface (Gerente et al. 2007). $CoFe_2O_4$-impregnated pine bark waste-based biochars showed low Pb(II) and Cd(II) removal and were observed to be highly dependent on pH (removal rose with a rise in pH from 2 to 8). At low pH, high concentrations of H^+ protonated the biochar surface making it positively charged, thereby reducing metal adsorption due to electrostatic repulsion. Correspondingly, metal adsorption increased at high pH (due to negatively charged surface).

The maximum Pb adsorption capacity of biochar-based composites varies from 4.9 to 367.7 mg/g, where ZnS/magnetic biochar composite demonstrated the best removal efficiency (Yan et al. 2015b). The removal could incorporate a number of mechanisms with the participation of (i) increased functional groups on the surface of composite, (ii) enhanced pore structure and surface area of the composite, and (iii) augmented active sites on the surface of composite (Gan et al. 2015). The impregnation of nano-materials helps increase O-containing functional groups (e.g., C=O, –OH and –COOH) on the surface of composite, which facilitates binding sites for metals and strengthens interaction with metals through surface complexation, cation-π bonding, electrostatic attraction and ionic exchange (Amen et al. 2020). However, nano-materials could also fill/block pores, thereby reducing the pore volume surface area of composite (Shaikh et al. 2022b). Magnetic biochar/ZnS composite was prepared by depositing ZnS nano-crystals on magnetic biochars in a polyol solution, where the composite showed super-paramagnetic properties for separation and excellent Pb(II) adsorption capacity (Wang et al. 2017a). Chitosan-impregnated magnetic biochar was fabricated by combining Fe_2O_3, chitosan, and biochar, which showed benefits of low operation cost, easy availability of biomass feedstock, magnetic property facilitation, an abundance of functional groups, and enhanced Cr(VI) removal efficiency (Mondal et al. 2017).

9.15 REMOVAL OF ORGANIC CONTAMINANTS AND EMERGING CONTAMINANTS

Aquatic systems are contaminated with organic contaminants such as antibiotics, pharmaceuticals, dyes, pesticides, halogenated flame retardants, surfactants, personal care products, other emerging contaminants, and polycyclic aromatic hydrocarbons, which shows detrimental impacts on environment (Basu et al. 2022, Shaikh et al. 2020a, 2021b, 2022b, c, Woo and Jhung 2021). It is crucial to remove the organic contaminants present in wastewaters before they are discharged into aquatic environments. Modified biochars have been used for removing organic contaminants. Attapulgite–biochar (lotus stalk biomass) composites removed 80.1% of norfloxacin at pH 2.0–11.0 (pH is largely involved in influencing the surface charge of the composite and molecular form of norfloxacin), while the attapulgite particles helped in augmenting adsorption capacity of norfloxacin (Li et al. 2017).

9.16 DYES

More than 0.7 million tons of dyestuff are produced annually with ~1 million commercial dyes present in the market, which are released as effluents in the environment (Rafatullah et al. 2010). Further, the dyes market is expected to rise above 37 billion US$ by 2023 with >5% annual growth, suggesting that the problem of dye-related contamination is expected to rise in the near future (Shaikh et al. 2021). Dyes are organic compounds containing auxochromic and chromophoric groups, which impart colour and exhibit high solubility in water, making conventional removal challenging (Shaikh et al. 2022c). Moreover, the dyes are resistant to degradation (biological, thermal, or photolytic) and persist for long in water bodies (Foroutan et al. 2021). The dyes damage aquatic environments irreparably by deterring water transparency, photosynthetic performance, and gas solubility, and are mutagenic and carcinogenic, thereby triggering health issues by damaging DNA apart from causing nasopharyngeal track damage, vertigo, hypertension, eyesight impairment, and skin irritation (Shaikh et al. 2018, 2020b). It is therefore crucial to minimize the concentrations of dyes from the environment and the associated adverse effects to contain the damages (Orts et al. 2018, Lellis et al. 2019). Pristine biochars are considered to possess limited adsorption capacity because of the comparatively lower surface area and functionality, making large-scale application difficult (Premarathna et al. 2019).

Different modification techniques have been used to augment dye removal efficiency of biochar, inclusive of physical treatment, chemical treatment, and metal/nano-material impregnation, which improves surface topography and morphology (Zhang et al. 2021, Kumar et al. 2022c) (Figure 9.3). Nanoparticles have received a lot of attention for biochar modification, enabled by their superior surface, thermal, and catalytic properties, while the nanoparticle-doped biochars are highly stable, cost-effective and eco-friendly showing enhanced target specificity and surface area (Kumar et al. 2021b, Shaikh et al. 2022c). Nano-materials like metals, metal oxides, or graphene oxide have been used for de-polluting wastewater and industrial effluents containing dyes (Mittal et al. 2013). Amongst metal nanoparticles, silver nanoparticles have become immensely popular as a nano-catalyst, enabled by high specificity, surface area, electro-thermal, optical, catalytic properties and adsorption capacity, which enhances its applicability as photo-catalyst and dopant material (Ishida et al. 2000). Therefore, biochar-based nano-composite materials have gained huge research interests because of the aforementioned superior properties, and their ability to possess unique properties of both nanoparticles and biochar eventually enhancing dye removal efficiency and demonstrating high de-colorization efficiency (Ye et al. 2019, Amen et al. 2020). In a study, >92% adsorption efficiency was reported for biochar-based nano-composite used for removing Rhodamine-B and Congo red, where adsorption was multi-layered dominated by chemisorption, with participation of van der Waals interaction, complexation, H-bonding, electrostatic attraction, pore-filling and surface adsorption (Shaikh et al. 2022b).

9.17 ANTIBIOTICS

There has been large-scale use and consumption of pharmaceutics, including antibiotics like azithromycin and hydroxychloroquine, especially after the emergence and spread of COVID-19 (SARS-CoV-2) globally as a pandemic (Calderón-Parra et al. 2021, Shaikh et al. 2023). Antibiotic consumption has escalated by > 65% globally since 2000, while >30% per capita increase in use was observed in India, with consumption increasing > 47% since 2010 (Sulis et al. 2021). Consequently, antibiotics have entered aquatic environments via wastewater and threatened drinking water quality, human health, and aquatic ecology (Qiu et al. 2021). Large amounts of untreated wastewaters are discharged from pharmaceutical industries and hospitals or as agricultural/urban runoffs. To consider an example, the antibiotic, tetracycline, is used for treating bacterial/microbial inflammation (El Messaoudi et al. 2022). When administered, the complete dose is not metabolized, and the residual fraction is released into the environment (in wastewater and solid waste) through urination/egestion (Balasubramani et al. 2020, Phoon et al. 2020). Moreover, there could be direct discharge of pharmaceuticals via leakage of septic tanks, during livestock breeding, or through untreated hospital/industry wastewater effluents (Russell and Yost 2021).

Correspondingly, it is crucial to develop eco-friendly sustainable solutions to minimize the contamination caused because of antibiotics prior to their entry into food chains. Methods such as advanced oxidation, reverse osmosis, and ion exchange have been used to treat antibiotic-loaded wastewaters but have been suggested to be inefficient in complete treatment and could be associated with higher costs (Kumar and Bhattacharya 2021, Shaikh et al. 2021). In this regard, biochar-based nano-composite has been reported to efficiently treat such wastewaters, since these composites are highly stable (thermally and mechanically) along with promising target specificity, enhanced reactivity, improved pore structure and large surface area (Sharma et al. 2019, Shaikh et al. 2020a, Kumar et al. 2021b). Moreover, these composites contain several functional groups, which helps in augmented electrostatic interaction through Van-der Waals interactions, H-bonding, and π–π stacking (Shaikh et al. 2022b, Sharma and Naushad 2020). Rajapaksha et al. prepared biochars using tea waste and burcucumber at 300–700°C and modified them with steam, where biochars produced at 700°C showed better porosity parameters, biochars produced from tea waste demonstrated superior surface areas (342 m^2/g for tea waste-based biochar vs 2.3 m^2/g for burcucumber-derived biochar) and steam treatment enabled increase in surface area of the adsorbent (576 m^2/g for tea waste-based biochar vs 7.1 m^2/g for burcucumber-derived biochar). However, adsorption capacity was higher for burcucumber-based biochars, possibly due to differences in ash composition (ash contents of 11% for tea waste-based biochar vs 44% for burcucumber-derived biochar and 16% for steam-treated tea waste-based biochar vs 71% for steam-treated burcucumber-derived biochar) and polarity (O/C molar ratio of 0.08 for tea waste-based biochar vs 0.26 for burcucumber-derived biochar and 0.11 for steam-treated tea waste-based biochar vs 0.67 for steam-treated burcucumber-derived biochar). The ash contents are involved in complexation, while hydrophilic groups participate in electrostatic interaction and hydrogen bonding.

9.18 ADSORPTION AND DEGRADATION OF ORGANIC CONTAMINANTS

Biochar-based composites containing oxidative/reductive and catalytic nano-particles spread on the surface have been used to treat organic contaminants simultaneously through adsorption and degradation (Shaikh et al. 2022a). Biochar serves as a substrate for oxidative/reductive and catalytic nanoparticles, consequently enhancing dispersion of the particles and escalating contaminant contact (Han et al. 2015b). Further, the abundant O-containing functional groups and large surface area promotes reactivity of oxidative/reductive and catalytic nanoparticles (Yan et al. 2015a). Importantly, biochar accelerates contaminant transfer from solution to composite surface through adsorption (Pi et al. 2015). The oxidative/reductive and catalytic nanoparticles spread on the surface of biochar could help regenerate biochar by degrading the adsorbed contaminants and prevent contaminant accumulation in biochar matrix, which eventually increases adsorption of contaminants (Pi et al. 2015). Graphitic C_3N_4 and zero-valent iron biochar-based composites are among the most commonly used composites for removing organic contaminants including trichloroethylene (Yan et al. 2015b), tar (Kastner et al. 2015), pentachlorophenol (Devi and Saroha 2015), methylene blue (Pi et al. 2015), methyl orange (Han et al. 2015a), and acid orange 7 (Wang et al. 2019).

9.19 BIOCHAR-BASED NANOCOMPOSITE FOR ENERGY STORAGE AND ENVIRONMENTAL PROTECTION APPLICATIONS

Biochar-based nano-composites have been largely used for applications including energy storage and environmental protection (Kumar et al. 2020). These biochar-based nano-composites have been used for catalysis, energy storage, and conversion, and in super-capacitors, primarily enabled by increase in surface area, pore structure, and surface chemistry (Cheng et al. 2017, Xiu et al. 2017) (Figure 9.3). In comparison to conventional carbon materials, biochar-based nano-composites have abundant O-rich functional groups, like OH, C=O, and COOH, which enables its

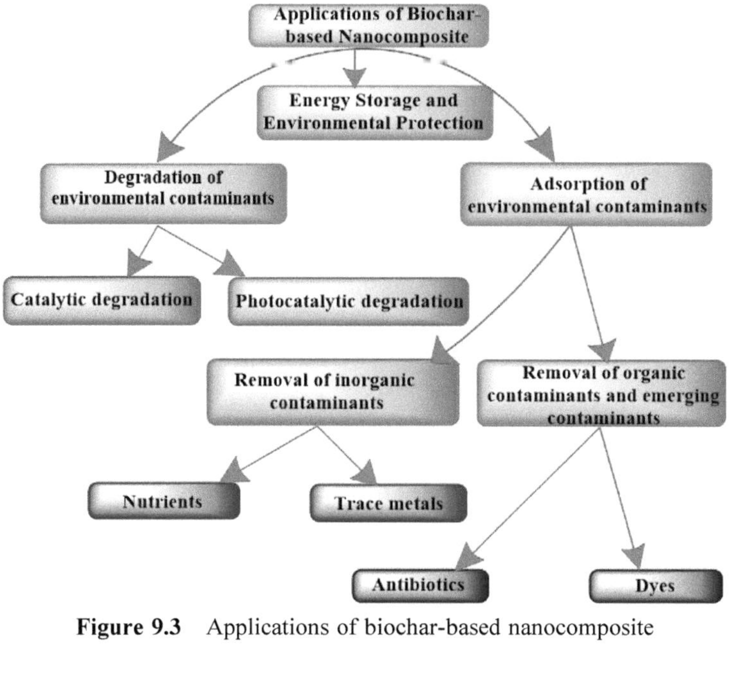

Figure 9.3 Applications of biochar-based nanocomposite

applicability as electrode in super-capacitors. Crucial properties could be fine-tuned by changing pyrolysis conditions, including temperature and heating rate, and judicious selection of catalyst and biomass feedstock that could greatly increase pseudo-capacitance performance and stored energy. Moreover, O-containing functional groups, which are electrochemically inert, could improve surface utilization and pore access, which eventually increases specific capacitance. Additionally, O-containing functional groups hinder further oxidation of carbon matrix, enhancing the cycle stability of electrodes. The abundant functionality and high surface area of biochar-based nano-composites also help in soil remediation by increasing pH, electrical conductivity, porosity, cation exchange capacity, microbial activities, nutrient profile, organic matter, hydraulic conductivity, and gas exchange properties (Gupta et al. 2022, Hafeez et al. 2022, Ramola et al. 2022). The biochar-based composites could also act as sink for greenhouse gases, like nitrous oxide, methane, and carbon dioxide, decreasing their concentration in the atmosphere and eventually minimizing global warming (Wang et al. 2018b, Yang et al. 2020, Rittl et al. 2021).

9.20 CONCLUSIONS AND FUTURE PERSPECTIVES

Modification of biochar influences surface properties including pore size, functional groups, surface charge, and surface area. Modifying agents such as metals, zero-valent ion, metal oxides, clays, graphene oxides, polymer, and other organic compounds have been used. The various modifiers mostly enable commendable increase in contaminant removal efficiency but negative effects have been observed sometimes. A lot of investigation is needed to identify suitable modifying agent for removing specific pollutants. The modification could be achieved by changing the physical and chemical properties of adsorbent. Further, Environmental conditions could also be changed to enhance removal efficiency such as manoeuvring pressure and temperature. Additionally, sate-of-the-art instruments like ICP and HPLC could be utilised to augment the detection limit of contaminants.

Biochar modification with metals or metal oxides has been the most utilised treatment method for preparing enhanced biochar to remove contaminants. Biochar composites have been reported with strengthened photo-catalytic activity. Biochar composites have not been observed to increase sorption of nutrients (like nitrates and phosphates) which are largely responsible for eutrophication and pollution of fresh water ecosystems. Greater research is needed in the direction of developing suitable adsorbents for reducing these nutrients. In the recent past, modified biochars have been successfully used for multi-faceted applications. Modification allows tuning of surface properties and augmenting of functional groups. Fabrication of modified biochar-based composites has been recommended for enhancing performance of different applications.

Most contaminant removal studies were performed in lab conditions. However, natural environment mostly differs from lab environment significantly, especially with continuous changes in temperature and pH, which imposes substantial impact on removal efficiency of adsorbents. Correspondingly, long-term pilot-scale experiments would be critical for assessing the stability and efficiency of biochars and their composites in contaminant remediation. Additionally, the limited literature related to large-scale production of modified biochars makes its documentation necessary. Life cycle assessment and cost analysis are key parameters necessary for estimating the profitability of biochar-based remediation when comparing it to other available methods. Municipal solid wastes have been increasing in quantity and therefore could be pyrolyzed with abundant clay materials to fabricate clay-biochar composites. Bio-oils and syngas generated during pyrolysis could be used for production of energy. Therefore, biochar-based composites have been used in wide-ranging applications inclusive of removal of various pollutants/contaminants from soils, water, and air, oxygen evolution and hydrogen production through photochemical reaction in photo-electrodes, and production and storage of energy.

References

Abad, M.K., Seyed Shahab, A., Javidan, P., Baghdadi, M. and Mehrdadi, N. 2022. Green synthesis of Pd@ biochar using the extract and biochar of corn-husk wastes for electrochemical Cr(Vi) reduction in plating wastewater. SSRN Electron. J. 11. 109911. https://doi.org/10.2139/ssrn.4265589.

Alam, M.A., Shaikh, W.A., Alam, M.O., Bhattacharya, T., Chakraborty, S., Show, B. and Saha, I. 2018. Adsorption of As(III) and As(V) from aqueous solution by modified *Cassia fistula* (golden shower) biochar. Appl. Water Sci. 8: 198. https://doi.org/10.1007/s13201-018-0839-y.

Almeelbi, T. and Bezbaruah, A. 2014. Aqueous phosphate removal using nanoscale zero-valent iron. Nanotechnol. Sustain. Dev. First Ed. 14: 197–210. https://doi.org/10.1007/978-3-319-05041-6_16.

Amen, R., Bashir, H., Bibi, I., Shaheen, S.M., Niazi, N.K., Shahid, M., Hussain, M.M., et al. 2020. A critical review on arsenic removal from water using biochar-based sorbents: the significance of modification and redox reactions. Chem. Eng. J. 396: 125195. https://doi.org/10.1016/j.cej.2020.125195.

Balasubramani, K., Sivarajasekar, N. and Naushad, M. 2020. Effective adsorption of antidiabetic pharmaceutical (metformin) from aqueous medium using graphene oxide nanoparticles: equilibrium and statistical modelling. J. Mol. Liq. 301: 112426. https://doi.org/10.1016/j.molliq.2019.112426.

Basu, S., Lakra, R., Kumari, R., Shaikh, W.A. and Chakraborty, S. 2022. Separation of congo red dye from water using AgNPs based hybrid UF membrane. pp. 1037–1046. *In:* Ratan, J.K., Sahu, D., Pandhare, N.N. and Bhavanam, A. (eds). Environmental Science and Engineering. Springer International Publishing, Cham. https://doi.org/10.1007/978-3-030-96554-9_68.

Calderón-Parra, J., Muiño-Miguez, A., Bendala-Estrada, A.D., Ramos-Martínez, A., Muñez-Rubio, E., Carracedo, E.F., Montes, J.T., et al. 2021. Inappropriate antibiotic use in the COVID-19 era: Factors associated with inappropriate prescribing and secondary complications. Analysis of the registry SEMI-COVID. PLoS One. 16: e0251340. https://doi.org/10.1371/journal.pone.0251340.

Chen, L., Chen, X.L., Zhou, C.H., Yang, H.M., Ji, S.F., Tong, D.S., Zhong, Z.K. et al. 2017. Environmental-friendly montmorillonite-biochar composites: facile production and tunable adsorption-release of ammonium and phosphate. J. Clean. Prod. 156: 648–659. https://doi.org/10.1016/j.jclepro.2017.04.050.

Chen, S., Zhou, M., Wang, H.F., Wang, T., Wang, X.S., Hou, H.B. and Song, B.Y. 2018. Adsorption of reactive brilliant Red X-3B in aqueous solutions on clay-biochar composites from bagasse and natural attapulgite. Water (Switzerland) 10. https://doi.org/10.3390/w10060703.

Cheng, B.H., Zeng, R.J. and Jiang, H. 2017. Recent developments of post-modification of biochar for electrochemical energy storage. Bioresour. Technol. 246: 224–233. https://doi.org/10.1016/j.biortech.2017.07.060.

Colmenares, J.C., Lisowski, P. and Łomot, D. 2013. A novel biomass-based support (Starbon) for TiO_2 hybrid photocatalysts: a versatile green tool for water purification. RSC Adv. 3: 20186–20192. https://doi.org/10.1039/c3ra43673j.

Creamer, A.E., Gao, B. and Zhang, M. 2014. Carbon dioxide capture using biochar produced from sugarcane bagasse and hickory wood. Chem. Eng. J. 249, 174–179. https://doi.org/10.1016/j.cej.2014.03.105.

Devi, P., Saroha and A.K. 2015. Simultaneous adsorption and dechlorination of pentachlorophenol from effluent by Ni-ZVI magnetic biochar composites synthesized from paper mill sludge. Chem. Eng. J. 271: 195–203. https://doi.org/10.1016/j.cej.2015.02.087.

El Messaoudi, N., El Khomri, M., Ablouh, E.H., Bouich, A., Lacherai, A., Jada, A. and Lima, E.C. 2022. Biosynthesis of SiO_2 nanoparticles using extract of *Nerium oleander* leaves for the removal of tetracycline antibiotic. Chemosphere. 287: 132453. https://doi.org/10.1016/j.chemosphere.2021.132453.

Faraji, A.R., Khoramdareh, N.B., Ahmadi, S., Gil, A., Farahanipour, A. and Hekmatian, Z. 2023. Synthesis of magnetic Mn/Fe oxides derived from drinking water sludges from a one-pot solvothermal process for the degradation of tetracycline and mixed dyes: synergistic effect and mechanism study. J. Environ. Chem. Eng. 11: 109668. https://doi.org/10.1016/j.jece.2023.109668.

Foroutan, R., Peighambardoust, S.J., Esvandi, Z., Khatooni, H. and Ramavandi, B. 2021. Evaluation of two cationic dyes removal from aqueous environments using CNT/MgO/$CuFe_2O_4$ magnetic composite powder: a comparative study. J. Environ. Chem. Eng. 9: 104752. https://doi.org/10.1016/j.jece.2020.104752.

Gai, X., Wang, H., Liu, J., Zhai, L., Liu, S., Ren, T. and Liu, H. 2014. Effects of feedstock and pyrolysis temperature on biochar adsorption of ammonium and nitrate. PLoS One. 9. https://doi.org/10.1371/journal.pone.0113888.

Gan, C., Liu, Y., Tan, X., Wang, S., Zeng, G., Zheng, B., Li, T., et al. 2015. Effect of porous zinc-biochar nanocomposites on Cr(VI) adsorption from aqueous solution. RSC Adv. 5: 35107–35115. https://doi.org/10.1039/c5ra04416b.

Gerente, C., Lee, V.K.C., Le Cloirec, P. and McKay, G. 2007. Application of chitosan for the removal of metals from wastewaters by adsorption—mechanisms and models review. Crit. Rev. Environ. Sci. Technol. 37: 41–127. https://doi.org/10.1080/10643380600729089.

Gupta, I., Singh, R., Batish, D.R., Singh, H.P., Raghubanshi, A.S. and Kohli, R.K. 2022. Engineered biochar as soil fertilizer. Eng. Biochar. 197–221. https://doi.org/10.1007/978-981-19-2488-0_11.

Hafeez, A., Pan, T., Tian, J. and Cai, K. 2022. Modified biochars and their effects on soil quality: a review. Environ. 9: 60. https://doi.org/10.3390/environments9050060.

Han, L., Xue, S., Zhao, S., Yan, J., Qian, L. and Chen, M. 2015a. Biochar supported nanoscale iron particles for the efficient removal of methyl orange dye in aqueous solutions. PLoS One. 10: e0132067. https://doi.org/10.1371/journal.pone.0132067.

Han, Y., Li, W., Zhang, J., Meng, H., Xu, Y. and Zhang, X. 2015b. Adsorption behavior of Rhodamine B on nanoporous polymers. RSC Adv. 5: 104915–104922. https://doi.org/10.1039/c5ra21130a.

Hu, Q., Jung, J., Chen, D., Leong, K., Song, S., Li, F., Mohan, B.C., et al. 2021. Biochar industry to circular economy. Sci. Total Environ. 757. https://doi.org/10.1016/j.scitotenv.2020.143820.

Inyang, M. and Dickenson, E. 2015. The potential role of biochar in the removal of organic and microbial contaminants from potable and reuse water: a review. Chemosphere. 134: 232–240. https://doi.org/10.1016/j.chemosphere.2015.03.072.

Irshad, A., Warsi, M.F., Agboola, P.O., Dastgeer, G. and Shahid, M. 2022. Sol-gel assisted Ag doped $NiAl_2O_4$ nanomaterials and their nanocomposites with g-C_3N_4 nanosheets for the removal of organic effluents. J. Alloys Compd. 902. https://doi.org/10.1016/j.jallcom.2022.163805.

Ishida, H., Campbell, S. and Blackwell, J. 2000. General approach to nanocomposite preparation. Chem. Mater. 12: 1260–1267. https://doi.org/10.1021/cm990479y.

Jung, K.W. and Ahn, K.H. 2016. Fabrication of porosity-enhanced MgO/biochar for removal of phosphate from aqueous solution: application of a novel combined electrochemical modification method. Bioresour. Technol. 200: 1029–1032. https://doi.org/10.1016/j.biortech.2015.10.008.

Jang, H.M. and Kan, E. 2019. A novel hay-derived biochar for removal of tetracyclines in water. Bioresour. Technol. 274: 162–172. https://doi.org/10.1016/j.biortech.2018.11.081.

Kastner, J.R., Mani, S. and Juneja, A. 2015. Catalytic decomposition of tar using iron supported biochar. Fuel Process. Technol. 130: 31–37. https://doi.org/10.1016/j.fuproc.2014.09.038.

Klüpfel, L., Keiluweit, M., Kleber, M. and Sander, M. 2014. Redox properties of plant biomass-derived black carbon (biochar). Environ. Sci. Technol. 48: 5601–5611. https://doi.org/10.1021/es500906d.

Kulaksiz, E., Gözmen, B., Kayan, B. and Kalderis, D. 2017. Adsorption of malachite green on Fe-modified biochar: Influencing factors and process optimization. Desalin. Water Treat. 74: 383–394. https://doi.org/10.5004/dwt.2017.20601.

Kumar, A., Bhattacharya, T., Mozammil Hasnain, S.M., Kumar Nayak, A. and Hasnain, M.S. 2020. Applications of biomass-derived materials for energy production, conversion, and storage. Mater. Sci. Energy Technol. 3: 905–920. https://doi.org/10.1016/j.mset.2020.10.012.

Kumar, A. and Bhattacharya, T. 2021. Biochar: a sustainable solution. Environ. Dev. Sustain. 23: 6642–6680. https://doi.org/10.1007/s10668-020-00970-0.

Kumar, A., Bhattacharya, T., Shaikh, W.A., Roy, A., Mukherjee, S. and Kumar, M. 2021a. Performance evaluation of crop residue and kitchen waste-derived biochar for eco-efficient removal of arsenic from soils of the Indo-Gangetic plain: a step towards sustainable pollution management. Environ. Res. 200: 111758. https://doi.org/10.1016/j.envres.2021.111758.

Kumar, A., Nagar, S. and Anand, S. 2021. Nanotechnology for sustainable crop production: recent development and strategies. pp. 31–47. *In*: Singh, P., Singh, R., Verma, P., Bhadouria, R., Kumar, A. and Kaushik, M. (eds). Plant-Microbes-Engineered Nano-particles (PM-ENPs) Nexus in Agro-Ecosystems. Advances in Science, Technology & Innovation. Springer, Cham. https://doi.org/10.1007/978-3-030-66956-0_3.

Kumar, A., Bhattacharya, T., Mukherjee, S. and Sarkar, B. 2022a. A perspective on biochar for repairing damages in the soil–plant system caused by climate change-driven extreme weather events. Biochar 4: 22. https://doi.org/10.1007/s42773-022-00148-z.

Kumar, A., Bhattacharya, T., Shaikh, W.A., Chakraborty, S., Owens, G. and Naushad, M. 2022b. Valorization of fruit waste-based biochar for arsenic removal in soils. Environ. Res. 213: 113710. https://doi.org/10.1016/j.envres.2022.113710.

Kumar, A., Bhattacharya, T., Shaikh, W.A., Chakraborty, S., Sarkar, D. and Biswas, J.K. 2022c. Biochar modification methods for augmenting sorption of contaminants. Curr. Pollut. Reports. 8: 519–555. https://doi.org/10.1007/s40726-022-00238-3.

Lellis, B., Fávaro-Polonio, C.Z., Pamphile, J.A. and Polonio, J.C. 2019. Effects of textile dyes on health and the environment and bioremediation potential of living organisms. Biotechnol. Res. Innov. 3: 275–290. https://doi.org/10.1016/j.biori.2019.09.001.

Lenka, S.P., Shaikh, W.A., Owens, G., Padhye, L.P., Chakraborty, S. and Bhattacharya, T. 2021. Removal of copper from water and wastewater using dolochar. Water. Air. Soil Pollut. 232: 167. https://doi.org/10.1007/s11270-021-05135-x.

Li, J., Liu, K., Gao, X., Yao, B., Huo, K., Cheng, Y., Cheng, X., et al. 2015. Oxygen- and nitrogen-enriched 3D porous carbon for supercapacitors of high volumetric capacity. ACS Appl. Mater. Interfaces 7: 24622–24628. https://doi.org/10.1021/acsami.5b06698.

Li, Y., Wang, Z., Xie, X., Zhu, J., Li, R. and Qin, T. 2017. Removal of norfloxacin from aqueous solution by clay-biochar composite prepared from potato stem and natural attapulgite. Colloids Surfaces A Physicochem. Eng. Asp. 514: 126–136. https://doi.org/10.1016/j.colsurfa.2016.11.064.

Lian, J., Xiong, L., Cheng, R., Pang, D., Tian, X., Lei, J., He, R., et al. 2019. Ultra-high nitrogen content biomass carbon supercapacitors and nitrogen forms analysis. J. Alloys Compd. 809: 151664. https://doi.org/10.1016/j.jallcom.2019.151664.

Lisowski, P., Colmenares, J.C., Mašek, O., Lisowski, W., Lisovytskiy, D., Kamińska, A. and Łomot, D. 2017. Dual functionality of TiO_2/biochar hybrid materials: photocatalytic phenol degradation in the liquid phase and selective oxidation of methanol in the gas phase. ACS Sustain. Chem. Eng. 5: 6274–6287. https://doi.org/10.1021/acssuschemeng.7b01251.

Liu, S.B., Tan, X.F., Liu, Y.G., Gu, Y.L., Zeng, G.M., Hu, X.J., Wang, H., et al. 2016a. Production of biochars from Ca impregnated ramie biomass (*Boehmeria nivea* (L.) Gaud.) and their phosphate removal potential. RSC Adv. 6: 5871–5880. https://doi.org/10.1039/c5ra22142k.

Liu, T., Gao, B., Fang, J., Wang, B. and Cao, X. 2016b. Biochar-supported carbon nanotube and graphene oxide nanocomposites for Pb(II) and Cd(II) removal. RSC Adv. 6: 24314–24319. https://doi.org/10.1039/c6ra01895e.

Lyu, H., Zhang, Q. and Shen, B. 2020. Application of biochar and its composites in catalysis. Chemosphere. 240: 124842. https://doi.org/10.1016/j.chemosphere.2019.124842.

Ma, F., Zhao, H., Zheng, X., Zhao, B., Diao, J. and Jiang, Y. 2023. Enhanced adsorption of cadmium from aqueous solution by amino modification biochar and its adsorption mechanism insight. J. Environ. Chem. Eng. 11: 109747. https://doi.org/10.1016/j.jece.2023.109747.

Mittal, A.K., Chisti, Y. and Banerjee, U.C. 2013. Synthesis of metallic nanoparticles using plant extracts. Biotechnol. Adv. 31: 346–356. https://doi.org/10.1016/j.biotechadv.2013.01.003.

Mondal, N.K., Samanta, A., Dutta, S. and Chattoraj, S. 2017. Optimization of Cr(VI) biosorption onto *Aspergillus niger* using 3-level box-behnken design: equilibrium, kinetic, thermodynamic and regeneration studies. J. Genet. Eng. Biotechnol. 15: 151–160. https://doi.org/10.1016/j.jgeb.2017.01.006.

Norouzi, O., Safari, F., Jafarian, S., Tavasoli, A. and Karimi, A. 2017. Hydrothermal gasification performance of *Enteromorpha intestinalis* as an algal biomass for hydrogen-rich gas production using Ru promoted Fe–Ni/γ-Al_2O_3 nanocatalysts. Energy Convers. Manag. 141: 63–71. https://doi.org/10.1016/j.enconman.2016.04.083.

Norouzi, O., Maria, F. Di and Dutta, A. 2020. Biochar-based composites as electrode active materials in hybrid supercapacitors with particular focus on surface topography and morphology. J. Energy Storage. 29: 101291. https://doi.org/10.1016/j.est.2020.101291.

Orts, F., del Río, A.I., Molina, J., Bonastre, J. and Cases, F. 2018. Electrochemical treatment of real textile wastewater: trichromy procion HEXL®. J. Electroanal. Chem. 808: 387–394. https://doi.org/10.1016/j.jelechem.2017.06.051.

Phoon, B.L., Ong, C.C., Mohamed Saheed, M.S., Show, P.L., Chang, J.S., Ling, T.C., Lam, S.S., et al. 2020. Conventional and emerging technologies for removal of antibiotics from wastewater. J. Hazard. Mater. 400: 124658. https://doi.org/10.1016/j.jhazmat.2020.122961.

Pi, L., Jiang, R., Zhou, W., Zhu, H., Xiao, W., Wang, D. and Mao, X. 2015. g-C_3N_4 Modified biochar as an adsorptive and photocatalytic material for decontamination of aqueous organic pollutants. Appl. Surf. Sci. 358: 231–239. https://doi.org/10.1016/j.apsusc.2015.08.176.

Pradhan, N., Pal, A. and Pal, T. 2002. Silver nanoparticle catalyzed reduction of aromatic nitro compounds. Colloids Surfaces A Physicochem. Eng. Asp. 196: 247–257. https://doi.org/10.1016/S0927-7757(01)01040-8.

Premarathna, K.S.D., Rajapaksha, A.U., Sarkar, B., Kwon, E.E., Bhatnagar, A., Ok, Y.S. and Vithanage, M. 2019. Biochar-based engineered composites for sorptive decontamination of water: a review. Chem. Eng. J. 372: 536–550. https://doi.org/10.1016/j.cej.2019.04.097.

Qiu, G., Chen, H., Srinivasa Raghavan, D.S. and Ting, Y.P. 2021. Removal behaviors of antibiotics in a hybrid microfiltration-forward osmotic membrane bioreactor for real municipal wastewater treatment. Chem. Eng. J. 417: 129146. https://doi.org/10.1016/j.cej.2021.129146.

Rafatullah, M., Sulaiman, O., Hashim, R. and Ahmad, A. 2010. Adsorption of methylene blue on low-cost adsorbents: a review. J. Hazard. Mater. 177: 70–80. https://doi.org/10.1016/j.jhazmat.2009.12.047.

Ramola, S., Mohan, D., Masek, O., Méndez, A. and Tsubota, T. (eds). 2022. Engineered Biochar: Fundamentals, Preparation, Characterization and Applications. Springer Nature Singapore, Singapore. https://doi.org/10.1007/978-981-19-2488-0.

Rittl, T.F., Oliveira, D.M.S., Canisares, L.P., Sagrilo, E., Butterbach-Bahl, K., Dannenmann, M. and Cerri, C.E.P. 2021. High application rates of biochar to mitigate N_2O emissions from a n-fertilized tropical soil under warming conditions. Front. Environ. Sci. 8: 611873. https://doi.org/10.3389/fenvs.2020.611873.

Russell, J.N. and Yost, C.K. 2021. Alternative, environmentally conscious approaches for removing antibiotics from wastewater treatment systems. Chemosphere. 263: 128177. https://doi.org/10.1016/j.chemosphere.2020.128177.

Salcedo, O.D.C., Vargas, D.P., Giraldo, L. and Moreno-Piraján, J.C. 2021. Study of mercury [Hg(II)] Adsorption from aqueous solution on functionalized activated carbon. ACS Omega. 6: 11849–11856. https://doi.org/10.1021/acsomega.0c06084.

Shaheen, S., Saeed, Z., Ahmad, A., Pervaiz, M., Younas, U., Mahmood Khan, R.R., Luque, R., et al. 2023. Green synthesis of graphene-based metal nanocomposite for electro and photocatalytic activity; recent advancement and future prospective. Chemosphere. 311: 136982. https://doi.org/10.1016/j.chemosphere.2022.136982.

Shaikh, W.A., Chakraborty, S. and Ul Islam, R. 2018. UV-assisted photo-catalytic degradation of anionic dye (Congo red) using biosynthesized silver nanoparticles: a green catalysis. Desalin. Water Treat. 130: 232–242. https://doi.org/10.5004/dwt.2018.23004.

Shaikh, W.A., Alam, M.A., Alam, M.O., Chakraborty, S., Owens, G., Bhattacharya, T. and Mondal, N.K. 2020a. Enhanced aqueous phase arsenic removal by a biochar based iron nanocomposite. Environ. Technol. Innov. 19. https://doi.org/10.1016/j.eti.2020.100936.

Shaikh, W.A., Chakraborty, S. and Islam, R.U. 2020b. Photocatalytic degradation of rhodamine B under UV irradiation using Shorea robusta leaf extract-mediated bio-synthesized silver nanoparticles. Int. J. Environ. Sci. Technol. 17: 2059–2072. https://doi.org/10.1007/s13762-019-02473-6.

Shaikh, W.A., Islam, R.U., Chakraborty, S. 2021. Stable silver nanoparticle doped mesoporous biochar-based nanocomposite for efficient removal of toxic dyes. J. Environ. Chem. Eng. 9: 104982. https://doi.org/10.1016/j.jece.2020.104982.

Shaikh, W.A., Chakraborty, S., Islam, R.U., Ghfar, A.A., Naushad, M., Bundschuh, J., Maity, J.P., et al. 2022a. Fabrication of biochar-based hybrid Ag nanocomposite from algal biomass waste for toxic dye-laden wastewater treatment. Chemosphere. 289: 133243. https://doi.org/10.1016/j.chemosphere.2021.133243.

Shaikh, W.A., Kumar, A., Chakraborty, S., Islam, R.U., Bhattacharya, T. and Biswas, J.K. 2022b. Biochar-based nanocomposite from waste tea leaf for toxic dye removal: from facile fabrication to functional fitness. Chemosphere. 291: 132788. https://doi.org/10.1016/j.chemosphere.2021.132788.

Shaikh, W.A., Kumar, A., Chakraborty, S., Naushad, M., Islam, R.U., Bhattacharya, T. and Datta, S. 2022c. Removal of toxic dye from dye-laden wastewater using a new nanocomposite material: Isotherm, kinetics and adsorption mechanism. Chemosphere. 308: 136413. https://doi.org/10.1016/j.chemosphere.2022.136413.

Shaikh, W.A., Chakraborty, S., Kumar, A., Biswas, J.K., Jha, A.K., Bhattacharya, T., Vithanage, M., et al. 2023. Tailor-made biochar-based nanocomposite for enhancing aqueous phase antibiotic removal. J. Water Process Eng. 55: 104215. https://doi.org/https://doi.org/10.1016/j.jwpe.2023.104215.

Sharma, G., Dionysiou, D.D., Sharma, S., Kumar, A., Al-Muhtaseb, A.H., Naushad, M. and Stadler, F.J. 2019. Highly efficient Sr/Ce/activated carbon bimetallic nanocomposite for photoinduced degradation of rhodamine B. Catal. Today. 335: 437–451. https://doi.org/10.1016/j.cattod.2019.03.063.

Sharma, G. and Naushad, M. 2020. Adsorptive removal of noxious cadmium ions from aqueous medium using activated carbon/zirconium oxide composite: Isotherm and kinetic modelling. J. Mol. Liq. 310: 113025. https://doi.org/10.1016/j.molliq.2020.113025.

Shi, J. 2013. On the synergetic catalytic effect in heterogeneous nanocomposite catalysts. Chem. Rev. 113: 2139–2181. https://doi.org/10.1021/cr3002752.

Sulis, G., Batomen, B., Kotwani, A., Pai, M. and Gandra, S. 2021. Research article sales of antibiotics and hydroxychloroquine in India during the covid-19 epidemic: an interrupted time series analysis. PLoS Med. 18: e1003682. https://doi.org/10.1371/journal.pmed.1003682.

Tan, X.-F., Liu, Y.-G., Gu, Y.-L., Xu, Y., Zeng, G.-M., Hu, X.-J., Liu, S.-B., et al. 2016. Biochar-based nano-composites for the decontamination of wastewater: a review. Bioresour. Technol. 212: 318–333. https://doi.org/10.1016/j.biortech.2016.04.093.

Thines, K.R., Abdullah, E.C., Mubarak, N.M. and Ruthiraan, M. 2017. Synthesis of magnetic biochar from agricultural waste biomass to enhancing route for waste water and polymer application: a review. Renew. Sustain. Energy Rev. 67: 257–276. https://doi.org/10.1016/j.rser.2016.09.057.

Tripathi, R.M., Kumar, N., Shrivastav, A., Singh, P. and Shrivastav, B.R. 2013. Catalytic activity of biogenic silver nanoparticles synthesized by Ficus panda leaf extract. J. Mol. Catal. B Enzym. 96: 75–80. https://doi.org/10.1016/j.molcatb.2013.06.018.

Varadavenkatesan, T., Vinayagam, R. and Selvaraj, R. 2019. Green synthesis and structural characterization of silver nanoparticles synthesized using the pod extract of *Clitoria ternatea* and its application towards dye degradation. Mater. Today Proc. 23: 27–29. https://doi.org/10.1016/j.matpr.2019.04.216.

Vithanage, M., Herath, I., Joseph, S., Bundschuh, J., Bolan, N., Ok, Y.S., Kirkham, M.B., et al. 2017. Interaction of arsenic with biochar in soil and water: a critical review. Carbon. 113: 219–230. https://doi.org/10.1016/j.carbon.2016.11.032.

Wang, G., Zhang, L. and Zhang, J. 2012. A review of electrode materials for electrochemical supercapacitors. Chem. Soc. Rev. 41: 797–828. https://doi.org/10.1039/c1cs15060j.

Wang, H., Gao, B., Wang, S., Fang, J., Xue, Y. and Yang, K. 2015a. Removal of Pb(II), Cu(II), and Cd(II) from aqueous solutions by biochar derived from $KMnO_4$ treated hickory wood. Bioresour. Technol. 197: 356–362. https://doi.org/10.1016/j.biortech.2015.08.132.

Wang, M.C., Sheng, G.D. and Qiu, Y.P. 2015b. A novel manganese-oxide/biochar composite for efficient removal of lead(II) from aqueous solutions. Int. J. Environ. Sci. Technol. 12: 1719–1726. https://doi.org/10.1007/s13762-014-0538-7.

Wang, S., Gao, B., Li, Y., Creamer, A.E. and He, F. 2017a. Adsorptive removal of arsenate from aqueous solutions by biochar supported zero-valent iron nanocomposite: batch and continuous flow tests. J. Hazard. Mater. 322: 172–181. https://doi.org/10.1016/j.jhazmat.2016.01.052.

Wang, X., Bayan, M.R., Yu, M., Ludlow, D.K. and Liang, X. 2017b. Atomic layer deposition surface functionalized biochar for adsorption of organic pollutants: improved hydrophilia and adsorption capacity. Int. J. Environ. Sci. Technol. 14: 1825–1834. https://doi.org/10.1007/s13762-017-1300-8.

Wang, L., Yan, W., He, C., Wen, H., Cai, Z., Wang, Z., Chen, Z., et al. 2018a. Microwave-assisted preparation of nitrogen-doped biochars by ammonium acetate activation for adsorption of acid red 18. Appl. Surf. Sci. 433: 222–231. https://doi.org/10.1016/j.apsusc.2017.10.031.

Wang, Y.Q., Bai, R., Di, H.J., Mo, L.Y., Han, B., Zhang, L.M. and He, J.Z. 2018b. Differentiated mechanisms of biochar mitigating straw-induced greenhouse gas emissions in two contrasting paddy soils. Front. Microbiol. 9. https://doi.org/10.3389/fmicb.2018.02566.

Wang, S., Zhao, M., Zhou, M., Li, Y.C., Wang, J., Gao, B., Sato, S., et al. 2019. Biochar-supported nZVI (nZVI/BC) for contaminant removal from soil and water: a critical review. J. Hazard. Mater. 373: 820–834. https://doi.org/10.1016/j.jhazmat.2019.03.080.

Wang, H., Wang, X., Liang, M., Chen, G., Kong, R.M., Xia, L. and Qu, F. 2020. A boric acid-functionalized lanthanide metal-organic framework as a fluorescence "turn-on" Probe for selective monitoring of Hg^{2+} and CH_3Hg^+. Anal. Chem. 92: 3366–3372. https://doi.org/10.1021/acs.analchem.9b05410.

Woo, H.C. and Jhung, S.H. 2021. Adsorptive removal of nitro- or sulfonate-containing dyes by a functional metal–organic framework: quantitative contribution of hydrogen bonding. Chem. Eng. J. 425: 130598. https://doi.org/10.1016/j.cej.2021.130598.

Wu, X.L., Wen, T., Guo, H.L., Yang, S., Wang, X. and Xu, A.W. 2013. Biomass-derived sponge-like carbonaceous hydrogels and aerogels for supercapacitors. ACS Nano. 7: 3589–3597. https://doi.org/10.1021/nn400566d.

Wu, Z., Yuan, X., Zhong, H., Wang, H., Jiang, L., Zeng, G., Wang, H., et al. 2017. Highly efficient adsorption of Congo red in single and binary water with cationic dyes by reduced graphene oxide decorated NH_2–MIL-68(Al). J. Mol. Liq. 247: 215–229. https://doi.org/10.1016/j.molliq.2017.09.112.

Xiu, S., Shahbazi, A. and Li, R. 2017. Characterization, modification and application of biochar for energy storage and catalysis: a review. Trends Renew. Energy. 3: 86–101. https://doi.org/10.17737/tre.2017.3.1.0033.

Xu, D., Tong, Y., Yan, T., Shi, L. and Zhang, D. 2017. N,P-Codoped Meso-/Microporous carbon derived from biomass materials via a dual-activation strategy as high-performance electrodes for deionization capacitors. ACS Sustain. Chem. Eng. 5: 5810–5819. https://doi.org/10.1021/acssuschemeng.7b00551.

Yan, J., Han, L., Gao, W., Xue, S. and Chen, M. 2015a. Biochar supported nanoscale zerovalent iron composite used as persulfate activator for removing trichloroethylene. Bioresour. Technol. 175: 269–274. https://doi.org/10.1016/j.biortech.014.10.103.

Yan, L., Kong, L., Qu, Z., Li, L. and Shen, G. 2015b. Magnetic biochar decorated with ZnS nanocrytals for Pb(II) removal. ACS Sustain. Chem. Eng. 3: 125–132. https://doi.org/10.1021/sc500619r.

Yan, J., Zuo, X., Yang, S., Chen, R., Cai, T. and Ding, D. 2022. Evaluation of potassium ferrate activated biochar for the simultaneous adsorption of copper and sulfadiazine: Competitive versus synergistic. J. Hazard. Mater. 424: 127435. https://doi.org/10.1016/j.jhazmat.2021.127435.

Yang, X., Zhang, S., Ju, M. and Liu, L. 2019. Preparation and modification of biochar materials and their application in soil remediation. Appl. Sci. 9. https://doi.org/10.3390/app9071365.

Yang, W., Feng, G., Miles, D., Gao, L., Jia, Y., Li, C. and Qu, Z. 2020. Impact of biochar on greenhouse gas emissions and soil carbon sequestration in corn grown under drip irrigation with mulching. Sci. Total Environ. 729: 138752. https://doi.org/10.1016/j.scitotenv.2020.138752.

Yao, Y. 2013. Sorption of phosphate and other contaminants on biochar and its environmental implications. ProQuest Diss. Theses. 177.

Yao, Y., Gao, B., Chen, J. and Yang, L. 2013. Engineered biochar reclaiming phosphate from aqueous solutions: mechanisms and potential application as a slow-release fertilizer. Environ. Sci. Technol. 47, 8700–8708. https://doi.org/10.1021/es4012977.

Ye, S., Yan, M., Tan, X., Liang, J., Zeng, G., Wu, H., Song, B., et al. 2019. Facile assembled biochar-based nanocomposite with improved graphitization for efficient photocatalytic activity driven by visible light. Appl. Catal. B Environ. 250: 78–88. https://doi.org/10.1016/j.apcatb.2019.03.004.

Yoon, K., Cho, D.W., Kwon, G., Rinklebe, J., Wang, H. and Song, H. 2023. Practical approach of As(V) adsorption by fabricating biochar with low basicity from $FeCl_3$ and lignin. Chemosphere. 329: 138665. https://doi.org/10.1016/j.chemosphere.2023.138665.

Zeng, K., Hachem, K., Kuznetsova, M., Chupradit, S., Su, C.H., Nguyen, H.C. and El-Shafay, A.S. 2022. Molecular dynamic simulation and artificial intelligence of lead ions removal from aqueous solution using magnetic-ash-graphene oxide nanocomposite. J. Mol. Liq. 347. https://doi.org/10.1016/j.molliq.2021.118290.

Zhang, M., Gao, B., Yao, Y., Xue, Y. and Inyang, M. 2012. Synthesis of porous MgO-biochar nanocomposites for removal of phosphate and nitrate from aqueous solutions. Chem. Eng. J. 210: 26–32. https://doi.org/10.1016/j.cej.2012.08.052.

Zhang, S. and Lu, X. 2018. Treatment of wastewater containing Reactive Brilliant Blue KN-R using TiO_2/BC composite as heterogeneous photocatalyst and adsorbent. Chemosphere. 206: 777–783. https://doi.org/10.1016/j.chemosphere.2018.05.073.

Zhang, S., Zhong, L., Wang, J., Tang, A. and Yang, H. 2021. Porous carbon-based $MgAlF5 \cdot 1.5H_2O$ composites derived from carbon-coated clay presenting super high adsorption capacity for Congo Red. Chem. Eng. J. 406: 126784. https://doi.org/10.1016/j.cej.2020.126784.

Zhou, Y., Gao, B., Zimmerman, A.R., Chen, H., Zhang, M. and Cao, X. 2014. Biochar-supported zerovalent iron for removal of various contaminants from aqueous solutions. Bioresour. Technol. 152: 538–542. https://doi.org/10.1016/j.biortech.2013.11.021.

Zhu, Z., Fan, W., Liu, Z., Yu, Y., Dong, H., Huo, P. and Yan, Y. 2018. Fabrication of the metal-free biochar-based graphitic carbon nitride for improved 2-Mercaptobenzothiazole degradation activity. J. Photochem. Photobiol. A Chem. 358: 284–293. https://doi.org/10.1016/j.jphotochem.2018.03.027.

Chapter 10

Removal of Toxic Azo Dyes from Wastewater Using Multi-functional Biochar-based Nanocomposite

Wasim Akram Shaikh*[1,2], Sukalyan Chakraborty*[2], Abhishek Kumar[2] and Tanushree Bhattacharya[2]

[1]School of Science and Technology, The Neotia University, Sarisha, South 24 Parganas, West Bengal, India–743368

[2]Dept. of Civil and Environmental Engineering, Birla Institute of Technology, Mesra, India–835215

Rafique Ul Islam[3]

[3]Department of Chemistry, School of Physical Sciences, Mahatma Gandhi Central University, Motihari, Bihar, India – 845401

Jayanta Kumar Biswas[4,5]

[4]Enviromicrobiology, Ecotoxicology and Ecotechnology Research Laboratory (3E-MicroToxTech Lab), Department of Ecological Studies, University of Kalyani, Kalyani, Nadia- 741235, West Bengal, India

[5]International Centre for Ecological Engineering, University of Kalyani, Kalyani- 741235, West Bengal, India

10.1 INTRODUCTION

Dye is a finely ground solid organic substance (dispersed in a liquid) that imparts colour to materials (cosmetics, leather, paper, textiles, and other materials), which does not washout by heat, light, or washing (Majumdar et al. 2022). Dyes are broadly categorized into two types viz, natural dye and synthetic dye. Natural dyes are derived from natural sources like vegetables, plants, trees, lichens, mollusks, and even from few insects. Synthetic dyes are manufactured from

*Corresponding Authors: sukalyanchakraborty@bitmesra.ac.in; wasim.bu12@gmail.com

minerals, petroleum by products, and organic chemicals (Shaikh et al. 2022b). Natural dyes are generally biodegradable, resulting in null or less environmental effects. However, the synthetic dyes and dye by-products are quite complex and highly toxic in nature creating tremendous threat to the aquatic ecosystem, environment, and human health, when discharged untreated into the environment (Chowdhury et al. 2020). The first synthetic dye, picric acid (2,4,6-trinitrophenol) was identified in 1771 by Peter Woulfe, which produced yellow colour on silk. However, the discovery of the first synthetic dye was recorded in 1856 by an 18-year-old chemistry student Sir William Perkin, during the fabrication of quinine, an anti-malarial drug, derived from coal tar. These synthetic dyes can affect aquatic plants by reducing sunlight transmission, can damage the central nervous system, brain, liver, reproductive system, and kidneys of human, and even can be mutagenic and carcinogenic to aquatic organisms and humans (Shaikh et al. 2020b).

The first synthetic dye, picric acid, was identified in 1771, which produced yellow colour on silk. Although, the discovery of the first synthetic dye was recorded in 1856 by an 18-year-old chemistry student Sir William Perkin, during the fabrication of quinine, an anti-malarial drug, derived from coal tar. He observed a black residue during experiments accidentally, which turned into a purple colour solution by dissolving methylated spirit, eventually worked well on silk. That purple-coloured compound (dye) was later named 'mauve' (a product of dichromate oxidation of impure aniline). Sometime between 1858 and 1859, a vivid purplish-red colour dye, 'fuchsia,' was developed by French chemist François-Emmanuel Verguin during the reaction of aniline ($C_6H_5NH_2$) with stannic chloride ($SnCl_4$), which he named 'fuchsine.' The discovery of fuchsine was the first synthetic triphenylmethane dye which triggered the second phase of the synthetic dye industry. Later some other types of synthetic dyes, including anthraquinone dyes, azo dyes, fluorescent brighteners, food dyes, phthalocyanine compounds, reactive dyes, quinacridone compounds, and xanthene dyes were also invented. The main advantages of synthetic dyes are cost and consistent lightfastness.

The main aim of the chapter is to summarize the advantages of biochar-based nanocomposites fabrication like generate new binding sites, and intensify overall surface area, which efficiently enhance the overall azo dye removal efficiency. The biochar-based nanocomposites were found to be techno-economically feasible and thermo-stable composites. Moreover, the utilization of waste biomasses would prevent them from ending up in landfill sites (reducing the pressure on landfill sites, minimizing release of leachates, and increasing levels of hygiene and sanitation), treat dye effluent efficiently, and not produce any toxic waste during synthesis or treatment.

10.1 CLASSIFICATION OF DYES

Synthetic dyes can be classified in several ways, including their colour, structural or functional groups, ionic charge dissolution in the aqueous medium, and application methods. Yagub et al. (2014) classified dyes based on chemical structure as presented in Table 10.1. However, the most commonly used classification is based on the ionic configuration and application due to the complexities of the colour nomenclature. Dyes are mainly classified as non-ionic and ionic (Figure 10.1). Ionic dyes are sub-classified as cationic and anionic, while the non-ionic dyes are sub-classified as two categories vat and disperse. Anionic dyes are further divided into three categories: acid dye, direct dye, and reactive dye (Table 10.2) (Panda et al. 2021). Several industries, including carpet, cosmetics, food paper, plastic, printing, and textile, use dyes to provide colour to their products for aesthetic view. However, after the dyeing process, most of the dyes and dyestuff are discharged in the environment as effluents or wastewater, which poses a higher risk to human health and environment because of their considerable reactivity, toxicity, and carcinogenicity. To maintain industrial production and sustainable development, it is thus necessary to treat the dye-laden effluent or wastewater before discharging into the environment.

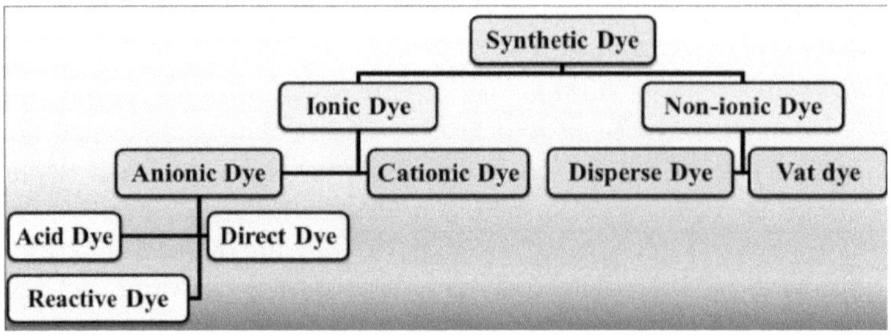

Figure 10.1 Classification of dyes

Table 10.1 Classification of dye based on structure (Yagub et al. 2014)

Sl. No.	Class	Chromospheres	Example
1	Azo dyes	–N=N–	Congo red
2	Anthraquinone dyes	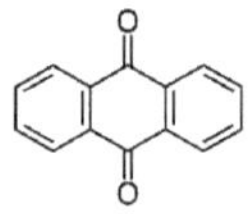	Reactive Blue 4
3	Indigoid dyes	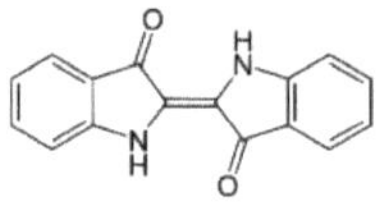	Acid Blue 71
4	Nitroso dyes	–N=O	Acid green 1
5	Nitro dyes	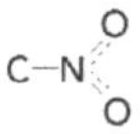	Acid Yellow 24
6	Triarylmethane dyes	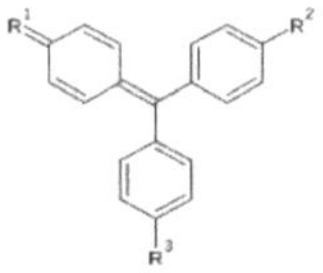	Basic Red 9
7	Xanthene dyes	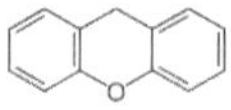	Rhodamine B

Table 10.2 Classification of dye based on the properties and application (Tan et al. 2015)

Classification of Dye	Properties	Examples	Applications	Toxicity
Anionic Dye	Water-soluble, Anionic	Acid red 10, sunset yellow, indigo carmine, and methyl orange	Leather, nylon, paper, silk, and wool colouring	Carcinogen
Cationic Dye	Water-soluble, Cationic	Basic green 5, Basic violet, Methylene blue, and Rhodamine B	Paper, nylons, polyacrylonitrile, polyesters colouring and indicator	Carcinogen
Disperse Dye	Water-insoluble, Non-ionic	Disperse orange, disperse red, and disperse yellow	Acrylic fibers, cellulose, nylon, and polyester colouring	Allergenic and carcinogen
Vat Dye	Low solubility in alkaline water	Vat blue 4, vat green 11, and vat orange 15	Cellulosic fibers colouring	Allergenic

10.2 WHAT ARE AZO DYES?

Azo dyes are a highly diverse groups of synthetic dyes. Azo dyes are organic in nature containing R–N=N–R′ functional group, where R and R′ represent aryl group. The aqueous solution of azo dyes is generally anionic due to the presence of 1–3 sulfonic acid groups. However, some azo dyes can also be found in cationic please. The tautomeric formation of azo dyes is found in the phenolic diazo dyes like Congo red. In general, the azo group is the backbone of azo dyes act as auxochrome, where chromophoric groups are the solubilizing groups. The auxochrome and associated chromophores determines the colour of the azo dyes by forming covalent bonds with substrates in textile. Azo dyes are classified into five types: Monozo, Disazo, Trisazo, Polyazo, and Azoic dye (Figure 10.2).

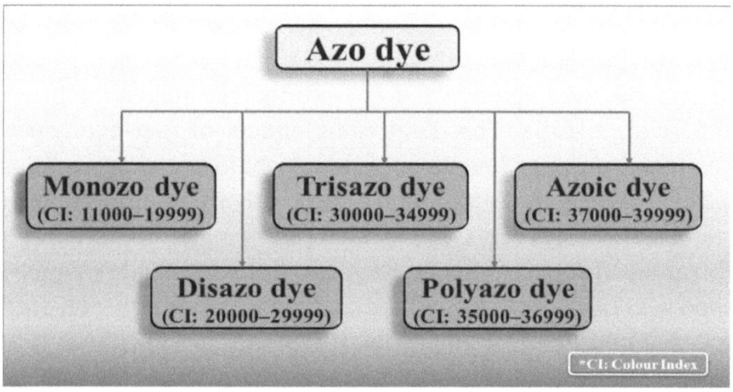

Figure 10.2 Classification of azo dyes.

10.3 SYNTHESIS OF AZO DYES

In 1858, the German chemist Johann Peter Griess developed a low-yield yellow compound having very similar dye-like properties from the reaction of nitrous acid (HONO) and aniline ($C_6H_5NH_2$). Later on, the yield problem was solved by reacting nitrous acid with aryl amines, resulting in higher yield reactive intermediates, which was proven to be the key step for the synthesis of ~50% of the commercially available synthetic dyes nowadays. However, the background chemistry of the reaction between aryl amines and nitrous acid was unclear at that time. In 1866, Kekulé proposed that the reaction of aryl group (Ar-NH_2) and nitrous acid produced an aryldiazonium ion (Ar-NN^+) which readily couples with anilines or phenols through a –N=N– bond (azo bond). That azo bond containing aryl intermediates (azo bond containing coloured aryl compounds) was termed azo dyes. Later diazo dye (dye with two azo groups) was synthesized through Diazotization, an electrophilic aromatic substitution reaction resulting in coupling primary aromatic amine (primarily m-phenylenediamine) into its diazonium salt. In general, azo dyes are synthesized through coupling reaction, through an electrophilic substitution of aryl diazonium cations with electron-donating aromatic coupling partners ($ArN_2^+ + Ar'H \rightarrow ArN = NAr' + H^+$).

Congo red which is one of the dyes used in the present study is also an azo dye, formerly known as a direct dye, is a conjugated diazo dye synthesized by the coupling reaction of 4-sulfo-1-naphthylamine ($C_{10}H_9NO_3S$) with 4,4′-Biphenylbis(diazonium) salt ($C_{12}H_8N_4^{+2}$) and found to be effective for cotton dyeing through simple immersion on cotton fabric in a hot aqueous bath. However, after the seventieth century (1970), the methyl- or methoxy-substituted benzidines derivatives (major group of diazo dyes) have not been produced in many countries due to their carcinogenic properties (Yaneva and Georgieva 2012).

10.4 SOURCES, USE, AND EFFECTS OF AZO DYES

Azo dyes are mainly used in the textile and pharmaceutics industries for dyeing garments, pH indicator, amyloidosis diagnosis, and histological stain for amyloid. Congo red (CR), a potential azo dye, is reported as a potent toxic dye causing neuro-degenerative disorders and associated diseases (Shaikh et al. 2022b). Azo dyes have been reported as a strong mutagen causing restricted cell division, reduced serum protein, aggregation of platelets, and blood clotting, reproductive disorders, gastrointestinal irritation, and respiratory problems (Shaikh et al. 2018). Cyclized-cyanine derivatives (ACC1 and ACC2) groups containing azo dyes are used to treatment of cancer in two-photon photodynamic therapy. Some of the azo dyes are also used in hypnotic medicine, which is used for the treatment of insomnia. In addition, benzidine and naphthoic acid derivatives of azo dyes are reported as co-carcinogens. Therefore, comprehending the toxic character of the dyes, it is important to search for feasible treatment options before discharging them into aquatic systems.

It is estimated that ~0.7 million tons of dye-related wastes are annually discharged as effluents, where the non-biodegradable dyes and dyestuff (by-products) threaten the biosphere, where ~60% are azo dyes (Shaikh et al. 2021b). The deleterious effects of azo dyes on aquatic ecosystems include minimized dissolved oxygen levels, diminished photosynthesis, reduced light penetration, and aesthetic damage(Li et al. 2020). Additionally, the synthetic azo dyes being non-biodegradable persist for long in aquatic environments and could get biomagnified after entering food chains, resulting in a significant adverse toxic effect on aquatic flora and fauna (Yagub et al. 2014). Even, soils can get impinged with azo dyes during their irrigation with dye-rich wastewater, triggering subsequent damage to the microbial communities and plant growth (Imran et al. 2015).

10.5 AZO DYE TREATMENT TECHNIQUES

10.5.1 Existing Treatment Techniques for azo Dye Laden (Waste) Water

Initially (early 1990's), the discolouration of dye-laden wastewater was carried out by equalization and sedimentation (Samsami et al. 2020). The development of discolouration or azo dye treatment technologies was however triggered after implementation of stricter effluent release standards in various countries across the world. To minimize the adverse environmental effects, several protecting agencies were established like Environment Agency (EA) of England and Wales, Scottish Environment Protection Agency (SEPTA) of Scotland, Ecological and Toxicological Association of the Dyestuffs Manufacturing Industry (ETAD) of Switzerland, and European Community (EC), etc. Moreover, few countries have their own regulation like the UK enforcing a law prohibiting discharge of synthetic effluents into marine environments; Australia, Canada, EU, USA have also set the maximum acceptable limits of different parameters for effluent discharge (Hao et al. 2000). However, in developing countries like India, Malaysia, Pakistan, the limits relating to effluent contamination have been recommended and not made mandatory (Samsami et al. 2020). To meet the standard regulation for azo dye laden effluent or wastewater discharge, four major types of the processes were followed viz, biological, electrochemical, chemical, and physical treatment technologies (Figure 10.3) (Miri et al. 2022, Choudhry et al. 2023, Kazak et al. 2023, Xia et al. 2023, Zhang et al. 2023).

The main objective of the discolouration was primarily to remove coloured substances or chemicals from synthetic solution or (azo)dye-laden wastewater. However, practically this process turned much more complex due to the presence of the high organic substances, formation of intermediate products, BOD, COD, heavy metals, and co-existing ions (Hao et al. 2000). The azo dyes and dyestuffs inhibit plant growth in the aquatic environment, enter the food chain, and bioaccumulate, thereby promoting toxicity, mutagenicity, and carcinogenicity in the aquatic organisms and humans (by consumption of the aquatic organisms). Indeed, these synthetic azo dyes and dyestuffs could trigger high alterations in ecological condition of aquatic flora and fauna, resulting from the toxicity and lack of their biodegradability (Berradi et al. 2019).

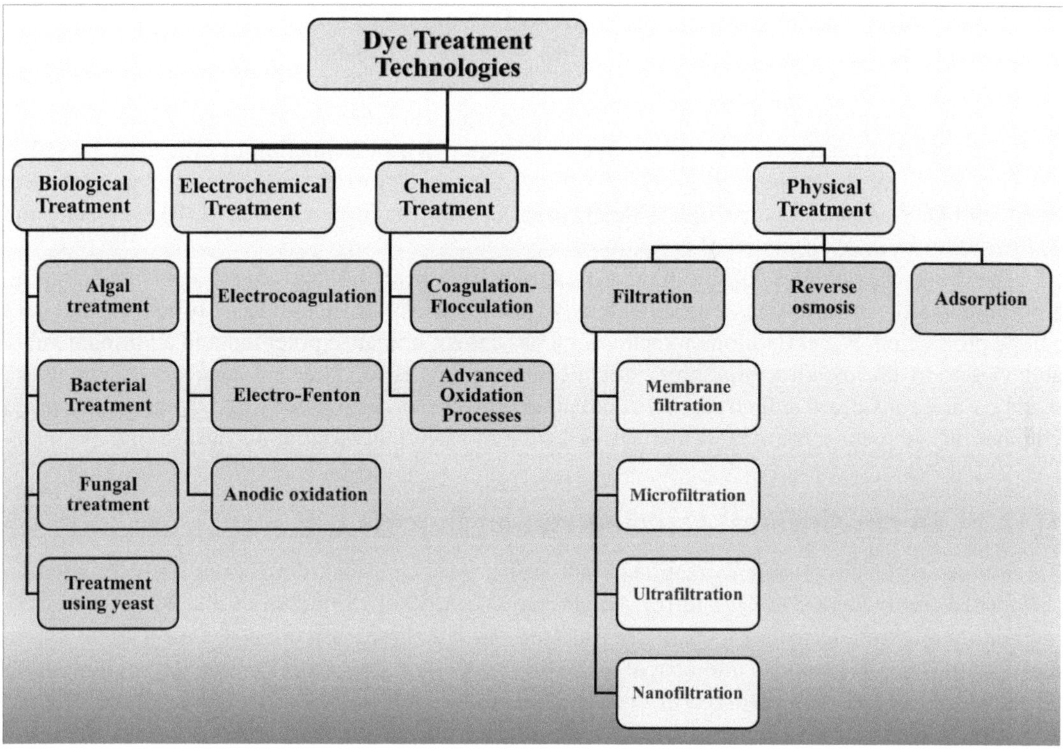

Figure 10.3 Different dye treatment technologies.

10.5.2 Biological Treatment of Synthetic azo Dyes

Biological treatment emerged as a cost-effective, environmentally friendly, and energy-saving process without the generation of toxic by-products and produces less sludge. There are two different pathways of azo dye decolourization using bio-organisms: (a) adsorption on the microbial biomass, and (b) biodegradation using microbes. Biodegradation has significant advantages to decolourize; basically, biodegradable azo dyes are converted into smaller non-toxic bi-products. Several bio-organisms were used to treat the azo dye, including algae, bacteria, fungi, and yeast. These bio-organisms can transform toxic azo dyes into by-products which are less harmful. The living organisms degrade the biodegradable azo dye through their metabolism or adsorb through electrostatic attraction between the cell wall (enabled by the availability of active functional groups like amino, carboxyl, hydroxyl, lipids, and (hetero)polysaccharides) and azo dye (Costa et al. 2018). The main bi-products of aerobic degradation are CO_2, water, dye fragments, and sludge biomass, where the anaerobic process produces the same except methane instead of carbon dioxide (Samsami et al. 2020).

Algal growth is primarily dependent on sunlight, CO_2, and nutrients, where nutrients were taken from wastewater, did not generate any secondary waste product, resulting in cheaper, non-toxic, and environment-friendly conventional treatment methods. Moreover, after degradation excess, algal biomass could also be used as production biomass and oil. Al-Fawwaz and Abdullah (2016) reported algae can metabolize dyes and also adsorb azo dyes and ions like: $-OH^-$, $RCOO^-$, $-NH_2$, and PO_4^{3-}. Bacteria, such as *Rhodococcus* sp., *Bacillus subtilis*, *Klebsiella* sp., *Pseudomonas luteola*, *Aeromonas* sp., and *Shigella* sp. are reported biodegrading azo dyes due to ease to culture and growing faster (Chang et al. 2001). The biodegradation of azo dye using bacterial took place due to the azo-reductase enzyme-assisted cleavage of azo bonds (–N=N–) (Chang et al. 2001).

Fungi and yeast can also decolourize azo dyes, but little research is available to support it due to the complex handling and growth in stressed conditions (Samsami et al. 2020). Fungi such as *Trametes versicolour* and *Cylindrocephalum aurelium* RY06 have been reported to degrade azo dyes like Methyl Red and Mordant Orange-1, showing an efficiency of about 91% and 85%, respectively (Martorell et al. 2018). *Ascomycetous* sp., *Pseudozyma rugulosa, Candida krusei* were the reported yeast species for azo dye degradation with the maximum efficiency of ~92% (Yu and Wen 2005). Mostafa et al. (2019) showed intra- and extra-cellular enzymes, including enzymes like manganese peroxidase (MnP), laccase, and, lignin peroxidase (LiP) accelerate the dye metabolism in environmental conditions.

The main advantages of biological treatment of synthetic azo dyes are cost-effective, environmentally friendly, less energy intensive techniques with no toxic bi-products, and less sludge production. Recently bionanotechnology has shown immense potential and additional value and vigour to the existing biological treatments (Pandey et al. 2021). However, though these methods are considered effective, environment friendly, the main disadvantages are incomplete elimination, time-consuming, and ineffective for large-scale industrial applications.

10.5.3 Electrochemical Decolourization Process

Three types of electrochemical treatments are being practiced, including: (a) Electrocoagulation (EC), (b) Electro-Fenton (EF), and (c) Anodic oxidation (AO) (Shindhal et al. 2021). In the EC method, two metal electrodes (Aluminium or Iron) with a direct power source were used, where metal anodes act as catalysts and coagulants. During operation, flocculated species begin floating on the surface by H_2 gas (evolved at the cathode). In situ generation of coagulants take place at the suitable electrode in electro-chemical reactors, responsible for the electrolytic oxidation and reduction reaction. EF basically hybridizes Fenton's degradation and electrolysis, and removes the colour substance by simultaneous oxidation and coagulation reaction (Bassyouni et al. 2017). AO degrades the azo dye by direct or indirect oxidations. In direct degradation, anode's surface absorb the azo dye molecule and then degrade, while indirect oxidation triggers electro-chemically produced strong oxidants (like O_3, OCl^-, $OH^\bullet$) (Aravind et al. 2023, Hamad et al. 2018).

There are two types of indirect oxidation like:

(a) **generation of active radicals:**

$$\text{Metal salt} + H_2O \rightarrow \text{Metal}(OH^\bullet) + H^+ + e^- \quad (1)$$

$$\text{Azo dye} + \text{Metal}(OH^\bullet) \rightarrow M^{+n} + CO_2 + H_2O \quad (2)$$

(b) **generation of chlorinated species:**

$$2Cl^- \text{(hypochlorite ions or hypochlorous acid)} \rightarrow Cl_2 + 2e^- \quad (3)$$

$$Cl_2 + H_2O \rightarrow HOCl + Cl^- + H^+ \quad (4)$$

$$HOCl \rightarrow OCl^- + H^+ \quad (5)$$

These active radicals and chlorinated ions degrade azo dyes to produce smaller fragments like CO_2 and H_2O.The main advantages of the EC method were: (a) operable in ambient conditions, (b) low sludge generation, (c) high reaction rate, and (d) environment friendly. The major disadvantages are the maintenance and optative cost, generation of toxic bi-products like OCl^- and not suitable for large-scale industrial application.

10.6 CHEMICAL PROCESSES FOR AZO DYE REMOVAL

Chemical treatment is one of the most effective treatment technologies where chemicals are used to degrade azo dyes. Though it is a very efficient treatment technology, this process is recently in the shade due to being more expensive than biological or physical treatments, with highly toxic by-products, requiring heavy equipment, costly maintenance, and requiring high energy. There are two popular chemical methods viz., (a) Coagulation-Flocculation and (b) Advanced Oxidation Processes (AOPs) (Faraji et al. 2023, Sachin et al. 2023).

Coagulation-flocculation is a highly popular chemical decolourization techniques where colour substances, including solid particles, are removed through sedimentation. Coagulants like metal salts (aluminium sulphate solution ($Al_2(SO_4)_3 \cdot 18H_2O$), alum (chrome alum: $KCr(SO_4)_2 \cdot 12H_2O$) and polymers (MO–KCl: *Moringa oleifera* Lam. Seed extract in KCl solution) has been reported to degrade azo dyes (Dotto et al. 2019).

AOPs are also efficient and widely adopted wastewater treatment technologies based on the in-situ generation of strong oxidizing agents and radicals like hydroxyl radicals ($OH^{\bullet}$), hydroperoxyl radical ($HO_2^{\bullet}$), superoxide radical anion ($^{\bullet}O_2^-$), etc.(Shaikh et al. 2021a, 2018, 2020b). Various AOPs were extensively used to dye laden wastewater treatment including, degradation, fenton, ozonation, photo-fenton, ultrasound, and UV-based. Among them, degradation is the most efficient and commonly practiced azo dye-laden wastewater treatment technology (Monga et al. 2022).

10.7 PHYSICAL TREATMENT TECHNIQUES FOR AZO DYE REMOVAL

The mass transfer mechanism is the basic principle of physical azo dye treatment technology. Applying this physical method ~86.8–99% of azo dyes (either single azo dye or a mixture of dyes) can be removed, making it the most efficient among these four (biological, electrochemical, chemical, and physical) methods (Shaikh et al. 2022c, Shindhal et al. 2021). The main advantage of the physical method is the low operational cost, simple design, high efficiency, easy operation, and most importantly, it can bypass toxic chemical hazards (Samsami et al. 2020). Several types of physical treatments have been developed and implemented to treat the azo dyes and azo dye-laden wastewater, including filtration (membrane filtration, microfiltration, ultrafiltration, nanofiltration), reverse osmosis (RO), and adsorption (Samsami et al. 2020).

Filtration is one of the cutting-edge physical dye treatment technologies. Different types of filtration methods are also used for azo dye treatment, like membrane filtration, where wastewater passes through the membrane. Typical microporous membranes are used in this microfiltration process, where the pore size ranges between 10 μm and 0.5 nm (Shindhal et al. 2021). In the reverse osmosis process, the contaminated water or effluent is passed through a semi-permeable membrane to remove dissolved solids, dyes, organic matters, ions, and microorganisms (Al-Bastaki 2004). It converts the wastewater to almost deionized water without thermal energy or chemicals. This process is extensively implemented in beverage industries, distillery spent wash, and groundwater treatment.

10.8 ADSORPTIVE AZO DYE REMOVAL

Adsorption is the most effective, low-cost, and eco-friendly physical separation process with a wide range of applications. Most importantly, adsorption has also been proven to be superior with respect to chemical, biological, and even other physical methods in terms of reusability, simplicity of design, and ease of operation without the generation of toxic by-products. Adsorption refers to a surface phenomenon in which liquid or gaseous phase adsorbates are attracted to a solid adsorbent surface by either physisorption or chemisorption (in some cases, both). Adsorption on porous carbon was documented in 1550 B.C. (in an ancient Egyptian papyrus) and later by Hippocrates and Pliny for their medicinal purpose (Gupta and Suhas 2009). The first scientific document on adsorption was reported by C.W. Scheele in 1773 for gas adsorption on carbon (Lennartson 2017). However, the term 'adsorption' was introduced by Kayser in 1881 (Gupta and Suhas 2009). The basic physisorption consisted of different types of forces like electrostatic and hydrophobic interactions, pore fillings, van der wall forces, and hydrogen bonding (Shaikh et al. 2022a, c).

10.9 DIFFERENT ADSORBENTS FOR AZO DYE AND THEIR LIMITATIONS

The most important (key) step of an adsorption study is the selection of adsorbents. Researchers used a deluge of adsorbents for the treatment of azo dyes from wastewater. Adsorbents are generally small porous materials that can effectively remove contaminants from the solid, liquid, or gaseous phase. Broadly, these adsorbents are categorized into two types, inorganic adsorbents and organic adsorbents (Adeyemo et al. 2017). Inorganic adsorbents like montmorillonite clay and bentonite were reported to remove malachite green and Congo red with maximum removal of 95% by Tahir et al. (2010). A study on aqueous methyl orange adsorption (>80% removal) was reported by Du et al. (2022) with quaternized kapok fibre adsorbent. Clay-mediated adsorption of methyl red onto clay was also a very efficient adsorbent for azo dye removal (Romdhane et al. 2020). Mariyam et al. (2021) reported Chrysoidine R adsorption onto mesoporous carbon achieving >95% decolourization efficiency. However, most of the adsorbents used in the adsorption are organic or bioinspired. For example, El Nemr et al. (2022) reported aqueous Direct Red 23 adsorption using mesoporous nano-activated carbon prepared from green alga *Ulva lactuca* biomass. Various types of adsorbents used for azo dye adsorption are summarized in Table 10.3.

10.10 MODIFIED ADSORBENTS FOR AZO DYE ADSORPTION

Although biochar-mediated azo dye adsorption showed several advantages, including wide ranges or applicability, non-toxic, eco-friendly, and carbon-rich adsorbent, it often fails to meet the desired large-scale efficiency due to the lower surface area and limited adsorption capacity (Abd-Elhamid et al. 2020). Therefore, various methods were adopted to modify and decorate biochar to enhance its properties and efficiency. Most extensive and efficient modifications are either chemical modifications or physical modifications. The chemical modifications are acid modification (by the treatment of hydrochloric acid, nitric acid, oxalic acid, and sulfuric acid) (Peng et al. 2016), alkali modification (by potassium and sodium hydroxide) (Rajapaksha et al. 2016), modification with oxidizing-agents (by hydrogen peroxide) (Wang et al. 2015), carbonaceous materials modification, and modification with metal and metal oxide (Tan et al. 2016), while the physical modification mainly includes steam and gas purging (Wang and Wang 2019).

10.11 BIOCHAR-BASED NANOCOMPOSITE

Biochar itself is a carbon rich efficient adsorbent with abundance of mesopores and/or micropores. Its role as an adsorbent can further be increased manifold by doping of nanoparticles (NPs) on its surface resulting in the formation of biochar-based nanocomposites. NPs adds additional unique properties including high surface area, active functional groups with decreased tendency

Table 10.3 Azo dye removal using difference adsorbents

Adsorbent	Target azo dye	Surface area ($m^2\,g^{-1}$)	Removal Efficiency; Adsorption capacity ($mg\,g^{-1}$)	Isotherms	Kinetics	Comments	References
Bottom ash	Acid Orange 7	NA	86%; NA	Langmuir and Freundlich	Film diffusion	Dye recovery using NaOH	Gupta et al. 2006
De-oiled soya	Acid Orange 7	NA	92%; NA	Langmuir and Freundlich	Film diffusion	Dye recovery using NaOH	Gupta et al. 2006
Calcined MgNiAl layered double hydroxides	Methyl orange	246.5	NA; 375	Langmuir and Redlich–Peterson	Pseudo-second-order	Endothermic process	Zaghouane-Boudiaf et al. 2012
Steam activated-wood waste biochar	Rhodamine B	776.46	100%; 189.93	Langmuir	Pseudo-second-order	–OH and –COOH groups involved in removal	Maneerung et al. 2016
Fe_3O_4/MgO nanoparticles	Amaranth	71.59	96%; 38.1	Langmuir	Pseudo-second-order	Endothermic process	Salem et al. 2016
Ethylenediamine-modified peanut husk	Sunset Yellow	NA	NA; 117.7	Langmuir	Intra-particle diffusion	Endothermic process	Song et al. 2016
$MgAlCO_3$	Amaranth	78	100%; 1.6 $mmol{\cdot}g^{-1}$	Langmuir	Pseudo-second-order	Adsorption on external particle surface	Abdellaoui et al. 2017
Calcined Mg/Fe and Ni/Fe layered double hydroxides	Methyl orange	NA	NA; 850.21, 889.28	Langmuir	Pseudo-second-order	Exothermic process	Elmoubarki et al. 2017
Calcined Mg/Fe and Ni/Fe layered double hydroxides	Methylene blue	NA	NA; 22.73, 35.37	Langmuir	Pseudo-second-order	Endothermic process	Elmoubarki et al. 2017
Biochar from wood waste	Indosol Black NF1200	350.4	99%; 185	Langmuir and Freundlich	Pseudo-second-order	High removal at pH 2	Kelm et al. 2019
H_3PO_4-activated *Ziziphus lotus* stones	Basic Red 46	NA	92%; 307	Langmuir	Pseudo-second-order	High removal at pH 8	Boudechiche et al. 2019
ZnO nanoparticles	Methyl orange, Amaranth	49.36	70.3%, 94.2%; 75.9, 65.2	Langmuir	Pseudo-second-order	R–SO^{3-} was active sites; electrostatic attraction between dye and ZnO–NPs^{+}	Zafar et al. 2019

Table 10.3 *Contd.*

Table 10.3 9 (*Contd.*) Azo dye removal using difference adsorbents

Adsorbent	**Target azo dye**	**Surface area (m^2 g^{-1})**	**Removal Efficiency; Adsorption capacity (mg g^{-1})**	**Isotherms**	**Kinetics**	**Comments**	**References**
Activated carbon cloth	Acid Orange 8, Acid Yellow 14, Acid Red 151	1870	NA	Freundlich	Pseudo-second-order	Adsorption and Electrosorption	Sahin et al. 2020
Zr(IV) encapsulated carboxymethyl cellulose-montmorillonite	Reactive red 2, Acid orange 7	16.84	96.5%, 97%; 35.9, 28.8	Freundlich	NA	Electrostatic attraction, hydrogen bonding and surface complexationinvolved in removal	Sirajudheen et al. 2020
Corn stalk activated carbon	Rhodamine B	NA	89.6%; 5.6	Freundlich	Pseudo-second-order	High removal at pH 3	Mousavi et al. 2021
Poly-diallyl-dimethyl-ammonium chloride-modified silica gel	New coccine	169.7	87%; NA	Langmuir	NA	Electrostatic interactioninvolved in removal	Pham et al. 2021
Bentonite	Basic Red 46	47	94%; 594	Langmuir	Pseudo-second-order	Higher removal in mild conditions (neutral pH, 25°C)	Paredes-Quevedo et al. 2021
Silver nanoparticle-doped sal leaf biochar	Rhodamine B	20.9	93.6%; 23.4	Freundlich	Pseudo-second-order	Pore-filling and electrostatic attractioninvolved in removal	Shaikh et al. 2021
Cetyl trimethyl ammonium chloride-modified coconut cellulose	Tartrazine, Congo red	NA	97%; 19.61, 19.99	Langmuir for Tartrazine, Freundlich for Congo red	Pseudo-second-order and Elovich	Symmetric deformation of the C_6H_6–Cl group in 1472 cm^{-1}	Tejada-Tovar et al. 2021
Polycation-Modified Nano-Alpha Alumina	New coccine	NA	95%; 3.17	Freundlich	Pseudo-second-order	$-SO^3$ group in NCC molecules, $-N^+(CH_3)_3$ group in PVBTAC molecules involved in removal	Doan et al. 2022
Silver nanoparticle-doped tea leaf biochar	Rhodamine B	27.03	95.9; 39.9	Freundlich	Pseudo-second-order	Electrostatic attraction, H-bonding, surface complexation, surface adsorption, weak van der waals interaction, pore-fillinginvolved in removal	Shaikh et al. 2022

of agglomeration and increase the stability of NPs (Amen et al. 2020). The most emerging fields of nanocomposite are however the graphene-based composite and biochar-based metal nanocomposite. Figure 10.4 showed the percentage wise dye removal research articles published with different biochar and modified biochar from 2010–2022.

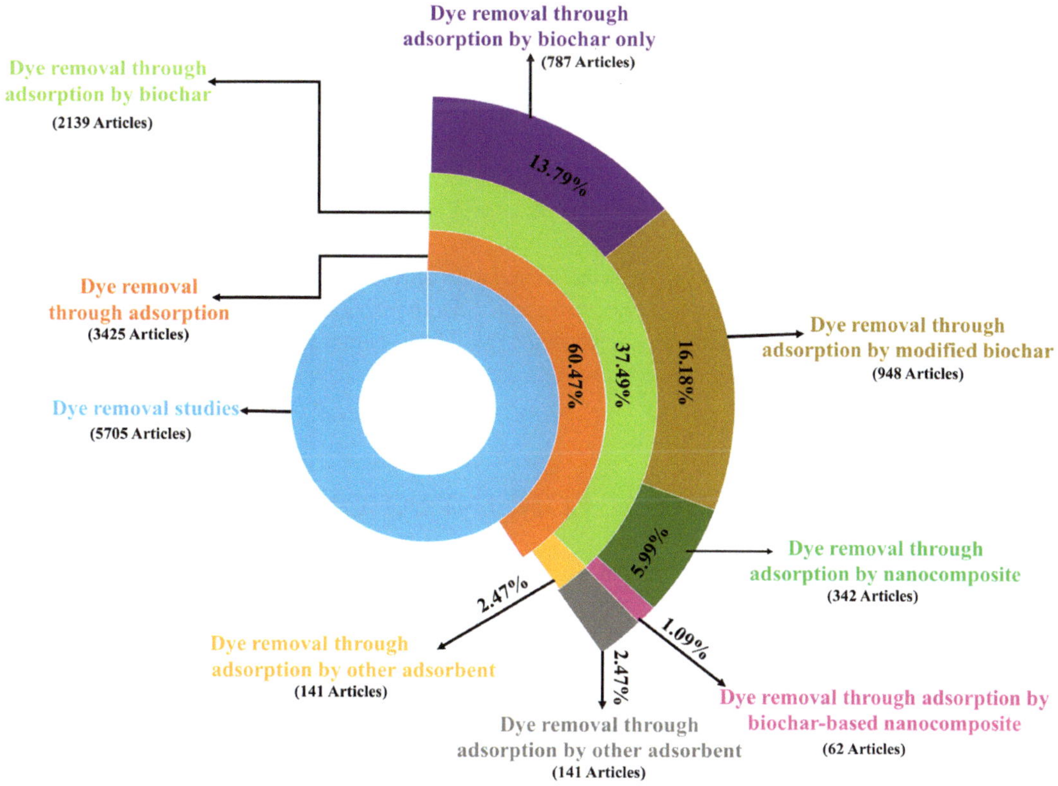

Figure 10.4 Percentage wise dye removal research articles by biochar and modified biochar from 1010-to 2022.

Metal nanoparticle-based nanocomposites are the most efficient techno-economic, eco-friendly, non-toxic, and low-cost adsorbent used for a wide range of environmental applications (Shaikh et al. 2022b). These metal nanoparticle-based nanocomposites are highly carbon-rich with little or no oxygen derived from pyrolyzed biomass (like animal litters, crop residues, solid wastes, and wood biomass) as a functional matrix for metal nanoparticles. Metal nanoparticles were then fabricated through a modified chemical-co precipitation method (Meyer et al. 2011). Silver and Iron nanoparticles (*n*ZVI) were the most commonly used metal nanoparticles to fabricate biochar-based metal nanocomposite for remediation of environmental contaminants (Pandey et al. 2021). Several feedstocks have been used as the supporting matrices of the biochar-based metal nanocomposite, including bamboo biomass (Zhou et al. 2013), chestnut biochar (Zhou et al. 2017), and pinewood biochar (Wang et al. 2017).

10.12 GRAPHENE-BASED COMPOSITE MATERIALS

Graphene sheets are the one-atom-thick 2D layers of sp^2-bonded carbon, having extraordinary electronic transport properties (Stankovich et al. 2006a). The most effective way to utilize this graphene is to incorporate graphene sheets in a composite material, which requires sufficient scale production and homogeneous distribution into various matrices. Graphene-polymer composites are one of the leading graphene-based composite materials, synthesized through chemically-modified graphene sheets within polymer hosts, a molecular-level dispersion method. In addition, modified stable graphite oxides can be prepared by the treatment with organic isocyanates, resulting in reducing the hydrophilic character of graphene oxide sheets due to the formation of amide and carbamate ester bonds (Stankovich et al. 2006b). Recently, Shi et al. (2022) has reported graphene oxide-chitosan aerogel composite fabricated and applied for methylene blue adsorption. Mahmood et al. (2022) reported the synthesis of novel graphene-based tin oxide composite and tested for aqueous phase Congo red adsorption owing to >85% removal efficiency.

10.13 ADSORPTION PROCESS PARAMETERS FOR AZO DYE REMOVAL

Different factors influence the dye removal efficiency and adsorption capacity (Table 10.3). These factors are inclusive of initial concentration, adsorbent dose, temperature, contact time, and solution pH (Alam et al. 2018, Tirkey et al. 2018). These parameters are optimized in studies to augment the dye removal for application at a large industrial scale and are discussed in the following sections.

10.14 INFLUENCE OF INITIAL CONCENTRATION

The impact of initial concentration of azodye on removal efficiency is closely connected to the concentration of adsorbate and active sites available on the surface of adsorbent (Huo et al. 2021). Generally, removal efficiency augments with rising azo dye concentration. The increase is contributed by the presence of binding sites on adsorbent surface and the strong driving force between these sites and dye concentration gradient (Ahmed et al. 2020). A rapid increase in removal is observed at lower concentrations due to the driving force provided by initial adsorbate concentration to ace the mass transfer resistance between adsorbent and azo dye (Shaikh et al. 2021b, Kumar et al. 2022). The continual saturation of active binding sites on adsorbent surface as adsorption proceeds after optimal removal decreases the number of available binding sites for azo dye interaction, thereby decreasing the removal efficiency minimally (Kumar et al. 2021). A silver nanoparticle-doped macroalgae-based biochar and silver nanoparticle-doped tea leaf-based biochar showed 77.63% and 86.32% removal efficiency for Congo red at 18 and 22.50 mg L^{-1} respectively (Adsorption capacity of 34.93 and 48.56 mg g^{-1} respectively) (Shaikh et al. 2022a, b).

10.15 INFLUENCE OF ADSORBENT DOSE

To assess the azo dye removal efficiency of adsorbent with respect to the adsorbent dose, overall costing, and minimize adsorbent wastage, experiments are performed to keep other parameters constant. Adsorption efficiency increases sharply with a rise in adsorbent dose, could be attributed to the increase in available active surface binding sites and adsorptive surface area of adsorbent. However, above the optimised adsorbent dose, removal efficiency decreases slightly, suggesting the saturation of binding sites participating in dye removal. On the contrary, adsorption capacity decreases with increase in adsorbent dose possibly due to aggregation of adsorbent, active binding sites getting overlapped and binding sites encountering a fixed number of adsorbate molecules. The impact of adsorbent dose gives an idea of the economic viability of the process of adsorption (Soltani et al. 2021).

10.16 INFLUENCE OF PROCESS TEMPERATURE

Studies suggested rise in temperature, removal efficiency and adsorption capacity increases till optimization (mostly at ambient temperature of 300 K). The rise in azodye removal with temperature is due to creation of surplus binding sites favouring augmented diffusion (by gaining kinetic energy), mass transfer, and dye mobility within adsorbent matrix (boosting the possibility of azo dye interaction with surface functional groups and azo dye entry into pores of the adsorbent (Kumar and Bhattacharya 2021, 2022). Rise in adsorption sites with rising temperature could be indicative of endothermic nature of azo dye adsorption. When the process temperature is increased beyond the optimized temperature, a decrease in azo dye removal efficiency and adsorption capacity is observed, possibly attributed to bond weakening and diffusion of pre-adsorbed dye molecules (Lenka et al. 2021, Shaikh et al. 2018). Formation of gases at higher process temperatures could create a gaseous barrier between azo dye and sorbent, blocking pores, consequently minimizing the azo dye removal efficiency (Yao et al. 2020). Decrease in adsorption capacity above optimization could be suggestive of exothermic adsorption beyond optimum removal.

To determine the kinetics of azo dye removal and equilibrium removal time, impact of contact time is studied. Initially, removal efficiency and adsorption capacity increases with a rise in contact time possibly attributable to surplus availability of pores and active binding sites, and a higher azo dye concentration gradient increasing its interaction with adsorbent (Nayak and Pal 2017). The increase in azo dye removal slows at higher contact times, which is resulting from the progressive saturation of pores and binding sites of adsorbent (Shaikh et al. 2021b). Finally, after achieving optimum removal, a minimal reduction in azo dye removal is observed possibly arising from the unavailability of pores and active binding sites of adsorbent and the competition between azo dye molecules and adsorbent, resulting in diffusion of pre-adsorbed azo dye molecules from the adsorbent matrix. A silver nanoparticle-doped macroalgae-based biochar and silver nanoparticle-doped tea leaf-based biochar showed 83.94% and 93.48% Congo red removal at 60 and 70 min respectively (Adsorption capacity of 21.58 and 30.05 mg g^{-1} respectively) (Shaikh et al. 2022a, b).

10.17 INFLUENCE OF SOLUTION pH

The increase or decrease of azo dye removal efficiency results from significant changes in azo dye solution properties, overall surface charge, ionization state of adsorbent binding groups, and dissociation of surface functional groups (Nautiyal et al. 2016). Litefti et al. (2019) suggested that ionic charge of aqueous azo dye is predominantly negative at pH above isoelectric point possibly because of anionic non-dissociated form. When pH of the final solution is lesser than pH_{zpc} (pH at point of zero charge),the overall surface charge is predominantly positive capable of electrostatically attracting negatively charged dye molecules and vice-versa (Shaikh et al. 2020). $pH>pH_{pzc}$ is helpful for adsorption of cationic dyes, while $pH<pH_{pzc}$ is needed for adsorption of anionic dyes. At alkaline pH, deprotonation of functional groups like hydroxyl and carboxyl group ($-COOH + OH^- \rightarrow COO^- + H_2O$) results in predominant negative charges on surface of adsorbent, significantly enhancing electrostatic interactions between positively charged azo dye molecules and negatively charged adsorbent (Maneerung et al. 2016). At neutral pH, there could be hydrophobic Van der Waals interactions and/or H-bonding between functional groups (like –SOOO–) of azo dye and functional groups (like –OH and $-NH_2$) of adsorbent (Shaikh et al. 2021b, Zhang et al. 2014). At acidic pH, electrostatic repulsion and protonation of functional groups like hydroxyl ($-OH + H^+ \rightarrow OH_2^{+179}$), carboxyl ($-COOH + H^+ \rightarrow COOH_2^+$), sulfonate ($-SO_3^- + H^+ \rightarrow -SO_3H$), or amino ($-NH_2 + H^+ \rightarrow -NH_3^+$) groups on adsorbent lead to a reduction in dye removal efficiency. Shaikh et al. (2022a, b) reported Ag nanoparticle-modified macroalgae-based biochar and Ag nanoparticle-doped tea leaf-based biochar showed 90.92% and 91.23%

Congo red removal efficiency at pH 6 and 7 respectively (adsorption capacity of 16.43 and 29.33 mg g^{-1} respectively).

10.18 AZO DYE ADSORPTION MECHANISM

The overall surface charge of composite, *n*Ag, surface functional groups on biochar synergistically lead to chemisorption, electrostatic attraction, H-bonding, and surface complexation with aqueous azo dye molecules (Koohi et al. 2021). The negatively charged SO_3^- and COO^- groups get attracted towards positively charged surface functional groups of nanocomposite (=NH, =CH_2, ≡CH, –NH_2 etc.) (Wu et al. 2017). Additionally, O-containing functional groups (NO, SO, C=O, and –OH) on surface of nanocomposite create surface complexation, van der Waal interactions, and H-bonding with S- and O-rich functional groups onazo dye molecules, thereby contributing to dye adsorption significantly (Figure 10.5) (Zhang et al. 2022).

Further, supramolecular azo dye molecules, the giant azo dye complex, could be involved in surface sorption on NPs and pores of nanocomposite, ionic interactions between hydroxyl O of stabilizing biomolecules, and surface complexation via π–π stacking between aromatic π-bond of dye and π-bonds of nanocomposite(Shaikh et al. 2022a). Moreover, functional groups containing π-bond like C=C in nanocomposite participate in π–π stacking with delocalized aromatic π-bonds of dye, thereby enhancing the removal efficiency (Costa et al. 2017) (Figure 10.5).

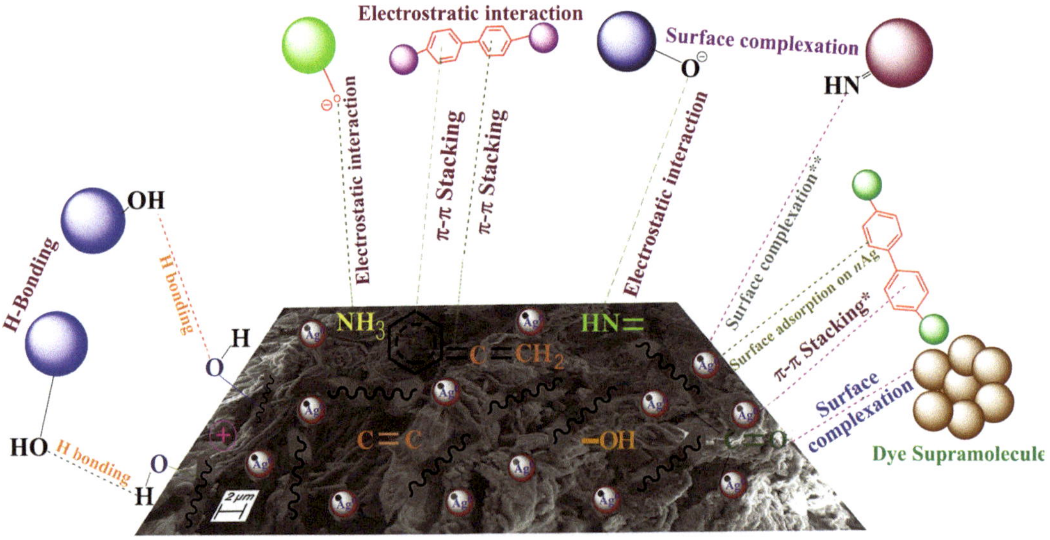

Figure 10.5 Dye adsorption mechanism using biochar-based metal nanocomposite.

10.19 ADVANTAGES AND DISADVANTAGES OF NANOCOMPOSITES

Biochar-based nanocomposite has been proved to be an efficient hybrid adsorbent used in different environmental fields. The main advantages are as follows:

(i) Nanocomposites or biochar-based nanocomposites have the potential to subside the problems related to yield and efficiency by the associated significantly high yields during production while utilizing minimum amount of nanoparticles.

(ii) Biochar-based nanocomposites could retain the crucial properties of both nanomaterials and biochar.

(iii) Improved porosity, functional groups, surface area, and electrical conductivity of the functional matrix.
(iv) New organic functional groups introduced onto the functional matrix.
(v) Hold excellent abilities to adsorb a range of aqueous environmental contaminants.
(vi) Improved adsorption mechanism through and van der Waal interaction, H-bonding, surface complexation, and electrostatic interaction.
(vii) Biochar-based nanocomposite have enormous potential in large-scale industrial applications for treating toxic dye-containing wastewater thereby promoting polluted water recycling.

10.20 SUSTAINABLE UTILIZATION AND MANAGEMENT OF THE SPENT BIOCHAR-BASED NANOCOMPOSITE THROUGH CIRCULAR ECONOMY

The biochar-based nanocomposites were found to be significantly remove toxic azo dyes could also contribute in proliferation of circular economy. Utilization of biomass waste to produce biochar would assist in waste minimization and associated pollution of soil and groundwater (Matrapazi and Zabaniotou 2020). Additionally, these waste biomasses would be prevented from ending up in landfill sites (reducing the solid biowaste). With regards to agriculture and horticulture, spent organic waste-derived biochar (with high nutrient content) could be used as a soil amendment to improve the nutritional status and fertility of soils (Hu et al. 2021). Improvement of soil fertility would also minimize the need for addition of costly fertilizers to nutrient-deprived soils.

Moreover, in terms of energy, biochar possesses a high calorific value of 24–36 MJ Kg^{-1} (comparable to non-renewable coal and peat), the spent sorbent could be utilized as an energy source (Kumar et al. 2020). The fabricated nanocomposite would be cheaper than industrial sorbents (such as activated charcoal), which could induce material cost savings for the industries. The preceding discussion (incorporating economic, environmental, and social dimensions) suggests a significant role of the composite in promoting and establishing circular economy. Nonetheless, some ecotoxicological risk of nanomaterials should be duly considered and adequate risk mitigation and management strategies must be put into practice from sustainability and environmental health perspectives (Shaikh et al. 2022b).

10.21 CONCLUSIONS AND FUTURE PERSPECTIVES

Although nanoparticles could degrade a significant amount of these toxic azo dyes, it was realized that the low yield of the NPs could be a hindrance or limitation for large-scale implementation. To address this inefficiency, the hybrid biochar-based nanocomposites (*n*C-BCs) were fabricated using biochar synthesized from biowaste materials like algal scum, sal leaf, husks, leaf litter, and discarded tea leaves as the substrate, which can overcome the inadequacy with multi-dimensional utility. The concept of biochar-based nanocomposites fabrication was to generate new binding sites, and intensify overall surface area, efficiently enhance the overall azo dye removal efficiency. Moreover, *n*C-BCs were techno-economically feasible and thermo-stable composites, in comparison to previously used composites. Physicochemical characteristics indicated the involvement of active surfaces binding sites, like C=O, CH, SO, NH, and –OH could be present, played the key role in the removal of azo dye. Shaikh et al. (2021a) reported CR adsorption on stable *n*Ag encompassed mesoporous heterogeneous nanocomposite where the adsorption mechanism was governed by surface complexation, H-bonding, π–π stacking interactions, van der Waal interaction, and electrostatic interaction. The *n*C-BCs also have the potential to contribute towards achieving the circular economy by helping restore and regenerate natural systems along with dye-laden wastewater treatment. This utilization of waste biomasses would prevent them from ending up in landfill sites (reducing the pressure on landfill sites, minimizing release of leachates, and increasing levels of hygiene and sanitation), treat dye effluent efficiently, and not produce any toxic waste during synthesis or treatment. In addition, the *n*C-BCs are reported to be techno-economically feasible which can treat ~2000 L of wastewater at the cost of ~7–7.5 US$. The preceding benefits suggested a crucial role of *n*C-BCs in promoting and establishing circular economy for a sustainable environment.

However, there are still many aspects of research to be unveiled. The effluent released from the industries are increasing, so efficient and continuous effluent treatment technique is required for the sustainable environmental and industrial development. The future scope of research lies in enhancing the dye uptake capacity, and batch experiment results can be further employed for column experiments. Continuous batch-mode pilot-scale treatment setup might also be one of the substantial options with the *n*C-BCs. This will result in direct implementation for in-situ dye-laden wastewater treatment.

References

Abd-Elhamid, A.I., Emran, M., El-Sadek, M.H., El-Shanshory, A.A., Soliman, H.M.A., Akl, M.A. and Rashad, M. 2020. Enhanced removal of cationic dye by eco-friendly activated biochar derived from rice straw. Appl. Water Sci. 10: 45. https://doi.org/10.1007/s13201-019-1128-0.

Abdellaoui, K., Pavlovic, I., Bouhent, M., Benhamou, A. and Barriga, C. 2017. A comparative study of the amaranth azo dye adsorption/desorption from aqueous solutions by layered double hydroxides. Appl. Clay Sci. 143: 142–150. https://doi.org/10.1016/j.clay.2017.03.019.

Adeyemo, A.A., Adeoye, I.O. and Bello, O.S. 2017. Adsorption of dyes using different types of clay: a review. Appl. Water Sci. 7: 543–568. https://doi.org/10.1007/s13201-015-0322-y.

Ahmed, D.N., Naji, L.A., Faisal, A.A.H., Al-Ansari, N. and Naushad, M. 2020. Waste foundry sand/MgFe-layered double hydroxides composite material for efficient removal of Congo red dye from aqueous solution. Sci. Rep. 10: 2042. https://doi.org/10.1038/s41598-020-58866-y.

Al-Bastaki, N. 2004. Removal of methyl orange dye and Na_2SO_4 salt from synthetic waste water using reverse osmosis. Chem. Eng. Process. Process Intensif. 43: 1561–1567. https://doi.org/10.1016/j.cep.2004.03.001.

Al-Fawwaz, A.T. and Abdullah, M. 2016. Decolorization of methylene blue and malachite green by immobilized *Desmodesmus* sp. isolated from North Jordan. Int. J. Environ. Sci. Dev. 7: 95–99. https://doi.org/10.7763/ijesd.2016.v7.748.

Alam, M.A., Shaikh, W.A., Alam, M.O., Bhattacharya, T., Chakraborty, S., Show, B. and Saha, I. 2018. Adsorption of As(III) and As(V) from aqueous solution by modified *Cassia fistula* (golden shower) biochar. Appl. Water Sci. 8: 198. https://doi.org/10.1007/s13201-018-0839-y.

Amen, R., Bashir, H., Bibi, I., Shaheen, S.M., Niazi, N.K., Shahid, M., Hussain, M.M., et al. 2020. A critical review on arsenic removal from water using biochar-based sorbents: The significance of modification and redox reactions. Chem. Eng. J. 396: 125195. https://doi.org/10.1016/j.cej.2020.125195.

Aravind, M., Amalanathan, M., Aslam, S., Noor, A.E., Jini, D., Majeed, S., Velusamy, P., et al. 2023. Hydrothermally synthesized Ag-TiO_2 nanofibers (NFs) for photocatalytic dye degradation and antibacterial activity. Chemosphere 321: 138077. https://doi.org/10.1016/j.chemosphere.2023.138077.

Bassyouni, D.G., Hamad, H.A., El-Ashtoukhy, E.S.Z., Amin, N.K. and El-Latif, M.M.A. 2017. Comparative performance of anodic oxidation and electrocoagulation as clean processes for electrocatalytic degradation of diazo dye Acid Brown 14 in aqueous medium. J. Hazard. Mater. 335: 178–187. https://doi.org/10.1016/j.jhazmat.2017.04.045.

Berradi, M., Hsissou, R., Khudhair, M., Assouag, M., Cherkaoui, O., El Bachiri, A. and El Harfi, A. 2019. Textile finishing dyes and their impact on aquatic environs. Heliyon 5(11): e02711. https://doi.org/10.1016/j.heliyon.2019.e02711.

Boudechiche, N., Fares, M., Ouyahia, S., Yazid, H., Trari, M. and Sadaoui, Z. 2019. Comparative study on removal of two basic dyes in aqueous medium by adsorption using activated carbon from *Ziziphus lotus* stones. Microchem. J. 146, 1010–1018. https://doi.org/10.1016/j.microc.2019.02.010.

Chang, J.S., Chou, C., Lin, Y.C., Lin, P.J., Ho, J.Y. and Hu, T.L. 2001. Kinetic characteristics of bacterial azo-dye decolorization by *Pseudomonas luteola.* Water Res. 35: 2841–2850. https://doi.org/10.1016/S0043-1354(00)00581-9.

Choudhry, A., Sharma, A., Siddiqui, S.I., Ahamad, I., Sajid, M., Khan, T.A. and Chaudhry, S.A. 2023. Origanum vulgare manganese ferrite nanocomposite: An advanced multifunctional hybrid material for dye remediation. Environ. Res. 220: 115193. https://doi.org/10.1016/j.envres.2022.115193

Chowdhury, M.F., Khandaker, S., Sarker, F., Islam, A., Rahman, M.T. and Awual, M.R. 2020. Current treatment technologies and mechanisms for removal of indigo carmine dyes from wastewater: a review. J. Mol. Liq. 318: 114061. https://doi.org/10.1016/j.molliq.2020.114061.

Costa, A.L., Gomes, A.C., Pillinger, M., Gonçalves, I.S., Pina, J. and Seixas de Melo, J.S. 2017. Insights into the photophysics and supramolecular organization of congo red in solution and the solid state. Chem. Phys. Chem. 18, 564–575. https://doi.org/10.1002/cphc.201601236.

Costa, A.F.S., Albuquerque, C.D.C., Salgueiro, A.A. and Sarubbo, L.A. 2018. Color removal from industrial dyeing and laundry effluent by microbial consortium and coagulant agents. Process Saf. Environ. Prot. 118: 203–210. https://doi.org/10.1016/j.psep.2018.03.001.

Doan, T.H.Y., Pham, H.A., Nguyen, N.H., Le, T.D., Nguyen, T.B. and Le, T.S. 2022. Adsorptive removal of azo dye new coccine using high-performance adsorbent-based polycation-modified nano-alpha alumina particles. J. Anal. Methods Chem. 2022: 9425334. https://doi.org/10.1155/2022/9425334.

Dotto, J., Fagundes-Klen, M.R., Veit, M.T., Palácio, S.M. and Bergamasco, R. 2019. Performance of different coagulants in the coagulation/flocculation process of textile wastewater. J. Clean. Prod. 208: 656–665. https://doi.org/10.1016/j.jclepro.2018.10.112.

Du, J., Wu, Y., Dong, Z., Zhang, M., Yang, X., Xiong, H. and Zhao, L. 2022. Single and competitive adsorption between Indigo Carmine and Methyl orange dyes on quaternized kapok fiber adsorbent prepared by radiation technique. Sep. Purif. Technol. 292. https://doi.org/10.1016/j.seppur.2022.121103.

El Nemr, A., Shoaib, A.G.M., El Sikaily, A., Ragab, S., Mohamed, A.E.D.A. and Hassan, A.F. 2022. Utilization of green alga *Ulva lactuca* for sustainable production of meso-micro porous nano activated carbon for adsorption of Direct Red 23 dye from aquatic environment. Carbon Lett. 32: 153–168. https://doi.org/10.1007/s42823-021-00262-1.

Elmoubarki, R., Mahjoubi, F.Z., Elhalil, A., Tounsadi, H., Abdennouri, M., Sadiq, M., Qourzal, S., et al. 2017. Ni/Fe and Mg/Fe layered double hydroxides and their calcined derivatives: Preparation, characterization and application on textile dyes removal. J. Mater. Res. Technol. 6: 271–283. https://doi.org/10.1016/j.jmrt.2016.09.007.

Faraji, A.R., Khoramdareh, N.B., Ahmadi, S., Gil, A., Farahanipour, A. and Hekmatian, Z. 2023. Synthesis of magnetic Mn/Fe oxides derived from drinking water sludges from a one-pot solvothermal process for the degradation of tetracycline and mixed dyes: synergistic effect and mechanism study. J. Environ. Chem. Eng. 11: 109668. https://doi.org/10.1016/j.jece.2023.109668.

Gupta, V.K., Mittal, A., Gajbe, V. and Mittal, J. 2006. Removal and recovery of the hazardous azo dye acid orange 7 through adsorption over waste materials: bottom ash and de-oiled soya. Ind. Eng. Chem. Res. 45: 1446–1453. https://doi.org/10.1021/ie051111f.

Gupta, V.K. and Suhas. 2009. Application of low-cost adsorbents for dye removal—a review. J. Environ. Manage. 90: 2313–2342. https://doi.org/10.1016/j.jenvman.2008.11.017.

Hamad, H., Bassyouni, D., El-Ashtoukhy, E.S., Amin, N. and Abd El-Latif, M. 2018. Electrocatalytic degradation and minimization of specific energy consumption of synthetic azo dye from wastewater by anodic oxidation process with an emphasis on enhancing economic efficiency and reaction mechanism. Ecotoxicol. Environ. Saf. 148: 501–512. https://doi.org/10.1016/j.ecoenv.2017.10.061.

Hao, O.J., Kim, H. and Chiang, P.C. 2000. Decolorization of wastewater. Crit. Rev. Environ. Sci. Technol. 30: 449–505. https://doi.org/10.1080/10643380091184237.

Hu, Q., Jung, J., Chen, D., Leong, K., Song, S., Li, F., Mohan, B.C., et al. 2021. Biochar industry to circular economy. Sci. Total Environ. 757: 143820. https://doi.org/10.1016/j.scitotenv.2020.143820.

Huo, J.B., Yu, G., Xu, L. and Fu, M.L. 2021. Porous walnut-like $La_2O_2CO_3$ derived from metal-organic frameworks for arsenate removal: a study of kinetics, isotherms, and mechanism. Chemosphere. 271: 129528. https://doi.org/10.1016/j.chemosphere.2020.129528.

Imran, M., Crowley, D.E., Khalid, A., Hussain, S., Mumtaz, M.W. and Arshad, M. 2015. Microbial biotechnology for decolorization of textile wastewaters. Rev. Environ. Sci. Biotechnol. 14: 73–92. https://doi.org/10.1007/s11157-014-9344-4

Kazak, O., Akkaya, G.K. and Tor, A. 2023. Sustainable and efficient removal of cationic, anionic and neutral dyes from water by pre-deposited vinasse biochar membrane. J. Environ. Chem. Eng. 11: 110042. https://doi.org/10.1016/j.jece.2023.110042.

Kelm, M.A.P., da Silva Júnior, M.J., de Barros Holanda, S.H., de Araujo, C.M.B., de Assis Filho, R.B., Freitas, E.J., dos Santos, D.R., et al. 2019. Removal of azo dye from water via adsorption on biochar produced by the gasification of wood wastes. Environ. Sci. Pollut. Res. 26: 28558–28573. https://doi.org/10.1007/s11356-018-3833-x.

Koohi, P., Rahbar-kelishami, A. and Shayesteh, H. 2021. Efficient removal of congo red dye using Fe_3O_4/NiO nanocomposite: synthesis and characterization. Environ. Technol. Innov. 23: 101559. https://doi.org/10.1016/j.eti.2021.101559.

Kumar, A., Bhattacharya, T., Mozammil Hasnain, S.M., Kumar Nayak, A. and Hasnain, M.S. 2020. Applications of biomass-derived materials for energy production, conversion, and storage. Mater. Sci. Energy Technol. 3: 905–920. https://doi.org/10.1016/j.mset.2020.10.012.

Kumar, A., Bhattacharya, T., Shaikh, W.A., Roy, A., Mukherjee, S. and Kumar, M. 2021. Performance evaluation of crop residue and kitchen waste-derived biochar for eco-efficient removal of arsenic from soils of the Indo-Gangetic plain: a step towards sustainable pollution management. Environ. Res. 200: 111758. https://doi.org/10.1016/j.envres.2021.111758.

Kumar, A. and Bhattacharya, T. 2021. Biochar: a sustainable solution. Environ. Dev. Sustain. 23: 6642–6680. https://doi.org/10.1007/s10668-020-00970-0.

Kumar, A. and Bhattacharya, T. 2022. Removal of arsenic by wheat straw biochar from soil. Bull. Environ. Contam. Toxicol. 108: 415–422. https://doi.org/10.1007/s00128-020-03095-2.

Kumar, A., Bhattacharya, T., Shaikh, W.A., Chakraborty, S., Owens, G. and Naushad, M. 2022. Valorization of fruit waste-based biochar for arsenic removal in soils. Environ. Res. 213: 113710. https://doi.org/10.1016/j.envres.2022.113710.

Lenka, S.P., Shaikh, W.A., Owens, G., Padhye, L.P., Chakraborty, S. and Bhattacharya, T. 2021. Removal of copper from water and wastewater using dolochar. Water. Air. Soil Pollut. 232: 167. https://doi.org/10.1007/s11270-021-05135-x.

Lennartson, A. 2017. The Chemical Works of Carl Wilhelm Scheele. Springer, Cham. https://doi.org/10.1007/978-3-319-58181-1_3.

Li, Z., Hanafy, H., Zhang, L., Sellaoui, L., Netto, M.S., Oliveira, M.L.S., Seliem, M.K., et al. 2020. Adsorption of Congo red and methylene blue dyes on an ashitaba waste and a walnut shell-based activated carbon from aqueous solutions: experiments, characterization and physical interpretations. Chem. Eng. J. 388: 124263. https://doi.org/10.1016/j.cej.2020.124263.

Litefti, K., Freire, M.S., Stitou, M. and González-Álvarez, J. 2019. Adsorption of an anionic dye (Congo red) from aqueous solutions by pine bark. Sci. Rep. 9: 16530. https://doi.org/10.1038/s41598-019-53046-z.

Mahmood, T., Noreen, U., Ali, R., Ullah, A., Naeem, A. and Aslam, M. 2022. Adsorptive removal of Congo red from aqueous phase using graphene–tin oxide composite as a novel adsorbent. Int. J. Environ. Sci. Technol. 19: 10275–10290. https://doi.org/10.1007/s13762-021-03867-1.

Majumdar, R., Shaikh, W.A., Chakraborty, S. and Chowdhury, S. 2022. A review on microbial potential of toxic azo dyes bioremediation in aquatic system. pp. 241–261. *In*: Das, S. and Dash, H.R. (eds). Microbial Biodegradation and Bioremediation: Techniques and Case Studies for Environmental Pollution, 2nd Ed. Elsevier. https://doi.org/10.1016/b978-0-323-85455-9.00018-7.

Maneerung, T., Liew, J., Dai, Y., Kawi, S., Chong, C. and Wang, C.H. 2016. Activated carbon derived from carbon residue from biomass gasification and its application for dye adsorption: kinetics, isotherms and thermodynamic studies. Bioresour. Technol. 200: 350–359. https://doi.org/10.1016/j.biortech.2015.10.047.

Mariyam, A., Mittal, J., Sakina, F., Baker, R.T. and Sharma, A.K. 2021. Adsorption behaviour of chrysoidine R dye on a metal/halide-free variant of ordered mesoporous carbon. Desalin. Water Treat. 223: 425–433. https://doi.org/10.5004/dwt.2021.27147.

Martorell, M.M., Rosales Soro, M. del M., Pajot, H.F. and de Figueroa, L.I.C. 2018. Optimization and mechanisms for biodecoloration of a mixture of dyes by trichosporon akiyoshidainum HP 2023. Environ. Technol. 39: 3169–3180. https://doi.org/10.1080/09593330.2017.1375024

Matrapazi, V.K. and Zabaniotou, A. 2020. Experimental and feasibility study of spent coffee grounds upscaling via pyrolysis towards proposing an eco-social innovation circular economy solution. Sci. Total Environ. 718: 137316. https://doi.org/10.1016/j.scitotenv.2020.137316.

Meyer, S., Glaser, B. and Quicker, P. 2011. Technical, economical, and climate-related aspects of biochar production technologies: a literature review. Environ. Sci. Technol. 45: 9473–9483. https://doi.org/10.1021/es201792c.

Miri Kafi Abad, S.S.A., Javidan, P., Baghdadi, M. and Mehrdadi, N. 2022. Green synthesis of Pd@Biochar using the extract and biochar of corn-husk wastes for electrochemical Cr(Vi) reduction in plating wastewater. SSRN Electron. J. 11: 109911. https://doi.org/10.2139/ssrn.4265589.

Monga, D., Shetti, N.P., Basu, S., Raghava Reddy, K., Badawi, M., Bonilla-Petriciolet, A. and Aminabhavi, T.M. 2022. Engineered biochar: a way forward to environmental remediation. Fuel. 311: 122510. https://doi.org/10.1016/j.fuel.2021.122510.

Mostafa, A.A.F., Elshikh, M.S., Al-Askar, A.A., Hadibarata, T., Yuniarto, A. and Syafiuddin, A. 2019. Decolorization and biotransformation pathway of textile dye by *Cylindrocephalum aurelium*. Bioprocess Biosyst. Eng. 42: 1483–1494. https://doi.org/10.1007/s00449-019-02144-3.

Mousavi, S.A., Kamarehie, B., Almasi, A., Darvishmotevalli, M., Salari, M., Moradnia, M., Azimi, F., et al. 2021. Removal of Rhodamine B from aqueous solution by stalk corn activated carbon: adsorption and kinetic study. Biomass Convers. Biorefinery. https://doi.org/10.1007/s13399-021-01628-1

Nautiyal, P., Subramanian, K.A. and Dastidar, M.G. 2016. Adsorptive removal of dye using biochar derived from residual algae after *in-situ* transesterification: alternate use of waste of biodiesel industry. J. Environ. Manage. 182: 187–197. https://doi.org/10.1016/j.jenvman.2016.07.063.

Nayak, A.K. and Pal, A. 2017. Green and efficient biosorptive removal of methylene blue by *Abelmoschus esculentus* seed: process optimization and multi-variate modeling. J. Environ. Manage. 200: 145–159. https://doi.org/10.1016/j.jenvman.2017.05.045.

Panda, S.K., Aggarwal, I., Kumar, H., Prasad, L., Kumar, A., Sharma, A., Vo, D.-V.N., et al. 2021. Magnetite nanoparticles as sorbents for dye removal: a review. Environ. Chem. Lett. 19: 2487–2525. https://doi.org/10.1007/s10311-020-01173-9.

Pandey, D., Singh, S., Dutta, K., Daverey, A. and Arunachalam, K. 2021. Biochar-based nanocomposites: a sustainable solution for water and wastewater treatment. pp. 619–639. *In*: Sarma, H., Joshi, S.J., Prasad, R. and Jampilek, J. (eds). Biobased Nanotechnology for Green Applications. Nanotechnology in the Life Sciences. Springer, Cham. https://doi.org/10.1007/978-3-030-61985-5_22.

Paredes-Quevedo, L.C., González-Caicedo, C., Torres-Luna, J.A. and Carriazo, J.G. 2021. Removal of a textile azo-dye (Basic Red 46) in water by efficient adsorption on a natural clay. Water. Air. Soil Pollut. 232: 4. https://doi.org/10.1007/s11270-020-04968-2.

Peng, P., Lang, Y.H. and Wang, X.M. 2016. Adsorption behavior and mechanism of pentachlorophenol on reed biochars: PH effect, pyrolysis temperature, hydrochloric acid treatment and isotherms. Ecol. Eng. 90: 225–233. https://doi.org/10.1016/j.ecoleng.2016.01.039.

Pham, T.D., Bui, V.P., Pham, T.N., Le, T.M.D., Nguyen, K.T., Bui, V.H. and Nguyen, T.D. 2021. Adsorptive removal of anionic azo dye new coccine using silica and silica-gel with surface modification by polycation. Polymers (Basel). 13: 1–13. https://doi.org/10.3390/polym13101536.

Rajapaksha, A.U., Chen, S.S., Tsang, D.C.W., Zhang, M., Vithanage, M., Mandal, S., Gao, B., et al. 2016. Engineered/designer biochar for contaminant removal/immobilization from soil and water: potential and implication of biochar modification. Chemosphere. 148: 276–291. https://doi.org/10.1016/j.chemosphere.2016.01.043.

Romdhane, D.F., Satlaoui, Y., Nasraoui, R., Charef, A. and Azouzi, R. 2020. Adsorption, modeling, thermodynamic, and kinetic studies of methyl red removal from textile-polluted water using natural and purified organic matter rich clays as low-cost adsorbent. J. Chem. 2020: 4376173. https://doi.org/10.1155/2020/4376173.

Sachin, J., Singh, N., Singh, R., Shah, K. and Pramanik, B.K. 2023. Green synthesis of zinc oxide nanoparticles using lychee peel and its application in anti-bacterial properties and CR dye removal from wastewater. Chemosphere. 327: 138497. https://doi.org/10.1016/j.chemosphere.2023.138497.

Sahin, E.M., Tongur, T. and Ayranci, E. 2020. Removal of azo dyes from aqueous solutions by adsorption and electrosorption as monitored with in-situ UV-visible spectroscopy. Sep. Sci. Technol. 55: 3287–3298. https://doi.org/10.1080/01496395.2019.1676786.

Salem, A.N.M., Ahmed, M.A. and El-Shahat, M.F. 2016. Selective adsorption of amaranth dye on Fe_3O_4/MgO nanoparticles. J. Mol. Liq. 219: 780–788. https://doi.org/10.1016/j.molliq.2016.03.084.

Samsami, S., Mohamadi, M., Sarrafzadeh, M.H., Rene, E.R. and Firoozbahr, M. 2020. Recent advances in the treatment of dye-containing wastewater from textile industries: overview and perspectives. Process Saf. Environ. Prot. 143: 138–163. https://doi.org/10.1016/j.psep.2020.05.034.

Shaikh, W.A., Chakraborty, S. and Ul Islam, R. 2018. UV-assisted photo-catalytic degradation of anionic dye (Congo red) using biosynthesized silver nanoparticles: a green catalysis. Desalin. Water Treat. 130: 232–242. https://doi.org/10.5004/dwt.2018.23004.

Shaikh, W.A., Alam, M.A., Alam, M.O., Chakraborty, S., Owens, G., Bhattacharya, T. and Mondal, N.K. 2020a. Enhanced aqueous phase arsenic removal by a biochar based iron nanocomposite. Environ. Technol. Innov. 19: 100936. https://doi.org/10.1016/j.eti.2020.100936.

Shaikh, W.A., Chakraborty, S. and Islam, R.U. 2020b. Photocatalytic degradation of rhodamine B under UV irradiation using Shorea robusta leaf extract-mediated bio-synthesized silver nanoparticles. Int. J. Environ. Sci. Technol. 17: 2059–2072. https://doi.org/10.1007/s13762-019-02473-6

Shaikh, W.A., Chakraborty, S., Owens, G. and Islam, R.U. 2021a. A review of the phytochemical mediated synthesis of AgNP (silver nanoparticle): the wonder particle of the past decade. Appl. Nanosci. 11: 2625–2660. https://doi.org/10.1007/s13204-021-02135-5.

Shaikh, W.A., Islam, R.U. and Chakraborty, S. 2021b. Stable silver nanoparticle doped mesoporous biochar-based nanocomposite for efficient removal of toxic dyes. J. Environ. Chem. Eng. 9: 104982. https://doi.org/10.1016/j.jece.2020.104982.

Shaikh, W.A., Chakraborty, S., Islam, R.U., Ghfar, A.A., Naushad, M., Bundschuh, J., Maity, J.P., et al. 2022a. Fabrication of biochar-based hybrid Ag nanocomposite from algal biomass waste for toxic dye-laden wastewater treatment. Chemosphere. 289: 133243. https://doi.org/10.1016/j.chemosphere.2021.133243.

Shaikh, W.A., Kumar, A., Chakraborty, S., Islam, R.U., Bhattacharya, T. and Biswas, J.K. 2022b. Biochar-based nanocomposite from waste tea leaf for toxic dye removal: from facile fabrication to functional fitness. Chemosphere. 291: 132788. https://doi.org/10.1016/j.chemosphere.2021.132788.

Shaikh, W.A., Kumar, A., Chakraborty, S., Naushad, M., Islam, R.U., Bhattacharya, T. and Datta, S. 2022c. Removal of toxic dye from dye-laden wastewater using a new nanocomposite material: isotherm, kinetics and adsorption mechanism. Chemosphere. 308: 136413. https://doi.org/10.1016/j.chemosphere.2022.136413.

Shi, Y., Song, G., Li, A., Wang, J., Wang, H., Sun, Y. and Ding, G. 2022. Graphene oxide-chitosan composite aerogel for adsorption of methyl orange and methylene blue: effect of pH in single and binary systems. Colloids Surf. A: Physicochemical and Engineering Aspects. 641: 128595. https://doi.org/10.1016/j.colsurfa.2022.128595.

Shindhal, T., Rakholiya, P., Varjani, S., Pandey, A., Ngo, H.H., Guo, W., Ng, H.Y., et al. 2021. A critical review on advances in the practices and perspectives for the treatment of dye industry wastewater. Bioengineered. 12: 70–87. https://doi.org/10.1080/21655979.2020.1863034.

Sirajudheen, P., Karthikeyan, P., Basheer, M.C. and Meenakshi, S. 2020. Adsorptive removal of anionic azo dyes from effluent water using Zr(IV) encapsulated carboxymethyl cellulose-montmorillonite composite. Environ. Chem. Ecotoxicol. 2: 73–82. https://doi.org/10.1016/j.enceco.2020.04.002.

Soltani, A., Faramarzi, M. and Parsa, S.A.M. 2021. A review on adsorbent parameters for removal of dye products from industrial wastewater. Water Qual. Res. J. 56: 181–193. https://doi.org/10.2166/wqrj.2021.023.

Song, Y., Xu, H. and Ren, J. 2016. Adsorption study for removal of sunset yellow by ethylenediamine-modified peanut husk. Desalin. Water Treat. 57: 17585–17592. https://doi.org/10.1080/19443994.2015.1086897.

Stankovich, S., Dikin, D.A., Dommett, G.H.B., Kohlhaas, K.M., Zimney, E.J., Stach, E.A., Piner, R.D., et al. 2006a. Graphene-based composite materials. Nature. 442: 282–286. https://doi.org/10.1038/nature04969

Stankovich, S., Piner, R.D., Nguyen, S.B.T. and Ruoff, R.S. 2006b. Synthesis and exfoliation of isocyanate-treated graphene oxide nanoplatelets. Carbon N.Y. 44: 3342–3347. https://doi.org/10.1016/j.carbon.2006.06.004.

Tahir, H., Hammed, U., Sultan, M. and Jahanzeb, Q. 2010. Batch adsorption technique for the removal of malachite green and fast green dyes by using montmorillonite clay as adsorbent. African J. Biotechnol. 9: 8206–8214. https://doi.org/10.5897/ajb10.911.

Tan, K.B., Vakili, M., Horri, B.A., Poh, P.E., Abdullah, A.Z. and Salamatinia, B. 2015. Adsorption of dyes by nanomaterials: recent developments and adsorption mechanisms. Sep. Purif. Technol. 150: 229–242. https://doi.org/10.1016/j.seppur.2015.07.009.

Tan, X.-F., Liu, Y.-G., Gu, Y.-L., Xu, Y., Zeng, G.-M., Hu, X.-J., Liu, S-B., et al. 2016. Biochar-based nano-composites for the decontamination of wastewater: a review. Bioresour. Technol. 212: 318–333. https://doi.org/10.1016/j.biortech.2016.04.093.

Tejada-Tovar, C., Villabona-Ortíz, Á. and Gonzalez-Delgado, Á.D. 2021. Adsorption of azo-anionic dyes in a solution using modified coconut (*Cocos nucifera*) mesocarp: kinetic and equilibrium study. Water. 13: 1–20. https://doi.org/10.3390/w13101382.

Tirkey, P., Bhattacharya, T. and Chakraborty, S. 2018. Optimization of fluoride removal from aqueous solution using Jamun (*Syzygium cumini*) leaf ash. Process Saf. Environ. Prot. 115: 125–138. https://doi.org/10.1016/j.psep.2017.10.022.

Wang, H., Gao, B., Wang, S., Fang, J., Xue, Y. and Yang, K. 2015. Removal of Pb(II), Cu(II), and Cd(II) from aqueous solutions by biochar derived from $KMnO_4$ treated hickory wood. Bioresour. Technol. 197: 356–362. https://doi.org/10.1016/j.biortech.2015.08.132.

Wang, S., Gao, B., Li, Y., Creamer, A.E. and He, F. 2017. Adsorptive removal of arsenate from aqueous solutions by biochar supported zero-valent iron nanocomposite: batch and continuous flow tests. J. Hazard. Mater. 322: 172–181. https://doi.org/10.1016/j.jhazmat.2016.01.052.

Wang, J. and Wang, S. 2019. Preparation, modification and environmental application of biochar: a review. J. Clean. Prod. 227: 1002–1022. https://doi.org/10.1016/j.jclepro.2019.04.282.

Wu, Z., Yuan, X., Zhong, H., Wang, H., Jiang, L., Zeng, G., Wang, H., et al. 2017. Highly efficient adsorption of Congo red in single and binary water with cationic dyes by reduced graphene oxide decorated NH_2-MIL-68(Al). J. Mol. Liq. 247: 215–229. https://doi.org/10.1016/j.molliq.2017.09.112.

Xia, C., Liang, Y., Li, X., Garalleh, H.A., Garaleh, M., Hill, J.M. and Pugazhendhi, A. 2023. Remediation competence of nanoparticles amalgamated biochar (nanobiochar/nanocomposite) on pollutants: a review. Environ. Res. 218: 114947. https://doi.org/10.1016/j.envres.2022.114947.

Yagub, M.T., Sen, T.K., Afroze, S. and Ang, H.M. 2014. Dye and its removal from aqueous solution by adsorption: a review. Adv. Colloid Interface Sci. 209: 172–184. https://doi.org/10.1016/j.cis.2014.04.002.

Yaneva, Z.L. and Georgieva, N.V. 2012. Insights into congo red adsorption on agro-industrial materials—spectral, equilibrium, kinetic, thermodynamic, dynamic and desorption studies: a review. Int. Rev. Chem. Eng. 4: 127–146.

Yao, X., Ji, L., Guo, J., Ge, S., Lu, W., Cai, L., Wang, Y., et al. 2020. Magnetic activated biochar nanocomposites derived from wakame and its application in methylene blue adsorption. Bioresour. Technol. 302: 122842. https://doi.org/10.1016/j.biortech.2020.122842.

Yu, Z. and Wen, X. 2005. Screening and identification of yeasts for decolorizing synthetic dyes in industrial wastewater. Int. Biodeterior. Biodegrad. 56: 109–114. https://doi.org/10.1016/j.ibiod.2005.05.006.

Zafar, Md. Nadeem, Dar, Q., Nawaz, F., Zafar, Md. Naveed, Iqbal, M. and Nazar, M.F. 2019. Effective adsorptive removal of azo dyes over spherical ZnO nanoparticles. J. Mater. Res. Technol. 8: 713–725. https://doi.org/10.1016/j.jmrt.2018.06.002.

Zaghouane-Boudiaf, H., Boutahala, M. and Arab, L. 2012. Removal of methyl orange from aqueous solution by uncalcined and calcined MgNiAl layered double hydroxides (LDHs). Chem. Eng. J. 187: 142–149. https://doi.org/10.1016/j.cej.2012.01.112.

Zhang, R., Zhang, J., Zhang, X., Dou, C. and Han, R. 2014. Adsorption of Congo red from aqueous solutions using cationic surfactant modified wheat straw in batch mode: kinetic and equilibrium study. J. Taiwan Inst. Chem. Eng. 45: 2578–2583. https://doi.org/10.1016/j.jtice.2014.06.009.

Zhang, X., Zhang, T., Guo, J., Ahmad, M., Yang, H., Su, X. and Huang, F., et al. 2022. Hierarchically porous tobacco midrib-based biochar prepared by a simple dual-templating approach for highly efficient Rhodamine B removal. Arab. J. Chem. 15: 103904. https://doi.org/10.1016/j.arabjc.2022.103904.

Zhang, S., Zhang, S., Liu, H., Li, L. and Guo, R. 2023. Fe-N-C-based cathode catalyst enhances redox reaction performance of microbial fuel cells: azo dyes degradation accompanied by electricity generation. J. Environ. Chem. Eng. 11: 109264. https://doi.org/10.1016/j.jece.2023.109264.

Zhou, Y., Gao, B., Zimmerman, A.R., Fang, J., Sun, Y. and Cao, X. 2013. Sorption of heavy metals on chitosan-modified biochars and its biological effects. Chem. Eng. J. 231: 512–518. https://doi.org/10.1016/j.cej.2013.07.036.

Zhou, Z., Liu, Y.-G., Liu, S.-B., Liu, H.-Y., Zeng, G.-M., Tan, X.-F., Yang, C.-P., et al. 2017. Sorption performance and mechanisms of arsenic(V) removal by magnetic gelatin-modified biochar. Chem. Eng. J. 314: 223–231. https://doi.org/10.1016/j.cej.2016.12.113.

Chapter 11

Nanostructured Materials for Removal of Microplastics from Water

Dhruba Jyoti Sarkar*[1], Santhana Kumar V[1] and Basanta Kumar Das[1]

[1]Aquatic Environmental Biotechnology and Nanotechnology Division, ICAR-Central Inland Fisheries Research Institute, Kolkata-700120, India

Soma Das Sarkar[2]

[2]Fisheries Resources Assessment and Informatics Division, ICAR-Central Inland Fisheries Research Institute, Kolkata-700120, India

Jayanta Kumar Biswas[3]

[3]Department of Ecological Studies, University of Kalyani, West Bengal-741235, India

11.1 INTRODUCTION

Worldwide, plastics have made their way into all commercial sectors viz. food packaging, farming, healthcare, construction, automobiles, electronic devices, etc. leading to the age of plastic (Plastics Europe 2019). The usage of plastic in every phase of life has increased many folds over the last three decades owing to faster lifestyles and changing consumption patterns (Sarkar et al. 2019). In the last 40 years, plastic use has expanded 25-fold due to its lightweight, durability, low cost, and elasticity (Sutherland et al. 2016). However, misuse of plastics and especially mismanagement of plastic waste has led them to sink into aquatic resources causing threats to numerous aquatic species and the balance of aquatic life. According to a recent report by WWF (WWF 2019), nearly 75% of the total plastic produced becomes waste and about 87% of mismanaged solid wastes from uncontrolled landfills are dripped into the natural environment contributing to plastic pollution. Out of 2.5 billion tonnes of solid waste generated globally, coastal countries generate nearly 275 million tonnes of plastic and simultaneously oceans receive 4.8–12.7 million tonnes (Jambeck et al. 2015). Not only marine system but freshwater rivers are also now found to be heavily

*Corresponding Authors: dhruba.sarkar@icar.gov.in; dhruba1813@gmail.com

contaminated with MPs (Sarkar et al. 2019, 2020a, b). After entering into the natural stream, these plastics are reported to be broken down to small pieces due to various environmental processes viz. photodegradation, oxidation, biodegradation, etc. (Sarkar et al. 2021c). Besides aquatic biota, these plastic pieces have also become a real threat to terrestrial species including humans due to exposure through drinking water. They are also reported to act as vectors for various other toxicants including, heavy metal, antibiotics, pesticides, etc. (Sarkar et al. 2021b). It was reported that approximately 5 g of plastic is consumed by an average person through food and beverage depending upon consumption practice whereas, consumption through drinking water (both tap and bottled) is as much as 1769 particles of plastic on a weekly basis (WWF 2019). These tiny plastic particles, christened as microplastics (MPs < 5 mm), are produced from larger plastic items through several environmental physical or chemical processes and have invaded every nook and corner of the planet.

Since bioaccumulation and ecological risk of MPs pollution have become a major concern for the public and scientific communities, MPs removal from their source point is the recent focal point of research. It was reported that most conventional water treatment plants can remove MPs to a certain extent (Sarkar et al. 2022a). But these treatment plants are not intentional for the removal of MPs, hence limitations still exists in achieving 100% removal efficiency in most cases (Sarkar et al. 2021a, 2022b). Acknowledging the toxic effects of MPs pollution, more research is going on to improve conventional water treatment technologies. In this aspect, nanotechnology possesses tremendous potential to improvising the MPs removal processes. Nanostructured materials (NSMs) due to their structural multi-functionalities and flexibility are now being researched to improve the efficiency of drinking water and wastewater treatment processes. Novel NSMs including nanoadsorbents, nanocatalysts, etc. due to their unique structural and chemical properties help in the quick separation or mineralization of MPs (Sarkar et al. 2022b). Hence this chapter particularly highlights the use of those NSMs in resolving the bottlenecks and improvising the existing processes used for MPs removal. The present chapter discusses these issues under a few subheadings based on the removal mechanisms viz. adsorption, filtration, magnetic separation, and catalytic oxidation. The details of the NSMs used in MPs removing strategies are summarised in Table 11.1.

11.2 ADSORPTION OF MPs

Adsorption is the mass transfer process in which a substance is transferred from a liquid phase to the surface of an adsorbent due to physical and/or chemical interactions. It was reported that among the various remediation technologies, adsorption is the most employed method for toxic pollutant remediation due to its effectiveness, cost efficiency and versatile nature (Anjum 2017). The adsorption technology is used for many environmentally important pollutants like heavy metals (Anjum 2017), pesticides (Rasool et al. 2022), POPs (Ighalo et al. 2022), etc. and very recently the process is being employed for the removal of MPs from water using latest nanostructured materials viz. graphene oxide (Yuan et al. 2020), metal-organic framework (Chen et al. 2020), etc.

11.2.1 Reduced Graphene Oxide

An alternative form of graphene oxide (GO) is reduced graphene oxide (RGO) which is prepared by several chemical, thermal or other methods to decrease the amount of oxygen in the GO sheet, thus making the material more stable (Papageorgiou et al. 2015). RGO was reported to be used for many applications, viz. energy storage (Okhay and Tkach 2021), composite materials (Huang et al. 2012), sensing (Reiner-Rozman et al. 2016), etc. Recently the material has been used for the removal of polystyrene (PS) MPs from water (Yuan et al. 2020). It was reported that three-

Table 11.1 Different nanomaterials tested for microplastic removal or complete degradation

Microplastics and dose	Nanomaterials and dose	Removal mechanism and time	Removal efficiency	References
PS and modified PS (Carboxylated and amine modified) (1 mg/ L)	Chitin based sponges material fabricated with graphene oxide (GO) and oxygen-doped carbon nitride, 1 cm^3	Adsorption, 48 hr	71.6–92.1%	Sun et al. 2021
PS (0.6 g/L)	3D Reduced graphene oxide, 0.75 mg/mL	Adsorption, 2 hr	> 50%	Yuan et al. 2020
PVDF, PS, PMMA, (0.001 g/mL)	Zirconium metal-organic framework-based foam materials (UiO-66-OH@Melamine foam), 1 cm thick	Membrane filtration, flow rate 1.5 L/h	95.5 ± 1.2%	Chen et al. 2020
PE, PP, PS and PET (0.5g/L)	Magnetic nano-Fe_3O_4 (0.5–1.5 g/L)	Adsorption and separation by magnetism, 150–250 min	62.83–86.87%	Shi et al. 2022
PE, PET, PS, polyurethane (PU), polyvinyl chloride (PVC), and PP (5–15 items/200 mL)	Hydrophobic Fe nanoparticles, 0.2 μg/mL to 10 mg/L	Adsorption and separation by magnetism, 5–30 min	92% (PE, PS; 10–20 μm; seawater) 93% (PE, PET, PS, PU, PVC and PP; > 1 mm; seawater) 84% (PE, PS, PU, PVC and PP; 200 μm –1 mm; fresh water) 78% (PE, PS, PU, PVC and PP; 200 μm –1 mm; Sediment)	Grbic et al. 2019
PE, PET, and polyamide (PA) (5 g/L)	Magnetic carbon nanotubes, 2–7 g/L	Adsorption and separation by magnetism, 300 min	100%	Tang et al. 2021
Cosmetic product-extracted MPs (5 g/L)	Magnetic spring-like carbon nanotubes, 0.2 g/L	Carbocatalytic oxidation and hydrothermal hydrolysis, 8 h	50	Kang et al. 2019
PE, PS	TiO_2 nanoparticle film	Photocatalytic mineralization, 12 h	98.4%	Nabi et al. 2020

dimensional RGO with a porous structure can adsorb PS MPs (Figure 11.1A). The adsorption mechanism is attributed to the strong π–π interaction between benzene ring of PS MPs and the carbon ring of three-dimensional RGO. Under optimal condition of pH 6.0, ionic concentration of 600 mg/L, time 120 min and temperature of 26°C, the maximum adsorption capacity of RGO for PS MPs obtained was 617.28 mg/g. The PS MPs removal efficiency of RGO was 56.08% in tap water with adsorption capacity of 448.60 mg/g and 53.85% in micro-polluted water with adsorption capacity of 430.78 mg/g (Figure 11.1B). These values are slightly lower than the spiked distilled water where 66.63% removal was observed with adsorption capacity of 533.06 mg/g. This reduced removal from the actual water sample was attributed to multiple interaction effects of metal ions, organic matter, algae secretions, ammonia nitrogen, and other pollutants with RGO. It was also reported that suspended particles in the real water sample can block the pore structure of RGO thus reducing active site for binding PS MPs. Very recently, antibiotic adsorption behaviour of GO/RGO nanofilled MPs (polypropylene (PP), polyamide (PA), and PS) was evaluated (Liu et al. 2022). The study reports that the presence of GO/RGO as nanofiller in the MPs could contribute to increase adsorption of antibiotics on the MPs surface. It was reported that the tetracycline adsorption capacity was increased to 100% in an aged PP-RGO system as compared to aged PP system.

11.2.2 Chitin-based Sponge Materials

One of the most abundant biomacromolecule in aquatic system is chitin which exhibits superiority in biocompatibility, biodegradability, antibacterial properties, etc. (Yan et al. 2019). It's wide availability and low cost makes it an excellent biomaterial in various forms including fiber (Huang et al. 2014, Yan et al. 2019), membrane (Nagahama et al. 2008), hydrogel (Chang et al. 2011), microgel (Arun Kumar et al. 2015), sponge (Chang et al. 2011), etc. with different application possibilities. In order to provide a green and sustainable approach to pollution remediation, chitin-based materials have been explored as adsorbent for various pollutants viz. heavy metal (Begum et al. 2021), dye (Wan Ngah et al. 2011), pesticides (Moradi Dehaghi et al. 2014), salt (EL-Gendi et al. 2014), etc. Very recently, chitin based sponge material, composed with GO and oxygen doped carbon nitride was tested for MPs removal (Sun et al. 2021). The material could effectively remove PS MPs along functionalized PS MPs viz. carboxylate-modified polystyrene (PS–COOH), and amine-modified polystyrene (PS–NH_2) (Figure 11.2). At test MPs (~1 µm) concentration of 1 mg/L, the MPs removal efficiency of sponge material could reach up to 71.6–92 %. The adsorption mechanism between chitin sponge and MPs involved were electrostatic interactions, hydrogen bond interactions, and π–π interactions. Due to excellent elasticity, mechanical durability and compressibility, the chitin sponge material can be reused for three cycles. It was found that the developed chitin sponge materials are biocompatible with algae and degradable in natural soil medium.

11.3 FILTRATION USING METAL–ORGANIC FRAMEWORKS

Metal-organic frameworks (MOFs) are a class of material consisting of metal ions coordinated to organic ligands to form 1, 2 or 3-dimensional structures and are also called as porous coordination polymers. They are recognised as a unique class of promising microporous materials whose structure can be designed according to target properties based on the geometric of organic ligands and coordination of metal ions or metal clusters (Zhu and Xu 2014). The key advantages of MOFs are high porosity (> 90%) and extremely high surface area (> 10000 m^2/g) for which they are finding applications in storage and separation (Murray et al. 2009, Sumida et al. 2012), sensing (Yanai et al. 2011, Kreno et al. 2012), proton conduction (Sadakiyo et al. 2012, Yamada et al. 2013) and drug delivery (Horcajada et al. 2012). Recently, a zirconium MOF-based

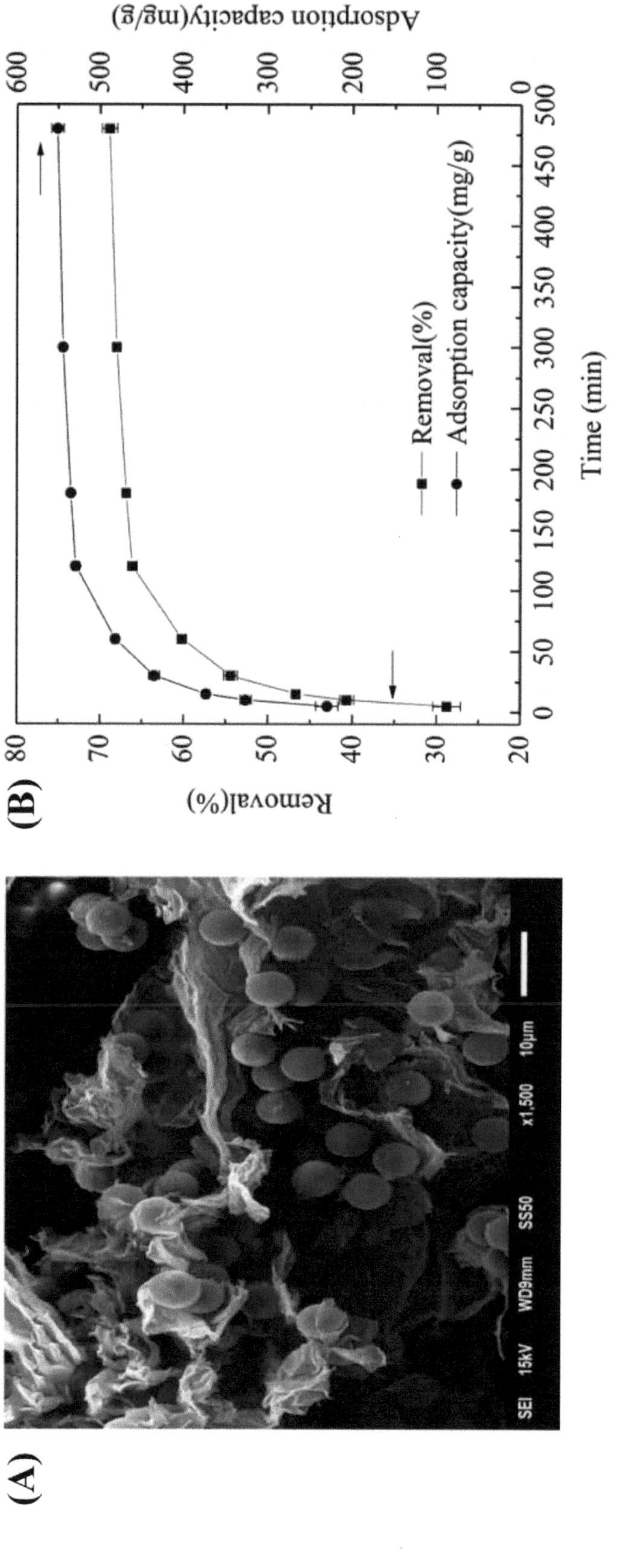

Figure 11.1 (A) SEM images of 3D RGO after adscrption of PS MPs; (B) Effect of time on the adsorption of PS MPs on 3D RGO (Yuan et al. 2020).

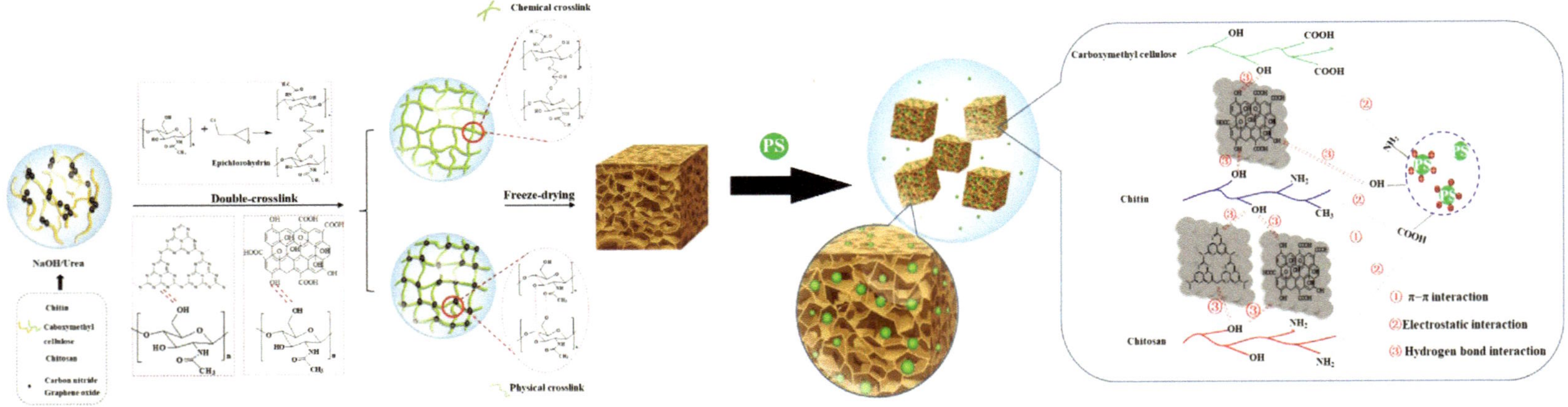

Figure 11.2 Schematic diagram of the synthesis process of chitin based sponges and the proposed adsorption mechanisms of PS, PSNH2, and PS-COOH on the chitin-based sponges. (Sun et al. 2021).

foam material was tested for the removal of poly(vinylidene fluoride) (PVDF, 273 nm) MPs from fresh water and seawater condition (Chen et al. 2020). In the study, a series of Zr-MOFs based foam materials (UiO-66-X@MF) were prepared by heating (100°C) a equimolar (0.39 mMol) acetonic solution of $ZnCL_4$ and 2,5-dihydroxyterephthalic acid (UiO-66-OH@MF) or terephthalic acid (UiO-66-H@MF) or 2-aminoterephthalic acid (UiO-66-NH_2@MF) or bromoterephthalic acid (UiO-66-Br@MF) or nitroterephthalic acid (UiO-66-NO_2@MF) in a Teflon lined autoclave in presence of the melamine foam (MF, thickness 1 cm) for 24 hrs followed by washing and drying (Figure 11.3A–C). The UiO-66-OH@MF exhibits the highest PVDF removal efficiency (90.1 ± 2.1%) followed by UiO-66-NH2@MF (75.3 ± 2.8%) > UiO-66@MF (65.6 ± 3.7%) > UiO-66-NO_2@MF 62.0 ± 4.2%) > UiO-66-Br@MF (61.8 ± 3.2%) (Figure 11.3D). The bare melamine foam (MF) showed 33.0 ± 2.6% removal efficiency for PVDF MPs. It was also found that melamine foam with maximum UiO-66-OH loading (25.4%) has the highest MPs removal efficiency. Besides, PVDF, other MPs like PMMA (325 nm) and PS (183 nm) were also tested and found that UiO-66-OH@MF can efficiently remove microplastics with efficiency up to 95.5 ± 1.2%.

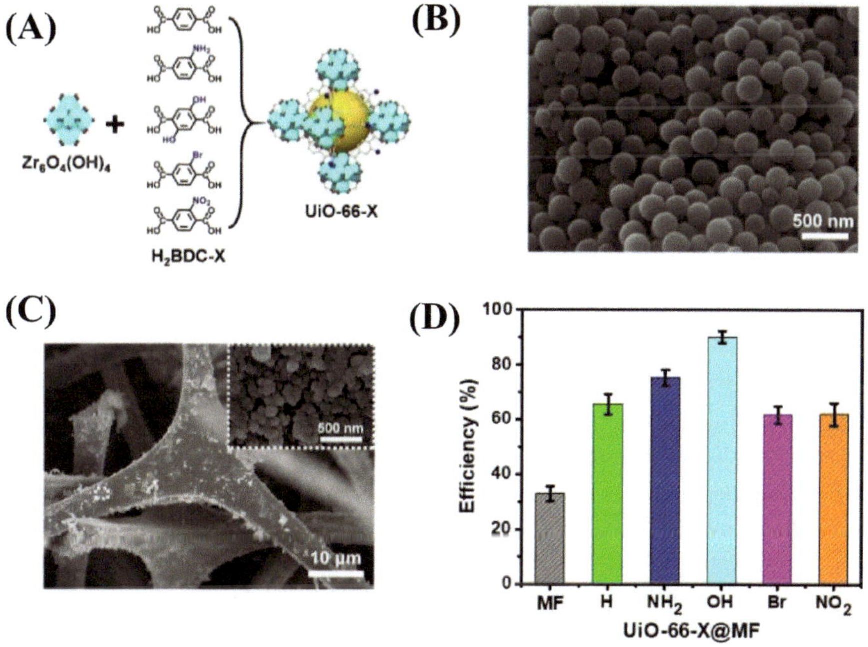

Figure 11.3 (A) The structures of UiO-66-X (X=H, NH_2, OH, Br and NO_2). (B) The SEM image of PVDF nanoparticles (Mw. 5.34 × 105 g mol^{-1}); (C) The SEM image of UiO-66-OH@MF-3 (inset is the zoomed-in image of the square area in C). (D) The PVDF removal efficiency of UiO-66-X@MF. (Chen et al. 2020).

11.4 MAGNETIC SEPARATION OF MPs

Like adsorption, the magnetic separation of pollutants requires a magnetic adsorbent which is composed of magnetic materials. It is possible to separate various pollutants such as heavy metals (Horst et al. 2016, Tamjidi et al. 2019), radioactive substances (Ebner et al. 2007, Nishimura et al. 2009), pesticides (Ma et al. 2018, Zhoua et al. 2021, Gamonchuang et al. 2022), etc. using magnetic adsorbent by changing the properties of surface functional molecules. Magnetic separation was reported to have several advantages over conventional adsorbents for water purification. They (i) integrate the process of capture and purification of pollutants (ii) have high throughput processing and (iii) have a low energy requirement (Schwaminger et al. 2019).

11.4.1 Magnetic Nano-Fe_3O_4

Among the various nanostructured materials, magnetic nanoparticle has drawn a lot of attention in bioanalytical techniques and biomedical applications. They are being used in a range of biomedical applications such as bioimaging (Jordan et al. 1999), biosensors (Sundar et al. 2020), separation (Li et al. 2016), targeted drug delivery (Li et al. 2005), etc. due to their biocompatibility, high saturation magnetization, high magnetic susceptibility and chemical stability. Magnetite (Fe_3O_4) nanoparticles are one of the most employed magnetic nanoparticles due to their distinctive electrical and magnetic characteristics (Chen et al. 2021). Due to the unique magnetic property, nano-Fe_3O_4 is being extensively used for separation studies for removal of pollutants like heavy metals (Iconaru et al. 2016), dye (Ahmadipouya et al. 2021), etc. Recently, a study used magnetic nano-Fe_3O_4 for the separation of various MPs (Shi et al. 2022). The average particle size of nano-Fe_3O_4 used in the study is 30 nm which is used to coat the MPs (Figure 11.4A, B). The study indicates that an optimum concentration of 1.3 g/L of nano-Fe_3O_4 for a period of 150 minutes can cause enough magnetization on the surface of MPs which could be removed by a presence of magnet. A concentration and time period higher than 1.3 g/L of nano-Fe_3O_4 and 150 mins did not cause significant difference in removal efficiency (Figure 11.4C, D). However different types of MPs showed different removal efficiency, for example, the average removal efficiency of PE, PP, PS and PET MPs (200–900 μm) from pure water was 86.87 ± 6.92%, 85.05 ± 4.70%, 86.11 ± 6.21%, and 62.83 ± 8.34%, respectively (Figure 11.4E). It was discussed in the study that among tested MPs, PET has the weakest hydrophobicity and crystallinity which led to low adsorption of nano-Fe_3O_4 and hence lower removal efficiency. Besides polymer type, the removal efficiency was found to be dependent on the size of MPs, type of water and amount of nano-Fe_3O_4 adsorbed on the MPs. Evaluation with environmental waters showed that an approach with nano-Fe_3O_4 could remove more than 80% MPs from river water (11.67 ± 1.44 items/L), domestic sewage (2488.67 ± 315.11 items/L), and natural seawater (8.33 ± 1.44 items/L).

11.4.2 Hydrophobic Iron Nanoparticles

To better extract small MPs (<20 μm) from environmental samples, a study used hydrophobic iron nanoparticles to magnetize MPs (Grbic et al. 2019). Hydrophobization of Fe nanoparticles was done by surface modification of Fe nanoparticles with silane to generate hydrophobic hydrocarbon tails. The method involves addition of methanolic solution of hexadecyltrimethoxysilane (HDTMS) to Fe nanoparticles (25 nm) which cause reaction of hydroxyl group in the oxide layer of Fe nanoparticle with the methoxysilane (–Si–O–CH_3) of HDTMS to form strong covalent siloxane bond (–Si–O–Si–) revealing alkyl chain of HDTMS on the surface of Fe nanoparticles (Figure 11.5). The developed hydrophobic Fe nanoparticles were used in the recovery experiment involving large MPs (from ~1 to ~8 mm), (2) medium MPs (from 200 μm to 1 mm), and (3) small MPs (<20 μm). For large-size MPs, the recoveries ranged from 74 to 105% from sea water. For medium-sized microplastics, the recoveries ranged from 59 to 100% for reverse osmosis water and from 49 to 90% for sediment. For the small MPs, the recoveries were 96 ± 13% for PE and 88 ± 17% for PS.

11.4.3 Magnetic Carbon Nanotube

Carbon nanotube (CNT) is reported as an excellent material for environmental pollution control due to its high surface area and a large number of adsorption sites. CNT are hydrophobic polymers with a diameter ~1 nm and length in the micrometer range. It was previously reported that CNT has high affinity and adsorption capacity for hydrophobic aromatic compounds. In many studies, CNT is magnetically modified with magnetic materials (e.g., iron oxide) for easy separation under a magnetic field. Magnetic separation was reported to be faster than centrifugation and filtration,

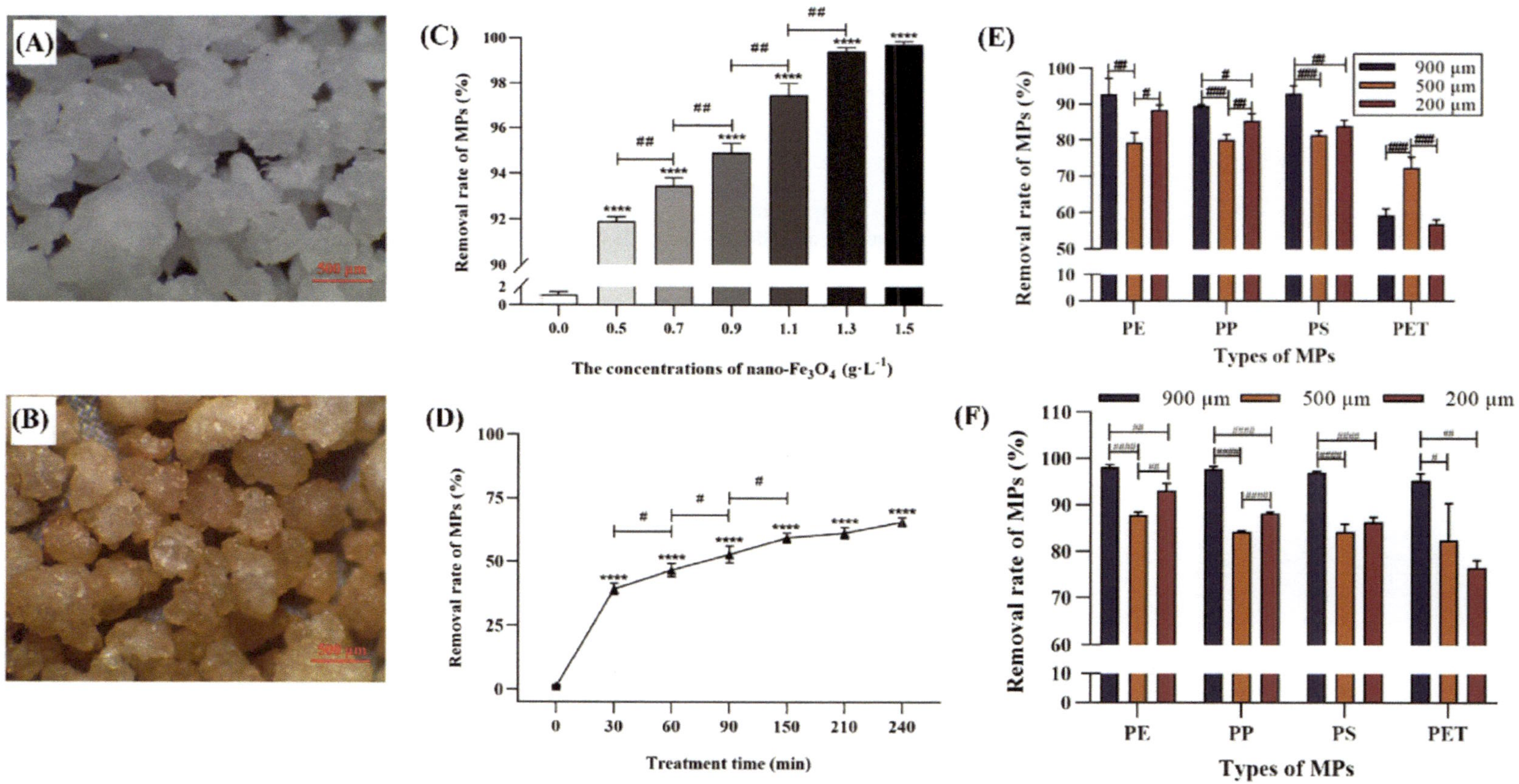

Figure 11.4 Optical microscopy depicting the change of surface morphology of PE MPs after adsorption of nano-Fe_3O_4. Virgin polyethylene MPs (A) and after adsorbed with nano-Fe_3O_4 (B); removal rates of MPs after magnetization by different concentrations of nano-Fe_3O_4 (C) and different time duration (D) Removal rates of PE, PP, PS, and PET MPs with different diameters in distilled water (E) and artificial seawater (F) (Shi et al. 2022).

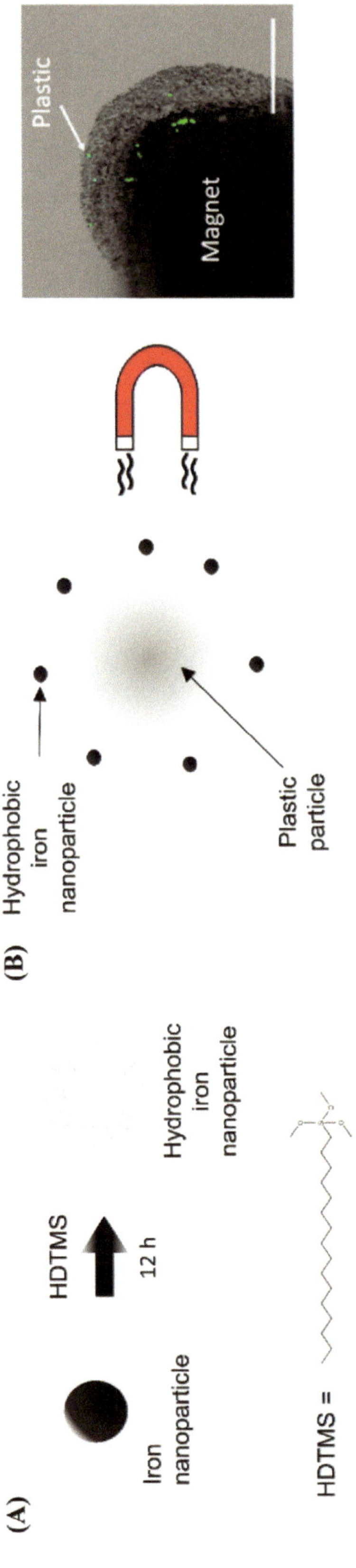

Figure 11.5 (A) Modification of Fe nanoparticles with HDTMS to create hydrophobic Fe nanoparticles; (B) Bound hydrophobic Fe nanoparticles allow magnetic recovery of MPs because of magnetic force. (Grbic et al. 2019).

and it also reduces the loss of nanomaterials from entering the natural environment. Very recently, the hydrophobicity of magnetic CNT (M-CNT) was used to remove MPs from water (Tang et al. 2021). In the study, the M-CNT was synthesised by addition of CNT (diameter, 1–2 nm; length, 1–3 μm) in a solution mixture of $FeCl_3$ and $FeSO_4$ followed by drop-wise addition of ammonium hydroxide to precipitate magnetic iron oxide nanoparticles which were separated, dried (80°C for 24 h) and heated (600°C for 2 h) (Figure 11.6A). The synthesised M-CNT effectively adsorbed PE, PET and Polyamide (PA) MPs when an optimal concentration of more than 5 g/L was maintained at 180 mins (Figure 11.6B). However, a residual concentration of 5–8 μg/L M-CNT was observed. The MPs adsorption capacity of M-CNT is best in neutral solution than in alkaline and acidic water. The MPs removal efficiency of M-CNT is nearly 100% and maintain the same in varied COD, ammonium and phosphate concentration.

11.5 CATALYTIC OXIDATIVE DEGRADATION OF MPs

Beside physical processes several chemical processes have also been developed to degrade the organic pollutants in situ through advanced oxidation processes in the presence of a catalyst or photocatalyst. Advanced oxidation process generates reactive hydroxyl radicals which cause the degradation of organic pollutants on the catalyst's surface. Photocatalytic semiconductor such as TiO_2 is extensively used for pesticide degradation under solar or UV irradiation (Jiménez et al. 2015, Farner Budarz et al. 2019). In photocatalysis, photon bombardment results in the generation of electron-hole pairs which leads to the degradation of organic pollutants. These photocatalytic semiconductors like TiO_2 are reported to have the advantages of being economical, non-poisonous, efficient radiation absorbers and having a prolonged life (Bano et al. 2021).

11.5.1 TiO_2 Nanoparticle Film

A recent study reports that photocatalytic mineralisation of MPs was studied using TiO_2 nanoparticle film under UV light irradiation (Nabi et al. 2020). It was observed that 98.4% PS microsphere can be mineralised using TiO_2 nanoparticle film in 12 hrs. Figure 11.7A–C shows the SEM images of photocatalytically degraded PS MPs under 365 nm UV light at 0 h, 6 h and 12 h. The process of preparation of TiO_2 showed a significant influence in the degradation process of PS. It was observed that TiO_2 film prepared using Triton-X100 performed best in reducing the diameter of PS microsphere (Figure 11.7D). Similarly, TiO_2 nanoparticle film using Triton-X100 had shown the highest MPs degradation efficiency (Figure 11.7E). The diameter of the test PS microsphere had also shown to affect the degradation process by TiO_2 film (Figure 11.7F). Under UV stimulation, TiO_2 produces electron in the conduction band (e^-_{CB}) and positive hole in the valence band (h^+_{VB}) resulting in the formation of two degradation pathways involving the formation of ROS such as $^{\bullet}OH$ and $O_2^{\bullet -}$ (Eq. 1). h^+_{VB} is a very powerful oxidizing agent resulting in CO_2 and H_2O as degradation products (Eq. 2). Further it also generates $^{\bullet}OH$, which due to its electrophilic nature can non-selectively oxidise all electron-rich organic compound resulting in their mineralisation to CO_2 and H_2O (Eq. 3 and 4) (Shang et al. 2003). Secondly, the e^-_{CB} reacts with oxygen forming $O_2^{\bullet -}$ which produces hydrogen peroxide that leads to degradation of MPs (Eq. 5 to 8) (Cho and Choi 2001).

$$TiO_2 \xrightarrow{hv} h^+_{VB} + e^-_{CB} \quad (1)$$

$$h^+_{VB} + \text{microplastic} \longrightarrow CO_2 + H_2O \quad (2)$$

$$h^+_{VB} + H_2O \longrightarrow {}^{\bullet}OH + H^+ \quad (3)$$

$${}^{\bullet}OH + \text{microplastic} \longrightarrow CO_2 + H_2O \quad (4)$$

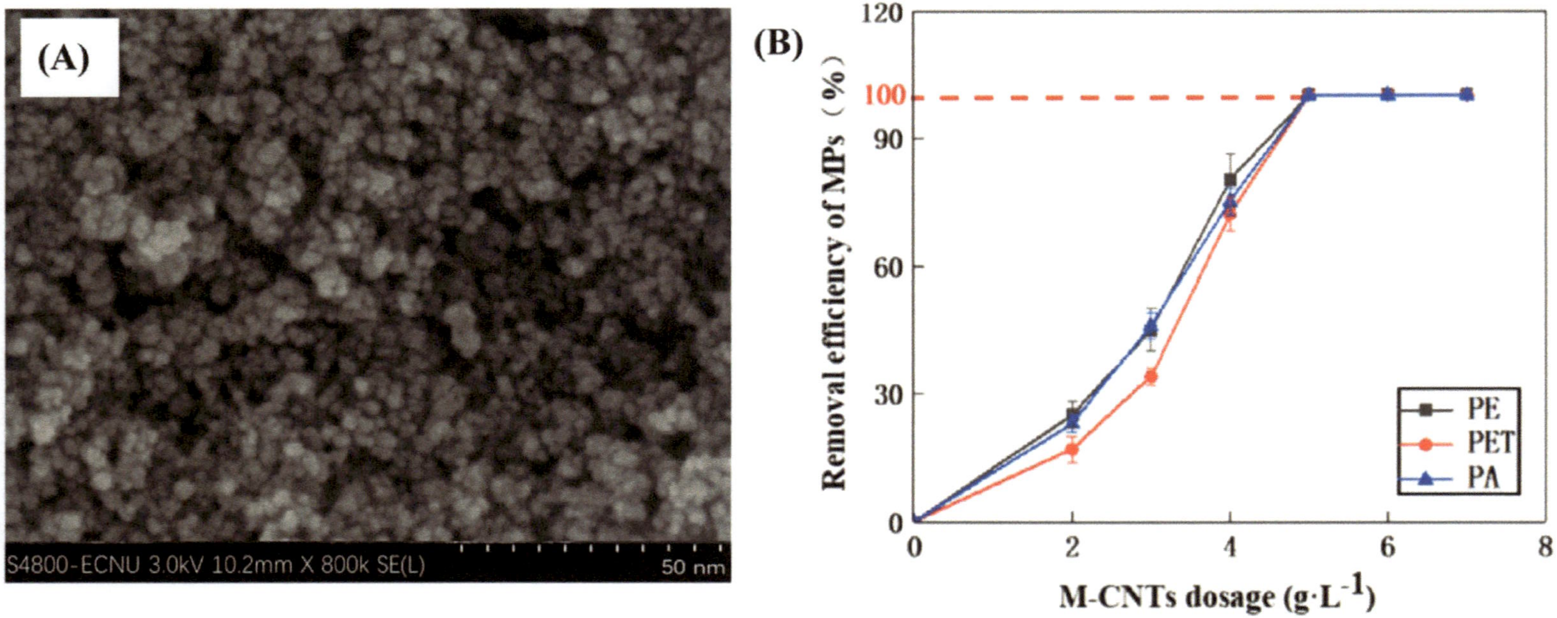

Figure 11.6 (A) SEM images for original M–CNTs; (B) Effect of M–CNTs dosage on MPs removal (pH = 7.0 ± 0.1, MPs dosage 5 g/L, contact time 300 min, 25°C). (Tang et al. 2021).

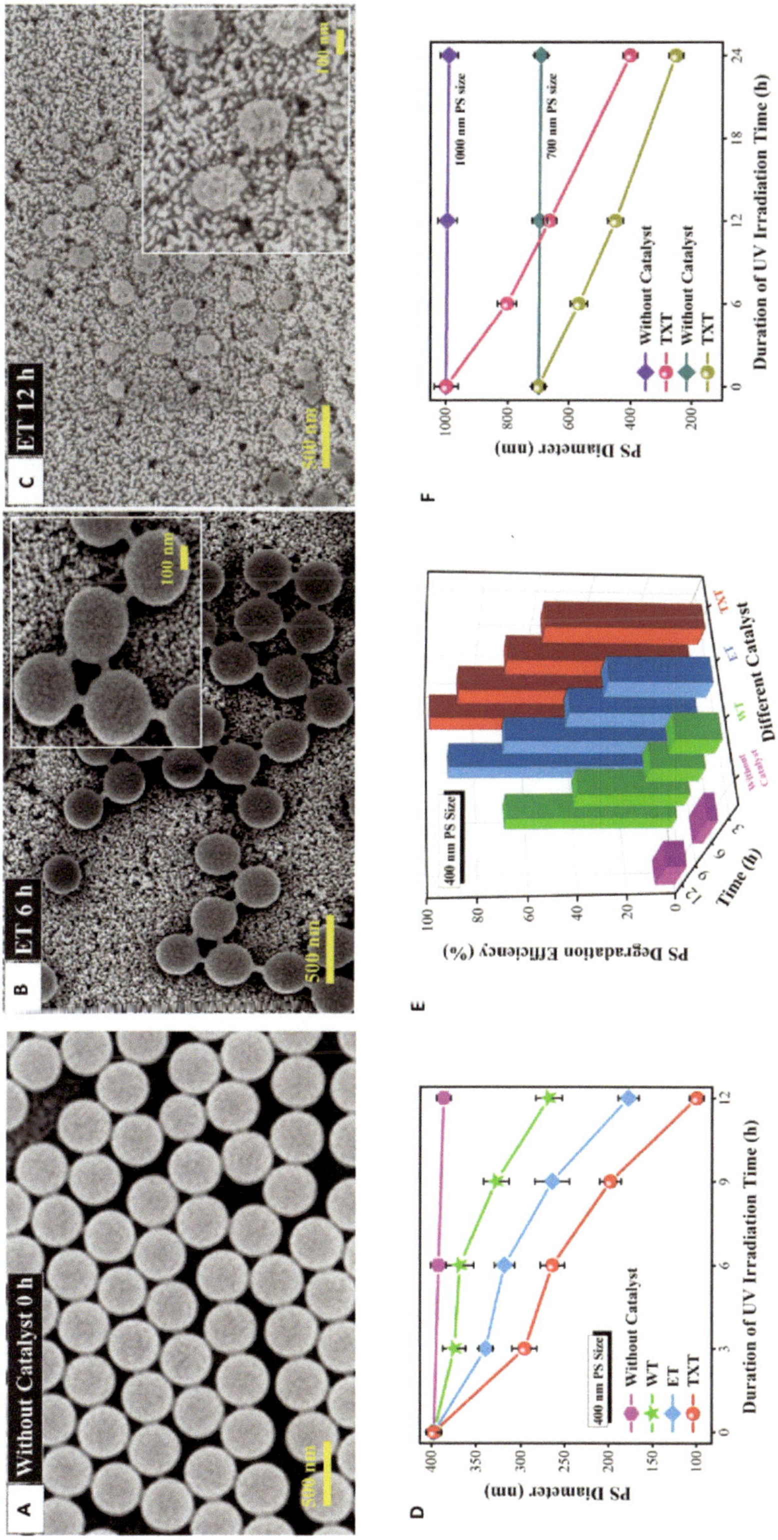

Figure 11.7 (A) SEM images of photocatalytically degraded PS MPs under 365 nm UV Light at 0 h; (B) 6 h and (C) 12 h; (D) Diameter change of 400-nm PS on TXT (Triton X-100), ET (ethanol), WT (water) films and without catalyst; (E) PS degradation efficiency (%) with (TXT, ET, WT) and without catalyst; (F) Diameter change of 1-µm and 700-nm PS on TXT film, and without catalyst. (Nabi et al. 2020).

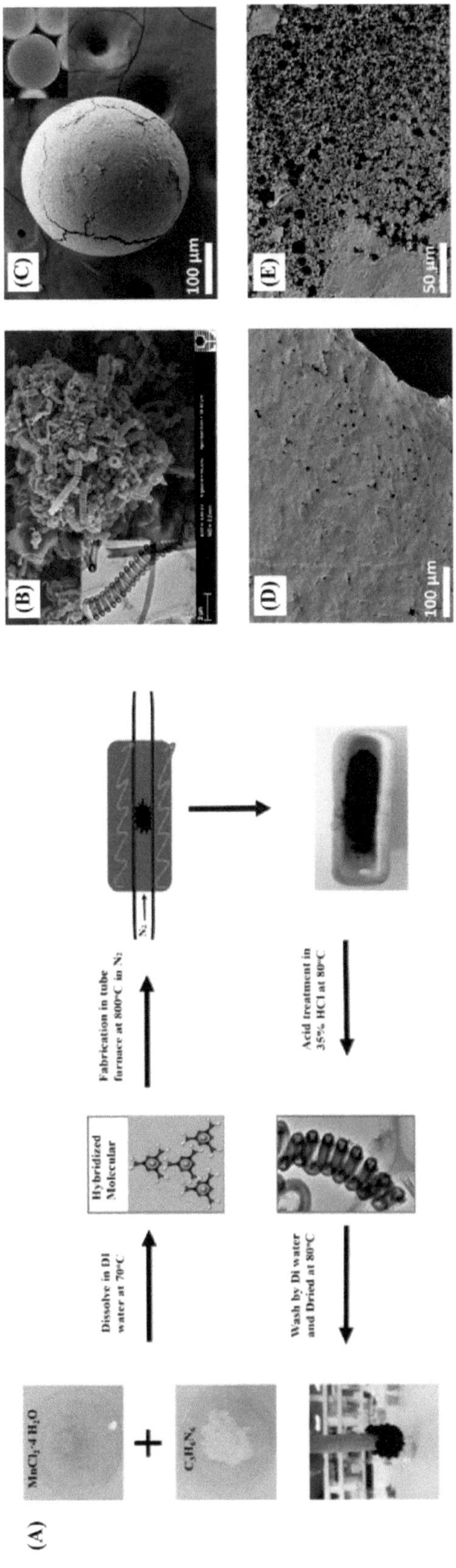

Figure 11.8 (A) Schematic illustration for preparation of Mn@NCNTs; (B) SEM image of Mn@NCNTs synthesized at 900°C; (C) MPs after 2hrs; (D) 6 hrs; (E) 8 hrs, of degradation using Mn@NCNTs. (Kang et al. 2019).

$$O_2 + e^-_{CB} \longrightarrow O_2^{\bullet-} \quad (5)$$
$$O_2^{\bullet-} + H_2O \longrightarrow {}^{\bullet}OOH + OH^- \quad (6)$$
$$2{}^{\bullet}OOH \longrightarrow O_2 + H_2O_2 \quad (7)$$
$$H_2O_2 \longrightarrow 2{}^{\bullet}OH \quad (8)$$

11.5.2 Carbon Nanostring

In a recent study carbon nanostring was developed to catalyse oxidative degradation of MPs. The carbon nanostring (Mn@NCNTs) was synthesised by encapsulation of manganese carbide nanoparticles in helically N-doped carbon nanotubes via one-pot pyrolysis (Kang et al. 2019). Briefly, to synthesis Mn@NCNTs, a mixed solution of melamine and $MnCl_2$ was dried and heated at 500°C for 2 hrs, then at 800°C for 6 hrs in N_2 atmosphere followed by washing of cooled product with Conc. HCl acid at 80°C for 24 hrs to remove the surface unstable species (Figure 11.8A). Synthesised Mn@NCNTs showed string like structure with surface area (Figure 11.8B). The oxidative degradation of microplastics was done in an autoclave in the presence of Mn@NCNTs and peroxymonosulfate (PMS), a persulphate salt which generates reactive oxygen species such as $SO_4^{\bullet}$ and hydroxyl radical (${}^{\bullet}OH$). Figure 11.8C–E shows SEM images of cosmetic product-extracted MPs after hydrothermal reaction in the presence of Mn@NCNTs and PMS. Fe, Co, Mn,

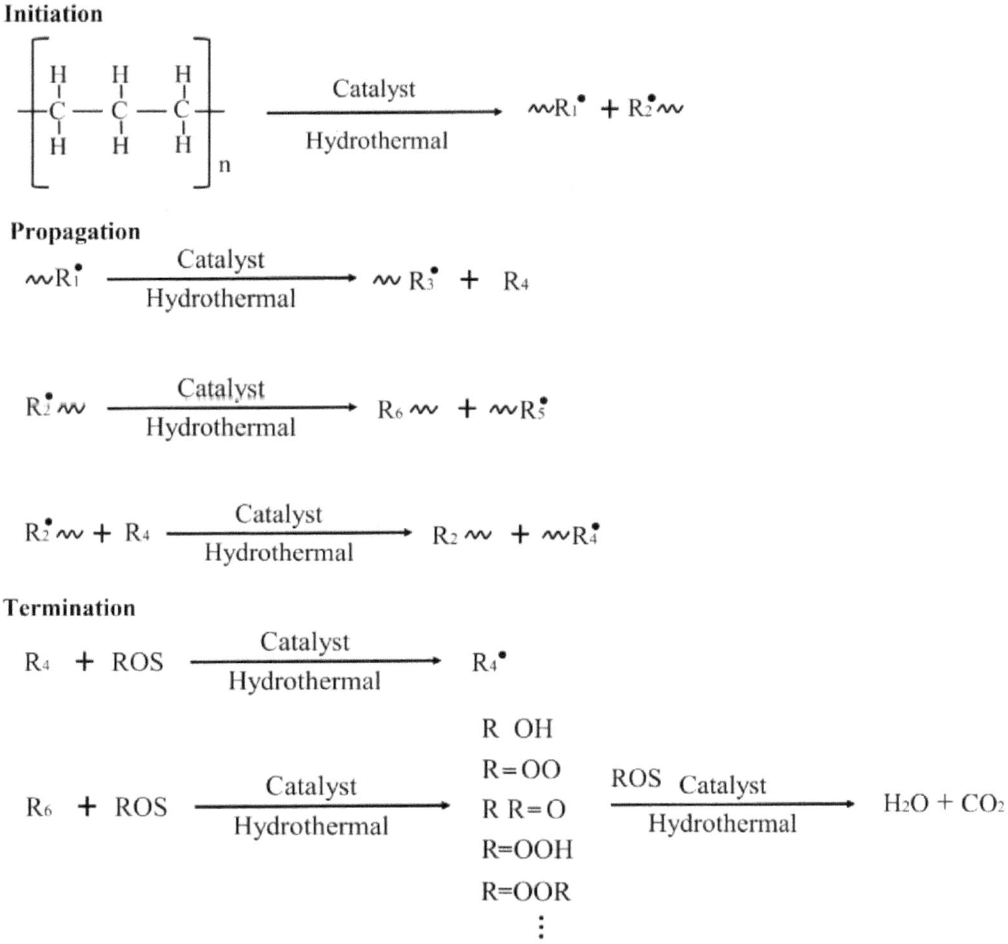

Figure 11.9 Schematic illustration of the catalytic oxidative degradation of MPs (Kang et al. 2019).

and their oxides were reported to act as activator for PMS to generate ROS; however, associated secondary pollution from metal leaching limits their application in catalytic oxidation. It was found that heteroatom doping in sp^2 carbons of CNT can create more active sites for enhanced catalytic oxidative activity for PMS (Duan et al. 2015). It was discussed in the study that in the oxidative degradation the long chain polyethylene polymers were shred in to subpolymers with shorter carbon chain followed by reaction with ROS to break into short chains with lower molecular weight (Figure 11.9).

11.6 CONCLUSION

MPs pollution has become a serious threat to environmental sustainability and human health. Establishment of remediation strategies for MPs pollution especially from aquatic environment is the need of the hour. Water treatment with especially formed nanostructured materials can remove MPs from the sources or degrade the same with good efficiency. The potential of nanostructured material in remediating MPs pollution through the mechanism of adsorption, magnetic separation, and catalytic oxidative degradation is huge. The limitations of conventional water treatment strategies can be improved using nanostructured materials while dealing particularly with MPs. However, at present these materials have the limitation of not being tested at large-scale treatment systems and lack of sufficient data on the effect on water quality, interaction with other pollutants, feasibility assessment with conventional water treatment systems, etc. As the concept of nanostructured materials in remediating MPs is new and material developments are at a nascent stage, more futuristic research is needed to develop robust technologies for eliminating MPs pollution from the environment.

References

Ahmadipouya, S., Haris, M.H., Ahmadijokani, F., Jarahiyan, A., Molavi, H., Moghaddam, F.M., Rezakazemi, M., et al. 2021. Magnetic Fe_3O_4@UiO-66 nanocomposite for rapid adsorption of organic dyes from aqueous solution. J. Mol. Liq. 322: 114910.

Anjum, A. 2017. Adsorption technology for removal of toxic pollutants. pp. 25–80. *In*: Rene, E., Sahinkaya, E., Lewis, A. and Lens, P. (eds). Sustainable Heavy Metal Remediation. Environmental Chemistry for a Sustainable World. Springer, Cham.

Arun Kumar, R., Sivashanmugam, A., Deepthi, S., Iseki, S., Chennazhi, K.P., Nair, S.V. and Jayakumar, R. 2015. Injectable chitin-poly(ε-caprolactone)/nanohydroxyapatite composite microgels prepared by simple regeneration technique for bone tissue engineering. ACS Appl. Mater. Interfaces. 7: 9399–9409.

Bano, K., Kaushal, S. and Singh, P.P. 2021. A review on photocatalytic degradation of hazardous pesticides using heterojunctions. Polyhedron. 209: 115465.

Begum, S., Yuhana, N.Y., Md. Saleh, N., Kamarudin, N.H.N. and Sulong, A.B. 2021. Review of chitosan composite as a heavy metal adsorbent: material preparation and properties. Carbohyd. Polym. 259: 117613.

Chang, C., Chen, S. and Zhang, L. 2011. Novel hydrogels prepared via direct dissolution of chitin at low temperature: structure and biocompatibility. J. Mater. Chem. 21: 3865–3871.

Chen, Y.J., Chen, Y., Miao, C., Wang, Y.R., Gao, G.K., Yang, R.X., Zhu, H.J., et al. 2020. Metal–organic framework-based foams for efficient microplastics removal. J. Mater. Chem. A. 8: 14644–14652.

Chen, F., Ilyas, N., Liu, X., Li, Z., Yan, S. and Fu, H. 2021. Size effect of Fe_3O_4 nanoparticles on magnetism and dispersion stability of magnetic nanofluid. Front. Energy Res. 9: 801.

Cho, S. and Choi, W. 2001. Solid-phase photocatalytic degradation of PVC-TiO_2 polymer composites. J. Photochem. Photobiol. A. 143: 221–228.

Duan, X., Sun, H., Wang, Y., Kang, J. and Wang, S. 2015. N-doping-induced nonradical reaction on single-walled carbon nanotubes for catalytic phenol oxidation. ACS Catalysis. 5: 553–559.

Ebner, A.D., Ritter, J.A. and Nuñez, L. 2007. High-gradient magnetic separation for the treatment of high-level radioactive wastes. Sep. Sci. Technol. 34: 1333–1350.

EL-Gendi, A., Deratani, A., Ahmed, S.A. and Ali, S.S. 2014. Development of polyamide-6/chitosan membranes for desalination. Egypt. J. Pet. 23: 169–173.

Farner Budarz, J., Cooper, E.M., Gardner, C., Hodzic, E., Ferguson, P.L., Gunsch, C.K. and Wiesner, M.R. 2019. Chlorpyrifos degradation via photoreactive TiO_2 nanoparticles: Assessing the impact of a multi-component degradation scenario. J. Hazard Mater. 372: 61–68.

Gamonchuang, J., Santaladchaiyakit, Y. and Burakham, R. 2022. Magnetic solid-phase extraction of carbamate pesticides using magnetic metal-organic frameworks derived from benzoate ligands, followed by digital image colorimetric screening and high-performance liquid chromatography analysis. ACS Omega. 7: 12202–12211.

Grbic, J., Nguyen, B., Guo, E., You, J.B., Sinton, D. and Rochman, C.M. 2019. Magnetic extraction of microplastics from environmental samples. Environ. Sci. Technol. Lett. 6: 68–72.

Horcajada, P., Gref, R., Baati, T., Allan, P.K., Maurin, G., Couvreur, P., Férey, G., et al. 2012. Metal-organic frameworks in biomedicine. Chem. Rev. 112: 1232–1268.

Horst, M.F., Alvarez, M. and Lassalle, V.L. 2016. Removal of heavy metals from wastewater using magnetic nanocomposites: analysis of the experimental conditions. Sep. Sci. Technol. 51: 550–563.

Huang, X., Qi, X., Boey, F. and Zhang, H. 2012. Graphene-based composites. Chem. Soc. Rev. 41: 666–686.

Huang, Y., Zhong, Z., Duan, B., Zhang, L., Yang, Z., Wang, Y. and Ye, Q. 2014. Novel fibers fabricated directly from chitin solution and their application as wound dressing. J. Mater. Chem. B. 2: 3427–3432.

Iconaru, S.L., Guégan, R., Popa, C.L., Motelica-Heino, M., Ciobanu, C.S. and Predoi, D. 2016. Magnetite (Fe_3O_4) nanoparticles as adsorbents for As and Cu removal. Appl. Clay Sci. 134: 128–135.

Ighalo, J.O., Yap, P.S., Iwuozor, K.O., Aniagor, C.O., Liu, T., Dulta, K., Iwuchukwu, F.U., et al. 2022. Adsorption of persistent organic pollutants (POPs) from the aqueous environment by nano-adsorbents: a review. Environ. Res. 212: 113123.

Jambeck, J.R., Geyer, R., Wilcox, C., Siegler, T.R., Perryman, M., Andrady, A., Narayan, R., et al. 2015. Plastic waste inputs from land into the ocean. Science. 347: 768–771.

Jiménez, M., Ignacio Maldonado, M., Rodríguez, E.M., Hernández-Ramírez, A., Saggioro, E., Carra, I. and Sánchez Pérez, J.A. 2015. Supported TiO_2 solar photocatalysis at semi-pilot scale: degradation of pesticides found in citrus processing industry wastewater, reactivity and influence of photogenerated species. J. Chem. Technol. Biotechnol. 90: 149–157.

Jordan, A., Scholz, R., Wust, P., Fähling, H. and Felix, R. 1999. Magnetic fluid hyperthermia (MFH): cancer treatment with AC magnetic field induced excitation of biocompatible superparamagnetic nanoparticles. J. Magn. Magn. Mater. 201: 413–419.

Kang, J., Zhou, L., Duan, X., Sun, H., Ao, Z. and Wang, S. 2019. Degradation of cosmetic microplastics via functionalized carbon nanosprings. Matter. 1: 745–758.

Kreno, L.E., Leong, K., Farha, O.K., Allendorf, M., van Duyne, R.P. and Hupp, J.T. 2012. Metal-organic framework materials as chemical sensors. Chem. Rev. 112: 1105–1125.

Li, Q., Xuan, Y. and Wang, J. 2005. Experimental investigations on transport properties of magnetic fluids. Exp. Therm. Fluid. Sci. 30: 109–116.

Li, P., Li, L., Zhao, Y., Sun, L. and Zhang, Y. 2016. Selective binding and magnetic separation of histidine-tagged proteins using Fe_3O_4/Cu-apatite nanoparticles. J. Inorg. Biochem. 156: 49–54.

Liu, N., Yu, F., Wang, Y. and Ma, J. 2022. Effects of environmental aging on the adsorption behavior of antibiotics from aqueous solutions in microplastic-graphene coexisting systems. Sci. Total Environ. 806.

Ma, J., Wu, G., Li, S., Tan, W., Wang, X., Li, J. and Chen, L. 2018. Magnetic solid-phase extraction of heterocyclic pesticides in environmental water samples using metal-organic frameworks coupled to high performance liquid chromatography determination. J. Chromatogr. A. 1553: 57–66.

Moradi Dehaghi, S., Rahmanifar, B., Moradi, A.M. and Azar, P.A. 2014. Removal of permethrin pesticide from water by chitosan–zinc oxide nanoparticles composite as an adsorbent. J. Saudi Chem. Soc. 18: 348–355.

Murray, L.J., Dinc, M. and Long, J.R. 2009. Hydrogen storage in metal–organic frameworks. Chem. Soc. Rev. 38: 1294–1314.

Nabi, I., Bacha, A.U.R., Li, K., Cheng, H., Wang, T., Liu, Y., Ajmal, S., et al. 2020. Complete photocatalytic mineralization of microplastic on TiO_2 nanoparticle film. iScience. 23: 101326.

Nagahama, H., Kashiki, T., Nwe, N., Jayakumar, R., Furuike, T. and Tamura, H. 2008. Preparation of biodegradable chitin/gelatin membranes with GlcNAc for tissue engineering applications. Carbohyd Polym. 73: 456–463.

Nishimura, K., Miura, O., Ito, D., Tsunasima, Y. and Wada, Y. 2009. Removal of radioactive heavy metal ions from solution by superconducting high-gradient magnetic separation with schwertmannite and zirconium-ferrite adsorbents. IEEE Trans. Appl. Superconduc. 19: 2162–2164.

Okhay, O. and Tkach, A. 2021. Graphene/reduced graphene oxide-carbon nanotubes composite electrodes: from capacitive to battery-type behaviour. Nanomaterials. 11(5): 1240.

Papageorgiou, D.G., Kinloch, I.A. and Young, R.J. 2015. Graphene/elastomer nanocomposites. Carbon. 95: 460–484.

Plastics Europe. 2019. Plastics e the Facts 2019. URL https://plasticseurope.org/wp-content/uploads/2021/10/2019-Plastics-the-facts.pdf (accessed 7.5.22).

Rasool, S., Rasool, T. and Gani, K.M. 2022. A review of interactions of pesticides within various interfaces of intrinsic and organic residue amended soil environment. Chem. Eng. J. Advances. 11: 100301.

Reiner-Rozman, C., Kotlowski, C. and Knoll, W. 2016. Electronic biosensing with functionalized rGO FETs. Biosensors. 6(2): 17.

Sadakiyo, M., Okawa, H., Shigematsu, A., Ohba, M., Yamada, T. and Kitagawa, H. 2012. Promotion of low-humidity proton conduction by controlling hydrophilicity in layered metal-organic frameworks. J. Am. Chem. Soc. 134: 5472–5475.

Sarkar, D.J., Sarkar, S.D., Das, B.K., Manna, R.K., Behera, B.K. and Samanta, S. 2019. Spatial distribution of meso and microplastics in the sediments of river Ganga at eastern India. Sci. Total Environ. 694: 133712.

Sarkar, D.J., Sarkar, S.D., Kumar Manna, R., Samanta, S. and Das, B.K. 2020a. Microplastics pollution: an emerging threat to freshwater aquatic ecosystem of India. J. Inland Fish Soc. India. 52: 5–15.

Sarkar, D.J., Sarkar, S.D., Manna, R.K. and Das, B.K. 2020b. Microplastic pollution: threat to Indian river system. Everyman's Science. 6: 376–382.

Sarkar, D.J., Sarkar, S.D., Das, B.K., Praharaj, J.K., Mahajan, D.K., Purokait, B., Mohanty, T.R., et al. 2021a. Microplastics removal efficiency of drinking water treatment plant with pulse clarifier. J. Hazard Mater. 413: 125347.

Sarkar, D.J., Sarkar, S.D., Das, B.K., Sahoo, B.K., Das, A., Nag, S.K., Manna, R.K., et al. 2021b. Occurrence, fate and removal of microplastics as heavy metal vector in natural wastewater treatment wetland system. Water Res. 192: 116853.

Sarkar, D.J., Sarkar, S.D., Mukherjee, S. and Das, B.K. 2021c. Impact and fate of microplastics in the riverine ecosystem. pp. 95–115. *In:* Kumar, M., Snow, D., Honda, R. and Mukherjee, S. (eds). Contaminants in Drinking and Wastewater Sources. Springer Transactions in Civil and Environmental Engineering. Springer, Singapore.

Sarkar, B., Dissanayake, P.D., Bolan, N.S., Dar, J.Y., Kumar, M., Haque, M.N., Mukhopadhyay, R., et al. 2022a. Challenges and opportunities in sustainable management of microplastics and nanoplastics in the environment. Environ. Res. 207: 112179.

Sarkar, D.J., Sarkar, S.D. and Das, B.K. 2022b. Occurrence, fate and removal of microplastics in wastewater treatment plants (WWTPs) and drinking water treatment plants (DWTPs). pp. 223–245. *In*: Sillanpää, M., Khadir, A. and Muthu, S.S. (eds). Microplastics Pollution in Aquatic Media. Environmental Footprints and Eco-design of Products and Processes. Springer, Singapore.

Schwaminger, S.P., Fraga-García, P., Eigenfeld, M., Becker, T.M. and Berensmeier, S. 2019. Magnetic separation in bioprocessing beyond the analytical scale: from biotechnology to the food industry. Front Bioeng Biotechnol. 7: 233.

Shang, J., Chai, M. and Zhu, Y. 2003. Photocatalytic degradation of polystyrene plastic under fluorescent light. Environ. Sci. Technol. 37: 4494–4499.

Shi, X., Zhang, X., Gao, W., Zhang, Y. and He, D. 2022. Removal of microplastics from water by magnetic nano-Fe_3O_4. Sci. Total Environ. 802: 149838.

Sumida, K., Rogow, D.L., Mason, J.A., McDonald, T.M., Bloch, E.D., Herm, Z.R., Bae, T.H., et al. 2012. Carbon dioxide capture in metal-organic frameworks. Chem. Rev. 112: 724–781.

Sun, C., Wang, Z., Zheng, H., Chen, L. and Li, F. 2021. Biodegradable and re-usable sponge materials made from chitin for efficient removal of microplastics. J. Hazard Mater. 420: 126599.

Sundar, S., Kwon, S.J. and Venkatachalam, G. 2020. Magneto-biosensor for the detection of uric acid using citric acid-capped iron oxide nanoparticles. J. Nanosci. Nanotechnol. 20: 2144–2153.

Sutherland, W.J., Broad, S., Caine, J., Clout, M., Dicks, L.V., Doran, H., Entwistle, A.C., et al. 2016. A horizon scan of global conservation issues for 2016. Trends Ecol. Evol. 31(1): 44–53.

Tamjidi, S., Esmaeili, H. and Moghadas, B.K. 2019. Application of magnetic adsorbents for removal of heavy metals from wastewater: a review study. Mater. Res. Express. 6: 102004.

Tang, Y., Zhang, S., Su, Y., Wu, D., Zhao, Y. and Xie, B. 2021. Removal of microplastics from aqueous solutions by magnetic carbon nanotubes. Chem. Eng. J. 406: 126804.

Wan Ngah, W.S., Teong, L.C. and Hanafiah, M.A.K.M. 2011. Adsorption of dyes and heavy metal ions by chitosan composites: a review. Carbohyd. Polym. 83: 1446–1456.

WWF, 2019. No plastic in nature: assessing plastic ingestion from nature to people. URL www.newcastle.edu.au/ (accessed 7.5.22).

Yamada, T., Otsubo, K., Makiura, R. and Kitagawa, H. 2013. Designer coordination polymers: dimensional crossover architectures and proton conduction. Chem. Soc. Rev. 42: 6655–6669.

Yan, L., Li, P., Zhou, W., Wang, Z., Fan, X., Chen, M., Fang, Y., et al. 2019. Shrimp shell-inspired antifouling chitin nanofibrous membrane for efficient oil/water emulsion separation with in situ removal of heavy metal ions. ACS Sustain. Chem. and Eng. 7: 2064–2072.

Yanai, N., Kitayama, K., Hijikata, Y., Sato, H., Matsuda, R., Kubota, Y., Takata, M., et al. 2011. Gas detection by structural variations of fluorescent guest molecules in a flexible porous coordination polymer. Nat. Mater. 10: 787–793.

Yuan, F., Yue, L., Zhao, H. and Wu, H. 2020. Study on the adsorption of polystyrene microplastics by three-dimensional reduced graphene oxide. Water Sci. Technol. 81: 2163–2175.

Zhoua, Q., Wu, Y., Sun, Y., Sheng, X., Tong, Y., Guo, J., Zhou, B., et al. 2021. Magnetic polyamidoamine dendrimers for magnetic separation and sensitive determination of organochlorine pesticides from water samples by high-performance liquid chromatography. J. Environ. Sci. 102, 64–73.

Zhu, Q.L. and Xu, Q. 2014. Metal–organic framework composites. Chem. Soc. Rev. 43: 5468–5512.

Chapter 12

Process Intensification of Fluoride-Containing Solution Using Nano-Adsorbent

Lopamudra Das*[1,2], Papita Das[1,2] and Avijit Bhowal[1,2]
[1]School of Advanced Studies on Industrial Pollution Control Engineering, Jadavpur University, Kolkata, India

Chiranjib Bhattacharjee[2]
[2]Department of Chemical Engineering, Jadavpur University, Kolkata, India

12.1 INTRODUCTION

Water is a significant constituent of the living body which is crucial for existence by improving the eminence of the lifecycle. But with growing technology, water is contaminated day by day by numerous toxic chemicals or waste runoffs from industries. Elemental Fluoride is one of the vital harmful pollutants which pollute groundwater as well as surface water and affect the heath of living beings. Fluoride initiates in ground-water because of natural leaching or suspension as of fluoride-rich rocks and volcanic ash such as fluorspar, cryolite, fluorapatite and sellaite, etc (Mohapatra et al. 2009, Wambu et al. 2015). Hydrofluoros is a disease caused by excess intake of fluoride-contaminated drinking water (permissible limit 1–1.5 mg/L) and hasbecome a global issue of great concern. Excess fluoride and contamination due to rapid industrial development is resulting in increased medical ailments among industrial personnel, inhabitants and domesticated cattle (Wambu et al. 2015, Choubisa and Choubisa 2016). Discharging effluents from numerous industrial activities for instance, coal-burning stations, steel production, aluminum, phosphorus, production of bricks, glass, plastic, cement, and hydrofluoric acid (HF acid) contains fluoride which affect the ecosystem and living beings by contaminating water, air, soil, crops and vegetables. Another source of fluoride pollution in potable water and surface water is because of the excessive

*Corresponding Author: lopa.che94@gmail.com

use of fertilizers (for instance superphosphate, NPK, potash, etc.) which also contain a notable amount of fluoride. Nearly 20 developing countries are facing hydrofluorosis diseases. Most Asian developing countries including India and African countries are severely affected (Choubisa and Choubisa 2016).

Agro-waste and byproducts/plant waste, for example, sugarcane bagasse, rice husk, groundnut shell, and tea waste; sawdust is furthermost extensively used biosorbent for the fluoride adsorption process. The nano-composite biosorbents comprising biomaterials, biopolymers and inorganic material have getting notable attention because of their integrated benefit. Aluminum-based nano-materials, magnetic nano-composites, clay, carbonaceous materials-based composites, biopolymers and polymeric nanocomposites also have been widely applied as biosorbents for defluoridation process (Mohapatra et al. 2009, Wambu et al. 2015).

Hence, the investigation envisioned eco-friendly, non-hazardous and efficient defluoridation of fluoride-containing wastewater-initiated intensification in the importance of nano-adsorbents. The aim of this chapter is to present a comprehensive and large-scale viewpoint on the topic of nano-biosorbents and their application for the removal of fluoride from wastewater along with the latest procedures. For this purpose, a brief enlightenment about the source of fluoride, the toxicity of fluoride on a living being, and a brief description of nano-biosorbent implemented in defluoridation techniques has been reviewed and discussed in the succeeding segments.

12.2 SOURCE OF FLUORIDE

12.2.1 Natural Source

Fluorine is a highly reactive and abundant element on earth's surface and its simplest electronegative ion is called fluoride (F^-) (Sivasankar et al. 2016). It shows a strong affinity to form compounds with other chemicals, that's why it can't occur in its natural elemental state. Fluoride exists in different amounts in the earth's crusts, rock-forming minerals, and volcanic ash. Fluoride concentration of groundwater is subject to the chemical, physical, geographical characteristics of the underground layer of water-bearing permeable sediment rock. It is also subject to the adjacent temperature, the intensity of weathering, deepness of underground water layer, permeability and acerbity of rocks and soil layer, presence of other chemical elements etc. (Kundu and Mandal 2009). Nowadays, huge variations of fluoride-bearing rock-forming minerals were recognized such as sellaite (MgF_2), fluorite (CaF_2), fluorapatite, cryolite (Na_3AlF_6), apatites, topaz, amphibole, mica, and phosphorite etc. (Datta et al. 1996, Mohapatra et al. 2009).

12.2.2 Manmade Source

Recently researchers reported that, fluoride is also emitted from effluents of the different manufacturing industries that employ fluorine-containing compounds (power stations, semiconductor industries, steel and aluminum industries, cosmetic industries, nickel production, glass, plastic, cement production industries) and as of several types of manmade activities. HF (Hydrogen Fluoride) is widely employed in the production of aluminum and CFCs (chlorofluorocarbons). Excessive usage of synthetic fertilizers including superphosphate, potash and NPK pollutes soil and fresh water by releasing significant quantities of fluoride into it. Precise addition of fluoride in dental appliance, foodstuffs, and pharmaceutical products also put-up fluoride accumulation in the environment. In addition, 'C6' short-chain fluorine-based formulations (fluorine-based durable water and oil textile repellents) are also responsible for fluoride contamination in water bodies and are also toxic and environmentally harmful (Choubisa and Choubisa 2016, Das et al. 2021b). Other fields where fluorine-containing compounds are utilized are electronic industries, tanning leathers, potable water fluoridation (Sodiumhexa fluorosilicate, sodium fluoride are generally used), insecticide production (sodium fluoride is extensively used), ceramics and fluorescent lamp industry (Calcium fluoride is widely used) (Bhatnagar et al. 2011).

Industrial fluorosis disease was first identified in cattle from a farm nearby to a superphosphate production industry (Choubisa and Choubisa 2016). India was introduced to the industrial fluorosis disease for the first time in Orissa. As per the report, the domestic cattle living nearby to an aluminum factory were suffering from this disease (Rao and Pal 1978).

12.3 TOXICITY OF FLUORIDE

Long-lasting fluoride exposure either by way of ingesting of fluoride-contaminated potable-water or fluoride released by the industries for a long time causes dental fluorosis (in the form of dental mottling, which is the prior pathognomonic sign of all types of long-lasting fluoride intoxication), and skeletal fluorosis (bone deformities) in living beings. More or less all type of foodstuffs and vegetables hold small amount of fluoride, depending on the source of soil and water. Fluoride contamination in drinking water is not the only principal cause of fluorosis but the overall consumption/intake of fluoride via all origins is responsible as well. In addition, non-skeletal fluorosis in soft tissues or body parts due to the fluoride endemic provinces (such as gastrointestinal discomforts, neurological disorders and irregular reproductive cycles, genotoxic effects, allergic reactions, polyuria, polydipsia, etc., have also been reported in human beings as well as in domestic cattle. (Choubisa 2012a).

12.4 TREATMENT METHODS

12.4.1 Adsorption Process for Defluoridation

There are different methods for fluoride remediation of wastewater which have been shown in Table 12.1.

Table 12.1 Fluoride removal techniques, its advantages and limitations (Roy et al. 2017b, Manna et al. 2018, Das et al. 2021a).

Methods	Advantages	Disadvantages
Adsorption	cheap, needs less space and easy to handle method, high efficiency, eco-friendly	Effective at specific condition
Reverse osmosis	Effective, work at wide pH range	Expensive, frequent cleaning required,
Precipitation /coagulation	Easy and efficient	high production of sludge and other ion precipitation
Ion exchange	Efficient color and odor intact	Expensive, effect of other ion concentration
Electrodialysis	efficient method, no chemical required, effective under wide pH ranges	Membrane fouling, higher power consumption
Electrochemical	Established and effective method	High energy consumption, Frequent replacement of anode required
Biological treatment	Effective towards Biocompatible and biodegradation process	High production of sludge, microbial activity is affected

Among all the wastewater treatment processes, adsorption is considered to be a reasonable technique because of its less operative cost, higher uptake capacity, recycle and regeneration capability of adsorbent, and it shows the high quality of water purification without generation of any secondary sludge. In recent years treatment of fluoride-contaminated water by adsorption technique has been extensively explored and importance of surface modification of adsorbent is rising to improve the removal proficiency and recyclability of adsorbent (Onyango et al. 2004). Adsorption process on to absorbent surface generally occurs through three stages (Mohapatra et al. 2009):

(a) Diffusion/transport of adsorbate (pollutant) molecule to the adsorbent's outer exterior from bulk solution across the boundary layer adjacent to the solid particle;
(b) Adsorption of adsorbate molecule on adsorbent surfaces;
(c) The adsorbed molecule about to replace with the structural elements within adsorbent or transported to the internal surfaces.

The principal undertakings of researchers and specialists are to formulate/develop inexpensive, effective, eco-friendly, and higher selective adsorbent material having remarkable operative adsorption capacity (Bhatnagar et al. 2011). In last few decades, an extensive variation of adsorbent materials and their chemical and physical modifications were experienced for defluoridation process such as activated rice husk (Roy et al. 2017a), activated carbons (Mohan et al. 2008), activated alumina (Maliyekkal et al. 2008), bauxite (Maliyekkal et al. 2008), fly ash, serpentine (Chidambaram et al. 2003), ceramics (Chen et al. 2011), chitin/chitosan and salt of alginate [95], ferric oxide/hydroxide (Shams et al. 2010), hydroxyapatite (Viswanathan et al. 2013), titanium-derived adsorbent (Wajima et al. 2009), cellulose fiber and cellulosic material reinforced composite (Zhao et al. 2008), clays (Çengeloğlu et al. 2002), zeolite (Nagendra Rao 2003) and magnesium-modified adsorbent (Mourabet et al. 2012).

12.5 NANOTECHNOLOGY APPROACHES FOR FLUORIDE ADSORPTION

Nanotechnology has an important position in the advancement of inventive techniques to replace available traditional production equipment to fabricate/reformulate novel materials/products and chemicals with enhanced performance ensuing with a reduction of energy and material consumption and improving the pollutant removal efficiency (Das et al. 2020c). New advanced methods are repeatedly being studied to enhance the existing conventional techniques for polluted water treatment. Therefore, it needs to be cost-effective and more efficient than existing accessible treatment methods for water purification process, in this review article nano-technological methods have been considered for this purpose (Ali Mansoori et al. 2008). A variety of classes of nano-materials have been reported to be proficient for treatment of wastewater like metal-based nano-materials, carbonaceous nano-materials, and polymeric nano-materials derived from agro/plant waste/byproduct materials. Though, because of the nano-size, the agglomeration tendency of nanomaterials is one serious limitation, which reduces the fluoride adsorption capacity. To overcome this problem, polymeric/carbonaceous materials were applied as the template to disperse the nanomaterials in the template for synthesis of the nanocomposite. One of the most important fields of nano-material is nano-composites having higher uptake capacity, high regeneration capability and high effectiveness in remediation of pollutants from contaminated water. In nano-composites, one of the phases is in nano-scale range and because of their small size and higher specific surface area, they differ as of conventional composite material with micro-scale structure. Nano-composites can be classified, in accordance with their structural and matrix variations, like polymer reinforce nano-composite (PVA-AC, nano-cellulose/PVA, chitosan/GO, etc.), metal based nano-composite ($Fe–Cr/Al_2O_3$, Al/CNT, etc), ceramic grafted nano-composite (Al_2O_3/SiO_2, Al_2O_3/TiO_2, etc.) (Das et al. 2020a, b). In this article, latest advances on nano-adsorbent development and their applications for treatment of fluoride contaminated wastewater have been reviewed. The most generally investigated nanosorbents depend on carbon (Roy et al. 2017b), metals and polymer materials holding significant importance to reduce the effect of harmfulness in wastewater treatment techniques.

The following segments of this review paper assemble most recent studies on the applications of various eco-friendly nano-adsorbent for defluoridation process.

12.6 CARBONACEOUS MATERIAL-BASED NANO-ADSORBENTS

The carbonaceous material, for instance, activated carbon, carbon nanotubes, graphene/graphene oxide, etc. because of its high surface area and hydrophobic surface property, requires it to be balanced out in composite synthesis to reduce agglomeration tendency. Carbonaceous materials have revealed immense effectiveness in wastewater treatment process due to their admirable adsorption characteristics and subsequent to the surface modifications of these materials, they can possibly be customized as needed. The chemical or gas activation of carbonaceous materials, for example, plant or mineral source was remarkably well-regarded. Usually, the raw materials employed for synthesis of activated carbon were woods, charcoal, seeds, coal, petroleum residues, synthetic polymers and also low-priced agro-waste materials such as sugarcane bagasses, rice-husk, sawdust, coconut shell, tea waste, fruit waste/ peels, etc are also used. Karthikeyan and Ilango have examined the fluoride uptake capacity of activated carbon derived from smoldering and carbonization of Morringa indica bark. They found the fluoride removal efficiency as 71% for 2 mg/l of the initial fluoride concentration at neutral pH and adsorption kinetics and thermodynamic study also have been evaluated (Karthikeyan and Ilango 2007). Karthikeyan and Elango have employed graphite for fluoride removal from its aqueous solution and they found notable efficiency of adsorbent (Karthikeyan and Elango 2008). Tchomgui-Kamga et al. have utilized $CaCl_2$ (calcium chloride) impregnated wood charcoal prepared by carbonization at different temperatures (500°C, 650°C, 900°C) and applied for defluoridation process in batch system. They reported fluoride removal efficiency >92% at neutral pH for initial F^- concentration of 10 mg/L (Tchomgui-Kamga et al. 2010). Li et al. investigated and reported on the fluoride remediation by graphene as adsorbent in batch reactor and they achieved maximum uptake capacity of adsorbent as 17.65 mg/g for 0.0025 g/L of the initial concentration of fluoride solution at 25°C temperature (Li et al. 2011). Zhu et al. prepared $AlCl_3$ modified cattle bone char and investigated its application in the defluoridation of water. They found 97% fluoride removal efficiency and adsorption capacity was 6.8 mg/g at neutral pH (Zhu et al. 2011). Yadav et al. evaluated fluoride adsorption potential of the wheat straw, sawdust, and activated carbon derived from sugarcane bagasse. They found the maximum fluoride uptake capacity as 1.15 mg/g (Yadav et al. 2013). Getachew et al. developed activated biosorbents by thermal activation using banana peel and coffee husk as raw materials and utilized them for the treatment of fluoride-contaminated water. They found that the maximum fluoride removal efficiency of banana peel and coffee husk were 85% and 86%, respectively for initial fluoride concentration of 10 mg/L at pH 2 (Getachew et al. 2015). Tang et al. developed composite containing hydroxyapatite decorated with carbon-nanotube and examined their application in the defluoridation process for treatment of wastewater. They achieved the upmost biosorption capacity of composite for fluoride remediation as 11.05 mg/g (Tang et al. 2018). Roy et al. studied the synthesis process and characterization of reduced graphene oxide and examined the fluoride removal potential of the prepared nano-adsorbent material or treatment of fluoride containing solution (Roy et al. 2017b). Zhang et al. developed zirconium-chitosan/graphene oxide composite and employed these adsorbents for fluoride biosorption from aqueous solution (Zhang et al. 2017). Jin et al. examined defluoridation of water onto amorphous alumina-modified expanded graphite. They achieved adsorption capacity as 1.18 mg/g and found highest biosorption efficiency to be approximately 94.4% (Jin et al. 2015).

12.7 BIOPOLYMER GRAFTED NANO-ADSORBENT

Polymer-based nano-composites are the combination of some materials such as fibers, whiskers, nanoparticles to support or fabricate with the polymer matrix for its longer stability, and higher efficiency. Cellulose, chitosan, chitin, alginate etc., are the most abundant renewable biopolymers which are very common and efficient raw materials available as template to synthesize polymeric nano-adsorbent for fluoride adsorption process (Das et al. 2020b, 2021a). Sairam Sundaram et al. developed bioorganic composite, nano-hydroxyapatite/chitosan and suggested that the prepared composite was effective for fluoride adsorption because of the low-priced, accessibility and high adsorption capacity (Sairam Sundaram et al. 2008). Sairam Sundaram et al. investigated the defluodation of fluoride contaminated water using magnesia/chitosan polymeric composite (Sairam Sundaram et al. 2009). Ammavasi and Mariappan, developed and examined the fluoride removal efficiency of magnetic iron oxide fabricated layered double hydroxide/cellulose composite for drinking water purification (Ammavasi and Mariappan 2018). Dongre prepared chitosan doped graphite composite and applied for the treatment of fluoride-contaminated water. They reported that the fluoride uptake capacity was achieved to be 37.9 mg/g at pH 6.5 and showed excellent regeneration capacity of the adsorbent (Dongre 2018). Shen et al. prepared and characterized the monetite bundles inlaid in chitosan beads. They also investigated its efficiency in defluoridation process and they obtained fluoride uptake of 50.01 mg/g (Shen et al. 2016). Chen et al. accounted the production of brushite nanosheets using chitosan as a template which prevents the agglomeration tendency of calcium phosphate. They utilized this nano-composite for defluoridation of water and found remarkable fluoride uptake capacity of 231.50 mg/g (Chen et al. 2017). Zhang et al. reported fluoride remediation process by Fe-impregnated chitosan. The fluoride removal ability of the adsorbent was 1.97 mg/g for 10 mg/L initial fluoride concentration (Zhang et al. 2015). Raghav and Kumar explored the defluoridation process using alginate based polymeric composite named Fe-Al-Ni/alginate. They also investigated thermodynamic and kinetic studies of fluoride adsorption study and maximum adsorption capacity was obtained 200 mg/g (Raghav and Kumar 2019). Valdez-Alegria et al. fabricated biopolymers grafted composites, chitosan-polyvinyl alcohol and investigated their capacityin fluoride remediation from aqueous solution (Valdez-Alegria et al. 2020). Wu et al. prepared sodium alginate/carboxymethyl cellulose composite impregnated with calcium and aluminum. The maximum fluoride removal capacity was calculated as of 35.98. mg/g at pH 2 and 298 K temperature (Wu et al. 2016). Sankararamakrishnan et al. reported the extraction of cellulose nanofibers from sugarcane bagasse, and prepared Zerovalent-iron/Zirconium doped cellulose nanofibers. They also evaluated the performance of the prepared composite in the treatment process of fluoride-contaminated water and fluoride uptake capacity was found to be 35.70 mg/g (Sankararamakrishnan et al. 2019). Das et al. investigated defluoridation study using extracted nanocellulose/PVA polymeric nanocomposite and highest fluoride remediation capacity at 25°C was found to be11.363 mg g^{-1} (Das et al. 2021a).

12.8 MAGNETIC NANO-PARTICLE BASED NANO-ADSORBENT

Xu et al. synthesized Al_2O_3–Fe_3O_4-expanded graphite nano-sandwich adsorbent and found maximum fluoride uptake capacity 3.35 mg/g (Xu et al. 2016). Dhillon et al. 2015 produced hydrous hybrid Fe–Ca–Zr oxide nanoadsorbent and investigated their applications and performance ability in fluoride remediation. The estimated adsorption capacity for fluoride removal was 250 mg/g (Dhillon et al. 2015). Abo Markeb et al. prepared Ce–Ti@Fe_3O_4 nano-composite and examined the remediation efficiency for defluoridation of polluted water. They achieved the fluoride uptake capacity of 91.04 mg/g at neutral pH (Abo Markeb et al. 2017). Zhao et al. developed Fe_3O_4@ $Al(OH)_3$ nano-adsorbent and analyzed the characterization of adsorbent which ensured the combined effect in terms of advantages of magnetic separation and aluminum oxide ($Al(OH)_3$). The investigation conveyed the excellent and efficient fluoride adsorption performance of prepared nano-composite (Zhao et al. 2010). Zhang et al. synthesized and characterized Fe–Ti bimetallic oxide coated Fe_3O_4 magnetic nanoparticle. They also investigated the defluoridation from aqueous solution and according to the report; fluoride adsorption capacity was calculated as 57.22 mg/g (Zhang et al. 2016). Liu et al. explored the defluoridation process using iron aluminum oxide magnetic nanoparticles grafted on graphene oxide for treatment of fluoride contaminated water. They established that prepared nano-adsorbent revealed superpara magnetism, higher adsorption efficiency for remediation of fluoride from polluted water (Liu et al. 2016). Yang et al. explored fluoride adsorption onto magnetic alumina aerogel from fluoride—contaminated water. They prepared magnetic composite by modifying alumina aerogel with Fe_3O_4 and the adsorption capacity was achieved to be 32.1 mg/g at pH 5 (Yang et al. 2016). Mohseni-Bandpi et al. fabricated chitosan impregnated Fe_3O_4 nano-adsorbents and evaluated the defluoridation of wastewater. They reported that the higher fluoride biosoroption proficiency of modified magnetic nano-adsorbent was 9.43 mg/g (Mohseni-Bandpiet al. 2015). Thakur et al. examined the treatment of fluoride containing water using Zirconium (IV)—polyacrylamide/Fe_3O_4 nano-composites and obtained excellent fluoride uptake capacity of 124.5 mg/g (Thakur et al. 2014). Ma et al. produced and investigated the characteristics of adsorption equilibrium of defluoridation study using magnetic-chitosan as nano-adsorbent material. They reported that the estimated fluoride removal capacity to be 20.93–23.98 mg/g (Ma et al. 2007). Wang et al. fabricated a magnetic nano-adsorbent Fe_3O_4/polypyrrole core-shell nano-composite and studied the potential of prepared nano-adsorbent in defluoridation process. They reported that it had an excellent electrochemical regeneration ability and showed higher fluoride uptake capacity (Wang et al. 2017). Sahoo and Hota, examined the adsorption ability of prepared magnetic nanocomposite, MgO–$MgFe_2O_4$/GO for fluoride remediation from polluted water and they reported that the fluoride uptake proficiency calculated as 34 mg/g (Sahoo and Hota, 2018). Kabir et al. synthesized and examined the characteristics of Mg–Al–Fe nanocomposite. They also investigated the performance ability of this nanomaterial in fluoride remediation process and derived value of fluoride uptake capacity was obtained as 90.68 mg/g (Kabir et al. 2018).

12.9 OTHER NOVEL BIO-COMPOSITE NANO-ADSORBENT

Hafshejani et al. prepared Al_2O_3 nanoparticles and examined their ability in defluoridation process as nano-adsorbent. They exhibited maximum removal capacity of 13.70 mg/g (Hafshejani et al. 2017). Dhillon et al. investigated the fluoride remediation study using Ca–Ce nanocomposite as nano-adsorbent and found fluoride uptake capacity of 83.33 mg/g (Dhillon et al. 2018). Mukherjee et al. 2020 developed ferrihydrite nano-composite grafted with cellulose nanofiber-polyaniline and examined the fluoride adsorption efficiency of this prepared nano-composite. Maximum uptake capacity was achieved as 50.80 mg/g (Mukherjee et al. 2020). Manna et al. investigated fluoride adsorption from polluted water on to neem oil phenolic resin treated lignocellulosic biosorbent in fixed-bed column system. They showed that prepared biosorbents were efficiently regenerable/reusable even after five cycles (Manna et al. 2018). Wei et al. developed and characterized reed biochar beads cross-linked with cerium alginate and utilized it for defluoridation process. They reported the highest fluoride uptake adsorption capacity as 34.86 mg/g (Wei et al. 2022). Kumari et al. synthesized aluminium fumarate-based polymeric composite using chitosan coated PVA support polymer matrix and examined fluoride adsorption efficiency of prepared composite from ground water. They reported that dipping 9 g of pouch-like tea bag of composite material can treat 1 L groundwater with fluoride concentration of 10–30 mg/L and reduced fluoride concentration to 1–1.5 mg/L after 100 min (Kumari et al. 2022). Shao et al. reported the feasibility and capacity of defluoridation by $CaSO_4 \cdot 2H_2O$ nanorods in batch mode and the investigation confirmed that fluoride adsorption was feasible in pH range of 5–11.32 (Shao et al. 20210). Zare et al. investigated the synthesis of the modified alginate beads modified and its application in defluoridation of wastewater, the uptake capacitiy was obtained as 51.02 mg/g (Zare et al. 2022).

12.10 CONCLUSION

From the literature review, wastewater purification techniques involving nanotechnology is an important filed in advance research work. This chapter summarizes the source of fluoride, its toxic effects on environment and human being, different types of recent and eco-friendly approaches involving nanotechnology. Focus has been given to efficient nano-materials including carbonaceous material-based nanoparticle, magnetic nano-composites, and polymeric nano-composites as nano-adsorbents. A review of the role of carbonaceous material-based nano-composites, polymeric nano-adsorbents and magnetic nano-particles shows high potential for fluoride removal from fluoride-contaminated water. A brief discussion on the application of the surface modified green organic material as biosorbent for the treatment of wastewater has also been discussed. Though, specified investigational results are applicable under definite experimental factors including pH, adsorbent mass, initial fluoride concentration, etc. Consequently, it can be concluded that nano-composites have a vast potential in fluoride remediation processes and the most proficient technique for defluoridation is adsorption because it is easy to operate, eco-friendly, cost-effective, easy disposal, and have no sludge generation problem.

ACKNOWLEDGEMENT

The authors are very thankful to the School of Advanced Studies on Industrial Pollution Control Engineering and Department of Chemical Engineering, Jadavpur University, Kolkata, India.

References

Abo Markeb, A., Alonso, A., Sánchez, A. and Font, X. 2017. Adsorption process of fluoride from drinking water with magnetic core-shell Ce-Ti@Fe_3O_4 and Ce-Ti oxide nanoparticles. Sci. Total Environ. 598: 949–958. https://doi.org/10.1016/j.scitotenv.2017.04.191.

Ali Mansoori, G., Bastami, T.R., Ahmadpour, A. and Eshaghi, Z. 2008. Environmental application of nanotechnology. pp. 439–493. In: Cao, G. (ed.). Annual Review of Nano Research, Volume 2. World Scientific. https://doi.org/10.1142/9789812790248 _0010.

Ammavasi, N. and Mariappan, R. 2018. Enhanced removal of hazardous fluoride from drinking water by using a smart material: Magnetic iron oxide fabricated layered double hydroxide/cellulose composite. J. Environ. Chem. Eng. 6: 5645–5654. https://doi.org/10.1016/j.jece.2018.08.071.

Bhatnagar, A., Kumar, E. and Sillanpää M. 2011. Fluoride removal from water by adsorption: a review. Chem. Eng. J. 171(3): 811–840. https://doi.org/10.1016/j.cej.2011.05.028.

Çengeloğlu, Y., Kir, E. and Ersöz, M. 2002. Removal of fluoride from aqueous solution by using red mud. Sep. Purif. Technol. 28: 81–86. https://doi.org/10.1016/S1383-5866(02)00016-3.

Chen, N., Zhang, Z., Feng, C., Li, M., Zhu, D. and Sugiura, N. 2011. Studies on fluoride adsorption of iron-impregnated granular ceramics from aqueous solution. Mater. Chem. Phys. 125: 293–298. https://doi.org/10.1016/j.matchemphys.2010.09.037.

Chen, Y., Shen, C., Rashid, S., Li, S., Ali, B.A. and Liu, J. 2017. Biopolymer-induced morphology control of brushite for enhanced defluorination of drinking water. J. Colloid Interface Sci. 491: 207–215. https://doi.org/10.1016/j.jcis.2016.12.032.

Chidambaram, S., Ramanathan, A.L. and Vasudevan, S. 2003. Fluoride removal studies in water using natural materials. Water SA. 29(3): 339–344.

Choubisa, S.L. 2012. Status of fluorosis in animals. Proc. Natl. Acad. Sci., India, Sect. B Biol.Sci. 331–339. https://doi.org/10.1007/s40011-012-0026-0.

Choubisa, S.L. and Choubisa, D. 2016. Status of industrial fluoride pollution and its diverse adverse health effects in man and domestic animals in India. Environ. Sci. Pollut. Res. 23: 7244–7254. https://doi.org/10.1007/s11356-016-6319-8.

Das, L., Das, P., Bhowal, A. and Bhattachariee, C. 2020a. Synthesis of hybrid hydrogel nano-polymer composite using Graphene oxide, Chitosan and PVA and its application in waste water treatment. Environ. Technol. Innov. 18: 100664. https://doi.org/10.1016/j.eti.2020.100664.

Das, L., Das, P., Bhowal, A. and Bhattachariee, C. 2020b. Treatment of malachite green dye containing solution using bio-degradable Sodium alginate/NaOH treated activated sugarcane baggsse charcoal beads: batch, optimization using response surface methodology and continuous fixed bed column study. J. Environ. Manage. 276: 111272. https://doi.org/10.1016/j.jenvman.2020.111272.

Das, L., Saha, N., Saha, P. Das, Bhowal, A. and Bhattacharya, C. 2020c. Application of synthesized nanocellulose material for removal of malachite green from wastewater. pp. 11–22. *In*: Ghosh, S., Saha, P. and Francesco Di, M. (eds). Recent Trends in Waste Water Treatment and Water Resource Management. Springer, Singapore. https://doi.org/10.1007/978-981-15-0706-9_2.

Das, L., Das, P., Bhowal, A. and Bhattacharjee, C. 2021a. Enhanced biosorption of fluoride by extracted nanocellulose/polyvinyl alcohol composite in batch and fixed-bed system: ANN analysis and numerical modeling. Environ. Sci. Pollut. Res. 28: 47107–47125. https://doi.org/10.1007/s11356-021-14026-x.

Das, L., Saha, N., Ganguli, A., Das, P., Bhowal, A. and Bhattacharjee, C. 2021b. Calcium alginate–bentonite/activated biochar composite beads for removal of dye and Biodegradation of dye-loaded composite after use: synthesis, removal, mathematical modeling and biodegradation kinetics. Environ. Technol. Innov. 24: 101955. https://doi.org/10.1016/j.eti.2021.101955.

Datta, P.S., Deb, D.L. and Tyagi, S.K. 1996. Stable isotope (^{18}O) investigations on the processes controlling fluoride contamination of groundwater. J. Contam. Hydrol. 24: 85–96. https://doi.org/10.1016/0169-7722(96)00004-6.

Dhillon, A., Nair, M., Bhargava, S.K. and Kumar, D. 2015. Excellent fluoride decontamination and antibacterial efficacy of Fe–Ca–Zr hybrid metal oxide nanomaterial. J. Colloid Interface Sci. 457: 289–297. https://doi.org/10.1016/j.jcis.2015.06.045.

Dhillon, A., Sapna, Choudhary, B.L., Kumar, D. and Prasad, S. 2018. Excellent disinfection and fluoride removal using bifunctional nanocomposite. Chem. Eng. J. 337: 193–200. https://doi.org/10.1016/j.cej.2017.12.030.

Dongre, R.S. 2018. Biosorption of fluoride from water by fabricated chitosan doped graphite novel composite. Res. Dev. Mater. Sci. 7: 1–9. https://doi.org/10.31031/rdms.2018.07.000654.

Getachew, T., Hussen, A. and Rao, V.M. 2015. Defluoridation of water by activated carbon prepared from banana (*Musa paradisiaca*) peel and coffee (*Coffea arabica*) husk. Int. J. Environ. Sci. Technol. 12: 1857–1866. https://doi.org/10.1007/s13762-014-0545-8.

Hafshejani, L.D., Tangsir, S., Daneshvar, E., Maljanen, M., Lähde, A., Jokiniemi, J., Naushad, M., et al. 2017. Optimization of fluoride removal from aqueous solution by Al_2O_3 nanoparticles. J. Mol. Liq. 238: 254–262. https://doi.org/10.1016/j.molliq.2017.04.104.

Jin, H., Ji, Z., Yuan, J., Li, J., Liu, M., Xu, C., Dong, J., et al. 2015. Research on removal of fluoride in aqueous solution by alumina-modified expanded graphite composite. J. Alloys Compd. 620: 361–367. https://doi.org/10.1016/j.jallcom.2014.09.143.

Kabir, H., Gupta, A.K. and Debnath, D. 2018. Synthesis, optimization and characterization of mesoporous Mg-Al-Fe tri-metal nanocomposite targeting defluoridation: synergistic interaction of molar ratio and thermal activation. J. Mol. Liq. 268: 376–385. https://doi.org/10.1016/j.molliq.2018.07.073.

Karthikeyan, G. and Ilango, S.S. 2007. Fluoride sorption using Morringa Indica-based activated carbon. Iran. J. Environ. Heal. Sci. Eng. 4: 21–28.

Karthikeyan, M. and Elango, K.P. 2008. Removal of fluoride from aqueous solution using graphite: a kinetic and thermodynamic study. Indian J. Chem. Technol. 15: 525–532.

Kumari, R., Kumar, A. and Basu, S. 2022. Aluminium fumarate-based polymer matrix composite for selective removal of fluoride from groundwater. Environ. Nanotechnol. Monit. Manage. 17: 100642. https://doi.org/10.1016/j.enmm.2022.100642.

Kundu, M.C. and Mandal, B. 2009. Assessment of potential hazards of fluoride contamination in drinking groundwater of an intensively cultivated district in West Bengal, India. Environ. Monit. Assess. 152: 97–103. https://doi.org/10.1007/s10661-008-0299-1.

Li, Y., Zhang, P., Du, Q., Peng, X., Liu, T., Wang, Z., Xia, Y., et al. 2011. Adsorption of fluoride from aqueous solution by graphene. J. Colloid Interface Sci. 363: 348–354. https://doi.org/10.1016/j.jcis.2011.07.032.

Liu, L., Cui, Z., Ma, Q., Cui, W. and Zhang, X. 2016. One-step synthesis of magnetic iron-aluminum oxide/graphene oxide nanoparticles as a selective adsorbent for fluoride removal from aqueous solution. RSC Adv. 6: 10783–10791. https://doi.org/10.1039/c5ra23676b.

Ma, W., Ya, F.Q., Han, M. and Wang, R. 2007. Characteristics of equilibrium, kinetics studies for adsorption of fluoride on magnetic-chitosan particle. J. Hazard. Mater. 143: 296–302. https://doi.org/10.1016/j.jhazmat.2006.09.032.

Maliyekkal, S.M., Shukla, S., Philip, L. and Nambi, I.M. 2008. Enhanced fluoride removal from drinking water by magnesia-amended activated alumina granules. Chem. Eng. J. 140: 83–192. https://doi.org/10.1016/j.cej.2007.09.049.

Manna, S., Saha, P., Roy, D., Adhikari, B. and Das, P. 2018. Fixed bed column study for water defluoridation using neem oil-phenolic resin treated plant bio-sorbent. J. Environ. Manage. 212: 424–432. https://doi.org/10.1016/j.jenvman.2018.02.037.

Mohan, D., Singh, K.P. and Singh, V.K. 2008. Wastewater treatment using low cost activated carbons derived from agricultural byproducts—a case study. J. Hazard. Mater. 152: 1045–1053. https://doi.org/10.1016/j.jhazmat.2007.07.079.

Mohapatra, M., Anand, S., Mishra, B.K., Giles, D.E. and Singh, P. 2009. Review of fluoride removal from drinking water. J. Environ. Manage. 91: 67–77. https://doi.org/10.1016/j.jenvman.2009.08.015.

Mohseni-Bandpi, A., Kakavandi, B., Kalantary, R.R., Azari, A. and Keramati, A. 2015. Development of a novel magnetite-chitosan composite for the removal of fluoride from drinking water: adsorption modeling and optimization. RSC Adv. 5: 73279–73289. https://doi.org/10.1039/c5ra11294j.

Mourabet, M., El Rhilassi, A., El Boujaady, H., Bennani-Ziatni, M., El Hamri, R. and Taitai, A. 2012. Removal of fluoride from aqueous solution by adsorption on Apatitic tricalcium phosphate using Box-Behnken design and desirability function. Appl. Surf. Sci. 258: 4402–4410. https://doi.org/10.1016/j.apsusc.2011.12.125.

Nagendra Rao, C.R. 2003. Fluoride and environment: a review. pp. 15–17. *In*: Proceedings of the Third International Conference on Environment and Health, Chennai, India.

Raghav, S. and Kumar, D. 2019. Comparative kinetics and thermodynamic studies of fluoride adsorption by two novel synthesized biopolymer composites. Carbohydr. Polym. 203: 430–440. https://doi.org/10.1016/j.carbpol.2018.09.054.

Rao D.N. and Pal, D. 1978 Cattle fluorosis problem around an aluminium factory. Environ. Physio. Eco. Plant. 2: 399–406.

Roy, S., Das, P. and Sengupta, S. 2017a. Treatability study using novel activated carbon prepared from rice husk: column study, optimization using response surface methodology and mathematical modeling. Process Saf. Environ. Prot. 105: 184–193. https://doi.org/10.1016/j.psep.2016.11.007.

Roy, S., Manna, S., Sengupta, S., Ganguli, A., Goswami, S. and Das, P. 2017b. Comparative assessment on defluoridation of waste water using chemical and bio-reduced graphene oxide: Batch, thermodynamic, kinetics and optimization using response surface methodology and artificial neural network. Process Saf. Environ. Prot. 111: 221–231. https://doi.org/10.1016/j.psep.2017.07.010.

Sahoo, S.K. and Hota, G. 2018. Surface functionalization of GO with $MgO/MgFe_2O_4$ binary oxides: a novel magnetic nanoadsorbent for removal of fluoride ions. J. Environ. Chem. Eng. 6: 2918–2931. https://doi.org/10.1016/j.jece.2018.04.054.

Sairam Sundaram, C., Viswanathan, N. and Meenakshi, S. 2008. Uptake of fluoride by nano-hydroxyapatite/chitosan, a bioinorganic composite. Bioresour. Technol. 99: 8226–8230. https://doi.org/10.1016/j.biortech.2008.03.012.

Sairam Sundaram, C., Viswanathan, N. and Meenakshi, S. 2009. Defluoridation of water using magnesia/chitosan composite. J. Hazard. Mater. 163: 618–624. https://doi.org/10.1016/j.jhazmat.2008.07.009.

Sankararamakrishnan, N., Srivastava, I. and Mishra, S. 2019. Studies on novel nano-bimetal doped cellulose nanofibers derived from agrowaste towards deflouridation. Int. J. Biol. Macromol. 128: 556–565. https://doi.org/10.1016/j.ijbiomac.2019.01.153.

Shams, M., Nodehi, R.N., Dehghani, M.H., Younesian, M. and Mahvi, A.H. 2010. Efficiency of granular ferric hydroxide (GFH) for removal of fluoride from water. Fluoride. 43(1): 61–66.

Shao, S., Ma, B., Chen, Y., Zhang, W. and Wang, C. 2021. Behavior and mechanism of fluoride removal from aqueous solutions by using synthesized $CaSO_4 \cdot 2H_2O$ nanorods. Chem. Eng. J. 426: 131364. https://doi.org/10.1016/j.cej.2021.131364.

Shen, C., Wu, L., Chen, Y., Li, S., Rashid, S., Gao, Y. and Liu, J. 2016. Efficient removal of fluoride from drinking water using well-dispersed monetite bundles inlaid in chitosan beads. Chem. Eng. J. 303: 391–400. https://doi.org/10.1016/j.cej.2016.05.103.

Sivasankar, V., Darchen, A., Omine, K. and Sakthivel, R. 2016. Fluoride: a world ubiquitous compound, its chemistry, and ways of contamination. pp. 5–32. *In*: Sivasankar, V. (eds). Surface Modified Carbons as Scavengers for Fluoride from Water. Springer, Cham. https://doi.org/10.1007/978-3-319-40686-2_2.

Tang, Q., Duan, T., Li, P., Zhang, P. and Wu, D. 2018. Enhanced defluoridation capacity from aqueous media via hydroxyapatite decorated with carbon nanotube. Front. Chem. 6: 104. https://doi.org/10.3389/fchem.2018.00104.

Tchomgui-Kamga, E., Ngameni, E. and Darchen, A. 2010. Evaluation of removal efficiency of fluoride from aqueous solution using new charcoals that contain calcium compounds. J. Colloid Interface Sci. 346: 494–499. https://doi.org/10.1016/j.jcis.2010.01.088.

Thakur, N., Kumar, S.A., Parab, H., Pandey, A.K., Bhatt, P., Kumar, S.D. and Reddy, A.V.R. 2014. A fluoride ion selective Zr(iv)-poly(acrylamide) magnetic composite. RSC Adv. 4: 10350–10357. https://doi.org/10.1039/c3ra46015k.

Valdez-Alegría, C.J., Fuentes-Rivas, R.M., García-Rivas, J.L., Arce, R.E.Z., Jiménez-Núñez, M.d.l.L. and García-Gaitán, B. 2020. Synthesis of chitosan-polyvinyl alcohol biopolymers to eliminate fluorides from water. Biomolecules. 10(1): 156. https://doi.org/10.3390/biom10010156.

Viswanathan, N., Prabhu, S.M. and Meenakshi, S. 2013. Development of amine functionalized co-polymeric resins for selective fluoride sorption. J. Fluor. Chem. 153: 143–150. https://doi.org/10.1016/j.jfluchem.2013.04.002.

Wajima, T., Umeta, Y., Narita, S. and Sugawara, K. 2009. Adsorption behavior of fluoride ions using a titanium hydroxide-derived adsorbent. Desalination. 249: 323–330. https://doi.org/10.1016/j.desal.2009.06.038.

Wambu, E.W., Ambusso, W.O., Onindo, C. and Muthakia, G.K. 2016. Review of fluoride removal from water by adsorption using soil adsorbents—an evaluation of the status. J. Water Reuse Desalin. 6(1): 1–29. https://doi.org/10.2166/wrd.2015.073.

Wang, M., Yan, W., Kong, W., Wu, Z., An, X., Wang, Z., Hao, X., et al. 2017. An electroactive and regenerable Fe_3O_4@Polypyrrole nanocomposite: fabrication and its defluorination in an electromagnetic coupling system. Ind. Eng. Chem. Res. 56(44): 12738-12744. https://doi.org/10.1021/acs.iecr.7b02601.

Wei, Y., Wang, L. and Wang, J. 2022. Cerium alginate cross-linking with biochar beads for fast fluoride removal over a wide pH range. Colloids Surf. A Physicochem. Eng. Asp. 636: 128161. https://doi.org/10.1016/j.colsurfa.2021.128161.

Wu, L., Lin, X., Zhou, X. and Luo, X. 2016. Removal of uranium and fluorine from wastewater by double-functional microsphere adsorbent of SA/CMC loaded with calcium and aluminum. Appl. Surf. Sci. 384: 466–479. https://doi.org/10.1016/j.apsusc.2016.05.056.

Xu, C., Li, J., He, F., Cui, Y., Huang, C., Jin, H. and Hou, S. 2016. Al_2O3–Fe_3O_4– expanded graphite nano-sandwich structure for fluoride removal from aqueous solution. RSC Adv. 6: 97376–97384. https://doi.org/10.1039/c6ra19390k.

Yadav, A.K., Abbassi, R., Gupta, A. and Dadashzadeh, M. 2013. Removal of fluoride from aqueous solution and groundwater by wheat straw, sawdust and activated bagasse carbon of sugarcane. Ecol. Eng. 52: 211–218. https://doi.org/10.1016/j.ecoleng.2012.12.069.

Yang, W., Dou, X., Li, Y., Mohan, D., Pittman, C.U. and Ok, Y.S. 2016. Performance and mass transfer of aqueous fluoride removal by a magnetic alumina aerogel. RSC Adv. 6: 112988–112999. https://doi.org/10.1039/c6ra23532h.

Zare, K., Banihashemi, A., Javanbakht, V. and Mohammadifard, H. 2022. Fluoride removal from aqueous solutions using alginate beads modified with functionalized silica particles. J. Mol. Struct. 1252: 132217. https://doi.org/10.1016/j.molstruc.2021.132217.

Zhang, J., Chen, N., Tang, Z., Yu, Y., Hu, Q. and Feng, C. 2015. A study of the mechanism of fluoride adsorption from aqueous solutions onto Fe-impregnated chitosan. Phys. Chem. Chem. Phys. 17: 12041–12050. https://doi.org/10.1039/c5cp00817d.

Zhang, C., Li, Y., Wang, T.J., Jiang, Y. and Wang, H. 2016. Adsorption of drinking water fluoride on a micron-sized magnetic Fe_3O_4@Fe-Ti composite adsorbent. Appl. Surf. Sci. 363: 507–515. https://doi.org/10.1016/j.apsusc.2015.12.071.

Zhang, J., Chen, N., Su, P., Li, M. and Feng, C. 2017. Fluoride removal from aqueous solution by Zirconium-Chitosan/Graphene oxide membrane. React. Funct. Polym. 114: 127–135. https://doi.org/10.1016/j.reactfunctpolym.2017.03.008.

Zhao, Y., Li, X., Liu, L. and Chen, F. 2008. Fluoride removal by Fe(III)-loaded ligand exchange cotton cellulose adsorbent from drinking water. Carbohydr. Polym. 72: 144–150. https://doi.org/10.1016/j.carbpol.2007.07.038.

Zhao, X., Wang, J., Wu, F., Wang, T., Cai, Y., Shi, Y. and Jiang, G. 2010. Removal of fluoride from aqueous media by Fe_3O_4@$Al(OH)_3$ magnetic nanoparticles. J. Hazard. Mater. 173: 102–109. https://doi.org/10.1016/j.jhazmat.2009.08.054.

Zhu, H., Wang, H., Wang, G. and Shi, L. 2011. Removal of fluorine from water by the aluminum-modified montmorillonite. pp. 6103–6105. *In:* 2011 International Conference on Electric Technology and Civil Engineering (ICETCE), Lushan, China. https://doi.org/10.1109/ICETCE.2011.5774229.

Section IV

Agriculture

Chapter 13

Nanomaterials for Controlled and Targeted Delivery of Agrochemicals for Cleaner Environment

Pramod U. Ingle[1], Sudhir S. Shende[1,2], Utkarsha Nimbalkar[1] and Aniket K. Gade*[1,3]

[1]Nanobiotechnology Laboratory, Department of Biotechnology, Sant Gadge Baba Amravati University, Amravati, Maharashtra, India–444602

Tatiana Minkina[2] and Vishnu D. Rajput[2]

[2]Academy of Biology and Biotechnology, Southern Federal University, Stachki 194/1, Rostov-on-Don, Russia–344090

Patrycja Golińska[3] and Mahendra Rai[1,3]

[3]Department of Microbiology, Nicolaus Copernicus University, Torun 87-100, Poland

Jayanta Kumar Biswas[4]

[4]Department of Ecological Studies & International Centre for Ecological Engineering, University of Kalyani, Kalyani, Nadia, West Bengal, India–741235

13.1 INTRODUCTION

The current century is facing a major challenge to feed the constantly growing global human population. The global human population could reach up to 8 billion by 2025 (McCalla 2001) and by 2050, this figure could be up to 9.6 billion (UNDESA 2019, Shende et al. 2022). At present, the global population is increasing at an annual rate of 1.2%, i.e., 77 million people added up every year (Carvalho 2006, De et al. 2014). India, Bangladesh, China, Indonesia, Nigeria, and Pakistan account for half of this worldwide yearly increment. Nonetheless, the world population requires food every day to stay alive. At the global level, considerable advancement has been

*Corresponding Author: aniketgade@sgbau.ac.in; aniketgade@gmail.com

made since 1960 towards improving nutrition as well as protecting the food for human beings. It can be safely assumed that the global gross agricultural production has developed more quickly than the worldwide population, through a normal positive enhancement in the per capita food production (Klassen 1995). Nevertheless, the gap between the quantity of food produced and the world population to feed is expected to augment till the year 2050. The global cereal grains production is anticipated at about 3,300 metric tons which is 60% extra than today (Gilland 2002).

The majority of the world population is directly or indirectly dependent on the agriculture sector (Mali et al. 2020, Rajput et al. 2020).A conventional method to improve crop production involves the use of agrochemicals like pesticides, fungicides, insecticides, fertilizers, etc. A huge variety of agrochemicals are employed to combat biotic and abiotic stresses in agriculture. However, these agrochemicals have a hostof side effects on soil health and crop quality. The release of great amounts of agrochemicals such as fertilizers, in the form of urea, ammonium salt, phosphate or nitrate compounds leads to worsening of the soil quality (Manjunatha et al. 2016, Mali et al. 2020). Most of the agrochemicals when applied to plants are not absorbed by the plants and are lost as run-off leading to water pollution. Haphazard application of agrochemicals such as fertilizers, herbicides, and pesticides result in the diminution of soil nutrients and fertility (Aktar et al. 2009). Agriculture is the backbone of the global economy and most developing countries rely on it for their livelihood. Hence, there is a calamitous need to develop a technology that may perhaps improve modern agriculture in a more dynamic, lucrative, productive, and environment-friendly manner. A variety of environment-friendly processes such as bioinsecticides and biopesticides are previously in use to control pests with non-harmful mechanisms (Mali et al. 2020).

In recent years, the important challenge has been not only to manufacture more food to feed the growing world population but also to ensure the security of the cultivated food crops. It is assumed that doing so will improve malnutrition and poverty and this could ultimately result in the advancement in general welfare and human health. This worldwide dispute for rising demand for per capita food production may perhaps be met by one of the several means, or a combined approach of the underneath: enhancing the water and soil management; increasing the agricultural land area; controlling pests for pre- and post-harvesting damages; improving the crop yield using organic fertilizers; encouraging the use of genetically modified organisms (GMOs) that are resistant to diseases and pests; using more productive plants and plant varieties that are resistant to the pests (De et al. 2014, Ingle 2023). Figure 13.1, below refers the various aspects of novel nano-based agrochemicals.

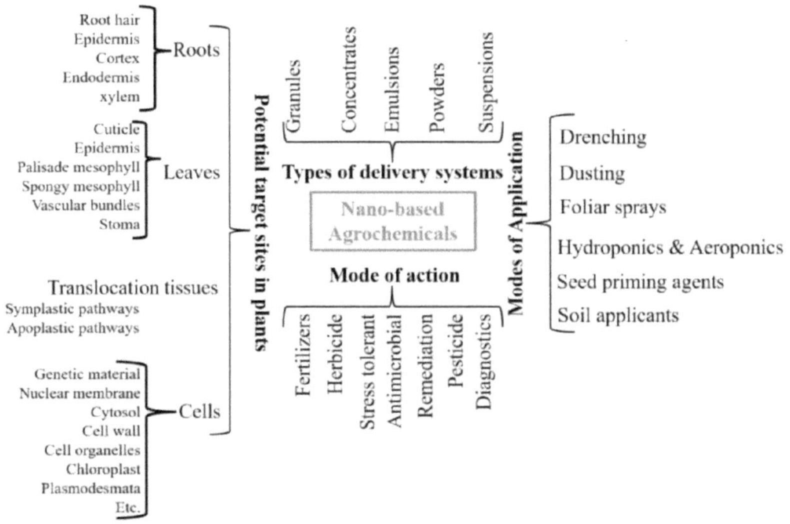

Figure 13.1 Various aspects of nano-based agrochemicals.

Nanotechnology (NT) is a novel, pioneering and systematic approach that directs to exploitation, design, and improvement of valuable nanomaterials (Rai and Biswas 2018). Nanotechnology produces materials in nanometer scale ranging in size from approx. 1–100 nm. Because of the smaller size, the surface area to volume ratio is improved in the nanoparticles/nanomaterials (compared with bulk counterparts), enhancing the biochemical reactivity and conferring remarkable and valuable physical properties (Ali et al. 2014). More than the last two decades, considerable research has been approved in the field of NT, stressing its plentiful applications in the agriculture sector (Chen et al. 2016, Prasad et al. 2017, Lv et al. 2018, Rai et al. 2018, Shang et al. 2019, Rajput et al. 2020, 2021, Shende et al. 2022). The application of NT in agriculture sector is escalating quickly to increase food values, improve nutrient content, reduce agricultural inputs, and enhance shelf life. Several nano-agricultural products are recently developed to diminish the employ of hazardous chemicals. Nanotechnology involves a number of aspects of food security, treatment of various plant diseases, new tools for the detection of pathogens, efficient delivery systems and packaging materials (Echiegu 2017). A variety of nanomaterials-based agrochemicals such as nanopesticides, nanoemulsions, nanoinsecticides, and nanoparticles were produced using NT (Table 13.1).

Table 13.1 Agriculture-related agro-nano-products and their applications (Mali et al. 2020)

Products	**Examples**	**Applications**	**References**
Plant protection products	Nanoemulsion from Neem oil (*Azadirachta indica*) as larvicidal agent	Nanoparticles,Nanocapsules, nanoemulsions and viral capsids as smart delivery systems of AIs for pest control and disease management in plants	Anjali et al. 2012
Fertilizers	Macronutrient Fertilizers Coated with Zinc Oxide nanoparticles and NPK controlled delivery Nano-coating of sulfur using Chitosan Nanoparticles	Nanoparticles and Nanocapsules for the nutrients enrichment absorption by plants and the delivery of nutrients to targeted/specific sites	Milani et al. 2015, Corradini et al. 2010
Soil enrichment	A nanoclay-based soil-enhancer product, f or water and nutrient retention and release	Water/liquid retention: Nanomaterials, e.g. zeolites and nanoclays, for water or liquid agrochemicals retention in the soil for their slow release to the plants	Parisi et al. 2015
Genetic material delivery	Gold nanoparticles (10–15 nm) Gold nanoparticles (5–25 nm) Starch (50–100)	DNA delivery	Torney et al. 2007, Vijayakumar et al. 2010, Liu et al. 2008
Nanosensors and diagnostic devices	Liposome-based nano-biosensor for the detection of pesticide in the soil and crops	Nanostructures and Nanomaterials (e.g. electrochemically active carbon nanotubes, fullerenes and nano-fibers) which are very responsive bio-chemical sensors to intimately monitor plant health and growth along with the environmental conditions	Vamvakaki and Chaniotakis 2007

There are several materials used to develop and coat nanomaterials, for example plant extracts, metal oxides, ceramics, polymers, silicates, lipids, and emulsions (Ruttkay-Nedecky et al. 2017). Nano-coated fertilizer particles or surface-coated nanomaterials hold the materials more powerfully to the plant because of higher surface tension than conventional surfaces. Furthermore, nanocoatings provide surface security for bigger particles. A nanocapsule is composed of a shell which has an active compound, such as a biological or chemical agent for the safety of plants against diseases and pests. The shells have different elements like polymers, lipids, nanoclays or viral capsids (Mali et al. 2020).

In this context, the focus has been given to the nanomaterials which are used for the controlled and targeted delivery of agrichemicals such as granules, nanoemulsions, powders, and other formulations. Various types of agrochemicals have been comprehensively discussed which could be used for sustainable development.

13.2 AGROCHEMICALS AND THEIR TYPES

In the agriculture sector, there are several agrochemicals used to enhance crop production and yield. These include fertilizers, pesticides, herbicides, insecticides, fungicides, and other forms. These agrochemicals are applied in various forms and formulations. The detailed account is given in the below sections.

13.2.1 Conventional Formulations for Delivery/Application of Agrochemicals

13.2.1.1 Granules

Granular formulations (GRF) are used for straight dissemination to the field frequently as pre-emergence herbicides or as soil insecticides (UNIDO 1983). The concentration of active ingredient (AI) is generally from 1 to 40%, and the size of granule mesh is usually between 250–1000 l. The granules are supposed to be free-flowing, non-caking, dust free, and must disintegrate in the soil so as to discharge the AI. Usually, granules are made either by coating a fine powder onto a substrate, for e.g. sand by using a sticker like PVP solution, or by solvent impregnation onto an absorbent carrier. Polymers or Resins might be sprayed onto the granules to control release rates (Knowles 2008). The granular carriers could be generally categorized into mineral and organic types. The mineral carriers involve limestone, sand, gypsum, kaolin, attapulgite, montmorillonite, and diatomite while, the organic granular carriers involve corncobs, peanut hulls, pecan shells, and recycled paper fibers (Collins et al. 1996). The absorptive capability of the carrier is a significant parameter and is a function of the crystalline structure and the available surface area of the carrier particles. Now granules are becoming less popular because of the increasing use of post-emergence herbicides.

13.2.1.2 Concentrates

Solution concentrates (SLs) are an aqueous solution of AI and is the simplest formulation to make, which merely requires dilution in a spray tank. Several pesticides that can be formulated in this way are limited by the hydrolytic stability and water solubility of the AI. Some of the SLsformulations contain a surfactant, which is usually a non-ionic ethylene oxide condensate to aid wetting onto the leaf surface (Knowles 2008). Solution concentrate formulations are generally very stable and, thus, have very few storage problems. Some problems do arise intermittently, like precipitation during dilution and deterioration of metal containers or spray applicators. Even though, these problems could be overcome by the use of suitable additives, like co-solvents, and corrosion inhibitors. The composition of a typical solution concentrates formulation is shown in Table 13.2.

Table 13.2 Approximate quantity of the excipients used in the various agrochemical formulations (De et al. 2014)

Excipients	Solution concentrate Quantity (%w/w)	Emulsion concentrate Quantity (%w/w)	Wettable powders Quantity (%w/w)	Suspension concentrate Quantity (%w/w)
Active ingredient	20–50	20–70	25–80	20–50
Antifreeze agent	5–10	–	–	5–10
Dispersing agent	–	–	2–5	2–5
Wetting agent	3–10	–	1–3	2–5
Anti-settling agent	–	–	–	0.2–2
Emulsifying agent	–	5–10	–	–
Solvents	–	q.s. 100	–	–
Inert filler	–	–	q.s. 100	–
Water	q.s. 100	–	–	q.s. 100

To enhance the bio-efficacy of SLs formulations, tallow amine ethoxylates or nonylphenol are generally used in tank mix wetters. On the other hand, to ensure the exact rate of wetting agent, these may be built into the formulation for optimizing the biological activity, for instance, with glyphosate and paraquat formulations. A significant amount of work has been done on novel surfactant-wetting agents for glyphosate formulations (Knowles 2008). In some cases, for preventing fungal growth or bacterial spoilage, preservatives must be added during long-term storage.

13.2.1.3 Emulsions

For many years, Emulsions or Emulsifiable Concentrate (EC) formulations have been very popular and signify the biggest volume of all agrochemical formulations in terms of worldwide consumption. Emulsifiable Concentrate is made from oily AIs or from low melting, waxy solid AIs, which are soluble in nonpolar hydrocarbon solvents like C9–C10 solvents, xylene, odorless kerosene, solvent naphtha, or other proprietary hydrocarbon solvents. Spontaneous emulsification through good emulsion stability properties has been ensured by adding surfactant emulsifiers to these formulations in the spray tank. Careful selection of a "balanced pair" emulsifier blend is normally required to make sure that the stability of emulsion dilution is preserved over extensively differing degrees of water hardness and climatic conditions. When the formulation is mixed with water, there is the production of emulsion droplets of 0.1–5 l. Over the last 20 years, the ECs formulation has been significantly facilitated by the commercial development of nonionic emulsifying agents in which the hydrophilic portion of the molecule consists of a polyethylene oxide chain. The nonionic surfactant, nonylphenol hydrophobic chain condensed with 12 or more ethylene oxide molecules is generally used (Knowles 2008). The other constituent of the balanced pair is usually an anionic surfactant like the oil-soluble calcium salt of dodecylbenzenesulfonic acid. Nevertheless, the nonylphenol ethoxylates have been assumed to have endocrine-modulating properties. Because of this potentially noxious effect there is a need of an alternative ethylene oxide condensate based on aliphatic alcohol hydrophobesis. The total concentration of the emulsifier blend is usually 5–10 % of the formulation. There are no specific rules to establish the ratio of anionic to nonionic surfactant in the mixed emulsifiers, but guidance could be obtained from the Hydrophilic-Lipophilic balance (HLB) system. In HLB system, a general thumb rule is that the higher the HLB value the more hydrophilic i.e. water-soluble is the surfactant. In general, the HLB range of 8–18, provides a good oil-in-water emulsion. The optimal ratio of anionic and nonionic surfactants is experimentally determined to provide spontaneous emulsification in water and to provide a stable emulsion by very little creaming and without coalescence of oil droplets. ECs are insufficient in the number of AIs for which they are appropriate. Many agrochemicals like pesticides need to be adequately soluble and are supplied economically in this form. Still, it might be plausible to improve the solubility of the AIs with the addition of a more polar solvent devoid of increasing the menace of crystallization in the spray tank. The composition of a typical EC formulation is given in Table 13.2.

The presence of solvents and emulsifiers in ECs formulations could sometimes provide improved biological competence compared with the other formulations. Many insecticides, for e.g. organophosphorus compounds and pyrethroids, are oil-soluble liquids or waxy solids and are readily formulated as ECs, and a few AIs necessitate to be prepared with solvents for optimal biological activity.

13.2.1.4 Powders

Like ECs, Wettable powder (WP) formulations of agrochemicals have also been known for many years and are made generally from solid AIs with high melting points that are appropriate for dry grinding through a mechanical grinder like a hammer- or pin-type mill or with air milling by a fluid energy micronizer. Air milling gives much finer particles (5–10 μ) than mechanical milling (20–40 μ) and can also be more suitable for AIs with lower melting points. Nevertheless, care should be taken to avoid suppressing or containing dust explosions that might arise if a source of ignition (like static energy) is present in both types of mills. Wettable powders generally include dry surfactants as powder wetting and dispersing agents and inert carriers or fillers. They usually contain more than 50% AIs, and the upper limit is normally determined by the amount of inert material like silica required for avoiding the AI particles from fusing together while processing in the dry grinding mills (Knowles 2008). For preventing the formulated product from aggregating or caking, an inert filler like talc or kaolin may also be used during storage. The excipients used in the preparation of wetting powders are given in Table 13.2.

13.2.1.5 Suspensions

Since the early 1970s, suspension concentrate (SC) technology has been progressively more applied to the formulation of many solid crystalline agrochemicals. Agrochemical particles could be suspended in an oil phase; however, it is much more suitable for SCs to be dispersions in water. In recent years, the production of aqueous SCs through high-energy wet grinding processes like bead milling received significant attention. The application of surfactants as dispersing and wetting agent has also led to a profusion of research on the colloidal and surface chemistry features of dispersion and stabilization of solid-liquid dispersions (Knowles 2008). The water-based SC formulations propose several advantages like a higher concentration of insoluble AIs, easiness in handling and application, safety to the machinist as well as environment, relatively lower cost, facilitate water-soluble adjuvants to be built-in for improved biological activity, and many others.

In general, farmers prefer SCs to WPs as these are non-dusty and easy to evaluate and pour into the spray tank. Nevertheless, there are several disadvantages; particularly the need to produce formulations that do not separate badly and also need to protect the product from freezing which might cause aggregation of the particles during storage. In the majority of cases, SCs are made by dispersing the AI powder in an aqueous solution of a dispersing and wetting agent using a high-shear mixer to provide a concentrated premix, followed by a wet grinding process in a bead mill to provide a particle size distribution in the range 1–10 l. The wetting or dispersing agent aids in the wetting of the powder into water and breaking of the agglomerates, aggregates, and single crystals into smaller particles. Additionally, the surfactant that becomes adsorbed onto the freshly produced particle surface during the grinding avoids reaggregation of the small particles and could guarantee the colloidal stability of the dispersion. There are several wetting/dispersing agents used in SCs formulations for e.g. sodium naphthalene sulfonate-formaldehyde condensates, sodium lignosulfonates, aliphatic alcohol ethoxylates, ethylene/propylene oxide block copolymers, tristyrylphenol ethoxylates, and esters are some of them. Recently, polymeric surfactants like "comb" surfactants are available that could adsorb strongly on particle surfaces and could provide significantly enhanced stabilization of SCs for long-term storage. A typical SC formulation is given in Table 13.2.

To increase the viscosity with an upsurge of three-dimensional network structures, there is an addition of anti-settling agent that prevents the separation of particles while storage for longer period. Swelling clay like bentonite (sodium montmorillonite) is generally used as anti-settling agent and perhaps mixed with water-soluble polymers which is generally cellulose derivatives, natural gums, or other types of polysaccharides, for eg. xanthan gum that are usually susceptible to microbial attack to offer synergistic rheological effects. To prevent degradation of the anti-settling agent, preservatives are generally added in SC formulations, which ensure that the long-term stability of the product is not damaged. In order to facilitate storage stability, a great deal of research has been carried out using rheological techniques so as to evaluate the forces acting between particles and polymers to be forecasted. Nevertheless, it is still very important to carry out the tests for long-term storage on variable temperatures so as to make sure that the particles do not agglomerate or separate irrevocably under usual storage conditions in the sales pack.

13.3 NEW GENERATION DELIVERY SYSTEMS FOR AGROCHEMICALS

There has been ever-increasing pressure from regulatory authorities as well as the government from the last few years on developing formulations that generally have very less impact on the environment. The most important issues that are being dealt with are security during use and manufacturing, feasibility for the user, alleviate disposal of the pack (or reuse), minimizing the amount of agrochemicals used along with reduced waste and all types of effluents. Therefore, the recent trends in the improvement of agrochemicals formulation are to eliminate solvents wherever feasible and employ aqueous emulsions or microemulsions (MEs), to develop multiple AI formulations where suitable, to replace dusty powders by SCs or water-dispersible granules, to build into the formulation bio-enhancing surfactant adjuvants, to develop novel formulations like gels or tablets, to organize discharge rate and targeting of agrochemicals by technique like encapsulation and seed treatment, to develop more effective spray tank adjuvants to reduce dosage and improve biological activity. The technical advances in surfactants and other formulation additives (mainly surfactant blends) collectively developed a more powerful dispersing agent that could meet the above complex requirements by obtaining a better understanding of the colloid/surface chemistry and rheology principles.

The ideal product seems to have characteristic such as, it has the maximum biological activity; it is free from volatile solvents, creating the least amount of pack disposal problems, and giving very less operator exposure hazard. Water-soluble granules (WPs) in water sachets could be added directly to a spray tank, which can meet these requirements a long way. The developmental work has been carried out by all the major agrochemical companies on these options. Conversely, it has to be noted that it has not been possible to prepare all the AIs like this and hence other alternatives are being comprehensively evaluated, together with ideas for application systems of closed transfer spray tanks and more expedient packaging (Knowles 2008). Aqueous-based agrochemical formulations are considered to be safe alternatives to the water-dispersible granule formulations which include SCs which have already been discussed along with oil-in-water emulsions (concentrated emulsions), suspo emulsions (SEs), microemulsions, multiple emulsions, microcapsules (Capsule suspensions) discussed in detail below.

13.3.1 Oil-in-water Emulsions

Now-a-days oil-in-water emulsions (EWs) are receiving significant attention as there is very less or negligible use of volatile organic solvents for safer handling. It can be due to their aqueous-based nature, EW could have noteworthy advantages over ECs in terms of safety and expenditure in manufacture, shipping, and application. Though, they need a careful selection of surfactant emulsifier to avoid creaming, flocculation, and coalescence of the oil droplets (Knowles 2008). To produce stable emulsions, nonionic surfactants are recently being used that block copolymers and other polymeric surfactants.

The nonionic surfactants are sometimes helpful to combine a high and a low HLB surfactant to provide an emulsifier mixture having an average HLB of 11–16 for the most advantageous emulsion stability (Knowles 2008). The size of droplet is also the best indication of stability and it should be below 21 volume mean diameter (VMD). The emulsions are generally thickened with polysaccharides like xanthan gum to avoid severance of the oil droplets. Occasionally polymers like polyvinyl alcohol are employed as emulsifiers as well as stabilizers/thickener.

13.3.2 Suspo Emulsions

Because of the convenience, mixed combination formulations are being most accepted. It needs to make sure that the farmer is supplied with accurate quantity of each component of agrochemical and conquers the troubles of tank mix incongruity (Knowles 2008). If one AI is solid and the other is liquid, then it is required to produce a suspo emulsion formulation that can distinguish in three phases, solid dispersed particles, liquid oil droplets, and continuous phase (generally water).

Suspo emulsions can, thus, be considered to be a combination of SCs and oil-in-water emulsions (EWs) with added surfactants (to prevent flocculation) and thickeners (to avoid separation of the dispersed phases). Surfactants used as dispersing agents for the solid phase can be comparable to those previously mentioned in SCs. Emulsifiers for the oily liquid phase can be comparable to those applied for EW. As these formulations are water-based and normally thickened with polysaccharides, it necessitates adding a preservative to avoid degradation of the thickener. To conquer the problem of hetero-flocculation between the oil droplets and the solid particles, there is a need to carefully select suitable emulsifying and dispersing agents. Thus, extensive storage testing for these formulations is essential.

13.3.3 Microemulsions

Microemulsions (MEs) are thermodynamically stable transparent emulsion systems, which are stable over a broad range of temperatures. It has a very fine droplet size of less than 0.051 (50 nm) and consists of three components, viz. solid dissolved in organic solvent or oily liquid, water, surfactant/cosurfactant system.

These components form a single phase having comparatively large "swollen micelles" in which the non-aqueous phase of the AI and solvent are solubilized or dissolved by the surfactant system. Two different types of surfactants, one water and one oil-soluble surfactants are required in the preparation of microemulsions. The water-soluble surfactant is generally nonionic or anionic with a very high HLB value, and the hydrophobic part of the molecule must match the oil. The cosurfactant must be oil soluble and must have a very low value of HLB (For e.g. hexanol). For a microemulsion, the total surfactant concentration could be as high as 10–30% or more (compared to about 5% for a typical oil-in-water emulsion). Microemulsions have comparatively lower AIs concentrations; however, the high surfactant content and solubilization of the AI could give rise to improved biological activity (Knowles 2008).

13.3.4 Multiple Emulsions

Multiple emulsions are another class of emulsions that could be oil-in-water-in-oil (O/W/O) or water-in-oil-in-water (W/O/W). These are complex formulations that necessitate a very careful selection of surfactant stabilizers and emulsifiers to conquer physical instability problems. Multiple emulsions are still in the research phase and may perhaps be of concern to lessen the toxicity of an AI by controlling it to the preliminary internal emulsion droplet phase (Knowles 2008). Conversely, because of the necessity to structure a second emulsion phase; there should be low AI content in the final product.

13.4 NANOBIOTECHNOLOGY: A NOVEL APPROACH FOR TARGETED DELIVERY OF AGROCHEMICAL

Nanobiotechnology primarily deals with the study of nanomaterials (NMs) having at least one dimension in the range of 1–100 nm and their application in life sciences and agriculture. Applications of nanomaterials are also continually expanding their horizon to various fields like chemical production, aerospace, electronics, etc. These nanostructured materials and engineered nanomaterials are mainly designed for their applications focused on efficiency and enhanced production strategies. Various polymers and liposomes are renewable, biodegradable, and environmentally friendly materials, usually used as organic carriers to encapsulate pesticides (Selyutina et al. 2017, 2020, An et al. 2022). Further research has revealed the potential of nanomaterials as an important agrochemical that can be used as formulated pesticides (de Oliveira et al. 2014, Huang et al. 2018) and their residual analysis (Kumar et al. 2020, Mahmoudpour et al. 2021), fertilizers (Taşkın et al. 2018, Adams et al. 2021), fungicides (Rai and Kratošová, 2015, Bawskar et al. 2021, Avila-Quezada et al. 2022) and plant growth promoters (Fincheira et al. 2020). NMs-based agrochemicals are gaining more attraction (Li et al. 2021) as they show beneficial properties such as increased penetration (Giraldo et al. 2014), higher uptake of active ingredients (Garcia-Gomez et al. 2018), and increased coverage (Su et al. 2020). Thus, NMs are proven for their promising applications in agrochemical delivery through various routes (Xu et al. 2020, Guha et al. 2020). But, the intensive research has raised some concerns (Malakar et al. 2020) regarding the use of NMs, such as unknown health risks and possible side effects on humans and environment, and difficulty in commercial production because of complex manufacturing procedures. This ultimately limits their use for agricultural purposes (An et al. 2022). This results in the necessity of controlled, regulated and well-monitored application of NMs in agriculture fields (Hussain 2018). Nanobiotechnology thus plays a vital role in developing an integrated system for the co-delivery of agrochemicals site specifically (Ji et al. 2021). It is thus necessary to understand nanomaterials as a delivery systems and their target sites in planta (Ingle 2023).

13.4.1 Cellular Delivery of Agrochemicals by Nanomaterials

In recent times, there has been a huge loss in the plant species including the crop plants due to the emergence of various bacterial, fungal and viral pathogens. In addition to this, other abiotic factors also raise the question of pollution, including the nonspecific use of various nanomaterials (Sebilo et al. 2013). This leads to major losses to farmers and agro based industries. Thus, plant diseases and their diagnosis has become a serious issue all over the world (Smith and Gilbertson 2018, Kulabhusan et al. 2022). To contribute towards solving these issues, the newly emerging field of nanotechnology can stand a sustainable solution. NMs are continuously being evaluated for their biocatalytic activity *in vitro* as well as *in vivo*. Although their fate inside the plant is not well understood, the NMs can be designed as per their desired site-specific activity and carry out a particular function within the plant body. This is only possible when the spatial path of the NMs *in vivo* is precisely known. Ultimately, this leads to the functionalization of NMs as per to perform a desirable activity after entering the plant body. This would help a controlled temporal as well as spatial distribution of NMs and the site-specific delivery of the cargo molecule they are carrying when used as a delivery system (Kulabhusan et al. 2022, Santana et al. 2020, Yin et al. 2023).

The literature review suggests that there are numerous locations within the plant body reported for which the NMs could be designed and applied for a specific purposeful function. To carry out the desired function NMs are supposed to be directed to a particular target site, where they can release the cargo molecule attached and fulfill the required activity within subcellular compartments. Santana et al. (2020) have recently demonstrated a mechanism for the delivery of NMs with a biochemical cargo that can be directed to the plant photosynthetic organelle i.e., chloroplast, with the help of a guiding peptide recognition motif. Authors used fluorescent quantum dots (QDs) functionalized with β-cyclodextrin molecular baskets. This enabled QDs to carry various biochemical compounds in the form of coating, ultimately targeting them to the chloroplast. Demirer et al. (2019), demonstrated efficient biomolecule delivery into intact plants of some unexplored species and chemically functionalized high aspect ratio nanomaterials. They observed the strong protein expression bypassing transgene interaction and efficient DNA delivery in plants like *Eruca sativa* (arugula), *Nicotiana benthamiana* (Nb), *Gossypium hirsutum* (cotton) *Triticum aestivum* (wheat) leaves, and arugula protoplasts. They reported the facilitated biomolecule transport into the plant without any DNA damage due to intracellular nucleases. The study paves the way towards the development of a new tool for passive and species-independent delivery of genetic material, escaping transgene interactions. Wang et al. (2019) have summarized some of agrochemicals and NMs used for agriculture purposes. Proteins can also be delivered to sub-cellular locations in plants using gold-functionalized mesoporous silica nanoparticles (Martin-Ortigosa et al. 2012). Zhang et al. (2019) reported the delivery of small interfering RNA (siRNA) molecules in *N. benthamiana*. Kwak and colleagues (Kwak et al. 2019) showed evidence of CNT-based plasmid DNA delivery and YFP transgene transient expression in the chloroplast based on fluorescence microscopy imaging. Various mechanisms triggered by the nano-based agrochemicals are shown in Figure 13.2, herewith.

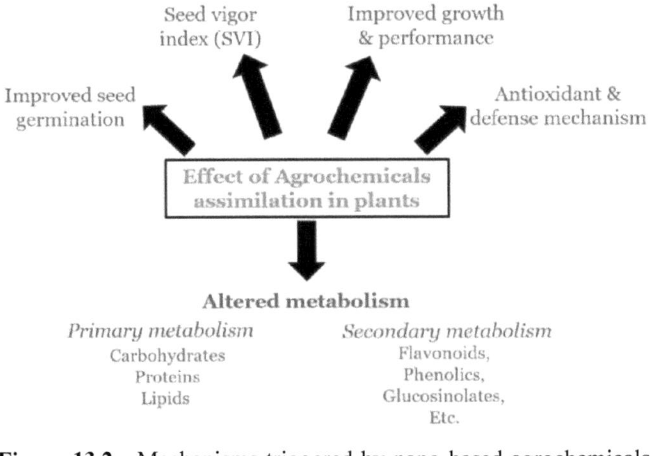

Figure 13.2 Mechanisms triggered by nano-based agrochemicals.

Nanocides are gaining more attention as a delivery system in agriculture and horticulture crops (Hussain 2017). Nano-enabled agrochemicals are getting widespread applications (Ghorbanpour and Shahid, 2022). Nanopesticides are also used for the controlled and differential action of pesticides in agricultural crops. These are also referred to as plant protection products (Kookana et al. 2014). Nanopesticides are beneficial in various ways as they show enhanced apparent solubility, faster decomposition in soil and/or plant (Jianhui et al. 2005, Guan et al. 2010), controlled release (Liu et al. 2006), targeted delivery (Pérez-de-Luque and Rubiales 2009), etc. ultimately the nanobased delivery systems are contributing to the development of crop biotechnology and sustainability (Nandini and Geetha 2021, Yin et al. 2023).

Even though chemical fertilizers are efficient in improving crops; bioaccumulation and biomagnification has led to their reduced usage. Chemical fertilizers are also causing deleterious effects on all forms of flora and fauna leading to eutrophication (Savci 2012, Chandini and Praksash 2019). This emphasizes the directed delivery of these chemicals or substituting them with other non-hazardous counterparts for sustainable agriculture. Nanofertilizers (NFs) in recent times have emerged as a smart and sustainable substitute for chemical fertilizers. They has also been explored as a novel delivery system in agriculture (Solanki et al. 2015). NFs are prepared by encapsulation of plant nutrients inside nanomaterials and delivering them as a nano-sized emulsion or colloid. NFs show higher utilization efficacy, absorption rate, and minimum losses. Stomatal openings and leaf pores in plants facilitate the uptake, penetration, and assimilation of NFs deep into the plant body. NFs are more efficiently transported into plants through plasmodesmata. This results in reduced nutrient loss, leading to higher production and nutrient quality of crops. However, availability and production of NFs, their adequate operative legislation, and allied risk management are the key limiting factors in their wide-ranging adoption as plant nutrient sources (Solanki et al. 2015, Iqbal 2019, Tarafder et al. 2020, Mejias et al. 2021, Basavegowda and Baek 2021, Mubeen et al. 2023).

Major losses i.e., yield and economic losses, to any agricultural crops are due to fungal infections (Godfray et al. 2016). Conventional fungicides tend to curb these losses but, on the contrary, non-specific living forms and biodiversity are targeted. Therefore, it is needed to look for replacement approaches to attain better meticulousness in fungal disease management (Patel et al. 2014). The development of nanofungicides as a new generation of fungicides could be an effective strategy for the management of fungal pathogens (Abd-Elsalam and Alghuthaymi 2015, Mittal et al. 2020). They show antimicrobial properties, circumventing their use in disinfection (Baker et al. 2005). Silver nanoparticles (AgNPs), CuNPs, ZnONPs, MgONPs, are fabricated and reported to be efficient and equal to chemical fungicides against the major fungal pathogens from a range of genera (Jo et al. 2009, Wani and Shah 2012, Ouda 2014). An MWCNT-g-PCA hybrid material encapsulated pesticides, zineb, and mancozeb, have been formulated for the management of *Alternaria* spp. (Sarlak et al. 2014). Nanocomposites functionalized with various metal, non-metal and polymeric nanomaterials are also shown to have a fungicidal effect on a huge range of crop pathogens (Alghuthaymi et al. 2021).

Thus, nanotechnology and nanomaterials are continuously advancing towards the development of site-specific and target-defined management strategies of various microbes like fungi, viruses, bacteria, and complex genomic portions (Das et al. 2021).

13.5 NANOBASED DELIVERY SYSTEMS FOR AGROCHEMICALS

The following section will discuss the various types of nano-based delivery systems that are reported for the targeted delivery of agrochemicals.

13.5.1 Metal and Metal Oxide Nanoparticles

Nanomaterials based on metals and non-metals themselves act as delivery systems of various micro- and macro-nutrients to the plants and soil. They can be further functionalized for the delivery of specific nutrients or cargo molecules at specific locations in the plant as desired. The basic mechanism of NPs interacting with the plants has been demonstrated by Rastogi et al. (2017). Metal nanoparticles include gold, copper, silver, platinum, titanium, nickel, and zinc nanoparticles, etc., (Yaqoob et al. 2020). As no further functionalization is required for their absorption, internalization, and activity inside the plants, these are more favored as compared to functionalized nanoparticles. But their intracellular localization cannot be controlled or monitored easily. The same is the case with metal oxide nanomaterials (Rastogi et al. 2017). Metal oxide NPs also affect plant metabolism, which is an attribute of their size, concentration and method of exposure. Titanium dioxide (TiO_2), silver, copper, copper oxide, zinc oxide, cerium dioxide, aluminum, nickel, and iron are most studied for their impacts on different plants as they are commonly used in industrial applications (Joner et al. 2008, Janson et al. 2019, Mora et al. 2019, Mittal et al. 2020).

13.5.2 Hydrogels NPs

Hydrogels are 3-dimensional, super absorbent polymeric nanoparticles that hold and retain water for a longer period of time. These are developed for the regions where there is either scarcity of water for agriculture or arid soil is present due to desertification. These properties make them suitable for various applications like medicine and agriculture as an alternative water resource (Vundavalli et al. 2015, Vega-Vásquez et al. 2020, Singh et al. 2021, Barrett-Catton et al. 2021, Nunes et al. 2022). Hydrogels can be used for conserving water in water deficit areas to reduce evaporation, prevent deep percolation and surface run-off, helping mitigation of water demands in rhizosphere of plant. Hydrogels thus can be used to coat the seeds along with impregnated nutrients, pesticides, additives and/or any other desired cargo molecule. It has the astonishing capacity in improving diverse hydrophysical, physicochemical, and biological properties of soil, concurrently lessening irrigation frequency, improving the water and nutrient use competences, and increasing the quantity and quality of the field, ornamental, and vegetable crops (Patra et al. 2022).

13.5.3 Dendrimers NPs

Dendrimers are 3D structures comprising of a central core, dendrons and cavities between these dendrons (Safari and Zarnegar 2014). They have even molecular weight and their 3D structure with outlying groups to determine solubility, which can be designed as per demands, smaller hydrodynamic volume and low molecular volume comparative to linear polymers of analogous molecular weight (Markowicz-Piasecka and Mikiciuk-Olasik 2016). Exposed groups can be functionalized to control their interactions with the other molecules in their proxy. Dendrimers can be used as an effective delivery system because of their size in nano range, ease of functionalization and functional stability (Vega-Vásquez et al. 2020). These are useful for functionalization with lipophilic as well as lipophobic cargo groups and are relatively resistant to hydrolysis making them a feasible alternative delivery system in various fields including medicine, veterinary, cosmetics, and agriculture (Kretzmann et al. 2017, Chauhan et al. 2020).

13.5.4 Polymeric NPs

Glyphosate acid is a broad-spectrum herbicide commonly used to kill unwanted plants in agriculture and nonagricultural landscapes. It is the active ingredient in numerous commercial herbicides for nonselective weed control. Glyphosate hinders the shikimic acid biosynthesis pathway in plants. Poly(alkylene-oxide) alkanol system acts as a carrier agent and glyphosate is trapped inside it to form a polymeric emulsion (Meyer and Scribner 2009, Chaud et al. 2021). Diazinon (Dimpylate) is a derivative of organothiophosphate and was used broadly as an agricultural and household insecticide. Cellulose ester/ PEG/ Inorganic salt was used as a polymeric coating by nanoencapsulation method. It has applications in agriculture to control insects in the soil, on ornamental plants, and on fruit, vegetable, nut, and field crops (Clarke 2007, NCBI-CID 2022). A composite composed of biopolymers (microcrystalline cellulose, chitosan, and alginate) and

ZnO NPs was developed to increase the utilization efficiency of Zn micronutrient for plant growth. ZnO-alginate beads exhibited the slowest controlled release of Zn (Martins et al. 2020). The controlled release of the NPK fertilizer sources such as urea, calcium phosphate, and potassium chloride was carried out by using polymeric chitosan NPs (Manjunatha et al. 2016). A system of silica NPs coated and functionalized with polystyrene and (trimethoxysilyl) propyl methacrylate was fabricated for delivery of the insecticide abamectin (Gao et al. 2019).

13.5.5 Solid- lipid NPs

Essential oil-loaded solid lipid nanoparticles can also be useful for the formulations of nanopesticides. Porous hollow silica nanoparticles (PHSNs) were prepared and applied as a controlled delivery system of water-soluble pesticide validamycin (Liu et al. 2006). Pyrethroids for example gamma-cyhalothrin (GCH) are broad-spectrum insecticides and acaricidal activities but have high and undesirable toxicity towards aquatic organisms. Compared with a conventional emulsifiable concentrate formulation of GCH; the solid lipid particle formulations reduced the toxicity towards aquatic organisms (Frederiksen et al. 2003).

Glyceryl tripalmitate (a solid present majorly in palm and coconut oil) is a carrier molecule for effective drug delivery. Carbendazim and tebuconazole are used in agriculture for the prevention and control of fungal diseases in plants. A reduced level of cytotoxicity was observed with the glyceryl tripalmitate-loaded pesticides than traditional pesticide formulations (Campos et al. 2015, Mubeen et al. 2023).

13.5.6 Micro and Nano Emulsion

Tamhane and colleagues have formulated a water-soluble plant protease inhibitor CanPI-7 as a Microemulsion biopesticide application. CanPI protein is synthesized by *Capsicum annuum* leaves and has 1–4 inhibitory domains with varied trypsin or chymotrypsin inhibitory sites. CanPIs are effective in reducing the fecundity of moths and fertility of eggs at low concentrations and can cause larval mortality at higher concentrations (Tamhane et al. 2012). Environment-friendly nanoemulsion systems for the water-soluble herbicide glyphosate isopropylamine mixed surfactant systems were prepared by mixing carbohydrate-derived alkyl poly glucoside and organosilicon surfactants. The nanoemulsions controlled growth of certain weeds such as Indian goosegrass (*Eleusine indica*), creeping foxglove (*A. gangetica*), slender buttonweed (*D. ocimifolia*), and buffalo grass (*P. conjugatum*) (De et al. 2014).

13.5.7 Micelles

The synthesized self-assembling amphiphile–picloram (SAP) conjugate has three extending arms: a lipophilic lauryl chain, a hydrophilic polyethylene glycol chain and the amphiphobic agrochemical active picloram (systemic herbicide used for general woody plant control). Upon contact with water, lipid-based conjugate form micelle. The interactions of SAP conjugates with the hydrophobic surface of crop plant leaves did not damage the epicuticular wax micromorphology hence decreasing both phytotoxicity and possible yield loss. SAP conjugates also prevent off-target environmental effects by minimizing off-target toxicity, less water requirement for spray application and low agrochemical runoff (Nadiminti et al. 2018). A system consisting of polyethylene glycol (PEG) and the pesticide dichlorphenoxyacetic acid (2,4-d) are used for delivery of micelle based pesticides. The photoactive molecule 3-nitro-4-bromomethylbenzoic acid (NBA) was used as a photo-trigger for the synthesis of above system. The 2,4-d-NBA-PEG system could self-assemble into core–shell micelles. 2,4-d is chemical analogs of auxin and is primarily used in agriculture to control weeds in corn and grain, forestry, and lawn care practices (Costa and Aschner 2014, Camara et al. 2019).

13.5.8 Carbon Nanomaterials

Graphene oxide films, a carbon-based nanomaterial, can prolong the process of potassium nitrate (K and N source) release, which amplifies the time of the function and minimizes dissipation by leaching (Shalaby et al. 2016). Carbon nanotubes (CNTs), are being used to deliver nucleic acid-based cargoes to plant species and tissues effectively in an almost species-independent manner (Demirer et al. 2019, Kwak et al. 2019). These are newly emerging nanomaterials and are less explored for their applications in biological systems. Thus, a detailed study is required for a thorough understanding and their applicability and compatibility with their biological counterparts (Pérez-de-Luque 2017).

13.6 ADVANTAGES OF NANOBASED DELIVERY SYSTEMS OVER CONVENTIONAL APPROACHES

Delivery systems in agriculture are important for application of pesticides and fertilizers as well as during genetic material-mediated plant improvement. Delivery systems for fertilizers face issues with bioavailability due to soil chelation, over-use, and run-offs while pesticides need to focus on efficacy enhancement and spray drift management. A feasible alternative for these complications is given by controlled delivery systems for pesticide and fertilizer application. Nano-based controlled delivery technique intends towards estimated release of essential and adequate quantities of agrochemicals over a period of time, to obtain the highest biological efficacy and to reduce the harmful effects. Controlled release of the active ingredient is accomplished due to the slow-release properties of the nanomaterial, the bonding of the ingredients to the material and the environmental conditions (Manjunatha et al. 2016). Developing nano-agrochemical delivery system through nanotechnology facilitates the enhancement of the stability and dispersion of active ingredients, contributes to the precise delivery of agrochemicals, decreases residual pollution, and reduces labor costs in different application circumstances (An et al. 2022).

13.7 CHALLENGES WITH NANO-BASED DELIVERY SYSTEMS

Nano-Enabled Agrochemicals in agriculture presents a targeted overview of the safe implementation of nanotechnologies within agricultural and horticultural settings, with the purpose of achieving enhanced production while maintaining ecological integrity. The growing global demand for agricultural crops and products requires high standards of quality and safety, which has stimulated the search for new technologies that preserve their quality and delay their decomposition. Nanotechnology may boost plant production by improving nutrient uptake/use efficiency with nanoformulations of fertilizers and agrochemicals for plant enhancement, detection and treatment of diseases, and host-parasite interactions at the molecular level using nanosensors. It also may improve plant disease diagnostics, removal of contaminants from soil and water, post-harvest management of vegetables and flowers, and reclamation of salt-affected soils.

Although the markets for nanoproducts and nanoformulations continue to increase, there are also growing concerns regarding the fate and behavior of nanomaterials in environmental systems. Exploring important topics related to nanotechnology and nanomaterials, the book includes the use of nanochemicals in insect pest management, such as nanofungicides, nanoherbicides, micronutrient supply, and nanosensors to monitor crop and soil health conditions, from detection of agrochemicals to their slow release of agrochemicals, and their impact on related environs (Ghorbanpour and Shahid 2022).

13.8 CONCLUSIONS AND FUTURE PROSPECTS

Nanotechnological applications in modern agriculture help in uplifting the global economy by providing support and advances in different ways. Considering various challenges posed by the increasing population, plant and animal diseases, and change in climatic conditions, the introduction of NPs contribute significantly to addressing these pressing issues, expanding the use of conventional pesticides and fertilizers to enhance crop production, ultimately causing harm to the natural environment. With the advent of NPs, the effectiveness and agronomic efficiency are highly improved compared to traditional resources. The better use of pesticides encapsulated in different nano-formulations provides better application, and controlled release protects the environment. Several nano pesticides and nano fertilizers are under development, and a few of them are already available in the market. The study of NP–plant interaction is an important point to consider in understanding the physiological and biochemical responses imposed by plants. Emerging techniques are technologically advanced to locate and quantify NPs through the plant system that supports the idea of their transformation and safety aspects in a complex system. Advances in agricultural sciences are critical for improving production efficiency and sustainability of agriculture. The rapid development of nanotechnology has provided potentially transformative opportunities to improve the performance of agrochemicals by exploring different agricultural applications of nanomaterials, including nano-fertilizers, nano-additives, nano-pesticides, nano-cleansers, and nano-sensors. In addition, nano-delivery systems have immense potential for delivering pesticides and/or nutrients into plants to eliminate pathogens or meet nutritional needs with maximum use efficiency and minimum side effects. However, despite the exciting outcomes of nanotechnology application in agriculture, nano-enabled agriculture is still in a juvenile stage with many challenges to overcome.

13.9 ACKNOWLEDGMENT

AKG and PG would like to acknowledge this research is part of the project No. UMO-2022/45/P/NZ9/01571 co-funded by the National Science Centre and the European Union Framework Programme for Research and Innovation Horizon 2020 under the Marie Skłodowska-Curie grant agreement No.945339. For the purpose of Open Access, the author has applied a CC-BY public copyright licence to any Author Accepted Manuscript (AAM) version arising from this submission.

References

Abd-Elsalam, K.A. and Alghuthaymi, M.A. 2015. Nanobiofungicides: is it the next-generation of fungicides? J. Nanotech. Mater. Sci. 2(2): 1–3. http://dx.doi.org/10.15436/2377-1372.15.013.

Adams, T., Ashworth, A.J. and Sauer, T. 2021. Soil CO_2 evolution is driven by forage species, soil moisture, grazing pressure, poultry litter fertilization, and seasonality in silvopastures. Agrosyst. Geosci. Env. 4: e20179. https://doi.org/10.1002/agg2.20179.

Aktar, W., Sengupta, D. and Chowdhury, A. 2009. Impact of pesticides use in agriculture: their benefits and hazards, Interdiscipl. Toxicol. 2: 1–12.

Alghuthaymi, M.A., Rajkuberan, C., Rajiv, P., Kalia, A., Bhardwaj, K., Bhardwaj, P., Abd-Elsalam, K.A., et al. 2021. Nanohybrid antifungals for control of plant diseases: current status and future perspectives. J. Fungi. 7(1): 48. doi: 10.3390/jof7010048. PMID: 33450851; PMCID: PMC7828323.

Ali, M.A., Rehman, I., Iqbal, A., Din S., Rao, A.Q., Latif, A. and Husnain, T. 2014. Nanotechnology, a new frontier in agriculture. Adv. Life Sci. 1: 129–138.

An, C., Sun, C., Li, N., Huang, B., Jiang, J., Shen, Y., Wang, C., et al. 2022. Nanomaterials and nanotechnology for the delivery of agrochemicals: strategies towards sustainable agriculture. J. Nanobiotech. 20(1): 11. https://doi.org/10.1186/s12951-021-01214-7.

Anjali, C.H., Sharma, Y., Mukherjee, A. and Chandrasekaran, N. 2012. Neem oil (*Azadirachta indica*) nanoemulsion: a potent larvicidal agent against *Culex quinquefasciatus*. Pest. Manag. Sci. 68: 158–163.

Avila-Quezada, G.D., Golinska, P. and Rai, M. 2022. Engineered nanomaterials in plant diseases: can we combat phytopathogens? App. Micro Biotech. 106(1): 117–129. 10.1007/s00253-021-11725-w.

Baker, C., Pradhan, A., Pakstis, L., Pochan, D.J. and Shah, S.I. 2005. Synthesis and antibacterial properties of silver nanoparticles. J. Nanosci. Nanotechnol. 5: 244–249. doi: 10.1166/jnn.2005.034.

Barrett-Catton, E., Ross, M.L. and Asuri, P. 2021. Multifunctional hydrogel nanocomposites for biomedical applications. polymers. 13: 856. https://doi.org/10.3390/polym13060856.

Basavegowda, N. and Baek, K.H. 2021. Current and future perspectives on the use of nanofertilizers for sustainable agriculture: the case of phosphorus nanofertilizer. 3 Biotech. 11(7): 357. doi: 10.1007/s13205-021-02907-4. Epub 2021 Jun 28. PMID: 34268065; PMCID: PMC8239106.

Bawskar, M., Bansod, S., Rathod, D., Santos, C., Ingle, P., Rai, M. and Gade, A. 2021. Silver nanoparticles as nanofungicide and plant growth promoter: evidences from morphological and chlorophyll 'a' fluorescence analysis. Adv. Mat. Lett. 12(10): 1–7. doi: 10.5185/aml.2021.15702.

Camara, M., Campos, E.V.R., Monteiro, R.A., do Espirito Santo Pereira, A., de Freitas Proença, P.L. and Fraceto, L.F. 2019. Development of stimuli-responsive nano-based pesticides: emerging opportunities for agriculture. J. Nanobiotech. 17(1): 100. https://doi.org/10.1186/s12951-019-0533-8.

Campos, E.V.R., de Oliveira, J.L., da Silva, C.M.G., Pascoli, M., Pasquoto, T., Lima, R., Abhilash, P.C., et al. 2015. Polymeric and solid lipid nanoparticles for sustained release of carbendazim and tebuconazole in agricultural applications. Scient. Rep. 5: 13809. https://doi.org/10.1038/SREP13809.

Carvalho, F.P. 2006. Agriculture, pesticides, food security and food safety. Environ. Sci. Policy. 9(7): 685–692.

Chandini., Kumar, R., Kumar, R. and Prakash, O. 2019. The impact of chemical fertilizers on our environment and ecosystem. pp. 71–85. In: Sharma, P. (ed.). Research Trends in Environmental Sciences, 2nd. New Delhi, India.

Chaud, M., Souto, E.B., Zielinska, A., Severino, P., Batain, F., Oliveira-Junior, J. and Alves, T. 2021. Nanopesticides in agriculture: benefits and challenge in agricultural productivity, toxicological risks to human health and environment. Toxics. 9(6): 131. https://doi.org/10.3390/toxics9060131.

Chauhan, A., Patil, C., Jain, P. and Kulhari, H. 2020. Dendrimer-based marketed formulations and miscellaneous applications in cosmetics, veterinary, and agriculture. pp. 325–334. *In*: Chauhan, A. and Kulhari, H. (eds). Pharmaceutical Applications of Dendrimers: A volume in Micro and Nano Technologies. Elsevier doi:10.1016/b978-0-12-814527-2.00014-7.

Chen, Y.W., Lee, H.V., Juan, J.C. and Phang, S.M. 2016. Production of new cellulose nanomaterial from red algae marine biomass *Gelidium elegans*. Carbohydr. Polym. 151: 1210–1219

Clarke, Z. 2007. Diazinon. pp. 1–4. *In*: Enna, S.J. and Bylund, D.B. (eds). xPharm: The Comprehensive Pharmacology Reference. Elsevier. https://doi.org/10.1016/B978-008055232-3.61621-6.

Collins, H.M., Hall, F.R. and Hopkinson, M. 1996. Pesticides ACE-o pesticide formulations and application systems. 15th. 1268. ASTM

Corradini, E., De Moura, M.R. and Mattoso, L.H.C. 2010. A preliminary study of the incorporation of NPK fertilizer into chitosan nanoparticles. Express. Polym. Lett. 4: 509–515.

Costa, L.G., and Aschner, M. 2014. Toxicology of pesticides. Ref. Mod. Biomed. Sci. https://doi.org/10.1016/B978-0-12-801238-3.00208-7. (https://www.sciencedirect.com/science/article/pii/B978012801238300208 7).

Das, K., Jhan, P.K., Das, S.C., Aminuzzaman, F. and Ayim, B.Y. 2021. Nanotechnology: past, present and future prospects in crop protection. *In:* Ahmad F. and Sultan M. (eds). Technology in Agriculture. IntechOpen. https://doi.org/10.5772/intechopen.98703.

De, A., Bose, R., Kumar, A., and Mozumdar, S. 2014. Targeted delivery of pesticides using biodegradable polymeric nanoparticles. pp. XXIII, 99. Series: Springer Briefs in Molecular Science. Springer New Delhi. DOI: 10.1007/978-81-322-1689-6.

de, Oliveira J.L., Campos E.V.R., Bakshi, M., Abhilash, P.C. and Fraceto, L.F. 2014. Application of nanotechnology for the encapsulation of botanical insecticides for sustainable agriculture: prospects and promises. Biotechnol. Adv. 32(8): 1550–1561. https:// doi. org/ 10. 1016/j. biotechadv.2014.10.010.

Demirer, G.S., Zhang, H., Matos, J.L., Goh, N.S., Cunningham, F.J., Sung, Y., Chang, R., et al. 2019. High aspect ratio nanomaterials enable delivery of functional genetic material without DNA integration in mature plants. Nat. Nanotechnol. 14(5): 456–464. doi: 10.1038/s41565-019-0382-5. Epub 2019 Feb 25. PMID: 30804481.

Echiegu, E.A. 2017. Nanotechnology applications in the food industry. pp. 153–171. *In:* Prasad, R., Kumar, V. and Kumar, M. (eds). Nanotechnology. Springer, Singapore. https://doi.org/10.1007/978-981-10-4678-0_9.

Fincheira, P., Quiroz, A., Medina, C., Tortella, G., Hermosilla, E., Diez, M.C. and Rubilar, O. 2020. Plant growth induction by volatile organic compound released from solid lipid nanoparticles and nanostructured lipid carriers. Coll. Surf. A: Physicochem. Engg. Asp. 596: 124739. doi:10.1016/j.colsurfa.2020.124739.

Frederiksen, H.K., Kristensen, H.G. and Pedersen, M. 2003. Solid lipid microparticle formulations of the pyrethroid gamma-cyhalothrin—incompatibility of the lipid and the pyrethroid and biological properties of the formulations. J. Controlled Release. 86(2–3): 243–252. https://doi.org/10.1016/S0168-3659(02)00406-6.

Gao, Y., Zhang, Y., He, S., Xiao, Y., Qin, X., Zhang, Y., Li, D., et al. 2019. Fabrication of a hollow mesoporous silica hybrid to improve the targeting of a pesticide. Chem. Engg. J. 364: 361–369. https://doi.org/10.1016/J.CEJ.2019.01.105.

Garcia-Gomez, C., Obrador, A., Gonzalez, D., Babin, M. and Fernandez, M.D. 2018. Comparative study of the phytotoxicity of ZnO nanoparticles and Zn accumulation in nine crops grown in a calcareous soil and an acidic soil. Sci. Total Environ. 644: 770–780. https:// doi. org/ 10. 1016/j. scitotenv.2018.06.356.

Ghorbanpour, M. and Shahid, M.A. (eds). 2022. Nano-enabled Agrochemicals in Agriculture. Elsevier Inc. USA. https://doi.org/10.1016/C2020-0-02858-6.

Gilland, B. 2002. World population and food supply: can food production keep pace with population growth in the next half-century? Food Policy. 27(1): 47–63. doi:10.1016/S0306-9192(02)00002-7.

Giraldo, J.P., Landry, M.P., Faltermeier, S.M., McNicholas, T.P., Iverson, N.M., Boghossian, A.A., Reuel, F.N., et al. 2014. Plant nanobionics approach to augment photosynthesis and biochemical sensing. Nat. Mater. 13(4): 400–8. https:// doi. org/ 10. 1038/ nmat3890.

Godfray, H.C., Mason-D'Croz, D. and Robinson, S. 2016. Food system consequences of a fungal disease epidemic in a major crop. Philos. Trans. R. Soc. Lond. B. Biol. Sci. 371: 20150467. doi: 10.1098/rstb.2015.0467.

Guan, H., Chi, D., Yu, J. and Li, H. 2010. Dynamics of residues from a novel nano-imidacloprid formulation in soybean fields. Crop Prot. 29: 942– 946.

Guha, T., Gopal, G., Kundu, R. and Mukherjee, A. 2020. Nanocomposites for delivering agrochemicals: a comprehensive review. J. Agric. Food Chem. 68(12): 3691–3702. https://doi.org/10.1021/acs.jafc.9b06982.

Huang, B., Chen, F., Shen, Y., Qian, K., Wang, Y., Sun, C., Zhao, X., et al. 2018. Advances in targeted pesticides with environmentally responsive controlled release by nanotechnology. Nanomaterials. 8(2): 102. https://doi.org/10.3390/nano8020102.

Hussain, T. 2017. Nanocides: smart delivery system in agriculture and horticultural crops. Adv. Plants Agric. Res. 6(6): 175. DOI: 10.15406/apar.2017.06.00233

Hussain, C.M. 2018. Handbook of Nanomaterials for Industrial Applications. Elsevier, Amsterdam.

Ingle, A. 2023. Nanotechnology in Agriculture and Agroecosystems. Elsevier, Netherlands.

Iqbal, M.A. 2019. Nano-fertilizers for sustainable crop production under changing climate: a global perspective. *In:* Hasanuzzaman, M., Filho, M.C.M.T., Fujita, M. and Nogueira, T.A.R. (eds). Sustainable Crop Production. Intech. Open. London. UK. https://doi.org/10.5772/intechopen.89089.

Janson, O., Gururaj, S., Pujari-Palmer, S., Ott, M.K., Strømme, M. and Engqvist, H. 2019. Titanium surface modification to enhance antibacterial and bioactive properties while retaining biocompatibility. Mater. Sci. Eng. C. 96: 272–279. doi: 10.1016/j.msec.2018.11.021.

Jianhui, Y., Kelong, H. Yuelong, W. and Suquin, L. 2005. Study on anti-pollution nano-preparation of dimethomorph and its performance. Chin. Sci. Bull. 50: 108–112.

Jo, Y.K., Kim, B.H. and Jung, G. 2009. Antifungal activity of silver ions and nanoparticles on phytopathogenic fungi. Plant Dis. 93: 1037–1043. doi: 10.1094/PDIS-93-10-1037.

Joner, E.J., Hartnik, T. and Amundsen, C.E. 2008. Environmental fate and ecotoxicity of engineered nanoparticles: norwegian pollution control authority report no. TA. 2304/2007. Bioforsk. 1–64.

Klassen, W. 1995. World food security up to 2010 and the global pesticide situation. *In:* Proceedings of the 8th International Congress on Pesticide Chemistry. 1–32. Am. Chem. Soc. Washington. D.C.

Knowles, A. 2008. Recent developments of safer formulations of agrochemicals. Environmentalist. 28(1): 35–44.

Kookana, R.S., Boxall, A.B.A., Reeves, P.T., Ashauer, R., Beulke, S., Chaudhry, Q., Cornelis, G., et al. 2014. Nanopesticides: guiding principles for regulatory evaluation of environmental risks. J. Agri. Food Chem. 62(19): 4227–4240. doi:10.1021/jf500232f.

Kretzmann J.A., Ho, D., Evans, C.W., Plani-Lam, J.H.C., Garcia-Bloj, B. Mohamed, A.E., O'Mara, M.L., et al. 2017. Synthetically controlling dendrimer flexibility improves delivery of large plasmid DNA. Chem. Sci. 8(4): 2923–2930. doi: 10.1039/c7sc00097a. Epub 2017 Jan 27. PMID: 28451358; PMCID: PMC5376716.

Kulabhusan, P.K., Tripathi, A. and Kant, K. 2022. Gold nanoparticles and plant pathogens: an overview and prospective for biosensing in forestry. Sensors. 22: 1259. https://doi.org/10.3390/s22031259.

Kumar, V., Vaid, K., Bansal S.A. and Kim, K.H. 2020. Nanomaterial-based immunosensors for ultrasensitive detection of pesticides/herbicides: current status and perspectives. Biosens. Bioelectron. 165: 112382. https:// doi.org/ 10. 1016/j. bios. 2020. 112382.

Kwak, S.Y., Giraldo, J.P., Wong, M.H., Koman, V.B., Lew, T.T.S., Ell, J., Weidman M.C., et al. 2017. A nanobionic light-emitting plant. Nano Lett. 17(12): 7951–7961. doi:10.1021/acs.nanolett.7b04369.

Kwak, S.Y., Lew, T.T.S., Sweeney, C.J., Koman, V.B., Wong, M.H., Bohmert Tatarev, K., Snell, K.D., et al. 2019. Chloroplast-selective gene delivery and expression in planta using chitosan-complexed single-walled carbon nanotube carriers. Nat. Nanotechnol. 14(14): 447–455. doi:10.1038/s41565-019-0375-4.

Li, P., Huang, Y., Fu, C., Jiang, S.X., Peng, W., Jia, Y., Peng, H., et al. 2021. Eco-friendly biomolecule-nanomaterial hybrids as next generation agrochemicals for topical delivery. EcoMat. 3(5): e12132. Mat. https:/doi.org/10.1002/eom2.

Liu, F., Wen, L.X., Li, Z.Z., Yu, W., Sun, H.Y. and Chen, J.F. 2006. Porous hollow silica nanoparticles as controlled delivery system for water-soluble pesticide. Mater. Res. Bull. 41(12): 2268–2275.

Liu, Y., Tong, Z. and Prud'homme, R.K. 2008. Stabilized polymeric nanoparticles for controlled and efficient release of bifenthrin. Pest. Manag. Sci. 64: 808–812.

Lv, M., Liu, Y., Geng, J.H., Kou, X.H., Xin, Z.H. and Yang, D.Y. 2018. Engineering nanomaterials-based biosensors for food safety detection. Biosens. Bioelectron. 106: 122–128.

Mahmoudpour, M., Karimzadeh, Z., Ebrahimi, G., Hasanzadeh, M. and Dolatabadi, J.E.N. 2021. Synergizing functional nanomaterials with aptamers based on electrochemical strategies for pesticide detection: current status and perspectives. Crit. Rev. Anal. Chem. https://doi.org/10.1080/10408347.2021.19199 87.

Malakar, A., Kanel, S.R., Ray, C., Snow, D.D. and Nadagouda, M.N. 2020. Nanomaterials in the environment, human exposure pathway, and health effects: a review. Sci. Total Environ. 759: 143470. https://doi.org/10.1016/j.scitotenv.143470.

Mali, S.C., Raj, S. and Trivedi, R. 2020. Nanotechnology a novel approach to enhance crop productivity. Biochem. Biophy. Rep. 24: 100821: 1–4. doi:10.1016/j.bbrep.2020.100821.

Manjunatha, S.B., Biradar, D.P. and Aladakatti, Y.R. 2016. Nanotechnology and its applications in agriculture: a review. J. Farm. Sci. 29(1): 1–13. https://www.researchgate.net/publication/303665019_Nanotechnology_and_its_applications_in_agriculture_A_review.

Markowicz-Piasecka, M. and Mikiciuk-Olasik, E. 2016. Dendrimers in drug delivery. Molecules. 23: E938. 10.1016/B978-0-323-42866-8.00002-2

Martin-Ortigosa, S., Valenstein, J.S., Lin, V.S.Y., Trewyn, B.G. and Wang, K. 2012. Gold functionalized mesoporous silica nanoparticle mediated protein and DNA codelivery to plant cells via the biolistic method. Adv. Funct. Mater. 22: 3576–3582.

Martins, N.R.C., Avellan, A., Rodrigues, S., Salvador, D., Rodrigues, S.M. and Trindade, T. 2020. Composites of biopolymers and ZnO NPs for controlled release of zinc in agricultural soils and timed delivery for maize. ACS Appl. Nano Mater. 3(3): 2134–2148. https://doi.org/10.1021/ACSANM.9B01492/SUPPL_FILE/AN9B01492_SI_001.PDF

McCalla, A.F. 2001. Challenges to world agriculture in the 21st century. Agri. Resour. Econom. 4(3): 1–4.

Mejias, J.H., Salazar, F., Pérez Amaro, L., Hube, S., Rodriguez, M. and Alfaro, M. 2021. Nanofertilizers: a cutting-edge approach to increase nitrogen use efficiency in grasslands. Front. Environ. Sci. 9: 635114. doi: 10.3389/fenvs.2021.635114.

Meyer, M.T. and Scribner, E.A. 2009. The evolution of analytical technology and its impact on water-quality studies for selected herbicides and their degradation products in water. pp. 289–313. *In*: Ahuja, S. (ed.). Handbook of Water Purity and Quality. Academic Press. https://doi.org/10.1016/B978-0-12-374192-9.00013-3.

Milani, N., Hettiarachchi, G.M., Kirby, J.K., Beak, D.G., Stacey, S.P. and McLaughlin, M.J. 2015. Fate of zinc oxide nanoparticles coated onto macronutrient fertilizers in an alkaline calcareous soil. PloS One. 10: e0126275.

Mittal, D., Kaur, G., Singh, P., Yadav, K. and Ali, S.A. 2020. Nanoparticle-based sustainable agriculture and food science: recent advances and future outlook. Front. Nanotechnol. 2: 579954. doi: 10.3389/fnano.2020.579954.

Mora, C.L.C., Mueller, A. and Janssen, A.P.G. 2019. Antibacterial medical product and method for producing same. Patent no. U.S. 2019/0124915 A1.

Mubeen, I., Mfarrej, M.F.B., Razaq, Z., Iqbal, S., Naqvi, S.A.H., Hakim, F., Mosa, W.F.A., et al. 2023. Nanopesticides in comparison with agrochemicals: outlook and future prospects for sustainable agriculture. Plant Physiol. Biochem. 198: 107670. https://doi.org/10.1016/j.plaphy.2023.107670.

Nadiminti, P.P., Liu, Q., Vanjari, L.K., Dong, Y.D., Boyd, B.J. and Cahill, D.M. 2018. Novel self-assembling conjugates as vectors for agrochemical delivery. J. Nanobiotech. 16(1): 94. https://doi.org/10.1186/s12951-018-0423-5.

Nandini, B. and Geetha, N. 2021. Smart delivery mechanisms of nanofertilizers and nanocides in crop biotechology. pp. 385–414. *In*: Sudisha, J., Harikesh, S., Leonardo, F.F. and Renata, L. (eds). Advances in Nano-Fertilizers and Nano-Pesticides in Agriculture. Elsevier. eBook. DOI: 10.1016/B978-0-12-820092-6.00015-X.

NCBI (National Center for Biotechnology Information). 2022. PubChem Compound Summary for CID 3017. Diazinon. https://pubchem.ncbi.nlm.nih.gov/compound/Diazinon.

Nunes, D., Andrade, S., Ramalho, M.J., Loureiro, J.A. and Pereira M.C. 2022. Polymeric nanoparticles-loaded hydrogels for biomedical applications: a systematic review on *in vivo* findings. Polymers. 14(5): 1010. doi: 10.3390/polym14051010. PMID: 35267833; PMCID: PMC8912535.

Ouda, S.M. 2014. Antifungal activity of silver and copper nanoparticles on two plant pathogens, *Alternaria alternata* and *Botrytis cinerea*. Res. J. Microbiol. 9: 34–42. doi: 10.3923/jm.2014.34.42.

Parisi, C., Vigani, M. and Rodríguez-Cerezo, E. 2015. Agricultural nanotechnologies: what are the current possibilities? Nano Today. 10: 124–127.

Patel, N., Desai, P., Patel, N., Jha, A. and Gautam, H.K. 2014. Agronanotechnology for plant fungal disease management: a review. Int. J. Curr. Microbiol. App. Sci. 3: 71–84.

Patra, S.K., Poddar, R., Brestic, M., Acharjee, P.U., Bhattacharya, P., Sengupta, S., Pal, P., et al. 2022. Prospects of hydrogels in agriculture for enhancing crop and water productivity under water deficit condition. 2022: 4914836. doi: https://doi.org/10.1155/2022/4914836.

Pérez-de-Luque, A. 2017. Interaction of nanomaterials with plants: what do we need for real applications in agriculture? Front Environ. Sci. 5: 12. doi: 10.3389/fenvs.2017.00012.

Pérez-de-Luque, A. and Rubiales, D. 2009. Nanotechnology for parasitic plant control. Pest Manage. Sci. 65(5): 540–545. https://doi.org/10.1002/ps.1732.

Prasad, R., Bhattacharyya A. and Nguyen Q.D. 2017. Nanotechnology in sustainable agriculture: recent developments, challenges, and perspectives. Front. Microbiol. 8: 1014.

Rai, M. and Kratošová, G. 2015. Management of phytopathogens by application of green nanobiotechnology: Emerging trends and challenges. Acta Agraria. Debreceniensis. DOI: 10.34101/actaagrar/66/1884.

Rai, M. and Biswas, J.K. (eds). 2018. Nanomaterials: Ecotoxicity, Safety, and Public Perception. Cham. Switzerland: Springer International Publishing.

Rai, M., Ingle, A.P., Pandit, R., Paralikar, P., Shende, S., Gupta, I., Biswas, J.K., et al. 2018. Copper and copper nanoparticles: role in management of insect-pests and pathogenic microbes. Nanotechnol. Rev. 7(4): 303–315.

Rajput, V., Minkina, T., Mazarji, M., Shende, S., Sushkova, S., Mandzhieva, S., et al. 2020. Accumulation of nanoparticles in the soil-plant systems and their effects on human health. Annals. Agri. Sci. 65: 137–143.

Rajput, V.D., Singh, A., Minkina, T.M., Shende, S.S., Kumar, P., Verma, K.K., et al. 2021. Potential applications of nanobiotechnology in plant nutrition and protection for sustainable agriculture. pp. 79–92. *In*: In: Ingle, A.P. (ed.). Nanotechnology in Plant Growth Promotion and Protection: Recent Advances and Impacts. John Wiley & Sons Ltd. USA.

Rastogi, A., Zivcak, M., Sytar, O., Kalaji, H.M., He, X., Mbarki, S. and Brestic, M. 2017. Impact of metal and metal oxide nanoparticles on plant: a critical review. Front Chem. 5: 78. doi: 10.3389/fchem.2017.00078

Ruttkay-Nedecky, B., Krystofova, O., Nejdl, L. and Adam, V. 2017. Nanoparticles based on essential metals and their phytotoxicity. J. Nanobiotechnol. 15: 33.

Safari, J. and Zarnegar Z. 2014. Advanced drug delivery systems: nanotechnology of health design a review. J. Saudi Chem. Soc. 18: 85–99. 10.1016/j.jscs.2012.12.009.

Santana, I., Wu, H., Hu, P. and Giraldo, J.P. 2020. Targeted delivery of nanomaterials with chemical cargoes in plants enabled by a biorecognition motif. Nature Commu. 11: 2045. doi:10.1038/s41467-020-15731-w.

Sarlak, N., Taherifar, A. and Salehi, F. 2014. Synthesis of nanopesticides by encapsulating pesticide nanoparticles using functionalized carbon nanotubes and application of new nanocomposite for plant disease treatment. J. Agric. Food Chem. 62: 4833–4838. doi: 10.1021/jf404720d.

Savci, S. 2012. Investigation of effect of chemical fertilizers on environment. APCBEE Procedia. 1: 287–292. doi: 10.1016/j.apcbee.2012.03.047.

Sebilo, M., Mayer, B., Nicolardot, B., Pinay, G. and Mariotti, A. 2013. Long-term fate of nitrate fertilizer in agricultural soils. Proc. Natl. Acad. Sci. USA. 110: 18185–18189.

Selyutina, O.Y., Apanasenko, I.E., Khalikov, S.S. and Polyakov, N.E. 2017. Natural poly-and oligosaccharides as novel delivery systems for plant protection compounds. J. Agric. Food Chem. 65(31): 6582–6587.

Selyutina, O.Y., Khalikov, S.S. and Polyakov, N.E. 2020. Arabinogalactan and glycyrrhizin based nanopesticides as novel delivery systems for plant protection. Environ. Sci. Pollut. Res.27: 5864–5872. https://doi.org/10.1007/s11356-019-07397-9.

Shalaby, T.A., Bayoumi, Y., Abdalla, N., Taha, H., Alshaal, T., Shehata, S., Amer, M., et al. 2016. Nanoparticles, soils, plants and sustainable agriculture. pp. 283–312. *In*: Ranjan, S., Dasgupta, N. and Lichtfouse, E. (eds). Nanoscience in Food and Agriculture 1. Sustainable Agriculture Reviews, vol 20. Springer, Cham. https://doi.org/10.1007/978-3-319-39303-2_10.

Shang, Y., Hasan, Md., K., Ahammed, G.J., Li, M., Yin, H. and Zhou, J. 2019. Applications of nanotechnology in plant growth and crop protection: a review. Molecules. 24: 2558. doi:10.3390/molecules24142558.

Shende, S., Rajput, V., Gade, A., Minkina, T., Fedorov, Y., Sushkova, S.N., Mandzhieva, S.S., et al. 2022. Metal-based green synthesized nanoparticles: boon for sustainable agriculture and food security. IEEE Trans. Nanobiosci. 21: 1. DOI:10.1109/TNB.2021.3089773.

Singh, N., Agarwal, S., Jain, A. and Khan, S. 2021. 3-Dimensional cross linked hydrophilic polymeric network “hydrogels”: an agriculture boom. Agri. Water Manage. 253: 106939. doi:10.1016/j.agwat.2021.106939.

Smith, A.M. and Gilbertson, L.M. 2018. Rational ligand design to improve agrochemical delivery efficiency and advance agriculture sustainability. ACS Sustain. Chem. Eng. 6: 13599–13610.

Solanki, P., Bhargava, A., Chhipa, H., Jain, N. and Panwar, J. 2015. Nano-fertilizers and their smart delivery system. pp. 81–101. *In:* Rai, M., Ribeiro, C., Mattoso, L. and Duran, N. (eds). Nanotechnologies in Food and Agriculture. Springer Int. New York. USA. doi:10.1007/978-3-319-14024-7_4.

Su, C., Ji, Y., Liu, S., Gao, S., Cao, S., Xu, X., Zhou, C., et al. 2020. Fluorescence-labeled abamectin nanopesticide for comprehensive control of pinewood nematode and *monochamus alternatus* hope. ACS Sustain. Chem. Eng. 8(44): 16555–16564. https://doi.org/10.1021/acssuschemeng.0c05771.

Tamhane, V.A., Dhaware, D.G., Khandelwal, N., Giri, A.P. and Panchagnula, V. 2012. Enhanced permeation, leaf retention, and plant protease inhibitor activity with bicontinuous microemulsions. J. colloid interf. sci. 383(1): 177–183. https://doi.org/10.1016/j.jcis.2012.06.025.

Tarafder, C., Daizy, M., Alam, M.M., Ali, M.R., Islam, M.J., Islam, R., Ahimmed, S., et al. 2020. Formulation of a hybrid nanofertilizer for slow and sustainable release of micronutrients. ACS Omega. doi:10.1021/acsomega.0c03233.

Taşkın, M.B., Şahin, O., Taskin, H., Atakol, O., Inal, A. and Gunes, A. 2018. Effect of synthetic nano-hydroxyapatite as an alternative phosphorus source on growth and phosphorus nutrition of lettuce (*Lactuca sativa* L.) plant. J. Plant Nutri. 41: 1148–1154. DOI: 10.1080/01904167.2018.1433836

Torney, F., Trewyn, B.G., Lin, V.S.Y. and Wang, K. 2007. Mesoporous silica nanoparticles deliver DNA and chemicals into plants. Nat. Nanotechnol. 2: 295–300.

UN Department of Economic and Social Affairs (UNDESA). Population Division. 2019. World Population Prospects. United Nations. New York, NY, USA, 2019.

UNIDO. 1983 Formulation of pesticides in developing countries. http://pdf.usaid.gov/pdf_docs/

Vamvakaki, V. and Chaniotakis, N.A. 2007. Pesticide detection with a liposome-based nano-biosensor. Biosens. Bioelectron. 22: 2848–2853.

Vega-Vásquez, P., Mosier, N.S. and Irudayaraj, J. 2020. Nanoscale drug delivery systems: from medicine to agriculture. Front. Bioeng. Biotechnol. 8: 79. doi: 10.3389/fbioe.2020.00079. PMID: 32133353; PMCID: PMC7041307.

Vijayakumar, P.S., Abhilash, O.U., Khan, B.M., Prasad, B.L. 2010. Nanogold-Loaded sharp-edged carbon bullets as plant gene carriers. Adv. Functional Mat. 20: 2416–2423.

Vundavalli, R., Vundavalli, S., Nakka, M. and Rao, D.S. 2015. Biodegradable nano-hydrogels in agricultural farming—alternative source for water resources. Procedia Mater. Sci. 10: 548–554. doi:10.1016/j.mspro.2015.06.005.

Wang, J.W., Grandio, E.G., Newkirk, G.M., Demirer, G.S., Butrus, S., Giraldo, J.P. and Landry, M.P. 2019. Nanoparticle mediated genetic engineering of plants. Mol. Plant. 12: 1037–1040.

Wani, A.H. and Shah, M.A. 2012. A unique and profound effect of MgO and ZnO nanoparticles on some plant pathogenic fungi. JAPS. 2(3): 40–44.

Xu, T., Ma, C., Aytac, Z., Hu, X., Ng, K.W., White, J.C. and Demokritou, P. 2020. Enhancing agrichemical delivery and seedling development with biodegradable, tunable, biopolymer-based nanofiber seed coatings. ACS Sustain. Chem. Eng. 8(25): 9537–9548. https:// doi. org/10.1021/acssuschemeng.0c02696.

Yaqoob, A.A., Ahmad, H., Parveen, T., Ahmad, A., Oves, M., Ismail, I.M.I., Qari H.A., et al. 2020. Recent advances in metal decorated nanomaterials and their various biological applications: a review. Front. Chem. 8: 341. doi: 10.3389/fchem.2020.00341.

Yin, J., Su, X., Yan, S. and Shen, J. 2023. Multifunctional nanoparticles and nanopesticides in agricultural application. Nanomaterials. 13: 1255. https://doi.org/10.3390/ nano13071255.

Zhang, H., Demirer, G.S., Zhang, H., Ye, T., Goh, N.S., Aditham, A.J., Cunningham, F.J., et al. 2019. DNA nanostructures coordinate gene silencing in mature plants. Proc. Natl. Acad. Sci. 116: 7543–7548.

Zobir, S.A.M., Ali, A., Adzmi, F., Sulaiman, M.R. and Ahmad, K.A. Review on nanopesticides for plant protection synthesized using the supramolecular chemistry of layered hydroxide hosts. Biology. 10: 1077. https://doi.org/ 10.3390/biology10111077.

Chapter 14

Nano-Enabled Sensors and Probes for Plant Pathogen Detection and Disease Control in Sustainable Agriculture

Avinash P. Ingle*[1], Shreshtha Saxena[1], Ashwini Savalkar[1] and Mangesh Moharil[1]

[1]Biotechnology Centre, Department of Agricultural Botany, Dr. PDKV, Akola, Maharashtra, India

Mahendra Rai[2]

[2]Department of Biotechnology, SGB Amravati University, Amravati-444602, Maharashtra, India

14.1 INTRODUCTION

Agriculture is the backbone of the economies of many countries and is the only primary source of food all over the globe. However, currently, agriculture is facing several challenges including the most important crop yield losses due to different biotic and abiotic factors (Ingle 2021, 2023). The loss in crop yield ultimately exerts pressure on the supply of sufficient quality of food to every individual present on the planet Earth. Moreover, the ever-increasing global population is also one of the major concerns of increased food demand. According to one of the reports, the global demand for food will continue to increase for the next 40 years if the human population will increase at the current pace. Similarly, it was estimated that an additional 70% of food production will be required by 2050 to meet the food demands (Fang and Ramasamy 2015). The scientific information available suggests that more than one billion populations are suffering from severe problems like malnutrition due to the insufficient supply of appropriate food.

Therefore, the production of enough food and its supply is of great importance. But, as discussed above biotic and abiotic factors are mainly responsible for these situations. Among the factors, mentioned-above biotic factors are mainly responsible for crop damage. It was observed

*Corresponding Author: ingleavinash14@gmail.com

that the losses in crop yield due to plant pathogen infections range between 20% and 40% (https://www.fao.org/news/story/en/item/1187738/icode/). In addition, post-harvest losses due to diseases and sub-standard quality are estimated to be 30–40% (Abass et al. 2014). Hence, to minimize the damage induced due to plant pathogens in crops during growth, harvest, and postharvest processing, as well as to enhance crop productivity ensuring agricultural sustainability, advanced disease detection, and prevention in crops is a need of the hour.

To date, immense attempts and path-breaking discoveries have been made in the field of agriculture for the effective management of issues like plant detection of plant pathogens and diseases and also to enhance crop production. Among these, development of synthetic agrochemicals is one of the important strategies. Agrochemicals like insecticides, pesticides, fertilizers, plant growth promoters, etc. are significantly helpful for the control of biotic factors or plant pathogens and also help to increase crop yield. However, extensive use of such synthetic agrochemicals has now became a major challenge for the agroecosystem. It was reported to cause several hazardous effects such as environmental pollution, contamination of water bodies, loss in soil fertility and beneficial soil microflora, toxicity to aquatic flora and fauna, etc.

Considering all these concerns, the development of novel, cost-effective and sustainable approaches for the detection and management of plant pathogens is highly necessary. In this context, it is believed that nanotechnology can play a very crucial role in sustainability because now days several nanomaterials have been promisingly used in the development of nanofertilizers, nanopesticides, nanoinsecticides, etc. In addition, considering the emerging role of nanotechnology particularly, nano-based sensors and probes in precision farming, it is believed that nano-based sensors can play a key role in the early detection and management of plant diseases. Figure 14.1 showed the possible application of nanomaterials in agriculture or plants. Hence, the main aim of the present chapter is to focus on the role of nanosensors and probes in the rapid and effective detection of plant pathogens and diseases.

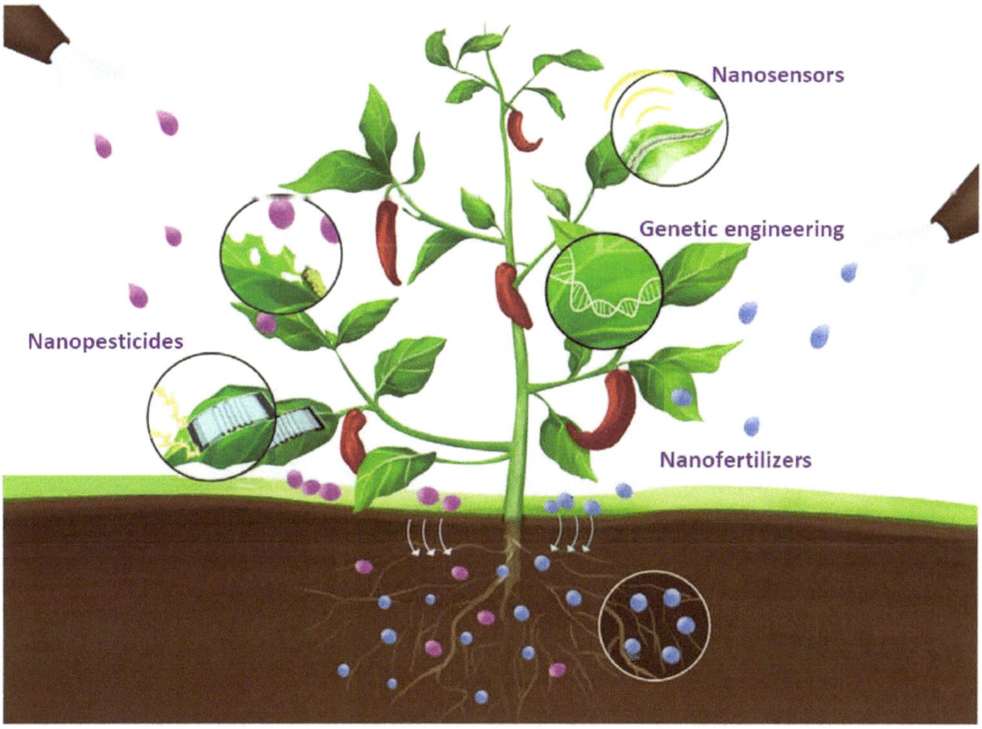

Figure 14.1 Possible applications of nanomaterials in agriculture [Adapted from Zhang et al. 2021, an open access article].

14.2 CONVENTIONAL METHODS FOR DETECTION OF PLANT PATHOGENS

Different conventional methods or approaches have been recommended for the early detection of plant pathogens which mainly include visual inspection, isolation of pathogenic organisms from infected tissue, and culturing it under laboratory conditions followed by its microscopy, serology, and molecular techniques based on polymerase chain reaction (PCR) i.e., quantitative real-time PCR (qPCR) and reverse transcription-PCR (RT-PCR) (Fang and Ramasamy 2015). Similarly, other techniques like DNA fingerprinting, isothermal amplification assays, nucleic acid blot assays, microarrays, and next-generation sequencing can also be used for the detection of a variety of plant pathogens (Buermans and den Dunnen 2014, Krawczyk et al. 2017, Ivanov et al. 2021). Moreover, various serological assays such as immunodiffusion tests, enzyme-linked immunosorbent assay, etc., and lateral flow devices (LF) are developed to detect plant pathogens depending on their specific antigens (Martinelli et al. 2015). It has been observed that monoclonal antibodies-based serological assays are found to have high specificity and are widely used for on-site field-level detection and diagnosis of pathogens from a large number of cereals and horticultural crops (Song et al. 2017). Apart from these techniques, several biosensors have been developed which are commonly used in the detection of plant pathogens. All such biosensors are described in the below section.

14.3 SENSORS IN PLANT PATHOGEN DETECTION AND MANAGEMENT

The most crucial time to identify illnesses is during the early growth phases of plants, and typically this is done by hand inspection. Plant diseases cause various symptoms to appear on leaves, stems, and fruits in plants (Dorthi et al. 2022). These symptoms are very helpful for visual inspection, which is the first step in the diagnosis of plant diseases, but they are ineffective for identifying the pathogen's presence in the early stages of infection when infections in the plant are asymptomatic (Mastin et al. 2022). Numerous methods have been employed in conjunction with a visual examination of plant symptoms in the field to diagnose plant disease issues, such as nucleic acid-based techniques and immunoassays for the determination of infectious protein and nucleic acid retrieved from diseased plant materials (Faizal Azizi and Lau 2022). The nucleic acid-based detection assays include numerous PCR-based and RNA-based technologies, as well as DNA-based approaches like fluorescence in-situ hybridization (FISH) and loop-mediated isothermal amplification (LAMP). These methods enable rapid and precise identification and quantification by overcoming inconsistent diagnoses or pathogen classifications (Martinelli et al. 2015, López-Goñi et al. 2008). As an alternative, immunoassays, commonly referred to as serological assays, have been used to detect pathogen antigens. These include the enzyme-linked immunosorbent assay (ELISA), lateral flow devices (LF), tissue print ELISA, and direct dot blot immunoassay (DTBIA) (Ali et al. 2022).

Traditional methods for pathogen detection (such as the nucleic acid-based detection method) often call for a research lab and skilled technician, and the process might take many hours or days. A sensing tool known as a biosensor possesses all of the additional qualities that traditional detection methods lack (Mohankumar et al. 2021). Consequently, employing biosensors to find biological indicators may be a good idea. A range of industries, particularly the food industry, agriculture, medicine, and biological research, have evolved as a result of recent advances in biosensor technology. As a result, the need for an in-field diagnostic procedure that is quicker, more dependable, exceptionally sensitive, and accurate has increased. Such "bedside-testing" might be created by utilizing the primary properties of biology, chemistry, and optics. This method adds certain benefits to bioassays by enabling quick reaction times without requiring the user to evaluate the data and low-cost on-site tests (Cassedy et al. 2020).

Several research disciplines, such as environmental monitoring, the monitoring of airborne pathogens, the detection of agrochemicals in foods and beverages, the real-time identification of biological components and biosensors for plant pathogens detection, have emerged as cutting-edge detection techniques (Liu et al. 2018). When a particular analyte of concern or pathogen comes into contact with a biosensor, then biosensors as a diagnostic tool, typically merge a biological sensing component and a physicochemical transducer to produce an electrical signal. A transducer then transforms a biomolecular connection into an output that is digital (Elmer and White 2018). Bioreceptors can be biological components like antibodies, DNA, enzymes, tissue types, complete cells, etc. The selective character of the biochemical interactions between these bioreceptors and the biosensor is what allows them to offer the biosensor with recognition specificity (Hui et al. 2022). A biosensor can be categorized as an electrochemical, thermal, optical, or piezoelectric biosensor depending on the type of transducer it uses (Kulkarni et al. 2022).

14.4 ELECTROCHEMICAL BIOSENSOR

The essential parts of an electrochemical biosensor are a molecular recognition layer as well as an electrochemical transducer, which transforms biological data resulting from a binding event into an electrical signal that is then shown on a readout device (Bakirhan et al. 2022). This category of biosensor may identify target diseases on plants in a range of locations, including greenhouses, fields, and post-harvest storage containers (Fang and Ramasamy 2015). DNA and Plant antibodies are used in point-of-care tests to identify plant infections because they are more favourable than other biological sensing elements that can be connected to a transducer (Ali et al. 2022). The majority of DNA-based electrochemical biosensors for detecting plant pathogens may use tag-dependent voltametric DNA hybridization detection (Patiti et al. 2022). The biomolecular connection between analyte and antibody that forms on the transducer surface is the underlying idea behind antibody-based immunosensors (Dyussembayev et al. 2021).

14.5 OPTICAL BIOSENSORS

A source of light, an optical carrier medium, an immobilized biorecognition component, and a signal recognizer are used in optical biosensors to quantify the association between the target analyte and ligand (Kumari and Saini 2022). The most commonly used optical biosensors for plant pathogen detection include colorimetric biosensors, surface plasmon resonance-based biosensors, and assays based on fluorescence. (Kulabhusan et al. 2022). The most widely used technologies for allowing the individual to immediately detect pathogens in a small number of samples merely within 10–15 min via a colour change are probably colorimetric biosensors. DNA and antibodies are used as bio-recognition elements in optical biosensors to identify plant pathogens (Dyussembayev et al. 2021). In crops like potato, maize, orchid, watermelon, and datura antibody-based optical biosensors are developed adopting the techniques like lateral flow immunoassay, surface plasmon resonance immunoassay, and microsphere immunoassay (Drygin et al. 2011, Charlermroj et al. 2013, Lin et al. 2014, Zhang et al. 2014, Feng et al. 2015). DNA-based optical biosensors were found to be effective in crops like banana, watermelon, citrulline, potato, banana, and cucumber when developed by adopting bridging flocculation, electrochemiluminescence-based DNA analysis, microfluidic microarray assembly method, and lateral flow assays (Tang et al. 2007, Wang and Li 2007, Zhao et al. 2011, Wei et al. 2014, Wee et al. 2015, Zhan et al. 2018).

The method by which the biological component, or bioreceptor layer, is bonded to the transducer is a key factor in the success of biosensors. The major objective is to create a strong bond between both the biomolecule and the sensing surface while retaining the stability of the biotic components (Zuo et al. 2015, Huang et al. 2021). Several essential parameters are being used to evaluate sensors like selectivity, high sensitivity, high resolution, repeatability (measurability), and high precision. Sometimes biosensors lack these features therefore, the use of nanostructures in the construction of biosensors nowadays for a variety of purposes has produced significant advancements in this field. It helps in raising the threshold needed to stabilize biomaterials to improve sensitivity, catalysis, the ability to react at low potentials, and the speed at which electrons are transferred from the catalytic reaction centre to the electrode surface (Huang et al. 2021).

Because of the large surface area of NPs, they act as potent transducers in biosensors. So, combining Nanomaterials (NMs) with electrical systems may lead to electrical transduction processes in smart nanoelectromechanical (NEMS) devices. Therefore, sensors based on NMs have become a potentially helpful analytical technique for plant pathogen detection because of their quick detection time, ease of use, and high sensitivity (Sharma et al. 2021).

14.6 NANOBIOSENSORS AND THEIR TYPES

14.6.1 Nanobiosensors

A system that uses at least one nanostructure to sense biological materials is referred to as a nano-biosensor. Usually, a typical nano-biosensors has three major components which mainly include recognition element, signal transducer and nanomaterial (Ku and Chung 2023). Figure 14.2 shows the schematic representation of different components of nanobiosensors. The goal of incorporating nanomaterials into biosensor structures is to raise the level necessary to stabilize biomaterials, which will increase sensitivity, catalyse the reaction, enable reactions to occur at low potentials, and speed up the exchange of electric charge from the active reaction centre to the surface of the electrode (Rai et al. 2020, Huang et al. 2021). By removing chemical intermediaries of electron transfer, the use of nanoparticles in the construction of biosensors can simplify the devices, which is crucial for the creation of third-generation biosensors (Li et al. 2011, Wang et al. 2013, Guo et al. 2014, Sun et al. 2019, Zhang and Wang 2019). Nanoparticles, nanotubes, nanowires, nanopores, self-adhesive monolayers, and nanocomposites are examples of nanostructures, and they share common properties with biological molecules like proteins and DNA (Xiong et al. 2020, Xue et al. 2021). Consequently, the interaction between biomaterials and nanostructures produces a biomaterial-nanostructure interaction with favourable characteristics and performance.

In the construction of electrochemical biosensors, nanoparticles play major roles (Kianfar et al. 2016). Mainly five categories are used to group these roles i.e. in the creation of sensors, nanostructures are employed for receiver stabilization, catalysis, enhanced electron transfer, labelling, and involvement in detecting reactions (Huang et al. 2021). Nanoparticles are capable of effectively adhering biomolecules to their surface and play a significant role in stabilizing biomolecules at the biosensor surface because of their highly effective surface area and free surface energy. Direct biomolecule adsorption on a bulk material's bare surface typically results in denatured proteins and loss of biological activity. The adhesion of the biomolecules on nanoparticle surfaces can retain their biological activity. This process is triggered by nanoparticles biocompatible nature (Kianfar et al. 2016). Because they have high surface energy, nanoparticles are often more active compared to bulk materials. Most nanoparticles, particularly metal nanoparticles, have substantial catalytic effects and are employed as catalysts in most processes due to their high surface activity (Liu et al. 2017, Wang et al. 2018a). Also, pesticides, fertilizers, and many other agronomic parameters are identified and controlled by nanosensors or nanobiosensors, which provide timely information for accurate decision-making and agricultural management (Duhan et al. 2017, Bhagat et al. 2023).

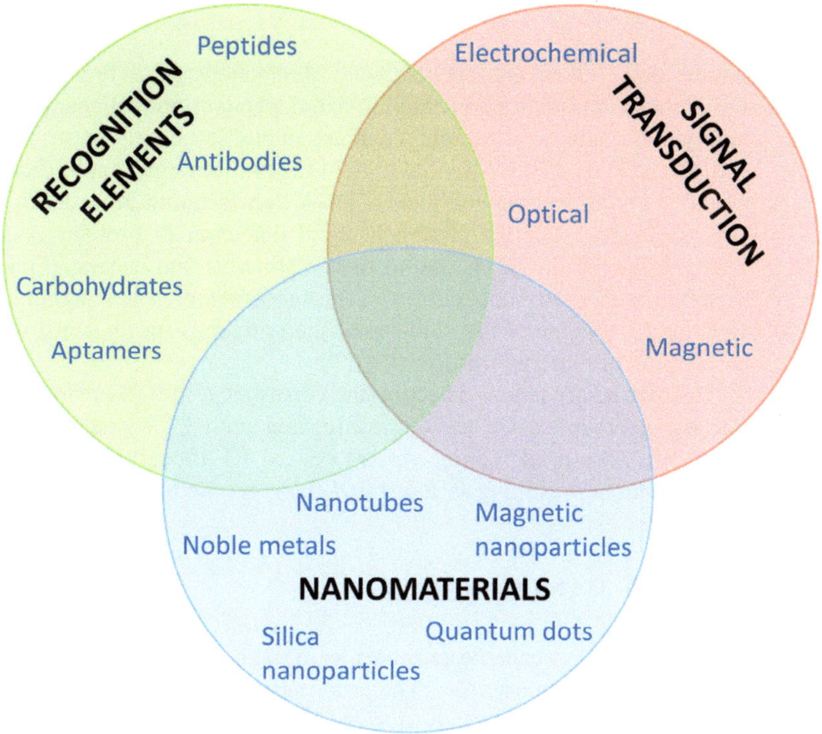

Figure 14.2 Schematic representation of nanobiosensors and its components [Adapted from Vikesland and Wigginton (2010) with copyright permission].

14.7 TYPES OF NANOBIOSENSORS

Recently, various biosensors based on nanoparticles have been developed and made available for commercial use in agriculture (Kundu et al. 2019).

14.8 METAL OR METALLOID NANOPARTICLES

A common category of nanosensors uses inorganic nanomaterials like Ag, Au, Si, and different metal oxide nanoparticles to detect and measure bioactive molecules of interest, using recognition ligands like antibodies or DNA oligonucleotides (Wu and Lin 2013, Darr et al. 2017). These nanoparticles can be utilized as special sensing platforms to examine interactions in vitro or in-vivo between both the nanomaterials and bioanalytes that are specifically relevant to plant diseases (Kwak et al. 2017). Matrix-assisted laser desorption ionization mass spectrometry (MALDI-MS) was considerably improved by the introduction of Pt nanosensors complexed with specific antibodies (IgG) to identify plant-associated microbes in soil and carrots (Ahmad et al. 2012). Surface-enhanced Raman spectroscopy has frequently been used to identify various phytopathogens or synthesized toxins utilizing Ag nanorods as sensor substrates (SERS) (Wu et al. 2012, Yüksel et al. 2015, Sanchez et al. 2019). A soil-borne species called *R. solanacearum* causes potato bacterial wilt, researchers utilized gold nanoparticles complexed with a particular single-stranded DNA to detect *R. solanacearum* genomic DNA at concentrations as low as 15 ng (Khaledian et al. 2017).

14.9 QUANTUM DOTS

Semiconductor quantum dots (QDs) are crystals with dimensions of a few nanometers that exhibit exceptional photophysical qualities, including size-dependent optical characteristics, high fluorescence quantum yields, and great stability against photobleaching. These characteristics eventually allow Multiplexed QDs as the potential optical tags for biological applications like cell sorting and tracking, the study of immunocomplexes, DNA hybridization procedures, biomedical imaging, and in vivo diagnostics. For the electrochemical detection of proteins or DNA, QDs can be utilized as labels. (Shao et al. 2011). Due to their distinctive and beneficial photophysical characteristics, QDs have been used successfully as biosensors for crop imaging and disease identification (Wang et al. 2016). One of the well-established methods for the construction of QD sensors is fluorescence resonance energy transfer (FRET).

A sophisticated FRET-based sensor for detecting the *Citrus tristeza* (CTV) virus is developed. In this, as a donor-acceptor complex, CdTe QDs conjugated with CTV coat protein (CP) and CP-labelled rhodamine dye were used. Target viruses caused CP-rhodamine to be replaced by free CP instead of CP-rhodamine, resulting in a restored fluorescence of QDs (Rafiq et al. 2022).

14.10 ARRAY-BASED NANOSENSORS

Another form of nanosensors that is capable of multiplexing and discriminating between distinct analytes are termed as array-based nanosensors. The array-based sensor assemblies, are structured similarly to the mammalian olfactory system and comprise many chromophores or artificial nanomaterials. The array-based technique is especially useful in recognizing differences between highly resembling analyte mixtures because of its cross-reactivity and nature to assess molecular fingerprints (Li et al. 2020). The analysis of the volatile chemical fingerprint of damaged plants is a crucial tool for detecting plant diseases (Cellini et al. 2017).

Recently a smartphone-integrated volatile organic compound (VOC) sensing platform made use of "plasmonic nanoparticles" for early diagnosis and to create the distinction between tomato late blight from other fungal infections that cause similar symptoms in tomato leaf (within 2 days of infection). To produce distinctive patterns of colour changes for each VOC, the gold nanoparticles have been functionalized with chemical ligands reactive to specific leafy VOCs but also printed on paper strips. After that, a smartphone microscope can be used to scan the data for digital pattern recognition and discrimination (Li et al. 2019).

14.11 FLUORESCENCE RESONANCE ENERGY TRANSFER (FRET) NANOSENSORS

In the fluorescence resonance energy transfer (FRET), an excited state donor 'D' (often a fluorophore) transmits energy to a nearby ground state acceptor 'A' through long-range dipole-dipole interactions (Sapsford et al. 2006). The acceptor must absorb the energy at the wavelength of the donor's emission, although it need not necessarily transfer the energy fluorescently. The FRET-based sensors have utilized a variety of chemical substances and nanoparticles as acceptors, including rhodamine, TAMRA, gold nanoparticles, graphene oxide nanoparticles, and carbon nanotubes (Shojaei et al. 2014, Taghdisi et al. 2014, Kalarestaghi et al. 2015).

Gold nanoparticles (AuNPs) serve as a decent basis for the FRET system. The distinctive characteristics of AuNPs include calorimetry, they have nonlinear optical characteristics, conductivity, and good building blocks for the creation of innovative bio-sensors. When it comes to the specific binding and identification of molecules and biological targets, AuNPs provide an ideal platform with a wide variety of chemical or biological fluorescent ligands. (Daniel and Astruc 2004). Nanomaterial surface energy transfer (NSET), which uses a gold metallic nanoparticle as an acceptor for energy transfer, differs from FRET in two key ways. Firstly, the measuring range is increased by the distance-dependent variations. Secondly, the single nanoparticle may quench dyes with a range of emission frequencies, from visible to near-infrared. NSET is therefore appropriate for studies where lengths above 10 nm are anticipated or if quenching of numerous dyes is required (Reinhard et al. 2005, Yun et al. 2005).

Plant pathology can benefit greatly from the detection of certain pathogens' DNA sequences. By using molecular beacons that have been tagged with AuNPs, hairpin FRET-based techniques for detecting DNA have been developed. The organic dye and nucleic acid sensor together produce a self-complementary hairpin structure on AuNPs, effectively quenching FRET fluorescence. Upon complementary hybridization only with target DNA, the hairpin shape transforms into a rod-like structure, increasing the dye's fluorescence (Dubertret et al. 2001, Ray et al. 2006).

14.12 SURFACE-ENHANCED RAMAN SCATTERING (SERS) NANOSENSORS

An optical approach called Surface-Enhanced Raman spectroscopy (SERS) combines traditional Raman spectroscopy with nanotechnology to provide an ultrasensitive, highly focused analytical tool (Schlücker 2014, Langer et al. 2020). SERS-based nanosensors offer the potential to mitigate the drawbacks of traditional approaches, especially those that integrate multifunctional features (Calderon et al. 2021).

SERS detection is combined with non-enzymatic DNA strand-displacement hybridization for the identification of specific nucleic acid targets, in a quick and selective "switch on" plasmonics-based nano biosensor. This SERS nano-biosensor offers a simple and effective tool for nucleic acid diagnostics. It is a target-label-free, quick nanoparticle-based biosensing system which uses a homogenous assay format (Wang et al. 2015). SERS has been often used for the broad detection and identification of bioactive compounds in plants (Du et al. 2018). It is practically used in food safety diagnostics, such as for pesticide detection using SERS Nanosensor based upon gold (Au) nanoparticles (Zhang et al. 2018, Li et al. 2022), as well as for plant hormones identification with the help of SERS nanosensor based upon Au colloidal nanoparticles and plasmonic silver nanoaggregates (Chen et al. 2017, Wang et al. 2017)

14.13 ELECTROCHEMICAL NANOSENSORS

Typically, electrochemical nanosensors have a working electrode, a counter electrode, and a reference electrode. This method has been demonstrated to be helpful in the qualitative research of plant pathogen sensing (Yotava and Yaneva 2013). Nanomaterial-designed electrodes have a comparatively large active surface area, which increases sensitivity (Shaw and Honeychurch 2022). Recent developments with electrochemical nanosensors are possible primarily due to metallic nanoparticles such as gold (Pingarrón et al. 2008, Yotava and Yaneva 2013, Shaw and Honeychurch 2022) and nanocarbon materials like carbon nanotubes and graphene-based materials due to their special electronic configuration (Pingarrón et al. 2008, Shaw and Honeychurch 2022).

Salicylic acid levels in oilseeds that have been infected with the fungus *Sclerotinia sclerotiorum* can be identified most precisely using an Au nanoelectrode complexed with Cu nanoparticles (Wang et al. 2010, Scott and Chen 2012, Singh et al. 2012, Dubey and Mailapalli 2016). Plant viruses such as *Pseudomonas syringae* can be determined by adopting the differential pulse voltammetry method of detection with electrochemical nanosensors based on gold nanoparticles (Lau et al. 2017).

14.14 PIEZOELECTRIC NANOSENSOR (PZN)

A piezoelectric sensor is a device that converts variations in pressure, acceleration, temperature, strain, or force into an electrical charge via the piezoelectric effect (Shaw and Honeychurch 2022). A pressure piezoelectric sensor with a nanometre size that can measure forces down to the nanonewton range has already been demonstrated (Wang et al. 2006).

A DNA-RNA hybridization-based piezoelectric DNA-sensor for *Cymbidium mosaic* virus (CymMV) and *Odontoglossum ringspot virus* (ORSV) is discovered for orchids. On a quartz crystal microbalance (QCM), specific oligonucleotide probes (having a mercaptohexyl group at the 5′-phosphate ends) are directly immobilized. These oligonucleotide probes are coated upon QCMs, which were then subjected to screen solutions containing viral RNA for hybridization. Both *CymMV* and *ORSV* were detectable by the appropriate nucleotide probe-coated QCM-based DNA sensors in concentrations as low as nearly 1 ng in purified RNA samples and 10 ng in the crude sap of diseased orchids (Eun et al. 2002).

Piezoelectric coupling sensors based on molecularly imprinted polymers for the quick and non-destructive detection of infested citrus with *Bactrocera dorsalis* are developed. The chief volatile gas in citrus 'd-limonene', changes significantly when the fruit becomes infected, making it easier to identify pest-infested citrus using information obtained from the gas. The one-month frequency drop variation was less than 5%, and the sensor's reproducibility and repeatability were both greater than 95%. The outcomes demonstrated the MIPs-QCM sensor's capability for quick, non-destructive detection of infested citrus (Wen et al. 2022).

14.15 ROLE OF NANOPARTICLE-BASED SENSORS IN PLANT DISEASE DETECTIONS

Plant diseases and their detection or identification are currently one of the serious global challenges which destroy the economy of farmers as well as industries depending on plant-based products (Kulabhusan et al. 2022, Giraldo and Kruss 2023). The accurate detection and diagnosis is an essential step for proper disease management and it helps execute a timely application of pesticides in the diseased field for proper disease management. Recognition of microscopic organisms like bacteria, fungi and viruses is the detection of a plant pathogen in quick reaction settings at the micro-propagation stage, nurseries and natural landscapes in infected plant tissues. Early detection is necessary and allows a farmer to take proper measurements to save the crop from complete failure. Biosensors technology play important role in plant disease detection and it has advantages like lessening the investigation time, sensitivity through automation as well as integration of multiple processes in a single piece of equipment (Patel 2021). Plants can emit an extensive set of chemical compounds which is a source of communicating vital information about the surrounding environment. A wireless network of nanosensors with numerous nanosensors inserted on or within plant leaves could detect the pathogens at early stages before morphological appearance which causes disease in the plant. The presence of crop diseases can be detected by using nanosensors due to their high sensitivity and shorter detection time (Kwak et al. 2017).

Bacterial, viral, and fungal infections cause plant diseases that affect production and economic losses in agriculture. There are lots of conventional methods available for the detection of plant diseases such as polymerase chain reaction (PCR), enzyme-linked immunosorbent assay (ELISA), immunofluorescence, and flow cytometry. But these conventional methods have some limitations like being time-consuming, laborious, complex and lack the necessary levels of detectability and specificity towards the target (Siddiquee et al. 2014). Therefore, new, rapid, sensitive and selective detection techniques are necessary to detect and control plant diseases. Thus, nanosensors have huge potential for the generation of multiple new methods to suppress diseases at early stages hence and enhancing the diagnosis of diseases (Yusof and Isha 2020).

A combination of portable imaging devices such as smartphones and nanostructure-supported non-invasive detection tools play important role in on-site and fast detection or diagnosis of plant diseases and long-term preservation of plant health conditions. Utilization of imaging techniques, such as hyperspectral imaging and thermographic imaging technique are helpful in the field for indirect detection of plant diseases (Li et al. 2020). Zhan et al. (2018) developed a nano biosensor by using a gold nanoparticle probe (AuNPs), target DNA, and capture probe. Here, sandwich type of hybridization reactions occurred in the presence of target DNA and it produces a characteristic red band, which indicates the presence of disease. This method is used for the early prediction of potato late blight disease and to investigate the management actions to reduce the risk of the development of an epidemic.

Li et al. (2019) developed a cost-effective smartphone-based volatile organic compound (VOC) fingerprinting platform which can be used for non-invasive detection of late blight caused due to *Phytophthora infestans* by monitoring leaf volatile emission in the field. Handheld optical scanning platform developed for digital quantification of relevant plant volatiles by integrating with disposable VOC sensor array with the smartphone camera module. This biosensor detects key plant volatiles at ppm level within 1 min of reaction. The novel colorimetric nano biosensor was developed by using localized surface plasmon resonance (LSPR) of AuNPs. The unamplified tomato yellow leaf curl virus (TYLCV) genome in infected plants was detected by using unmodified gold nanoparticles on a nanosensor. The presence of target virus infections was detected by a change in colour in samples. The colour change appears visually after the addition of salt and is confirmed through UV-Vis spectroscopy. This nanosensor was found to be very sensitive and efficient in the detection of TYLCV (Razmi et al. 2019).

The ATR-LSPR (attenuated total reflection-localized surface plasmon refractive) optical biosensor is used to detect chili leaf curl virus (ChiLCV). It requires a very small volume (20–40 μL) of a sample. In this biosensor, a complementary ssDNA sequence is used as a receptor (Das et al. 2021). *Pseudocercospora fijiensis* causes black Sigatoka in the banana plantation which results in a significant loss in both yield and product quality. The detection of *P. fijiensis* in leaf extract at an early stage is possible through SPR immunosensor (Luna-moreno et al. 2019). Razo et al. (2019) used AuNPs to fabricate lateral flow test strips for the diagnosis of *Ralstonia solanacearum* which causes potato brown rot. Due to the reduction of tetrachloroauric anion on the surface of AuNPs, the signal generated in the test strips test zone and the size of AuNPs increases approximately up to 100 nm. This technique is not complicated and it requires very less time about 3 min. The limit of detection (LOD) of *R. solanacearum* was 3×10^1 cells/mL and it allows detection of latent bacterial infection in potato tubers.

The nanoparticle-based electrochemical biosensor was 10,000 times more sensitive as compared to a conventional polymerase chain reaction (PCR)/gel electrophoresis. This biosensor can detect *Pseudomonas syringe* infected plant samples even before the visibility of disease symptoms. For sensitive and rapid identification of plant pathogen DNA on disposable screen-printed carbon electrodes, a nanoparticle-based electrochemical biosensor was developed. Rapid isothermal amplification of target pathogen DNA sequences through recombinase polymerase amplification (RPA) followed by AuNPs-based electrochemical assessment with differential pulse voltammetry (DPV) requires 60 min. Concerning sensitivity, portability, speed as well as simplicity this biosensor has the potential for a solution to rapid management of diseases for various applications in agriculture diagnostics (Lau et al. 2017).

Uda et al. (2020) developed Harumanis mango in the Perlis State of Malaysia which is seasonal. When plant diseases became a serious problem, it significantly disturbs not only the quality but also the quantity of production of Harumanis mango. The symptoms of Harumanis mango disease appear at a later stage, which becomes difficult to control or take any further action for its management or prevention. Interdigitated electrodes (IDE) nano biosensor plays a very important role in the detection of Harumanis mango disease as early as possible. Utilization of a label-free and simple sensing platform with IDE was helpful in the detection of any pathogen obtained in Harumanis disease. Pre-warning signals by using such promising tools are not only beneficial to farmers but also to researchers which protect the plant from harmful pathogens and diseases before getting worst.

14.16 ROLE OF NANOPARTICLE-BASED SENSORS IN PLANT DISEASE MANAGEMENT

For disease management or control as well as plant protection, simultaneous detection of all present phyto-pathogenic microbes with high accuracy as well as quick detection is essential in all areas related to agriculture and environmental safety. Effective mitigation, as well as management measures of plant diseases, depends on their not only early but also accurate detection of disease-causing agents, which is mainly initiated by initial detection and recognition of symptoms in the field before the appearance and severe attack on plants. Management of plant diseases in various conditions like in the greenhouse, field and at the country borders plant pathogen identification is necessary as the first step in the management of diseases (Khater et al. 2017). Early detection of plant pathogens plays an important role in the management and prevention of the spread of diseases. Nanotechnology-based biosensors are beneficial in crop disease management which is helpful for the improvement of productivity as well as the quality of the crop (Amjad and Serajuddin 2022).

The presence of fungus in oranges is identified or detected by a new whole cell-based biosensor. A whole cell-based biosensor was developed to manage or monitor the citrus field. Usually, in citrus post-harvest losses are predominant and these are mostly caused due to pathogen such as *Penicillium digitatum*. To manage or monitor rot the whole cell biosensor acts as a new tool. This technology is helpful in the reduction of food losses by enabling more effective crop management throughout postharvest treatment, storage, and transport. The detection of activity of fungus in the oranges was after the third day of infection, before the appearance of visible symptoms or signs on the fruit surface. This biosensor plays important role in the efficient management of postharvest orange (Chalupowicz et al. 2020). *Phytophthora palmivora* is a notorious pathogen that causes severe loss in cacao plants. For electrochemical detection and management of *P. palmivora,* a nano biosensor was developed (Franco et al. 2019).

With the help of hyperspectral imaging technology, a new OR-AC-GAN model is developed to solve the problems of disease control by early detection of disease in plants. Hyperspectral imaging is a powerful tool in monitoring diseases by following early-stage detection. This model is used to detect tomato spotted wilt virus (TSWV) which ultimately results in controlling and spread of such diseases (Wang et al. 2018b). One of the most dangerous plant bacteria worldwide is *Xylella fastidiosa* which causes several diseases resulting in a negative impact on the environment as well as agriculture. It can infect more than 350 species worldwide and its early detection is difficult to eliminate or control. So, the first detection is more important for managing diseases. The airborne imaging spectroscopy and thermography were used to diagnose *X. fastidiosa* infection in olive trees before occurring of symptoms on trees (Zarco-Tejada et al. 2018).

The false smut of rice was caused due to plant pathogen *Ustilaginoidea virens* and for its detection paper-based sensor device was developed and oxidized graphene has been used to improve efficiency and detection limit. False smut is an important fungal disease that significantly decreases rice production. This oxidized graphene-based geno biosensor is sensitive, simple and selective for the detection of false smut of rice. The cyclic voltammetry and linear sweep voltammetry can be used to measure the hybridization of target ssDNA and probe ssDNA. The limit of detection of the developed biosensor was 10 fM of target ssDNA. This technique was cost-effective as well as efficient. If once disease and disease-causing pathogens are identified or detected then they can be cured by either using of NPs, nano pesticides, or nanoinsecticides with the help of a nano biosensor (Rana et al. 2021).

Zulkifli et al. (2016) demonstrated that cucumber mosaic virus (CMV) can be detected and managed through a direct self-assembled monolayer (SAM) DNA sensor via thiol SAM which is a very promising tool. This electrochemical DNA sensor using a screen-printed gold electrode was developed for the diagnosis of the cucumber mosaic virus. The hybridization reaction of DNA which happens on the surface of the electrode was then measured by using cyclic voltammetric analysis.

14.17 CONCLUSIONS

Over the last few years, nanotechnology has demonstrated a great potential to revolutionize the entire scenario of varied fields including agriculture. To date, several path-breaking discoveries and innovations have been done that not only enhance crop productivity in a sustainable way but also effectively help the detection and management of plant pathogens/diseases. In this line, the development of nanobiosensors for plant disease monitoring and management is a major breakthrough. Nanobiosensors could possibly play a vital role in sustainable agriculture for the effective detection of plant pathogens, in addition to the detection of soil condition parameters like nutrient level, organic resources, soil moisture, pH, etc. Nanobiosensors can serve as next-generation biosensors and their integration with the GPS system may contribute to precision farming and smart agriculture.

References

Abass, A.B., Ndunguru, G., Mamiro, P., Alenkhe, B., Mlingi, N. and Bekunda, M. 2014. Post-harvest food losses in a maize-based farming system of semi-arid savannah area of Tanzania. J. Stored Prod. Res. 57: 49–57.

Ahmad, F., Siddiqui, M.A., Babalola, O.O. and Wu, H.F. 2012. Biofunctionalization of nanoparticle assisted mass spectrometry as biosensors for rapid detection of plant associated bacteria. Biosen. Bioelect. 35(1): 235–242.

Ali, Q., Zheng, H., Rao, M.J., Ali, M., Hussain, A., Saleem, M.H., Nehela, Y., et al. 2022. Advances, limitations, and prospects of biosensing technology for detecting phytopathogenic bacteria. Chemosphere. 296: 133773.

Amjad, S. and Serajuddin, M. 2022. Role of nanosensors and bionanosensors in crop abiotic stress. pp. 1–12. *In:* Ansari, S.A., Ansari, M.I. and Husen, A. (eds). Augmenting Crop Productivity in Stress Environment. Springer. Singapore.

Bakirhan, N.K., Topal, B.D., Ozcelikay, G., Karadurmus, L. and Ozkan, S.A. 2022. Current advances in electrochemical biosensors and nanobiosensors. Crit. Rev. Anal. Chem. 52(3): 519–534.

Bhagat, R., Ingle, A.P. and Chen, H. 2023. Nanosensors and nanobiosensors for sustainable agriculture. pp. 93–112. *In:* Ingle, A.P. (ed.). Nanotechnology in Agriculture and Agroecosystems. Elsevier. UK.

Buermans, H.P.J. and den Dunnen J.T. 2014. Next generation sequencing technology: advances and applications. Biochim. Biophys. Acta, Mol. Basis Dis. 1842(10): 1932–1941.

Calderon, I., Guerrini, L. and Alvarez-Puebla, R.A. 2021. Targets and tools: nucleic acids for surface-enhanced raman spectroscopy. Biosensors. 11(7): 230.

Cassedy, A., Mullins, E. and O'Kennedy, R. 2020. Sowing seeds for the future: the need for on-site plant diagnostics. Biotechnol. Adv. 39: 107358.

Cellini, A., Blasioli, S., Biondi, E., Bertaccini, A., Braschi, I. and Spinelli, F. 2017. Potential applications and limitations of electronic nose devices for plant disease diagnosis. Sensors. 17(11): 2596.

Chalupowicz, D., Veltman, B., Droby, S. and Eltzov, E. 2020. Evaluating the use of biosensors for monitoring of *Penicillium digitatum* infection in citrus fruit. Sen. Actuators B: Chem. 311: 127896. https://doi.org/10.1016/j.snb.2020.127896.

Charlermroj, R., Himananto, O., Seepiban, C., Kumpoosiri, M., Warin, N., Oplatowska, M., Gajanandana, O., et al. 2013. Multiplex detection of plant pathogens using a microsphere immunoassay technology. Plos One. 8(4): e62344.

Chen, M., Zhang, Z., Liu, M., Qiu, C., Yang, H. and Chen, X. 2017. In situ fabrication of label-free optical sensing paper strips for the rapid surface-enhanced Raman scattering (SERS) detection of brassinosteroids in plant tissues. Talanta. 165: 313–320.

Daniel, M.C. and Astruc, D. 2004. Gold nanoparticles: assembly, supramolecular chemistry, quantum-size-related properties, and applications toward biology, catalysis, and nanotechnology. Chem. Rev. 104(1): 293–346.

Darr, J.A., Zhang, J., Makwana, N.M. and Weng, X. 2017. Continuous hydrothermal synthesis of inorganic nanoparticles: applications and future directions. Chem. Rev. 117(17): 11125–11238.

Das, S., Agarwal, D.K., Mandal, B., Rao, V.R. and Kundu, T. 2021. Detection of the chili leaf curl virus using an attenuated total reflection-mediated localized surface-plasmon-resonance-based optical platform. ACS Omega. 6(27): 17413–17423.

Dorthi, K., Reddy, S.N. and Ramya, G. 2022. Plant disease prediction using image processing and deep convolutional neural network. pp. 455–462. *In:* Bhateja, V., Khin Wee, L., Lin, J.CW., Satapathy, S.C. and Rajesh, T.M. (eds). Data Engineering and Intelligent Computing. Lecture Notes in Networks and Systems, vol 446. Springer, Singapore.

Drygin, Y.F., Blintsov, A.N., Grigorenko, V.G., Andreeva, I.P., Osipov, A.P., Varitzev, Y.A., Uskov, A.I., et al. 2011. Highly sensitive field test lateral flow immunodiagnostics of PVX infection. Appl. Microbiol. Biotechnol. 93(1): 179–189.

Du, S., Yu, C., Tang, L. and Lu, L. 2018. Applications of SERS in the detection of stress-related substances. Nanomaterials. 8(10): 757.

Dubertret, B., Calame, M. and Libchaber, A.J. 2001. Single-mismatch detection using gold-quenched fluorescent oligonucleotides. Nat. Biotechnol. 19(4): 365–370.

Dubey, A. and Mailapalli, D.R. 2016. Nanofertilisers, nanopesticides, nanosensors of pest and nanotoxicity in agriculture. Sustain. Agri. Rev. 19: 307–330.

Duhan, J.S., Kumar, R., Kumar, N., Kaur, P., Nehra, K. and Duhan, S. 2017. Nanotechnology: the new perspective in precision agriculture. Biotechnol. Rep. 15: 11–23.

Dyussembayev, K., Sambasivam, P., Bar, I., Brownlie, J.C., Shiddiky, M.J. and Ford, R. 2021. Biosensor technologies for early detection and quantification of plant pathogens. Front. Chem. 9: 636245. doi: 10.3389/fchem.2021.636245.

Elmer, W. and White, J.C. 2018. The future of nanotechnology in plant pathology. Ann. Rev. Phytopathol. 56: 111–133.

Eun, A.J.C., Huang, L., Chew, F.T., Li, S.F.Y. and Wong, S.M. 2002. Detection of two orchid viruses using quartz crystal microbalance-based DNA biosensors. Phytopathology. 92(6): 654–658.

Faizal Azizi, M.M. and Lau, H.Y. 2022. Advanced diagnostic approaches developed for the global menace of rice diseases: a review. Canadian J. Plant Pathol. doi: 10.1080/07060661.2022.2053588.

Fang, Y. and Ramasamy, R.P. 2015. Current and prospective methods for plant disease detection. Biosensors. 5(3): 537–561.

Feng, M., Kong, D., Wang, W., Liu, L., Song, S. and Xu, C. 2015. Development of an immunochromatographic strip for rapid detection of *Pantoeastewartii* subsp. *stewartii.* Sensors. 15(2): 4291–4301.

Franco, A.J.D., Merca, F.E., Rodriguez, M.S., Balidion, J.F., Migo, V.P., Amalin, D.M., Alocilja, E.C., et al. 2019. DNA-based electrochemical nano biosensor for the detection of *Phytophthora palmivora* (Butler) Butler, causing black pod rot in cacao (*Theobroma cacao* L.) pods. Physiol. and Mol. Plant. Path. 10(7): 14–20.

Giraldo, J.P. and Kruss, S. 2023. Nanosensors for monitoring plant health. Nat. Nanotechnol. 18: 107–108.

Guo, C., Huo, H., Han, X., Xu, C. and Li, H. 2014. Ni/CdS bifunctional Ti@TiO_2 core-shell nanowire electrode for high-performance nonenzymatic glucose sensing. Anal. Chem. 86(1): 876–883.

Huang, X., Zhu, Y. and Kianfar, E. 2021. Nanobiosensors: properties, applications, and electrochemical techniques. J. Mater. Res. Technol. 12: 1649–1672.

Hui, Y., Huang, Z., Alahi, M., Nag, A., Feng, S. and Mukhopadhyay, S.C. 2022. Recent advancements in electrochemical biosensors for monitoring the water quality. Biosensors. 12(7): 551.

Ingle, A.P. 2021. Nanotechnology in Plant Growth Promotion and Protection. John Wiley & Sons Ltd. Publisher, UK.

Ingle, A.P., Moharil, M.P., Saxena, S., Patil, V. and Borse, S. 2023. Nanotechnology: the new perspective in agroecosystem and sustainable agriculture. pp. 1–24. *In:* Ingle, A.P. (ed.). Nanotechnology in Agriculture and Agroecosystems. Elsevier. UK.

Ivanov, A.V., Safenkova, I.V., Zherdev, A.V. and Dzantiev, B.B. 2021. The potential use of isothermal amplification assays for in-field diagnostics of plant pathogens. Plants. 10: 2424. https:// doi.org/10.3390/plants10112424.

Kalarestaghi, A., Bayat, M., Hashemi, S.J. and Razavilar, V. 2015. Highly sensitive fret-based fluorescence immunoassay for detecting of aflatoxin B1 using magnetic/silica core-shell as a signal intensifier. Iranian J. Biotechnol. 13(3): 25–31.

Khaledian, S., Nikkhah, M., Shams-bakhsh, M. and Hoseinzadeh, S. 2017. A sensitive biosensor based on gold nanoparticles to detect *Ralstonia solanacearum* in soil. J. Gen. Plant Pathol. 83(4): 231–239.

Khater, M., de la Escosura-Muñiz, A. and Merkoçi, A. 2017. Biosensors for plant pathogen detection. Biosen. Bioelect. 93: 72–86.

Kianfar, M., Kianfar, F. and Kianfar, E. 2016. The effect of nano-composites on the mechanic and morphological characteristics of NBR/PA6 blends. American J. Oil Chem. Technol. 4(1): 29–44.

Krawczyk, K., Uszczyńska-Ratajczak, B., Majewska, A. and Borodynko-Filas, N. 2017. DNA microarray-based detection and identification of bacterial and viral pathogens of maize. J. Plant. Dis. Prot. 124: 577–583.

Ku, C.A and Chung, C.K. 2023. Advances in humidity nanosensors and their application: review. Sensors. 23: 2328. https://doi.org/10.3390/s23042328.

Kulabhusan, P.K., Tripathi, A. and Kant, K. 2022. Gold nanoparticles and plant pathogens: an overview and prospective for biosensing in forestry. Sensors. 22(3): 1–18.

Kulkarni, M.B., Ayachit, N.H. and Aminabhavi, T.M. 2022. Biosensors and microfluidic biosensors: from fabrication to application. Biosensors. 12(7): 543.

Kumari, S. and Saini, S.M. 2022. Optical biosensors for diagnostic applications. pp. 155–193. *In*: Samantara, A.K., Raj, S. and Ratha, S. (eds). Nanomaterials-Based Sensing Platforms: Towards the Efficient Detection of Biomolecules and Gases. Apple Academic Press. https://doi.org/10.1201/9781003199304.

Kundu, M., Krishnan, P., Kotnala, R.K. and Sumana, G. 2019. Recent developments in biosensors to combat agricultural challenges and their prospects. Trends Food Sci. Technol. 88. 157–178.

Kwak, S.Y., Wong, M.H., Lew, T., Bisker, G., Lee, M.A., Kaplan, A., Dong, J., et al. 2017. Nanosensor technology applied to living plant systems. Ann. Rev. Aanal. Chem. 10(1): 113–140.

Langer, J., Jimenez de Aberasturi, D., Aizpurua, J., Alvarez-Puebla, R.A., Auguié, B. and Baumberg, J.J. 2020. Present and future of surface-enhanced Raman scattering. ACS Nano. 14(1): 28–117.

Lau, H.Y., Wu, H., Wee, E.J.H., Trau, M., Wang, Y. and Botella, J.R. 2017. Specific and sensitive isothermal electrochemical biosensor for plant pathogen DNA detection with colloidal gold nanoparticles as probes. Sci. Rep. 7: 38896; doi: 10.1038/srep38896.

Li, Y., Zhao, Y., Cheng, H., Hu, Y., Shi, G., Dai, L., Qu, L., et al. 2011. Nitrogen-doped graphene quantum dots with oxygen-rich functional groups. J. Am. Chem. Soc. 134(1): 15–18.

Li, Z., Paul, R., Tis, T.B., Saville, A.C., Hansel, J.C., Yu, T., Ristaino, J.B. et al. 2019. Non-invasive plant disease diagnostics enabled by smartphone-based fingerprinting of leaf volatiles. Nat. Plants. 5(8): 856–866.

Li, Z., Yu, T., Paul, R., Fan, J., Yang, Y. and Wei, Q. 2020. Agricultural nanodiagnostics for plant diseases: recent advances and challenges. Nanoscale Adv. 2(8): 3083–3094.

Li, H., Merkl, P., Sommertune, J., Thersleff, T. and Sotiriou, G.A. 2022. SERS hotspot engineering by aerosol self-assembly of plasmonic Ag nanoaggregates with tunable interparticle distance. Adv. Sci. 9(22): 2201133.

Lin, H.Y., Huang, C.H., Lu, S.H., Kuo, I.T. and Chau, L.K. 2014. Direct detection of orchid viruses using nanorod-based fibre optic particle plasmon resonance immunosensor. Biosen. Bioelect. 51: 371–378.

Liu, M., Zhao, H., Quan, X., Chen, S. and Fan, X. 2010. Distance-independent quenching of quantum dots by nanoscale-graphene in self-assembled sandwich immunoassay. Chem. Commun. 46(42): 7909–7911.

Liu, G., Ren, G., Zhao, L., Cheng, L., Wang. C. and Sun, B. 2017. Antibacterial activity and mechanism of bifidocin a against *Listeria monocytogenes*. Elsevier. Food. Control. 73: 854–861.

Liu, C., Xu, C., Xue, N., Sun, J.H., Cai, H., Li, T., Liu, Y., et al. 2018. Enzyme biosensors for point-of-care testing. *In:* Yellampalli, S. (ed.). MEMS sensors: design and application. Intech. Open. UK. https://doi.org/10.5772/intechopen.73249.

López-Goñi, I., García-Yoldi, D., Marín, C.M., de Miguel, M.J., Muñoz, P.M., Blasco, J.M., Jacques, I., et al. 2008. Evaluation of a multiplex PCR assay (Bruce-ladder) for molecular typing of all *Brucella* species, including the vaccine strains. J. Clin. Microbiol. 46(10): 3484–3487.

Luna-moreno, D., Araceli, S.Á., Islas-flores, I., Canto-canche, B., Carrillo-pech, M., Villarreal-chiu, J.F. and Rodr, M. 2019. Early detection of the fungal banana black sigatoka pathogen *Pseudocercospora fijiensis* by an SPR immunosensor method. Sensors. 19(3): 1–12.

Martinelli, F., Scalenghe, R., Davino, S., Panno, S., Scuderi, G., Ruisi, P., Villa, P., et al. 2015. Advanced methods of plant disease detection: a review. Agron. Sustain. Dev. 35: 1–25.

Mastin, A.J., Van. Den. Bosch, F., Bourhis, Y. and Parnell, S. 2022. Epidemiologically-based strategies for the detection of emerging plant pathogens. Sci. Rep. 12(1): 1–13.

Mohankumar, P., Ajayan, J., Mohanraj, T. and Yasodharan, R. 2021. Recent developments in biosensors for healthcare and biomedical applications: a review. Measurement. 167: 108293.

Patel, P. 2021. A review on plant disease diagnosis through biosensor. Int. J. Biosen. Boelect. 7(2): 50–62.

Patiti, C., Sfragano, P.S., Laschi, S., Pillozzi, S., Boddi, A., Crociani, O., Bernini, A., et al. 2022. Chip-based and wearable tools for isothermal amplification and electrochemical analysis of nucleic acids. Chemosensors. 10(7): 278.

Pingarrón, J.M., Yáñez-Sedeño, P. and González-Cortés, A. 2008. Gold nanoparticle-based electrochemical biosensors. Electrochimica. Acta. 53(19): 5848–5866.

Rafiq, A., Tahir, M., Zia, R., Nazir, K., Nayab, N. and Shaheen, A., Mansoor, S., et al. 2022. Virus detection using nanobiosensors. pp. 547–572. *In:* Denizli, A., Nguyen, T.A., Rajendran, S., Yasin, J. and Nadda, A.K. (ed.). Nanosensors for Smart Agriculture. Elsevier. UK.

Rai, M., Reshetilov, A., Plekhanova, Y. and Ingle, A.P. 2020. Macro, Micro, and Nano-Biosensors: Potential Applications and Possible Limitations. Springer Nature. Switzerland.

Rana, K., Mittal, J., Narang, J., Mishra, A. and Pudake, R.N. 2021. Graphene-based electrochemical DNA biosensor for detection of false smut of rice (*Ustilaginoidea virens*). Plant Pathol. J. 37(3): 291–298.

Ray, P.C., Fortner, A. and Darbha, G.K. 2006. Gold nanoparticle-based FRET assay for the detection of DNA cleavage. J. Phys. Chem. 110(42): 20745–20748.

Razmi, A., Golestanipour, A., Nikkhah, M., Bagheri, A., Shamsbakhsh, M. and Malekzadeh-Shafaroudi, S. 2019. Localized surface plasmon resonance biosensing of tomato yellow leaf curl virus. J. Virol. Meth. 267: 1–7.

Razo, S.C., Panferova, N.A., Panferov, V.G., Safenkova, I.V., Drenova, N.V., Varitsev, Y.A., Zherdev, A.V., et al. 2019. Enlargement of gold nanoparticles for sensitive immunochromatographic diagnostics of potato brown rot. Sensors 19(1): 153. https://doi.org/10.3390/s19010153.

Reinhard, B.M., Siu, M., Agarwal, H., Alivisatos, A.P. and Liphardt, J. 2005. Calibration of dynamic molecular rulers based on plasmon coupling between gold nanoparticles. Nano Letters. 5(11): 2246–2252.

Sanchez, L., Farber, C., Lei, J., Zhu-Salzman, K. and Kurouski, D. 2019. Non-invasive and non-destructive detection of cowpea bruchid within cowpea seeds with a hand-held Raman spectrometer. Anal. Chem. 91(3): 1733–1737.

Sapsford, K.E., Berti, L. and Medintz, I.L. 2006. Materials for fluorescence resonance energy transfer analysis: beyond traditional donor-acceptor combinations. Angew. Chemie. 45(28): 4562–4589.

Schlücker, S. 2014. Surface-enhanced Raman spectroscopy: concepts and chemical applications. Angew. Chemie. 53(19): 4756–4795.

Scott, N. and Chen, H. 2012. Nanoscale science and engineering for agriculture and food systems. Indus. Biotechnol. 8(6): 340–343

Shao, L., Gao, Y. and Yan, F. 2011. Semiconductor quantum dots for biomedical applications. Sensors 11(12): 11736–11751.

Sharma, P., Kumar, S., Patel, A., Datta, B. and DeLong, R.K. 2021. Nanomaterials for agricultural and ecological defence applications: active agents and sensors. Wiley Interdisciplinary Reviews: Nanomed. Nanobiotech. 13(5): 1713.

Shaw, D.S. and Honeychurch, K.C. 2022. Nanosensor applications in plant science. Biosensors. 12(9): 675.

Siddiquee, S., Rovina, K., Yusof, N.A., Rodrigues, K.F. and Suryani, S. 2014. Nanoparticle-enhanced electrochemical biosensor with DNA immobilization and hybridization of *Trichoderma harzianum* gene. Sens. Bio-Sens. Res. 2: 16–22.

Singh, S., Srivastava, A., Oh, H.M., Ahn, C.Y., Choi, G.G. and Asthana, R.K. 2012. Recent trends in development of biosensors for detection of microcystin. Toxicon. 60(5): 878–894.

Song, G., Wu, J.Y., Xie, Y., Liu, Y., Qian, Y.J., Zhou, X.P. and Wu, J.X. 2017. Monoclonal antibody-based serological assays for detection of Potato virus *S* in potato plants. J. Zhejiang Univ. Sci. B. 18(12): 1075–1082.

Sun, L., Li, C., Zhang, C., Liang, T. and Zhao, Z. 2019. The strain transfer mechanism of fiberbragg grating sensor for extra-large strain monitoring. Sensors. 19(8): 1851.

Taghdisi, S.M., Emrani, S.S., Tabrizian, K., Ramezani, M., Abnous, K. and Emrani, A.S. 2014. Ultrasensitive detection of lead (II) based on fluorescent aptamer-functionalized carbon nanotubes. Environ. Toxicol. Pharma. 37(3): 1236–1242.

Tang, Y.B., Xing, D., Zhu, D.B., and Liu, J.F. 2007. An improved electrochemiluminescence polymerase chain reaction method for highly sensitive detection of plant viruses. Anal Chim. Acta. 582(2): 275–280.

Uda, M.N.A., Gopinath, S.C.B., Hashim, U., Hakimi, A., Uda, M.N.A., Anuar, A., Bakar, M.A.A., et al. 2020. Harumanis mango: Perspectives in disease management and advancement using interdigitated electrodes (IDE) nano-biosensor. IOP Conf. Ser.: Mater. Sci. Eng. 864: 012180. doi: 10.1088/1757-899X/864/1/012180.

Vikesland, P.J. and Wigginton, K.R. 2010. Nanomaterial enabled biosensors for pathogenmonitoring—a review. Environ. Sci. Technol. 44: 3656–3669.

Wang, X., Zhou, J., Song, J., Liu, J., Xu, N. and Wang, Z.L. 2006. Piezoelectric field effect transistor and nanoforce sensor based on a single ZnO nanowire. Nano Letters. 6(12): 2768–2772.

Wang, L. and Li, P.C.H. 2007. Flexible microarray construction and fast DNA hybridization conducted on a microfluidic chip for greenhouse plant fungal pathogen detection. J. Agri. Food Chem. 55(26): 10509–10516,

Wang, Z., Wei, F., Liu, S.Y., Xu, Q., Huang, J.Y., Dong, X.Y., Yu, J.H., et al. 2010. Electrocatalytic oxidation of phytohormone salicylic acid at copper nanoparticles-modified gold electrode and its detection in oilseed rape infected with fungal pathogen *Sclerotinia sclerotiorum.* Talanta 80(3): 1277–1281.

Wang, T., Zhu, H., Zhuo, J., Zhu, Z., Papakonstantinou, P., Lubarsky, G., Lin, J., et al. 2013. Biosensor based on ultrasmall MoS_2 nanoparticles for electrochemical detection of H_2O_2 released by cells at the nanomolar level. Anal. Chem. 85(21): 10289–10295.

Wang, H.N., Fales, A.M. and Vo-Dinh, T. 2015. Plasmonics-based SERS nanobiosensor for homogeneous nucleic acid detection. Nanomed: Nanotechnol. Biol. Med. 11(4): 811–814.

Wang, P., Lombi, E., Zhao, F.J. and Kopittke, P.M. 2016. Nanotechnology: a new opportunity in plant sciences. Trends Plant Sci. 21(8): 699–712.

Wang, F., Gu, X., Zheng, C., Dong, F., Zhang, L., Cai, Y., You, Z., et al. 2017. Ehrlich reaction evoked multiple spectral resonances and gold nanoparticle hotspots for Raman detection of plant hormone. Anal. Chem. 89(17): 8836–8843.

Wang, X., Wang, J., Sun, X., Wei, S., Cui, L., Yang, W., Liu, J., et al. 2018a. Hierarchical coral-like NiMoS nanohybrids as highly efficient bifunctional electrocatalysts for overall urea electrolysis. Nano Res. 11(2): 988–996.

Wang, D., Vinson, R., Holmes, M., Seibel, G., Bechar, A., Nof, S., Luo, Y., et al. 2018b. Early tomato spotted wilt virus detection using hyperspectral imaging technique and outlier removal auxiliary classifier generative adversarial nets (OR-AC-GAN). ASABE Ann. Int. Meet. 1800660.

Wee, E.J.H., Lau, H.Y., Botella, J.R. and Trau, M. 2015. Re-purposing bridging flocculation for on-site, rapid, qualitative DNA detection in resource-poor settings. Chem. Commun. 51(27): 5828–5831.

Wei, J., Liu, H., Liu, F., Zhu, M., Zhou, X. and Xing, D. 2014. Miniaturized paper-based gene sensor for rapid and sensitive identification of contagious plant virus. ACS Appl. Mater. Interf. 6(24): 22577–22584.

Wen, T., Nie, Q., Han, L., Gong, Z., Li, D., Ma, Q., Wang, Z., et al. 2022. Molecularly imprinted polymers-based piezoelectric coupling sensor for the rapid and non-destructive detection of infested citrus. Food Chem. 387: 132905.

Wu, X., Gao, S., Wang, J.S., Wang, H., Huang, Y.W. and Zhao, Y. 2012. The surface-enhanced Raman spectra of aflatoxins: spectral analysis, density functional theory calculation, detection, and differentiation. Analyst. 137(18): 4226–4234.

Wu, S.H. and Lin, H.P. 2013. Synthesis of mesoporous silica nanoparticles. Chem. Soc. Rev. 42(9): 3862–3875.

Xiong, Z., Xiao, N., Xu, F., Zhang, X., Xu, Q. and Zhang, K., Ye, C., et al. 2020. An equivalent exchange-based data forwarding incentive scheme for socially aware networks. J. Signal Proc. Systems. 93(2–3): 249–263.

Xue, Q., Zhu, Y. and Wang, J. 2021. Joint distribution estimation and naïve Bayes classification under local differential privacy. IEEE Trans. Emer. Topics Comp. 9(4): 2053–2063.

Yotova, L. and Yaneva, S. 2013. Biomimetic nanosensors for determination of toxic compounds in food and agricultural products. J. Chem. Technol. Metall. 48(3): 215–227.

Yüksel, S., Schwenkbier, L., Pollok, S., Weber, K., Cialla-May, D. and Popp, J. 2015. Label-free detection of *Phytophthora ramorum* using surface-enhanced Raman spectroscopy. Analyst. 140(21): 7254–7262.

Yun, C.S., Javier, A., Jennings, T., Fisher, M., Hira, S., Peterson, S., Hopkins, B. et al. 2005. Nanometal surface energy transfer in optical rulers, breaking the FRET barrier. J. Am. Chem. Soc. 127(9): 3115–3119.

Yusof, N.A. and Isha, A. 2020. Nanosensors for early detection of plant diseases. pp. 407–419. *In:* Husen, A. and Jawaid, M. (eds). Nanomaterials for Agriculture and Forestry Applications, Elsevier. UK.

Zarco-Tejada, P.J., Camino, C., Beck, P.S.A., Calderon, R., Hornero, A., Hernández-Clemente, R., Kattenborn, T., et al. 2018. Previsual symptoms of *Xylella fastidiosa* infection revealed in spectral plant-trait alterations. Nat. Plants. 4(7): 432–439.

Zhan, F., Wang, T., Iradukunda, L. and Zhan, J. 2018. A gold nanoparticle-based lateral flow biosensor for sensitive visual detection of the potato late blight pathogen, *Phytophthora infestans*. Anal. Chimica. Acta. 1036: 153–161.

Zhang, F., Zou, M., Chen, Y., Li, J., Wang, Y., Qi, X., Xue, Q., et al. 2014. Lanthanide-labeled immunochromatographic strips for the rapid detection of *Pantoeastewartii subsp. stewartii*. Biosen. Bioelect. 51: 29–35.

Zhang, P., Wang, L.M., Zheng, D.W., Lin, T.F., Wei, X.D., Liu, X.Y., Wang, H.Q., et al. 2018. Surface-enhanced Raman spectroscopic analysis of N^6-benzylaminopurine residue quantity in sprouts with gold nanoparticles. J. Environ. Sci. Health. 53(9): 561–566.

Zhang, C. and Wang, H. 2019. Swing vibration control of suspended structure using active rotary inertia driver system: parametric analysis and experimental verification. Appl. Sci. 9(15): 3144.

Zhang, Q., Ying, Y. and Ping, J. 2021. Recent advances in plant nanoscience. Adv. Sci. 2103414. doi: 10.1002/advs.202103414.

Zhao, W., Lu, J., Ma, W., Xu, C., Kuang, H. and Zhu, S. 2011. Rapid on-site detection of *Acidovorax avenae* subsp. *Citrulli* by gold-labeled DNA strip sensor. Biosen. Bioelect. 26(10): 4241–4244.

Zulkifli, H., Salam, F., Saad, S.M., Rahman, R.A., Rani, R.M., Karim, M.S.A. and Ishak, Z. 2016. Preliminary study of electrochemical DNA Sensor for cucumber mosaic virus. Procedia Chem. 20: 98–101.

Zuo, C., Chen, Q., Tian, L., Waller, L. and Asundi, A. 2015. Transport of intensity phase retrieval and computational imaging for partially coherent fields: the phase space perspective. Optics Lasers Eng. 71: 20–32.

Section V

Energy and Conservation

Chapter 15

Applications of Nanomaterials for Energy Generation and Storage

Akshita Singh[1]
[1]Centre for Nanotechnology, Indian Institute of Technology Roorkee, Roorkee, 247667, Uttarakhand, India

Vimal Chandra Srivastava*[1,2]
[2]Department of Chemical Engineering, Indian Institute of Technology Roorkee, Roorkee, 247667, Uttarakhand, India

15.1 INTRODUCTION

The promise of nanoscience is due to the fact that at the nanoscale, new properties and phenomena arise, as compared to those at the larger scale. Along with minimization and compactness, it offers the precise handling of the atoms to design new properties or modify the existing ones. This ability to understand and control matter at the nanoscale could be used to enhance existing products or develop new ones. Some of the beneficial aspects of the nanoscale are shown in Figure 15.1.

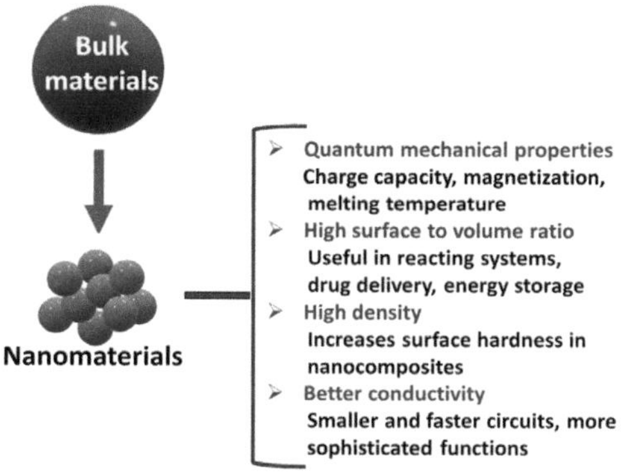

Figure 15.1 Properties of nanomaterials.

*Corresponding Authors: vimalcsr@yahoo.co.in; vimal.srivastava@ch.iitr.ac.in

Applications of nanoscience are too vast to be mentioned in the space of this write-up. Some of the fields are shown in Figure 15.2, however here, it is concentrated on their application in the energy sector. The importance of energy in human life is well known to us. Because of its ever-increasing demand, it is necessary to look for modern technologies to make existing energy sources more efficient and environmental friendly or develop new ones. The intersection of nanotechnology with energy has the potential to change the way energy was until now being generated, converted, transmitted, stored, and used. One can definitely say that each part of the value-added chain in the energy sector is strongly impacted by nanotechnology. The examples include both conventional energy sources as well as renewable energy sources. Researchers in the field of nanoscience are involved in finding ways to establish cleaner and more cost-friendly energy sources along with reducing their consumption. Nanotechnology, through improved catalysis, helps in increasing the efficiency of fuel production from raw petroleum materials (Applications of Nanotechnology|National Nanotechnology Initiative). Wires containing carbon nanotubes are used in electric grids in order to reduce transmission power loss. Nanostructured solar panels are more efficient, cheaper, and easier to install. New kinds of batteries are being developed using nanotechnology that is more efficient, have high power density, low charging time, and longer lifetimes. Other than the batteries, one can also witness the contribution made by nanotechnology in improving the design and functioning of supercapacitors. This write-up briefly discusses the impact of nanotechnology in enhancing the performance of fuel cells, supercapacitors, and solar cells.

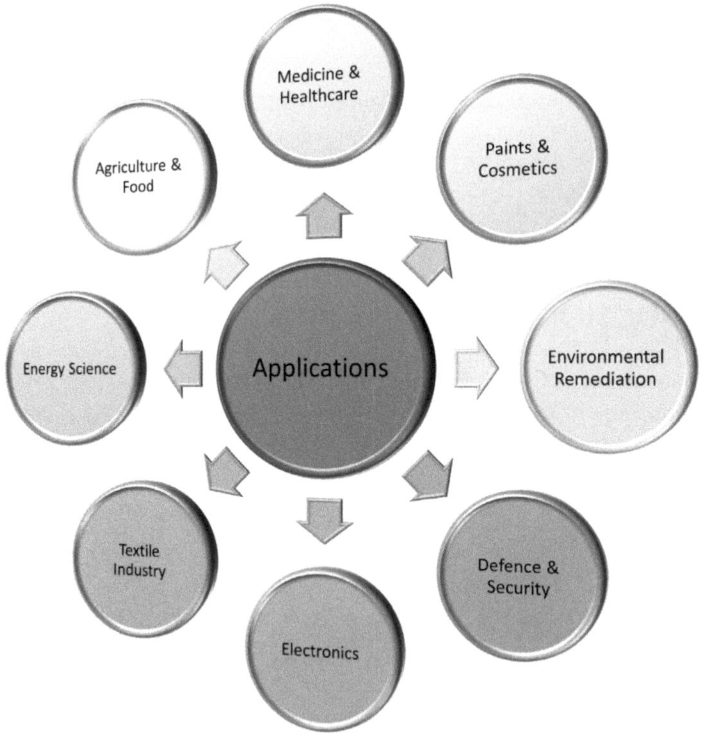

Figure 15.2 Applications of nanomaterials.

15.2 FUEL CELLS

The conversion of chemical energy into electrical energy, heat, and water takes place by the means of fuel cells, which are electrochemical devices. It offers features like high efficiency, compact structure, low emission, silent operation, long lifetime, and high-power density. Considering that the fuel cell's waste heat is also used, they can be highly efficient, i.e., up to 90%. It is similar to a battery, except it does not require periodic recharging. It continues to generate electricity as long as the fuel source is provided, and thus there is a continuous supply of reactants to be consumed. It consists of a cathode and an anode, each with a catalyst, separated by an electrolyte. Oxidation to fuel such as hydrogen, methane, methanol, etc., takes place at the electropositive cathode, whereas reduction to air, oxygen, etc., appears at the electronegative anode. There are special types of electrolytes that determine the type of fuel cell. They can be liquid, like phosphoric acid, alkaline, molten carbonate, or solid, like solid oxide and proton exchange membrane. The working principle of all the fuel cells is somewhat similar. Taking the example of a cell where hydrogen is used as fuel and oxygen is the oxidizing agent, the working is as follows: hydrogen supplied at the anode dissociates into electrons and protons; electrons, through an external circuit, travel from the anode to the cathode; positive hydrogen ions also travel from the anode to the cathode, but via electrolyte; separately, oxygen is pumped at the cathode which catalytically reacts with electrons and protons reached there, to form heat and water. The schematic of this process is shown in Figure 15.3.

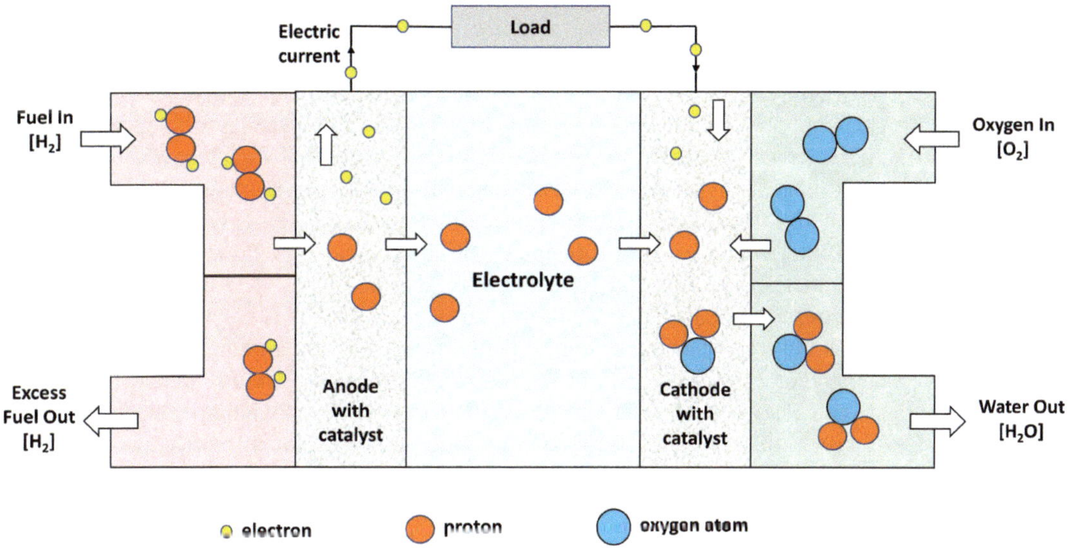

Figure 15.3 Working principle of PEM fuel cell.

Fuel cells include polymer electrolyte membrane fuel cells (PEMFCs), alkaline fuel cells (AFCs), direct methanol fuel cells (DMFCs), phosphoric acid fuel cells (PAFCs), molten carbonate fuel cells (MCFCs), and solid oxide fuel cells (SOFCs). The first four are low temperature (ca. <200°C) fuel cells, while the latter two are high temperature (ca. >400°C) fuel cells. This article will mainly focus on PEMFCs, DMFCs among low-temperature fuel cells, and SOFCs in high-temperature fuel cells. They use solid electrolytes that make their maintenance easy, however, their commercial exploitation is still limited because of their high cost and poor durability. High cost in PEMFC and DMFC arises due to the expensive Pt catalyst and Nafion membrane used. At the same time, SOFC results from high operating temperature, which requires expensive sealing materials and interconnects between the cells. Durability issues in PEMFC and DMFC arise due to degradation in the performance of the membrane, catalyst, and catalyst support. In SOFC, it is mainly because of carbon deposition, poisoning of catalyst, and decline in the microstructure. Nanotechnology has played an essential role in addressing these problems by developing high-performance materials with novel preparation methods and designs. Here the application of nanomaterials in the electrodes (Table 15.1) and electrolytes of the PEMFCs, DMFCs belonging to the low-temperature fuel cell, and SOFCs belonging to the high-temperature fuel cell will be discussed.

Table 15.1 Nanostructured-based different anode catalysts in the fuel cell

Catalyst	Experimental composition	Mass/specific activity	I_F/I_B	References
Pt/graphene	0.5 M H_2SO_4, 1 M CH_3OH	—	0.83	Li et al. 2010a
PtSn/graphene	0.5 M H_2SO_4, 1 M C_2H_5OH	407 mA mg^{-1}	1.10	Wang et al. 2018
Pd/MnO_2/CNT	0.5 M NaOH, 1 M CH_3OH	432 mA mg^{-1}	18.3	Zhao et al. 2010a
Ni@Pd/MWCNT	0.5 M NaOH, 1 M CH_3OH	770 mA mg^{-1}	1.99	Zhao et al. 2010b
Pt/Mn_3O_4–MWCNT	0.5 M H_2SO_4, 1 M CH_3OH	970 mA mg^{-1}	1.02	Yang et al. 2012
Pd/GS	1 M KOH, 1 M CH_3OH	222 mA mg^{-1}	—	Guo et al. 2013
Pd/NG	1 M KOH, 1 M CH_3OH	19.7 mA cm^{-2}	1.63	Kiyani et al. 2017
Pt/OMC	0.5 M H_2SO_4, 0.5 M CH_3OH	250 mA mg^{-1}	—	Su et al. 2010

Reduced graphene oxide (RGO); Graphene nanosheets (GS); Nitrogen-doped graphene (NG); Ordered mesoporous carbon (OMC); Carbon nanotube (CNT)

15.2.1 Low-temperature Fuel Cell

Electrodes: Fuel cells operating on low-temperature are mainly used in portable applications because of their fast operation. They prefer hydrogen as fuel because of its high activity and environment-friendly nature. In H_2-PEMFCs, Pt is generally used as an electrocatalyst in electrodes for better electrochemical performance. However, its high cost and low availability led researchers to find ways to either decrease Pt loading or use non-Pt electrocatalysts. One way to do so is to use binary and ternary Pt alloy as hydrogen oxidation reaction (HOR) anode material. Alcaide et al. (2010) replaced Pt/C with PtPd/C leading to 35 wt.% reduction in Pt loading. Other includes using Ru, Sn (Gözt and Wendt 1998), Co, Cr (Brouzgou et al. 2012), Mn, Ag, and Au to form different alloys. For oxygen reduction reaction (ORR) cathode in H_2-PEMFCs, Pd-Pt bimetallic catalysts and metal oxide electrocatalysts have been the most examined as they offer electrocatalytic activity comparable to Pt. It is observed that core-shell structured catalysts are gaining much attention to reduce the Pt-content. Examples include hollow-core mesoporous shell carbon doped with Pt-Pd bimetallic catalysts and Pt-rich shell with low Pt content Pt-Co core (Fıçıcılar et al. 2009, Zhou et al. 2010). Zeis et al. (2007) prepared Pt-coated nanoporous gold leaf to be used as a catalyst in PEMFC. Pt-coated nanoporous gold leaf along with Nafion membrane electrode assembly generated a maximum power density of 0.25 W cm^{-2} with a low Pt loading of 0.03 mg cm^{-2} in the unoptimized test cell. Lu et al. (2016) reviewed the 1D nanostructured electrocatalysts for HOR anode and ORR cathode in PEMFCs. In the review, Pt-based nanostructures, non-Pt-based non-precious metal nanostructures, as well as non-Pt-based precious metal nanostructures have been reported to act as active and durable electrocatalysts. Non-platinum electrocatalysts like Ir-V/C nanoparticles have also been examined for both HOR and ORR reactions in PEMFC and were found to give good stability and improved catalytic activity (Li et al. 2009, Qiao et al. 2010). Researchers are even modifying the trivial support materials by preparing Pt layers on 1-D, 2-D and 3-D metal nanostructures (Morozan et al. 2011). Dsoke et al. (2011) prepared a composite by mixing Pt-Co/C with $Cs_{2.5}H_{0.5}PW_{12}O_{40}$ salt for the electrode having ORR reaction and found it acting as an effective support for Pt nanoparticles with low loading of ~14 μg cm^{-2}. Also, the adoption of novel carbon support materials like multi-walled CNTs and single-walled CNTs with nanoscale features like large surface area, greater stability, and better conductivity has been proven to decrease the Pt loading in H_2-PEMFCs (Antunes et al. 2011).

In the case of DMFCs, other than the low electrocatalytic activity for ORR and methanol oxidation reaction (MOR), drawbacks include carbon monoxide poisoning of the anode electrocatalyst. Like H_2-PEMFCs, Pt alloys have also been much examined for MOR anode in DMFCs. Among them, Pt-Ru exhibits better electrocatalytic activity by reducing CO poisoning. Ando et al. (2009) developed thin Pt-Ru nanoplatelets on carbon nanoparticles as catalysts for MOR and found it to give much better mass activity instead of low Pt loading of 0.32 μg_{Pt} cm^{-2}. Core-shell structured bimetallic nanoparticles also exhibit excellent heterogeneous catalytic activity. For example, core-shell structured CdSe@Pt nanocomposites synthesized by Yang et al. (2011) showed better electrocatalytic activity than commercial Pt/C catalyst, with loading as low as 20 μg_{Pt} cm^{-2} and a current density of 142.5 mA cm^{-2}. Liu et al. (2010a) also reported nanocomposite of iron phthalocyanine, platinum, carbon black, and Nafion as catalytic cathode exhibiting enhanced activity for ORR. In DMFCs also, carbon nanostructures like MWCNTs used as support materials for catalysts like Pt-Ru nanoparticles (Prabhuram et al. 2007) and Pt-Co hollow nanospheres (Guo and Cui 2009) were found to improve their electrocatalytic performance. Another example of carbon support includes ordered porous carbons with a hollow core-mesoporous wall supporting a Pt-Ru catalyst that is found to exhibit high specific activity in the MOR (Chai et al. 2004). The incorporation of metal oxides in catalysts changes its surface electronic properties and also provides reactive oxygen species (ROS) to react with CO, which reduces catalytic poisoning. Other examples include Pt-rod-shaped CeO_2/graphene sheet and Pt/TiO_2 nanofiber catalyst which was found to show effective electrocatalysis in DMFC (Wang et al. 2015, Chen et al. 2015). Few MOR catalysts used in DMFCs have been listed in Table 15.2.

Table 15.2 Summary of the MOR catalysts in DMFCs

Composition	Structure	Specific surface area (m^2 g^{-1})	Mass normalized current densities toward MOR (mA $mg_{Pt/Pd}^{-1}$)	References
Pt	Nanostar	72	590	Wang et al. 2012a
Pd/C	Nanoparticles	43.4	5.5	Li et al. 2017
Pd/GS	Nanoparticles	99.4	222	Guo et al. 2013
Pt-Pd	Hollow porous nanospheres	23.3	204.6	Qiu et al. 2016
Pt/PANI-HPMo-GS	Nanoparticles	108.9	322	Cui et al. 2013

Polyaniline (PANI); $H_3PMo_{12}O_{48}$ (HPMo); Graphene nanosheets (GS)

Electrolyte: Another essential component that impacts the performance of fuel cells is the electrolyte. As discussed earlier, PEMFCs and DMFCs use solid electrolytes. The commonly used polymer electrolyte membrane material in these cells is Nafion. It belongs to a class of material called perfluorosulphonic acid polymers. Its structure consists of two major parts—(i) a Teflon backbone which is hydrophobic in nature, and (ii) side chains of vinyl-ether terminated in hydrophilic sulphonic acid groups in a cluster. On being hydrated, the sulphonic acid clusters in the membrane form channels for the transport of H^+ ions resulting in excellent proton conductivity. Thus, membrane performance is dependent on water content, and this requirement restricts the operating temperature (Devanathan 2008). Other than membrane dehydration, Nafion has drawbacks like high cost and methanol crossover. Also, at high temperatures, there is shrinkage in H^+ transporting channels which decreases the conductivity. However, as we know, at high temperatures, catalysts perform better, so we need to design a membrane having high conductivity at such temperatures. Nanotechnology is being used to do so by either modifying conventional Nafion membrane or developing some alternative. Researchers have studied the incorporation of carbon nanomaterial-based fillers like CNT (Yin et al. 2018), graphene (Sahu et al. 2016), and graphene oxide (Lee et al. 2014), which resulted in improved chemical and thermal stability along with better fuel-blocking capability in membranes. Nanoceramic oxide hybrid electrolyte membranes for PEMFCs are also reviewed to find enhanced hydration and conductivity (Xu and Mu 2014).

15.2.2 High-temperature Fuel Cell

Electrodes: Though solid-oxide fuel cells have significant advantages in terms of high performance and efficiency, their high operating temperature (700–900°C) is a serious drawback. Thus, researchers have been continuously involved in trying to bring the operating temperature to an intermediate range of 400–700°C. This uplift is accompanied by another problem, i.e., an increase in resistance of fuel cell components. An effort is being made to tackle these problems through the involvement of nanomaterials. Designing electrode materials at the nanoscale also led to the introduction of the quantum confinement effect, which favored the robustness and reliability of SOFCs at even reduced temperatures and lower material costs. Some of the basic features to be addressed before making the choice of the anode and cathode material are; high ionic and electronic conductivity, chemical compatibility and thermal stability, high porosity with an organized structure, and high catalytic activities. Nanotechnology played a great role in addressing some of these requirements. Nanostructured materials have been useful in improving the electrocatalytic activities for REDOX reactions (Zhuiykov 2014), minimizing the catalyst loading and lowering the cost, reducing the operating temperature (Chockalingam and Basu 2011), maintaining sufficient thermal stability, inter-connected networks (porosity) for rapid gas transport, and reducing carbon deposition and sulfur poisoning, which results in increased durability (Zhuiykov 2014). Some of the nanomaterials used in electrodes of SOFCs are listed in Table 15.3.

Table 15.3 Nanostructured-based electrodes for SOFCs.

Anode	Cathode	Electrolyte	Temperature (°C)	Reference
Ni	LSM-YSZ	ScSZ	700	Myung et al. 2015
Ni–SDC	PBSCF	YSZ	650	Zhang et al. 2021
Pt	Pt	YSZ	350–500	Wang et al. 2008
Ni–GDC	LSCF–GDC (LSC–LG)	GDC	550–650	Huang et al. 2018
Ni–YSZ	LSC	YSZ	650	Shin et al. 2019
CMZO–NSDC	CMZO–NSDC	NSDC	400–550	Abbas et al. 2017
$SFT_{0.2}$	$SFT_{0.2}$	LSGM	700	dos Santos-Gómez et al. 2015
LST–SDC	LSC–SDC	LSGM	650-800	Zhang et al. 2014a

$La_{0.6}Sr_{0.4}Co_{0.2}Fe_{0.8}O_{3-\delta}$–$Ce_{0.9}Gd_{0.1}O_{2-\delta}$ (LSCF–GDC); Gadolinium doped ceria (GDC $Sr_{0.98}Fe_{0.8}Ti_{0.2}O_{3-\delta}$ ($SFT_{0.2}$); $La_{0.8}Sr_{0.2}CoO_3$-Sm-doped ceria (LSC-SDC); Mg/Sr-doped lanthanum gallate (LSGM); Lanthanum Strontium Manganite-Yttria Stabilized Zirconia (LSM-YSZ); Scandia-stabilized zirconia (ScSZ); $PrBa_{0.5}Sr_{0.5}Co_{1.5}Fe_{0.5}O_{5+\delta}$ (PBSCF); $Pr_{0.6}Sr_{0.4}MnO_{3\pm\delta}$(PSM); La-doped strontium titanate-Sm-doped ceria (LST-SDC); Copper manganese zinc oxide (CMZO); Sodium carbonated calcium doped ceria (NSDC).

Electrolyte: In SOFCs, mostly solid oxide or ceramic materials with high oxide ion/proton conductivity and low electronic conductivity act as an electrolyte. Also, it should be chemically, mechanically, and thermally stable over a wide range of temperatures. Yttria-stabilized zirconia (YSZ) is the most common electrolyte used in SOFCs, however, it requires a high operating temperature (800–1000°C). So, there are electrolytes like Gadolinia doped ceria (GDC), Samarium doped ceria (SDC), and Samarium doped ceria carbonate (SDCC) for low temperatures (<600°C). Also, their nano-sized powders reduced the firing temperature in the fabrication of cells. Large surface area and grain boundaries in nanostructured-based systems increase the number of mobile defects, resulting in improved electrochemical behavior compared to bulk structures. When one reduces the operating temperature, the contribution of grain boundary increases which is beneficial (Guo and Waser 2006) but since in these electrolytes, the ion-transport mechanism is a thermally activated process, so reduced temperature results in increased electrolyte resistivity. One of the approaches to address this problem is decreasing electrolyte thickness at the nanoscale (Hao et al. 2017). Another way is the use of alternative nanomaterials with comparatively better ionic conductivity than YSZ in the intermediate temperature range. Also, since the property of nanomaterials depends on their morphology, so it can be tuned as required. For example, higher ionic conductivity is expected for 1D nanostructures compared to their analogs due to the presence of continuous grain boundaries. Thus, the performance of nanostructured electrolytes in SOFCs depends on their composition, shape, and size.

15.3 SUPERCAPACITORS

Supercapacitors act as the bridge between batteries and conventional capacitors. Not only do they have high power density in comparison to batteries but also have greater energy storing capacity than the conventional capacitors. Also, unlike batteries, they have a relatively long cycle life and work well in extreme temperatures. Supercapacitors store energy electrostatically in a way similar to conventional capacitors rather than chemical reactions as in batteries. It also has two plates separated by a thin insulator and soaked in an electrolyte. The schematic of a supercapacitor is shown in Figure 15.4. The huge difference in the energy storage capacity comes from the relatively large surface area of electrodes with a small distance between them. They are made up in the form of a thin coating of porous substances like activated charcoal, nanocomposites, and metal oxide nanoparticles applied on a metallic current collector. Some of the properties of electrodes are good conductivity, high-temperature stability, inertness, corrosion resistance, and preferably being environment friendly and cheaper. There is a separator that prevents the short circuit between the two electrodes by direct contact. Electrolyte provides the ionic conductive connection between the two electrodes. It should be chemically inert with low viscosity and high decomposition voltage. Other than the capacitance of the supercapacitor, operating voltage, temperature range, and equivalent series resistance also depends on the type of electrolyte.

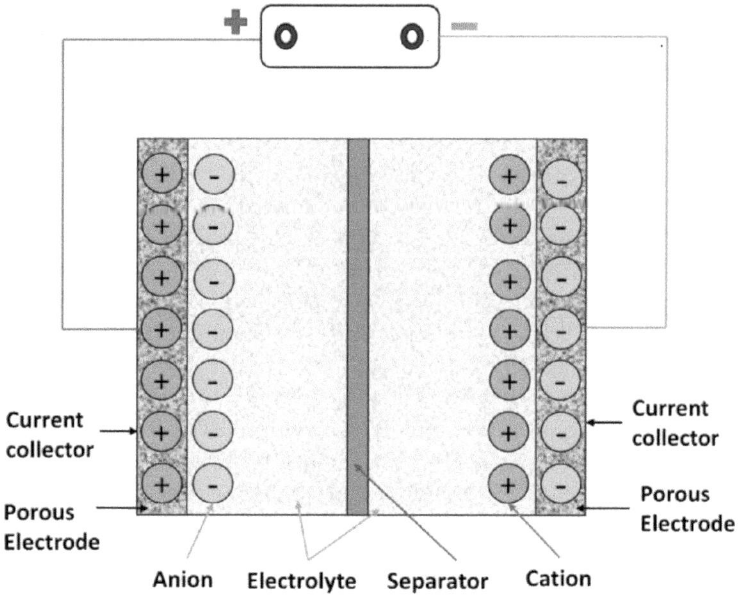

Figure 15.4 Design of a supercapacitor.

Electrodes make up the core component of electrochemical supercapacitors. As a result, improving the electrochemical activity of electrode materials is crucial in enhancing the complete performance of the supercapacitor. Recent studies found nano-structured materials to be a much better option for electrodes. They offer large active surface areas and also facilitate the transport of electrolytes. Generally, supercapacitors are classified into three different types: Electrical double-layer capacitors (EDLC), Pseudo-capacitors, and Hybrid supercapacitors. The kinds of nanomaterials used and their advantages in these three supercapacitors will now be discussed.

15.3.1 Electrical Double-Layer Capacitors

Electrical energy in EDLCs is stored electrostatically through double-layer capacitance, which is achieved by the separation of electrons and ions at the interface between electrode surface and electrolyte. Due to this purely physical mechanism, EDLCs have highly reversible operations and thus can bear a large number of charging/discharging cycles. Also, since the rate of charging/discharging depends mainly on the physical motion of charged particles, they are much faster in operation. Carbon-based nanostructures are prevalently used for electrode materials in EDLCs due to their properties like specific surface area, pore size, and distribution, electrical conductivity, etc.

The first material to be used for electrodes of EDLCs was activated carbon. Due to their porous nature, they provide a high specific surface area. Although their electrical conductivity is much less than those of metals, it's sufficient for application in supercapacitors. For typically activated carbons, the BET-specific surface area generally ranges between 1000–2000 m^2 g^{-1}. In a two-electrode cell system, the specific capacitance of activated carbon is typically less than 350 F g^{-1} (Sevilla and Mokaya 2014). It has been found that the distortion and partial removal of the ion solvation shell in carbon material having a pore size less than 1 nm, leads to an unusual rise in capacitance (Merlet et al. 2013). So narrow size distribution and interconnection between the porous structure of such activated carbons enhance the energy density of supercapacitors. Niu et al. (2017) prepared porous nanosheets of activated carbon from biomass sugarcane pith. It had a specific surface area of 1786.1 m^2 g^{-1} and exhibited a capacitance of 339 F g^{-1} at a current density of 0.25 A g^{-1}. Other than activated carbon powders, carbon aerogels can also be used as electrode material. They also have a continuous porous 3D structure but are more conductive than activated carbon powders and have low mass density. For such electrodes, the gravimetric capacitance of 104 F g^{-1} has been observed. Companies have already been producing aerogel supercapacitors having specific power of 20 W g^{-1} and specific energy as high as 90 Wh kg^{-1} (Lerner 2004). Lv et al. (2020) synthesized 3D porous carbon aerogels and evaluated their performance as the supercapacitor electrode material. It had a high surface area of 1303 m^2 g^{-1} and exhibited a specific capacitance of 260 F g^{-1} in the three-electrode system at a current density of 1 A g^{-1}.

CNTs can also enhance supercapacitor performance as a result of their good chemical and thermal stability, highly wettable surface area, high electrical conductivity, and excellent mechanical strength. An et al. (2001) studied the behavior of SWCNT-based capacitors and found that they have a specific capacitance of 180 F g^{-1} along with an energy density and a power density of 6.5 Wh kg^{-1}, and 20 kW kg^{-1} respectively. Aval et al. (2018) fabricated a symmetric paper supercapacitor comprising CNT-based electrodes which exhibited a specific capacitance of 411 F g^{-1} at 20 mV s^{-1}. Also, they obtained an energy density of 91 Wh kg^{-1} and a power density of 13.6 kW kg^{-1}. Graphene has also been widely investigated as an electrode material because of its large specific surface area and good electrical conductivity. Theoretically, graphene can have a capacity of up to 550 F g^{-1}, which is the highest of all carbon-based materials. But the stacking of graphene layers due to strong van der Waals force between the adjacent sheets decreases the specific surface area and utilization rate of electrolyte resulting in a far less actual specific capacitance value as compared with the theoretical one. Researchers have used curved graphene sheets instead of flat ones to prevent their restacking and achieved a high specific energy density of 85.6 Wh kg^{-1} (Liu et al. 2010b). Behzadi Pour et al. (2018) fabricated a flexible supercapacitor using graphene sheets as electrode material and PVA as electrolyte. The specific capacitance of this assembly came out to be 343 F g^{-1} at a scan rate of 20 mV s^{-1}. Another way is to make nanocomposites by combining graphene or its oxides with conducting polymers or metal oxides to introduce pseudocapacitance and improve electrochemical performance (Wu et al. 2010).

Teli et al. (2022) synthesized Mn_2O_3/GO nanocomposite thin film using the electrospinning method to use it as a supercapacitor electrode material. It exhibited an excellent specific capacitance of 588 F g^{-1} at a current density of 2.5 A g^{-1} with 98% retention capacity after 1000 cycles. Few other examples of carbon-based nanomaterials used in EDLCs are listed in Table 15.4.

15.3.2 Pseudocapacitors

It works by transferring charge originating from adsorbed and de-solvated ions between the electrode and the electrolyte. There is no chemical reaction involved in this charge transfer. It could be electrosorption, a redox reaction, or intercalation by adsorbed ions on the electrode surface. Pseudocapacitors most often use conducting polymers, transition metal oxides/sulfides/selenides, and their composites as electrode materials. RuO_2 was the first material to be used for this purpose. It has excellent electrical conductivity and a high theoretical capacitance of 2000 F g^{-1}. Its cyclic voltammetry curve is also similar to that of a double electrode layer, i.e., an almost perfect rectangle. Asim et al. (2019) prepared electrodes of RuO_2 nanorods decorated CNTs via chemical vapor deposition for supercapacitor application. The designed material was found to exhibit a high specific capacitance of 176 F g^{-1} with 97% retention after 10,000 cycles. Other than RuO_2, one can use metal oxides like MnO_2, Fe_3O_4, IrO_2, Co_3O_4, NiO, and conducting polymers (CP) like polyaniline, polypyrrole, polyacetylene, and polythiophene.

Non-toxic behavior, high theoretical capacitance, high abundance, and thus low cost make MnO_2 a potential candidate. Wei et al. (2013) designed a 2D β-MnO_2 network for use as electrode material. It not only exhibited a high specific capacitance of 453 F g^{-1} but also 97% retention capacity after 1800 cycles. Also, a three-dimensional core-shell composite electrode material was synthesized by Cai et al. (2020) which consisted of hollow N-doped carbon@MnO_2. It was found to exhibit 247.9 F g^{-1} specific capacitance at 0.5 A g^{-1}current density along with 82.9% retention capacity after 2000 cycles. Co_3O_4 also has a high theoretical specific capacitance i.e., 3560 F g^{-1}, good redox property, and offers great reversibility. Feng et al. (2015) reported the large-scale fabrication of atomic layers of Co_3O_4 nanofilms for application as electrode material and achieved an excellent specific capacitance of 1400 F g^{-1} at 1 A g^{-1}. It also showed a retention capacity with less than a 3% decrease after 1500 cycles. Co_3O_4 nanoparticles synthesized by the sol-gel method were bound with carbon using polyvinylidene fluoride and tested as electrode material, giving a specific capacitance of 761.25 F g^{-1} at 11 mA cm^{-2} (Priyadharsini et al. 2020). Another option for pseudocapacitor electrodes is using metal oxide nanocomposites. Manibalan et al. (2021) synthesized a nanocomposite of CeO_2-ZnO by chemical precipitation method and found it to give a high specific capacitance of 424 F g^{-1} with 94.6% cyclic stability after 1000 cycles. Similarly, nanostructures of pure CP, CP/carbon binary composites, CP/metal oxide binary composites, and CP/metal oxide/carbon ternary composites are also found to be widely studied as electrode materials in supercapacitors (Meng et al. 2017). Sivakkumar et al. (2007) evaluated the electrochemical behavior of polyaniline nanofibers in a two-electrode system. They found it to exhibit capacitance of 554 F g^{-1} at a current density of 1 A g^{-1}. A composite of polypyrrole nanotube/MnO_2, prepared by Ji et al. (2015) was found to show a specific capacitance of 403 F g^{-1} at 1 A g^{-1} with a cycling performance of 88.6% retention after 800 cycles. Zhang et al. (2014b) fabricated polythiophene/MWCNT composite which exhibited a specific capacitance of 216 F g^{-1} at 1 A g^{-1}. Some other examples of nanomaterials used in pseudocapacitors are given in Table 15.5.

15.3.3 Hybrid-supercapacitors

They are asymmetric supercapacitors that comprise two different types of electrodes in a cell. One of them is said to be a battery-type electrode that exhibits electrochemical capacitance and

Table 15.4 Carbon-based nanomaterials used in EDLCs

Electrode material	Surface area (m^2 g^{-1})	Electrolyte	Energy density (Wh kg^{-1})	Power density (W kg^{-1})	Potential (V)	Specific capacitance (F g^{-1})	Retention capacity (%)/ number of cycles	Reference
Porous interconnected carbon nanosheets	2200	$TEABF_4$/ACN	30	13000	2.7	150	92/10000	Sevilla and Fuertes 2014
N-doped mesoporous carbon spheres	1602	KOH	20.5	17600	1.2	326	–	Du et al. 2019a
Biomass-derived activated carbon	2115	PVA/KOH	14	440	1.5	294	97.6/5000	Du et al. 2019b
Graphene nanosheets	–	Li_2SO_4/ethylene glycol	3.16	–	1	91	–	Zhang et al. 2019
Graphene nanofibers	–	$EMIMBF_4$/ϒ-BL	–	–	3.7	196	100/8000	Tian et al. 2018
Activated carbon derived from Moroccan anthracite	2934	$EMIMBF_4$	72.51	~1312	3.5	41	~100/2000	Boujibar et al. 2019
Carbon nanofiber	588	H2SO4	–	–	0.9	150	94.1/35000	Wang et al. 2017

Tetraethylammonium tetrafluoroborate/acetonitrile ($TEABF_4$/ACN); Poly (vinyl alcohol); 1-ethyl-3-methylimidazolium tetrafluoroborate/ϒ-butyrolactone ($EMIMBF_4$/ϒ-BL) stores charge by means of redox reactions, whereas the other is a carbon-based electrode that stores charge electrostatically using an electrical double layer. These types of supercapacitors can exert the advantages of both EDLCs and pseudocapacitances, which resulted in improved energy density and increased operating voltage window. Rani et al. (2020) reported a gravimetric specific capacitance of 223.2 F g^{-1} by using activated carbon as material for cathode and a composite of MnO_2/reduced graphene oxide nanoscrolls as material for the anode in a hybrid supercapacitor. Another example is the all-solid-state flexible hybrid supercapacitor reported by Zhang et al. (2014c) which used CNT/graphene as anode and MnO_2/graphene as the cathode. Operated in a voltage range of 0–1.8 V, it exhibited 31.8 Wh kg^{-1} of energy density. A high energy density of 80 Wh kg^{-1} was obtained in a lithium-ion hybrid supercapacitor designed by Sennu et al. (2016) by utilizing activated carbon as the cathode and $Li_4Ti_5O_{12}$ as the battery type anode. In another report, Teli et al. (2022) used Mn_2O_3/GO and Li-chip as electrodes to assemble a coin cell hybrid device which showed an excellent performance with a specific capacitance of 740 F g^{-1}, energy density of 925 Wh kg^{-1}, a power density of 3.3 kWh kg^{-1} and 90% capacitance retention after 5000 cycles.

Thus, a large number of studies proved nanostructured materials to be very promising for application in supercapacitors. A variety of nanomaterials like carbon-based, metal oxides, nanocomposites, and conducting polymers have already been studied and reported. However, there are still many challenges that arise using nanostructured materials, so further research is needed in order to resolve them and find suitable supercapacitors according to their different applications.

15.4 SOLAR CELLS

Solar cells are electronic devices that convert the vast amount of solar energy radiated to the earth into electricity. Among all renewable energies, sunlight is the biggest energy source in the world, so it becomes important to harvest it. Most of the solar energy is in the visible range of the electromagnetic spectrum, so semiconductor (e.g., silicon) materials are used in designing a solar cell in order to absorb them. The most basic construction is shown in Figure 15.5. Sunlight falls on an n-type Si layer, and photons carry their energy through the layer inside. On reaching the p-type

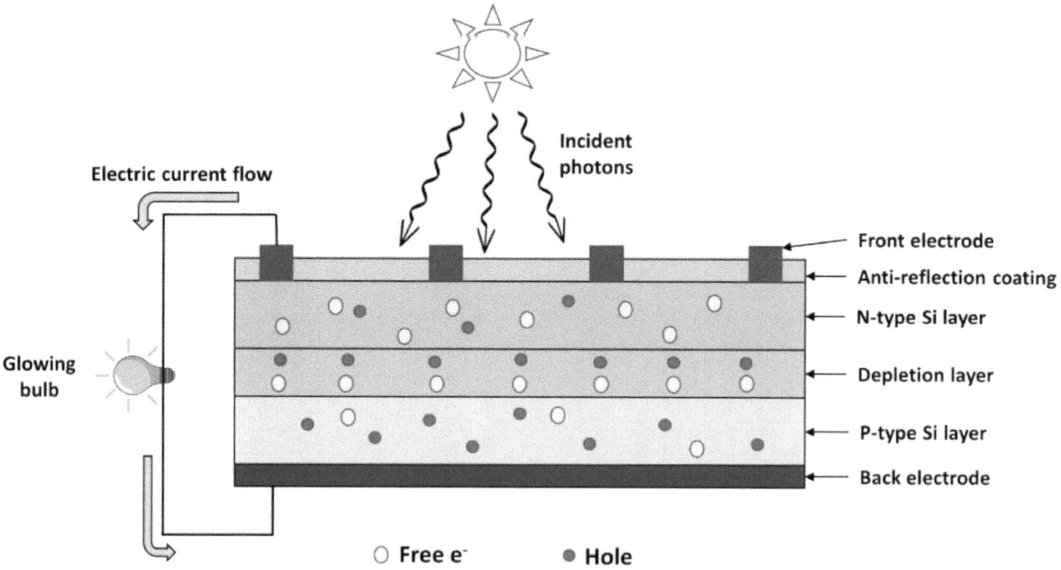

Figure 15.5 Working of a conventional Si-based solar cell.

Table 15.5 Nanostructured materials used in pseudocapacitors

Electrode	Electrolyte	Energy density (Wh kg^{-1})	Power density (W kg^{-1})	Specific capacitance (F g^{-1})	Number of cycles	Retention capacity (%)	References
NiS_2/ZnS nanocomposite	KOH	28.0	478900	1198	1000	87	Li et al. 2016
$MoSe_2$	H_2SO_4	–	–	257.38	5000	95	Vattikuti et al. 2020
V_2O_5/PEDOT	Na_2SO_4	85	250	614	50,000	122	Bi et al. 2018
NiO@nanofiber	Poly (vinyl alcohol)-H_3PO_4	22	–	161	5000	96	Kwak et al. 2020
IrO_2@Mn_3O_4 nanocomposite	Na_2SO_4	81	714	1027	10000	88	Beknalkar et al. 2021
$NiFe_2O_4$ nanostructure	KOH	–	–	541	1000	~77	Bhojane et al. 2017
Nd_2O_3/Mn_3O_4	Na_2SO_4	8.11	1750	205	500	67	Kubra et al. 2020
PANi/MoO_3/GNP	H_2SO_4	99.68	550.4	734	1000	92.4	Das et al. 2015
PPy/CNT/MnO_2	Na_2SO_4	38.42	100	529.3	1000	98.5	Zhou et al. 2015

Poly(3,4-ethylenedioxythiophene) (PEDOT); Polyaniline (PANi); Graphene nanoplatelets (GNP); Polypyrrole (PPy)

layer, that energy is used to knock out the electron present in the ground state to the excited state. Directed by the built-in electric field formed at the junction, these electrons jump across the barrier to the n-type layer and flow in the outer circuit. This is how electricity is generated.

Two major problems encountered while working with solar cells are: (i) less efficiency and (ii) high manufacturing cost. Though sunlight is available to us at zero cost, its conversion is not cost-effective for us till now. Also, humans are presently not at a stage where they can utilize the maximum amount of solar energy radiated to the earth. Let's take the case of silicon cells, where the incident photon should have definite energy equal to the bandgap of material to knock out the electrons. If the incident energy is greater than the bandgap, then extra energy will be lost as heat, and if it's lesser than the bandgap, it will just pass through without knocking out electrons. These two result in around 60–70% loss of solar energy incidents on the cell. Thus, cost reduction and improved efficiency of the solar cell are two main targets of researchers involved in this study. As shown in Figure 15.6, there are four generations of solar cells. It will be now discussed what role nanotechnology plays in solving the each of the above-discussed problems.

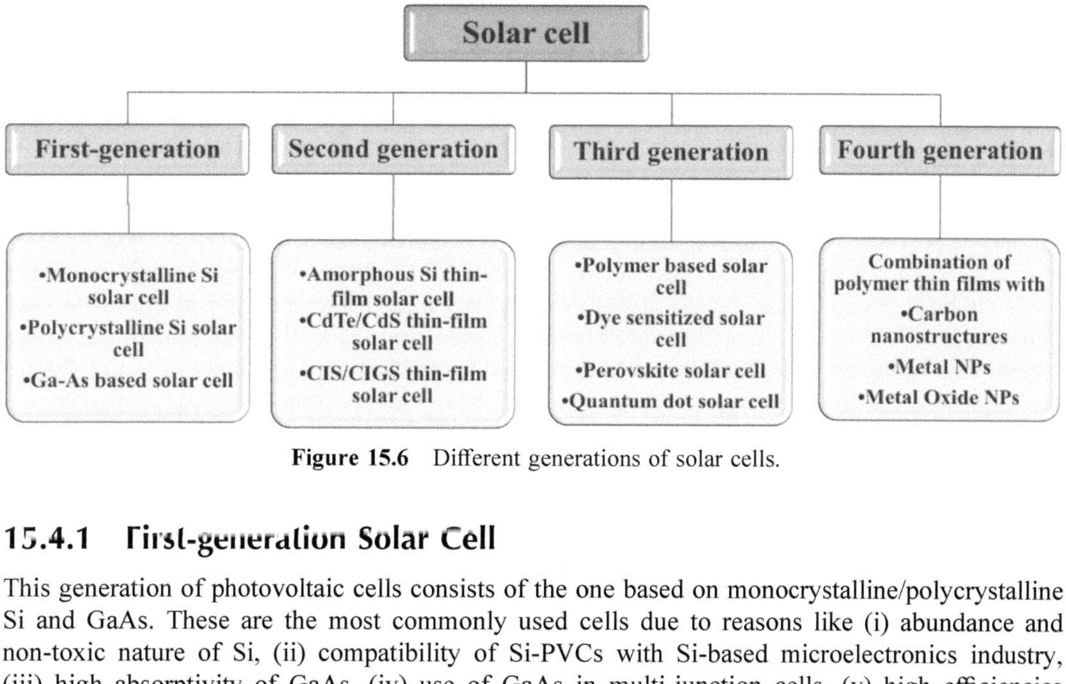

Figure 15.6 Different generations of solar cells.

15.4.1 First-generation Solar Cell

This generation of photovoltaic cells consists of the one based on monocrystalline/polycrystalline Si and GaAs. These are the most commonly used cells due to reasons like (i) abundance and non-toxic nature of Si, (ii) compatibility of Si-PVCs with Si-based microelectronics industry, (iii) high absorptivity of GaAs, (iv) use of GaAs in multi-junction cells, (v) high efficiencies offered by both Si- and GaAs-based cells. This initial concept of the solar cell was microstructured based. Gradually modifications began to take place, and the incorporation of nanotechnology, which began in the second generation, continued till the fourth generation. These advancements will be seen further.

15.4.2 Second-generation Solar Cell

This generation comprises thin-film technology-based solar cells. Some common materials used for this purpose are amorphous silicon, cadmium sulfide (CdS), cadmium telluride (CdTe), copper indium gallium diselenide (CIGS), copper indium diselenide (CIS), etc. Thin film technology is advantageous as they require lesser amount of material and hence are cheaper. However, one of the problems related to these traditional thin-film solar cells is the low efficiency it offers as compared to wafer-based technology. This is the reason why it has not been able to completely replace first-generation solar cells. Nanotechnology has some role to play in improving their performance. In nano-structured thin-film solar cells, the absorber layer is of the order of a few hundred nanometers as compared to several micrometers in the conventional thin-film solar cell. In such thin films, holes, and electrons generated have to travel much shorter distances which is beneficial for reducing recombination losses. Another application of nanotechnology in thin-film solar cells is modifying the bandgap of the material in the absorber and window layer by altering the size of the nanoparticles. Also, in nanostructured thin-film solar cells, the film thickness is much smaller than the effective optical path for absorption. These thin films having a width of a few nanometers can be assembled to make tandem solar cells that offer a wide bandgap for the conversion of different parts of the solar spectrum. Another way to improve the performance is by embedding different nanostructures in the planar thin-film solar cell. Fan et al. studied the CdS nanopillar/CdTe thin-film hybrid solar cell and found it to give better conversion efficiency than that of the planar CdS/CdTe cell (Yu et al. 2012). Tonpe et al. (2019) reported PANI/ZnO hybrid thin-film solar cell by utilizing the 1D morphology of ZnO nanorods for ultrafast charge transfer. All these factors help in improving the low efficiency as was seen in traditional thin-film solar cells.

15.4.3 Third-generation Solar Cell

These generation cells are designed to improve the efficiency obtained in second-generation thin film deposition along with employing new organic and nano-structured materials. They are based on phenomena such as multiple exciton generation and hot carrier collection. These generation cells are suitable for large-scale production with high efficiencies, but their cost per watt of electricity is still an issue to be addressed. They include quantum dot solar cells, organic and polymer solar cells, dye-sensitized solar cells, and multijunction solar cells. Dye-sensitized solar cells consist of a transparent conductive oxide anode with nano-crystalline mesoporous oxide deposited on it, a charge transfer dye, an electrolyte, and a catalyst coated crystal as the cathode. Common photoanodes used are TiO_2 and ZnO. They have been used in different nanostructures like nanowires, nanotubes, nanosheets, core-shell and 3D hierarchical nanostructures (Ye et al. 2015). These nanostructures help in improving the efficiency of DSSCs due to their unique properties like excellent electron transport and charge collection in 1D structures and large surface area in 3D structures (Macak et al 2005, Martinson et al. 2007, Jiang et al. 2007). Plasmonic nanoparticles are also infused in metal oxide film for increasing the device's efficiency (Kandasamy et al 2022). Even for counter electrodes in DSSCs, nobel metals like Pt, Au, and Ag are trying to be replaced by carbon-based nanostructures in order to reduce the cost (Wang and Hu 2012b, Hwang et al. 2015). Rahman et al. (2017) studied the performance of DSSC by investigating the effect of graphene oxide content in preparing reduced graphene oxide films for the counter electrode in the cell. Perovskite solar cell also consists of a transparent conductive oxide anode (generally FTO) and counter electrode along with an absorber layer sandwiched b/w electron transport mesoporous layer and hole transport layer. They are thin, low cost, highly efficient, and lightweight, however, their stability is the main challenge. In order to minimize the degradation, absorber material can be encapsulated within a carbon nanotube/polymer composite (Habisreutinger et al. 2014). Carbon-based nanomaterials like carbon nanotube, graphene, and graphene oxide have also been used as hole transport material and counter electrode material in perovskite solar cells (Moore and Wei 2021). The Quantum confinement effect seen in low dimensional perovskite solar cells not only reduces the losses but also increases the efficiency via carrier multiplication and is also beneficial in the formation of tandem devices (McDonald et al. 2019).

15.4.4 Fourth-generation Solar Cell

This generation of solar cells is a result of the combination of polymer thin films and inorganic nanomaterials. They are designed with a view to utilizing the flexibility and low cost of polymers along with improved charge transport and device lifetime due to nanostructures. The commonly used nanomaterials are metal nanoparticles, metal oxide nanoparticles, carbon nanotubes, graphene, and graphene oxide. These are used in different parts of a polymer solar cell like transparent conducting electrodes (TCE), charge transport layer, interlayer, and active layer. Indium tin oxide (ITO) is the most preferred TCE in solar cells due to its low sheet resistance and high visible light transmission. Novel technologies used for this purpose include metal films of nanometer thickness (Ghosh et al. 2010), silver nanowires (De et al. 2009), and carbon-based nanostructures (Wu et al. 2004, Sun and Shi 2013). Graphene can be used as TCE in polymer solar cells (PSC) since it is lightweight, flexible, and has good conductivity and transparency. The use of graphene in TCE showed a slight increment in the efficiency of PSC as compared to that of ITO (Sun et al. 2015). It can even be incorporated in a small amount in the host matrix of TCE to increase transparency. Nanostructures are also used in designing the active layer of PSCs. Metallic nanoparticles like Au and Ag are incorporated into the active layer in order to increase the absorption of incident rays owing to the longer optical path lengths they produce (Wang et al. 2011a, Wang et al. 2011b). The use of doped multiwalled carbon nanotubes (MWCNTs) in the active layers also led to increased efficiency from 3% to 4.1% in organic solar cells (Lee et al. 2011). Nanomaterials are even widely used in charge transport layers. For hole-transport layers (HTL), poly (3,4-ethylene dioxythiophene)-poly (styrene sulfonate) (PEDOT: PSS) have been the most commonly used, but devices made with this have less lifetime. So thin films of transition metal oxides like tungsten oxide (Han et al 2009), vanadium oxide (Li et al. 2006), and nickel-oxide (Park et al. 2010) have the advantage of high work functions, air stability high transparency, and an excellent ohmic contact with electrodes have been tested. CNTs have high thermal and chemical stability along with excellent energy level matching for hole injection, so they can be used to form composites with polymer matrix for HTLs (Díez-Pascual et al. 2010). Also, graphene oxide has been tested as an effective material for HTLs owing to its stable nature and high vertical conductivity, which allows tunneling between planar overlapping sheets (Li et al. 2010b). One can tune the energy levels of carbon-based nanomaterials through chemical modification so that they can even act as an electron transport layer. N-doped MWCNTs had a lower work function (4.6 eV) than pristine nanotubes (4.8 eV) which made them suitable for electron injection instead of hole injection (Lee et al. 2011). Similarly, esterification of -COOH groups of graphene oxide to -COOC groups also decreased the work function from 4.7 eV to 4.0 eV to be used in ETLs (Chang et al. 2012).

15.5 CONCLUSION AND FUTURE PERSPECTIVES

It is evident that nanotechnology has a great role to play in enhancing performance and reducing the cost of energy generation and storage devices. Other than those discussed in this write-up, there are many more energy-related promising applications of nanostructured materials like their use in batteries, hydrogen production and storage, biogas production, etc. However, it should be noted that there are many challenges that arise using nanostructured materials in different devices, so research is also needed in order to resolve them.

In fuel cells, intensive work is still required to solve the issues related to hydrogen generation and storage, the high cost of generator production and materials, and the short lifetime in order to commercialize properly. Research to minimize catalyst poisoning will lead to a durable cell which in turn also reduces the present effective cost. Though many nanostructured electrocatalysts have been studied for fuel cell applications, continuous research is still required in order to find the material with a better balance between performance to cost and thus wholly replace conventional platinum on a large scale. A study of the functional principles of nanostructured supports and reaction mechanisms of multimetal catalysts is also required. Nafion membrane used in PEMFCs suffers from drawbacks like excessive fuel crossover and reduction in conductivity of proton at higher temperatures. One can either look for an alternative or modify the Nafion with filler materials. Carbon nanomaterials with properly tuned properties in optimized quantity could be the best option for this.

Research in supercapacitors aims to improve energy density, maintain high power, reduce cost, and increase lifetime. The primary way to do so is by synthesizing new electrode and electrolyte materials. The cost of a supercapacitor majorly (40–60%) depends on the cost of electrodes used. Recently, several studies have focused on the use of high surface area carbon nanomaterials obtained from biomass as it's a sustainable and low-cost approach. However, there are problems associated with it, like difficulty in tuning properties of biomass obtained carbon and their unknown consequence on the electrochemical behavior of supercapacitors. Therefore, research is needed in this direction to understand the fundamental behavior of these materials. One can also develop theoretical models to predict the capacitive properties of biomass-derived carbon. Aqueous and ionic electrolyte materials should also be researched in order to compete with the organic ones in terms of wide decomposition voltage window and lower cost, respectively. For commercialization purposes, techniques should be developed to have better control over pore size distribution.

Though nanotechnology has been applied at a greater level in developing new generations of solar cells, its implementation at an industrial scale is something to be worked upon. Thin-film cells at present have not been developed to be efficient enough in order to replace wafer-based first-generation cells. Third- and fourth-generation cells offer efficiency better than second-generation; however, their short lifetime restricts commercialization. So, addressing these drawbacks in future research on photovoltaic cells is important. Also, the improvement of nanomaterials can enhance the performance of thermophotovoltaic cells, which use the IR wavelength for energy harvesting. Another problem in the long run use of solar cells is that the materials used in them are mostly toxic to the environment, so bio-based solar cells can be used for clean production.

It would be fair enough to say that nanotechnology has the power to solve the future energy crisis issue, and thus one can expect a bright future as these technologies continue to be developed. This write-up can act as a mini-review for readers, which concludes that the need of the hour is to continue exploring new nanomaterials to be utilized and new nanostructured-based designs for devices to be worked upon.

References

Abbas, G., Raza, R., Ashfaq Ahmad, M., Ajmal Khan, M., Jafar Hussain, M., Ahmad, M., Aziz, H., et al. 2017. Electrochemical investigation of mixed metal oxide nanocomposite electrode for low temperature solid oxide fuel cell. Int. J. Mod. Phys. B. 31(27): 1750193.

Alcaide, F., Alvarez, G., Cabot, P.L., Miguel, O. and Querejeta, A. 2010. Performance of carbon-supported PtPd as catalyst for hydrogen oxidation in the anodes of proton exchange membrane fuel cells. Int. J. Hydrog. Energy. 35(20): 11634–11641.

An, K.H., Kim, W.S., Park, Y.S., Choi, Y.C., Lee, S.M., Chung, D.C., Bae, D.J., et al. 2001. Supercapacitors using single-walled carbon nanotube electrodes. Adv. Mater. 13(7): 497–500.

Ando, Y., Sasaki, K. and Adzic, R. 2009. Electrocatalysts for methanol oxidation with ultra low content of Pt and Ru. Electrochem Commun. 11(6): 1135–1138.

Antunes, R.A., De Oliveira, M.C., Ett, G. and Ett, V. 2011. Carbon materials in composite bipolar plates for polymer electrolyte membrane fuel cells: a review of the main challenges to improve electrical performance. J. Power Sources. 196(6): 2945–2961.

Applications of Nanotechnology | National Nanotechnology Initiative. Available at: https://www.nano.gov/about-nanotechnology/applications-nanotechnology (Accessed: August 19, 2022).

Asim, S., Javed, M.S., Hussain, S., Rana, M., Iram, F., Lv, D., Hashim, M., et al. 2019. RuO_2 nanorods decorated CNTs grown carbon cloth as a free standing electrode for supercapacitor and lithium ion batteries. Electrochim. Acta. 326: 135009.

Aval, L.F., Ghoranneviss, M. and Pour, G.B. 2018. High-performance supercapacitors based on the carbon nanotubes, graphene and graphite nanoparticles electrodes. Heliyon. 4(11): e00862.

Behzadi Pour, G., Fekri Aval, L. and Mirzaee, M. 2018. Flexible graphene supercapacitor based on the PVA electrolyte and $BaTiO_3$/PEDOT: PSS composite separator. J. Mater. Sci: Mater. Electron. 29(20): 17432–17437.

Beknalkar, S.A., Teli, A.M., Harale, N.S., Shin, J.C. and Patil, P.S. 2021. Construction of IrO_2@Mn_3O_4 core-shell heterostructured nanocomposites for high performance symmetric supercapacitor device. J. Alloys Compd. 887: 161328.

Bhojane, P., Sharma, A., Pusty, M., Kumar, Y., Sen, S. and Shirage, P. 2017. Synthesis of ammonia-assisted porous nickel ferrite ($NiFe_2O_4$) nanostructures as an electrode material for supercapacitors. J. Nanosci. Nanotechnol. 17(2): 1387–1392.

Bi, W., Wu, Y., Liu, C., Wang, J., Du, Y., Gao, G., Wu, G., et al. 2018. Gradient oxygen vacancies in V_2O_5/PEDOT nanocables for high-performance supercapacitors. ACS Appl. Energy Mater. 2(1): 668–677.

Boujibar, O., Ghamouss, F., Ghosh, A., Achak, O. and Chafik, T. 2019. Activated carbon with exceptionally high surface area and tailored nanoporosity obtained from natural anthracite and its use in supercapacitors. J. Power Sources. 436: 226882.

Brouzgou, A., Song, S.Q. and Tsiakaras, P. 2012. Low and non-platinum electrocatalysts for PEMFCs: current status, challenges and prospects. Appl. Catal. B. 127: 371–388.

Cai, W., Kankala, R.K., Xiao, M., Zhang, N. and Zhang, X. 2020. Three-dimensional hollow N-doped ZIF-8-derived carbon@MnO_2 composites for supercapacitors. Appl. Surf. Sci. 528: 146921.

Chai, G.S., Shin, I.S. and Yu, J.S. 2004. Synthesis of ordered, uniform, macroporous carbons with mesoporous walls templated by aggregates of polystyrene spheres and silica particles for use as catalyst supports in direct methanol fuel cells. Adv. Mater. 16(22): 2057–2061.

Chang, C.Y., Cheng, Y.J., Hung, S.H., Wu, J.S., Kao, W.S., Lee, C.H. and Hsu, C.S. 2012. Combination of molecular, morphological, and interfacial engineering to achieve highly efficient and stable plastic solar cells. Adv. Mater. 24(4): 549–553.

Chen, H., Wang, S.X., Zhao, W.L., Zhang, N.N., Zheng, Y.P. and Sun, Y.M. 2015. Preparation of Pt/TiO_2 nanofibers and their electrocatalytic activity towards methanol oxidation. Acta Phys. Sin. 31(2): 302–308.

Chockalingam, R. and Basu, S. 2011. Impedance spectroscopy studies of Gd–CeO_2–(LiNa)CO_3 nano composite electrolytes for low temperature SOFC applications. Int. J. Hydrog. Energy. 36(22): 14977–14983.

Cui, Z., Guo, C.X. and Li, C.M. 2013. Self-assembled phosphomolybdic acid–polyaniline–graphene composite-supported efficient catalyst towards methanol oxidation. J. Mater. Chem. A. 1(22): 6687–6692.

Das, A.K., Karan, S.K. and Khatua, B.B. 2015. High energy density ternary composite electrode material based on polyaniline (PANI), molybdenum trioxide (MoO_3) and graphene nanoplatelets (GNP) prepared by sono-chemical method and their synergistic contributions in superior supercapacitive performance. Electrochim. Acta. 180: 1–15.

De, S., Higgins, T.M., Lyons, P.E., Doherty, E.M., Nirmalraj, P.N., Blau, W.J., Boland, J.J., et al. 2009. Silver nanowire networks as flexible, transparent, conducting films: extremely high DC to optical conductivity ratios. ACS Nano. 3(7): 1767–1774.

Devanathan, R. 2008. Recent developments in proton exchange membranes for fuel cells. Energy Environ. Sci. 1(1): 101–119.

Díez-Pascual, A.M., Naffakh, M., González-Domínguez, J.M., Ansón, A., Martínez-Rubi, Y., Martínez, M.T., Simard, B., et al. 2010. High performance PEEK/carbon nanotube composites compatibilized with polysulfones-I. Structure and thermal properties. Carbon. 48(12): 3485–3499.

dos Santos-Gómez, L., Porras-Vázquez, J.M., Losilla, E.R. and Marrero-López, D. 2015. Ti-doped $SrFeO_3$ nanostructured electrodes for symmetric solid oxide fuel cells. RSC Adv. 5(130): 107889–107895.

Dsoke, S., Moretti, A., Giuli, G. amd Marassi, R. 2011. Rotating disc electrode study of Pt-Co-$Cs_{2.5}PW_{12}O_{40}$ composite electrodes toward oxygen reduction reaction. Int. J. Hydrog. Energy. 36(13): 8098–8102.

Du, J., Liu, L., Yu, Y., Zhang, Y., Lv, H. and Chen, A. 2019a. N-doped ordered mesoporous carbon spheres derived by confined pyrolysis for high supercapacitor performance. J. Mater. Sci. Technol. 35(10): 2178–2186.

Du, W., Zhang, Z., Du, L., Fan, X., Shen, Z., Ren, X., Zhao, Y., et al. 2019b. Designing synthesis of porous biomass carbon from wheat straw and the functionalizing application in flexible, all-solid-state supercapacitors. J. Alloys. Compd. 797: 1031–1040.

Feng, C., Zhang, J., He, Y., Zhong, C., Hu, W., Liu, L. and Deng, Y. 2015. Sub-3 nm Co_3O_4 nanofilms with enhanced supercapacitor properties. ACS Nano. 9(2): 1730–1739.

Fıçıcılar, B., Bayrakçeken, A. and Eroğlu, İ. 2009. Effect of Pd loading in Pd-Pt bimetallic catalysts doped into hollow core mesoporous shell carbon on performance of proton exchange membrane fuel cells. J. Power Sources. 193(1): 17–23.

Ghosh, D.S., Chen, T.L. and Pruneri, V. 2010. High figure-of-merit ultrathin metal transparent electrodes incorporating a conductive grid. Appl. Phys. Lett. 96(4): 041109.

Götz, M. and Wendt, H. 1998. Binary and ternary anode catalyst formulations including the elements W, Sn and Mo for PEMFCs operated on methanol or reformate gas. Electrochim. Acta. 43(24): 3637–3644.

Guo, X. and Waser, R. 2006. Electrical properties of the grain boundaries of oxygen ion conductors: acceptor-doped zirconia and ceria. Prog. Mater. Sci. 51(2): 151–210.

Guo, D.J. and Cui, S.K. 2009. Hollow PtCo nanospheres supported on multi-walled carbon nanotubes for methanol electrooxidation. J. Colloid. Interface Sci. 340(1): 53–57.

Guo, S., Li, S., Hu, T., Gou, G., Ren, R., Huang, J., Xie, M., et al. 2013. Graphene decorated with Pd nanoparticles via electrostatic self-assembly: a highly active alcohol oxidation electrocatalyst. Electrochim. Acta. 109: 276–282.

Habisreutinger, S.N., Leijtens, T., Eperon, G.E., Stranks, S.D., Nicholas, R.J. and Snaith, H.J. 2014. Carbon nanotube/polymer composites as a highly stable hole collection layer in perovskite solar cells. Nano Lett. 14(10): 5561–5568.

Han, S., Shin, W.S., Seo, M., Gupta, D., Moon, S.J. and Yoo, S. 2009. Improving performance of organic solar cells using amorphous tungsten oxides as an interfacial buffer layer on transparent anodes. Org. Electron. 10(5): 791–797.

Hao, S.J., Wang, C., Liu, T.L., Mao, Z.M., Mao, Z.Q. and Wang, J.L. 2017. Fabrication of nanoscale yttria stabilized zirconia for solid oxide fuel cell. Int. J. Hydrog. Energy. 42(50): 29949–29959.

Huang, Y.L., Hussain, A.M. and Wachsman, E.D. 2018. Nanoscale cathode modification for high performance and stable low-temperature solid oxide fuel cells (SOFCs). Nano Energy. 49: 186–192.

Hwang, S., Batmunkh, M., Nine, M.J., Chung, H. and Jeong, H. 2015. Dye-sensitized solar cell counter electrodes based on carbon nanotubes. Chem. Phys. Chem. 16(1): 53–65.

Ji, J., Zhang, X., Liu, J., Peng, L., Chen, C., Huang, Z., Li, L., et al. 2015. Assembly of polypyrrole nanotube@ MnO_2 composites with an improved electrochemical capacitance. Mater. Sci. Eng. B. 198: 51–56.

Jiang, C.Y., Sun, X.W., Lo, G.Q., Kwong, D.L. and Wang, J.X. 2007. Improved dye-sensitized solar cells with a ZnO-nanoflower photoanode. Appl. Phys. Lett. 90(26): 263501.

Kandasamy, M., Selvaraj, M., Kumarappan, C. and Murugesan, S. 2022. Plasmonic Ag nanoparticles anchored ethylenediamine modified TiO_2 nanowires@graphene oxide composites for dye-sensitized solar cell. J. Alloys. Compd. 902: 163743.

Kiyani, R., Parnian, M.J. and Rowshanzamir, S. 2017. Investigation of the effect of carbonaceous supports on the activity and stability of supported palladium catalysts for methanol electro-oxidation reaction. Int. J. Hydrog Energy. 42(36): 23070–23084.

Kubra, K.T., Sharif, R., Patil, B., Javaid, A., Shahzadi, S., Salman, A., Siddique, S., et al. 2020. Hydrothermal synthesis of neodymium oxide nanoparticles and its nanocomposites with manganese oxide as electrode materials for supercapacitor application. J. Alloys Compd. 815: 152104.

Kwak, C.S., Ko, T.H., Lee, J.H., Kim, H.Y. and Kim, B.S. 2020. Flexible transparent symmetric solid-state supercapacitors based on NiO-decorated nanofiber-based composite electrodes with excellent mechanical flexibility and cyclability. ACS Appl. Energy Mater. 3(3): 2394–2403.

Lee, D.C., Yang, H.N., Park, S.H. and Kim, W.J. 2014. Nafion/graphene oxide composite membranes for low humidifying polymer electrolyte membrane fuel cell. J. Membr. Sci. 452: 20–28.

Lee, J.M., Park, J.S., Lee, S.H., Kim, H., Yoo, S. and Kim, S.O. 2011. Selective electron-or hole-transport enhancement in bulk-heterojunction organic solar cells with N-or B-doped carbon nanotubes. Adv. Mater. 23(5): 629–633.

Lerner, E.J. 2004. Less is more with aerogels. Industrial Physicist. 10: 26–32.

Li, G., Chu, C.W., Shrotriya, V., Huang, J. and Yang, Y. 2006. Efficient inverted polymer solar cells. Appl. Phys. Lett. 88(25): 253503.

Li, B., Qiao, J., Yang, D., Zheng, J., Ma, J., Zhang, J. and Wang, H. 2009. Synthesis of a highly active carbon-supported Ir–V/C catalyst for the hydrogen oxidation reaction in PEMFC. Electrochim. Acta. 54(24): 5614–5620.

Li, Y., Gao, W., Ci, L., Wang, C. and Ajayan, P.M. 2010a. Catalytic performance of Pt nanoparticles on reduced graphene oxide for methanol electro-oxidation. Carbon. 48(4): 1124–1130.

Li, S.S., Tu, K.H., Lin, C.C., Chen, C.W. and Chhowalla, M. 2010b. Solution-processable graphene oxide as an efficient hole transport layer in polymer solar cells. ACS Nano. 4(6): 3169–3174.

Li, G.C., Liu, M., Wu, M.K., Liu, P.F., Zhou, Z., Zhu, S.R., Liu, R., et al. 2016. MOF-derived self-sacrificing route to hollow NiS_2/ZnS nanospheres for high performance supercapacitors. RSC Adv. 6(105): 103517–103522.

Li, X., Niu, X., Zhang, W., He, Y., Pan, J., Yan, Y. and Qiu, F. 2017. One-pot anchoring of Pd nanoparticles on nitrogen-doped carbon through dopamine self-polymerization and activity in the electrocatalytic methanol oxidation reaction. Chem. Sus. Chem. 10(5): 976–983.

Liu, Y., Zheng, N., Chao, W., Liu, H. and Wang, Y. 2010a. A novel nanocomposite catalytic cathode for direct methanol fuel cells. Electrochim. Acta. 55(20): 5617–5623.

Liu, C., Yu, Z., Neff, D., Zhamu, A. and Jang, B.Z. 2010b. Graphene-based supercapacitor with an ultrahigh energy density. Nano Lett. 10(12): 4863–4868.

Lu, Y., Du, S. and Steinberger-Wilckens, R. 2016. One-dimensional nanostructured electrocatalysts for polymer electrolyte membrane fuel cells: a review. Appl. Catal. B. 199: 292–314.

Lv, Y., Ding, L., Wu, X., Guo, N., Guo, J., Hou, S., Tong, F., et al. 2020. Coal-based 3D hierarchical porous carbon aerogels for high performance and super-long life supercapacitors. Sci. Rep. 10(1): 1–11.

Macák, J.M., Tsuchiya, H., Ghicov, A. and Schmuki, P. 2005. Dye-sensitized anodic TiO_2 nanotubes. Electrochem. Commun. 7(11): 1133–1137.

Manibalan, G., Rajesh Kumar, M., Murugadoss, G., Yesuraj, J., Mohan Kumar, R. and Jayavel, R. 2021. Novel chemical route for synthesis of CeO_2–ZnO nanocomposite towards high electrochemical supercapacitor application. J. Mater. Sci.: Mater. Electron. 32(7): 8746–8755.

Martinson, A.B., Elam, J.W., Hupp, J.T. and Pellin, M.J. 2007. ZnO nanotube based dye-sensitized solar cells. Nano Lett. 7(8): 2183–2187.

McDonald, C., Ni, C., Maguire, P., Connor, P., Irvine, J.T., Mariotti, D. and Svrcek, V. 2019. Nanostructured perovskite solar cells. Nanomaterials. 9(10): 1481.

Meng, Q., Cai, K., Chen, Y. and Chen, L. 2017. Research progress on conducting polymer based supercapacitor electrode materials. Nano Energy. 36: 268–285.

Merlet, C., Péan, C., Rotenberg, B., Madden, P.A., Daffos, B., Taberna, P.L., Simon, P., et al. 2013. Highly confined ions store charge more efficiently in supercapacitors. Nat. Commun. 4(1): 1–6.

Moore, K. and Wei, W. 2021. Applications of carbon nanomaterials in perovskite solar cells for solar energy conversion. Nano Mater. Sci. 3(3): 276–290.

Morozan, A., Jousselme, B. and Palacin, S. 2011. Low-platinum and platinum-free catalysts for the oxygen reduction reaction at fuel cell cathodes. Energy Environ. Sci. 4(4): 1238–1254.

Myung, J.H., ho Shin, T., Kim, S.D., Park, H.G., Moon, J. and Hyun, S.H. 2015. Optimization of Ni–zirconia based anode support for robust and high-performance 5×5 cm^2 sized SOFC via tape-casting/co-firing technique and nano-structured anode. Int. J. Hydrog. Energy. 40(6): 2792–2799.

Niu, Q., Gao, K., Tang, Q., Wang, L., Han, L., Fang, H., Zhang, Y., et al. 2017. Large-size graphene-like porous carbon nanosheets with controllable N-doped surface derived from sugarcane bagasse pith/chitosan for high performance supercapacitors. Carbon. 123: 290–298.

Park, S.Y., Kim, H.R., Kang, Y.J., Kim, D.H. and Kang, J.W. 2010. Organic solar cells employing magnetron sputtered p-type nickel oxide thin film as the anode buffer layer. Sol. Energy Mater. Sol. Cells. 94(12): 2332–2336.

Prabhuram, J., Zhao, T.S., Liang, Z.X. and Chen, R. 2007. A simple method for the synthesis of PtRu nanoparticles on the multi-walled carbon nanotube for the anode of a DMFC. Electrochim. Acta. 52(7): 2649–2656.

Priyadharsini, C.I., Marimuthu, G., Pazhanivel, T., Anbarasan, P.M., Aroulmoji, V., Siva, V. and Mohana, L. 2020. Sol–Gel synthesis of Co_3O_4 nanoparticles as an electrode material for supercapacitor applications. J. Solgel. Sci. Technol. 96(2): 416–422.

Qiao, J., Lin, R., Li, B., Ma, J. and Liu, J. 2010. Kinetics and electrocatalytic activity of nanostructured Ir–V/C for oxygen reduction reaction. Electrochim. Acta. 55(28): 8490–8497.

Qiu, X., Dai, Y., Zhu, X., Zhang, H., Wu, P., Tang, Y. and Wei, S. 2016. Template-engaged synthesis of hollow porous platinum–palladium alloy nanospheres for efficient methanol electro-oxidation. J. Power Sources. 302: 195–201.

Rahman, M.Y.A., Sulaiman, A.S., Umar, A.A. and Salleh, M.M. 2017. Dye-sensitized solar cell (DSSC) utilizing reduced graphene oxide (RGO) films counter electrode: effect of graphene oxide (GO) content. J. Mater. Sci.: Mater. Electron. 28(2): 1674–1678.

Rani, J.R., Thangavel, R., Kim, M., Lee, Y.S. and Jang, J.H. 2020. Ultra-high energy density hybrid supercapacitors using MnO_2/reduced graphene oxide hybrid nanoscrolls. Nanomaterials. 10(10): 2049.

Sahu, A.K., Ketpang, K., Shanmugam, S., Kwon, O., Lee, S. and Kim, H. 2016. Sulfonated graphene–nafion composite membranes for polymer electrolyte fuel cells operating under reduced relative humidity. J Phys. Chem. C. 120(29): 15855–15866.

Sennu, P., Choi, H.J., Baek, S.G., Aravindan, V. and Lee, Y.S. 2016. Tube-like carbon for Li-ion capacitors derived from the environmentally undesirable plant: prosopis juliflora. Carbon. 98: 58–66.

Sevilla, M. and Mokaya, R. 2014. Energy storage applications of activated carbons: supercapacitors and hydrogen storage. Energy Environ Sci. 7(4): 1250–1280.

Sevilla, M. and Fuertes, A.B. 2014. Direct synthesis of highly porous interconnected carbon nanosheets and their application as high-performance supercapacitors. ACS Nano. 8(5): 5069–5078.

Shin, S.S., Kim, J.H., Li, G., Lee, S.Y., Son, J.W., Kim, H. and Choi, M. 2019. A highly activated and integrated nanoscale interlayer of cathodes in low-temperature solid oxide fuel cells via precursor-solution electrospray method. Int. J. Hydrog. Energy. 44(9): 4476–4483.

Sivakkumar, S.R., Kim, W.J., Choi, J.A., MacFarlane, D.R., Forsyth, M. and Kim, D.W. 2007. Electrochemical performance of polyaniline nanofibres and polyaniline/multi-walled carbon nanotube composite as an electrode material for aqueous redox supercapacitors. J. Power Sources. 171(2): 1062–1068.

Su, F., Poh, C.K., Tian, Z., Xu, G., Koh, G., Wang, Z., Liu, Z., et al. 2010. Electrochemical behavior of Pt nanoparticles supported on meso-and microporous carbons for fuel cells. Energy Fuels. 24(7): 3727–3732.

Sun, Y. and Shi, G. 2013. Graphene/polymer composites for energy applications. J. Polym Sci B: Polym Phys. 51(4): 231–253.

Sun, Y., Zhang, W., Chi, H., Liu, Y., Hou, C.L. and Fang, D. 2015. Recent development of graphene materials applied in polymer solar cell. Renew. Sustain. Energy. Rev. 43: 973–980.

Teli, A.M., Beknalkar, S.A., Karade, V.C., Mane, S.M., Go, J., Patil, P.S. and Shin, J.C. 2022. Electrospun deposited Mn_2O_3/GO nanofiber composite electrode for hybrid coin cell supercapacitor devices. J. Mater. Sci.: Mater. Electron. 33(11): 8844–8857.

Tian, J., Cui, C., Xie, Q., Qian, W., Xue, C., Miao, Y., Jin, Y., et al. 2018. EMIMBF 4–GBL binary electrolyte working at−70°C and 3.7 V for a high performance graphene-based capacitor. J. Mater. Chem. A. 6(8): 3593–3601.

Tonpe, D.A., Gattu, K.P., Kutwade, V.V., Sonawane, M.E., Dive, A.S. and Sharma, R. 2019. Development of organic/inorganic PANI/ZnO 1D nanostructured hybrid thin film solar cell by soft chemical route. J. Mater. Sci.: Mater. Electron. 30(17): 16056–16064.

Vattikuti, S.P., Devarayapalli, K.C., Nagajyothi, P.C. and Shim, J. 2020. Microwave synthesized dry leaf-like mesoporous $MoSe_2$ nanostructure as an efficient catalyst for enhanced hydrogen evolution and supercapacitor applications. Microchem. J. 153: 104446.

Wang, X., Huang, H., Holme, T., Tian, X. and Prinz, F.B. 2008. Thermal stabilities of nanoporous metallic electrodes at elevated temperatures. J. Power Sources. 175(1): 75–81.

Wang, D.H., Kim, D.Y., Choi, K.W., Seo, J.H., Im, S.H., Park, J.H., Park, O.O., et al. 2011a. Enhancement of donor–acceptor polymer bulk heterojunction solar cell power conversion efficiencies by addition of au nanoparticles. Angew. Chem. Int. Ed. 123(24): 5633–5637.

Wang, D.H., Park, K.H., Seo, J.H., Seifter, J., Jeon, J.H., Kim, J.K., Park, J.H., et al. 2011b. Enhanced power conversion efficiency in PCDTBT/$PC_{70}BM$ bulk heterojunction photovoltaic devices with embedded silver nanoparticle clusters. Adv. Energy Mater. 1(5): 766–770.

Wang, L., Imura, M. and Yamauchi, Y. 2012a. Tailored design of architecturally controlled Pt nanoparticles with huge surface areas toward superior unsupported Pt electrocatalysts. ACS Appl. Mater. Interfaces. 4(6): 2865–2869.

Wang, H. and Hu, Y.H. 2012b. Graphene as a counter electrode material for dye-sensitized solar cells. Energy Environ. Sci. 5(8): 8182–8188.

Wang, W., Lu, X., Zhu, M., Cao, Z., Li, C., Gao, Y., Li, L. et al. 2015. Rod-shaped CeO_2 intercalated Graphene for supporting Pt composite as anode catalysts for DMFCs. Electrochim. Acta. 176: 1338–1342.

Wang, L., Zhang, G., Zhang, X., Shi, H., Zeng, W., Zhang, H., Liu, Q., et al. 2017. Porous ultrathin carbon nanobubbles formed carbon nanofiber webs for high-performance flexible supercapacitors. J. Mater. Chem. A. 5(28): 14801–14810.

Wang, L., Li, Y., Miao, X., Qin, H., Ren, J., Wang, X., Sun, Z., et al. 2018. Three-dimensional graphene-carbon nanotubes composite-supported ptsn catalysts with a tunable microstructure to enhance electrocatalytic activity for ethanol oxidation. Int. J. Electrochem. Sci. 13: 11081–11095.

Wei, C., Pang, H., Zhang, B., Lu, Q., Liang, S. and Gao, F. 2013. Two-dimensional β-MnO_2 nanowire network with enhanced electrochemical capacitance. Sci. Rep. 3(1): 1–5.

Wu, Z., Chen, Z., Du, X., Logan, J.M., Sippel, J., Nikolou, M., Kamaras, K., et al. 2004. Transparent, conductive carbon nanotube films. Science. 305(5688): 1273–1276.

Wu, Q., Xu, Y., Yao, Z., Liu, A. and Shi, G. 2010. Supercapacitors based on flexible graphene/polyaniline nanofiber composite films. ACS Nano. 4(4): 1963–1970.

Xu, F. and Mu, S. 2014. Nanoceramic oxide hybrid electrolyte membranes for proton exchange membrane fuel cells. J. Nanosci. Nanotechnol. 14(2): 1169–1180.

Yang, J., Chen, X., Ye, F., Wang, C., Zheng, Y. and Yang, J. 2011. Core-shell CdSe@Pt nanocomposites with superior electrocatalytic activity enhanced by lateral strain effect. J. Mater. Chem. 21(25): 9088–9094.

Yang, X., Wang, X., Zhang, G., Zheng, J., Wang, T., Liu, X., Shu, C., et al. 2012. Enhanced electrocatalytic performance for methanol oxidation of Pt nanoparticles on Mn_3O_4-modified multi-walled carbon nanotubes. Int. J. Hydrog. Energy. 37(15): 11167–11175.

Ye, M., Wen, X., Wang, M., Iocozzia, J., Zhang, N., Lin, C. and Lin, Z. 2015. Recent advances in dye-sensitized solar cells: from photoanodes, sensitizers and electrolytes to counter electrodes. Mater. Today. 18(3): 155–162.

Yin, C., Li, J., Zhou, Y., Zhang, H., Fang, P. and He, C. 2018. Enhancement in proton conductivity and thermal stability in nafion membranes induced by incorporation of sulfonated carbon nanotubes. ACS Appl. Mater. Interfaces. 10(16): 14026–14035.

Yu, M., Long, Y.Z., Sun, B. and Fan, Z. 2012. Recent advances in solar cells based on one-dimensional nanostructure arrays. Nanoscale. 4(9): 2783–2796.

Zeis, R., Mathur, A., Fritz, G., Lee, J. and Erlebacher, J. 2007. Platinum-plated nanoporous gold: an efficient, low Pt loading electrocatalyst for PEM fuel cells. J. Power Sources. 165(1): 65–72.

Zhang, S.L., Li, C.X. and Li, C.J. 2014a. Chemical compatibility and properties of suspension plasma-sprayed $SrTiO_3$-based anodes for intermediate-temperature solid oxide fuel cells. J. Power Sources. 264: 195–205.

Zhang, H., Hu, Z., Li, M., Hu, L. and Jiao, S. 2014b. A high-performance supercapacitor based on a polythiophene/multiwalled carbon nanotube composite by electropolymerization in an ionic liquid microemulsion. J. Mater Chem. A. 2(40): 17024–17030.

Zhang, Z., Xiao, F., Qian, L., Xiao, J., Wang, S. and Liu, Y. 2014c. Facile synthesis of 3D MnO_2–graphene and carbon nanotube–graphene composite networks for high-performance, flexible, all-solid-state asymmetric supercapacitors. Adv. Energy Mater. 4(10): 1400064.

Zhang, Q., Yang, H., Lang, X., Zhang, X. and Wei, Y. 2019. 1-Ethyl-2,3-dimethylimidazolium tetra-fluoroborate ionic liquid mixture as electrolyte for high-voltage supercapacitors. Ionics. 25(1): 231–239.

Zhang, W., Zhou, Y., Hussain, A.M., Song, D., Miura, Y., Chen, Y., Luo, Z., et al. 2021. High-performance, thermal cycling stable, coking-tolerant solid oxide fuel cells with nanostructured electrodes. ACS Appl. Mater. Interfaces. 13(4): 4993–4999.

Zhou, W.P., Sasaki, K., Su, D., Zhu, Y., Wang, J.X. and Adzic, R.R. 2010. Gram-scale-synthesized Pd_2Co-supported Pt monolayer electrocatalysts for oxygen reduction reaction. J. Phys. Chem. C. 114(19): 8950–8957.

Zhao, Y., Zhan, L., Tian, J., Nie, S. and Ning, Z. 2010a. MnO_2 modified multi-walled carbon nanotubes supported Pd nanoparticles for methanol electro-oxidation in alkaline media. Int. J. Hydrog Energy. 35(19): 10522–10526.

Zhao, Y., Yang, X., Tian, J., Wang, F. and Zhan, L. 2010b. Methanol electro-oxidation on Ni@Pd core-shell nanoparticles supported on multi-walled carbon nanotubes in alkaline media. Int. J. Hydrog. Energy. 35(8): 3249–3257.

Zhou, J., Zhao, H., Mu, X., Chen, J., Zhang, P., Wang, Y., He, Y., et al. 2015. Importance of polypyrrole in constructing 3D hierarchical carbon nanotube@MnO_2 perfect core–shell nanostructures for high-performance flexible supercapacitors. Nanoscale. 7(35): 14697–14706.

Zhuiykov, S. 2014. Nanostructured Semiconductor Oxides for the Next Generation of Electronics and Functional Devices. Woodhead Publishing. 321–373.

Chapter 16

Passageway into a Clean World with the Magical Nanomaterial TiO_2

Tanmay Dagar, Sudeshna Sarkar, Biswarup Pal Choudhury and Swati De*

Department of Chemistry, University of Kalyani, West Bengal, Pin Code- 741 235, INDIA

16.1 INTRODUCTION

One of the major problems confronting us today is environmental pollution (Dozzi 2012). The human race had inherited a clean atmosphere, abundant water resources and fertile soil, all of which has served them endlessly. However, all these resources have been ravaged by different anthropogenic activities leading to air, water and soil pollution. Another potential threat to human health arises from the deposition of microbes on furniture, doors and windows and other surfaces. Prolonged exposure to such microbes and pollutants pose severe health hazards (Flannigan 2002). It is no wonder that the prime focus of present-day scientific research is searching for sustainable, renewable and environment-friendly technologies. In the quest for new materials to be used in such sustainable technologies, it was found that TiO_2 serves as a "*Magical Material*". Scientific research over the past few decades has produced a wealth of literature on TiO_2 materials. Towards the end of the last century, the development of nanomaterials further enriched this field. In this chapter, the interesting nanomaterial TiO_2 and its role in providing a "*Passageway into an Environmentally Clean World*" has been reviewed in detail.

Among the various applications of TiO_2 nanomaterials relevant to the environment, its photocatalytic activity, antimicrobial action, potential for use in solar light harvesting, self-cleaning and anti-fogging action, among others bear strong relevance to the context of this chapter. Microbial growth and deposition on surfaces of everyday use can be prevented by using photocatalytic coatings comprising TiO_2 nanoparticles (Parkin and Palgrave 2005). Undoped TiO_2 photocatalysts have shown promise in the decomposition of organic contaminants using clean solar energy without yielding any harmful by-products (Chun et al. 2009, Ziolli and Jardim 2002).

*Corresponding Authors: deswati1@gmail.com; sdechem1@klyuniv.ac.in

Contemporary research has seen rapid progress in the field of semiconductor-based photocatalysts, their mechanism of action, enhancing the photocatalytic efficiency and using them in mitigating the effects of environmental pollution. Significant applications include environmental remediation (Luo et al. 2019), generation of fuel via CO_2 photo reduction (Kumaravel et al. 2020), photocatalytic water splitting (Li et al. 2015), H_2 production (Corredor et al. 2019, Ganguly et al. 2019) and bacterial disinfection (Zhang and Xu 2019). There are several semiconductor-based photocatalytic materials, out of which TiO_2 is widely accepted due to its photostability, intrinsic electronic and surface properties, non-toxicity, low cost and environment friendliness (Fox and Dulay 1993, Hoffmann et al. 1995, Su et al. 2006, Wang et al. 2009). Recent years have witnessed exciting breakthroughs on TiO_2 materials. The first major advance was in 1972 when Fujishima and Honda reported the photoelectrochemical splitting of water using a TiO_2 anode and a Pt counter electrode (Fujishima and Honda 1972). TiO_2 photocatalyst was first used for environmental remediation in 1977 when the reduction of CN^- in water was reported (Frank and Bard 1977). This triggered further research on TiO_2 materials, as the potential for water and air purification using "free" solar energy was realized (Hoffmann et al. 1995, Fox and Dulay 1993, Zhao et al. 1998). Another significant finding was the discovery of TiO_2 surfaces with anti-fogging and self-cleaning abilities as reported by Wang et al. (1997). This was attributed to the superhydrophilic properties of the photoexcited TiO_2 surfaces (Wang et al. 1997). Another important work was the use of TiO_2 nanoparticles in the dye sensitized solar cell (DSSC), reported by Graetzel and O'Regan in 1991 (O'Regan et al. 1991)

However, TiO_2 materials have certain disadvantages *viz.* low visible light photoconversion efficiency, low quantum yield and high rate of electron-hole recombination. Such drawbacks limit their usage in practical and commercial applications (Guo et al. 2019). It was only natural that research on TiO_2 materials in the recent past focused on enhancing the sunlight harvesting ability of TiO_2 and on prevention of recombination of the photogenerated electron-hole pair. In this context, the most effective strategy to overcome the limitations of bulk TiO_2 has been the use of TiO_2 nanomaterials.

Several researchers have used different methods to improve the performance of TiO_2 photocatalysts by using nanomaterials. Liu et al. (2016) manufactured novel 3D mesoporous black $TiO_2/MoS_2/TiO_2$ nanosheets to achieve this. However, although the photocatalyst activity and recyclability improved, the cost of the process and the small quantity of TiO_2 produced restricted its application in industry.

In an alternative development, floating TiO_2-based photocatalysts were used to decompose oil (Wang et al. 2017b), *Floating photocatalysts* employ lightweight materials and thus can float on the water surface (Figure. 16.1). The main advantage is that the *floating photocatalyst* minimizes loss and prevents long-term contact between the photocatalyst and the pollutants (Xing et al. 2018).

In this book chapter, the reader will be led through a step-by-step journey into: (1) the history of development of TiO_2 materials; (2) the structure and polymorphic forms of TiO_2; (3) the semiconducting behavior of TiO_2; (4) the need for developing materials for environmental remediation; (5) comprehensive treatise on the TiO_2 nanomaterials i.e. synthesis, characterization and their versatile applications and finally (6) the important conclusions from the review and (7) the future perspective of the directions of research on design of TiO_2 nanomaterials for environmental applications.

16.2 HISTORY OF TiO_2 MATERIALS

Titanium oxide (TiO_2) is a semiconductor which has been commercially used as a white paint since the early 20th century. The photoactivity of TiO_2 was discovered when prolonged exposure to sunlight bleached the TiO_2 paint. This phenomenon was first reported during 1932–1934. It was also observed that in the presence of TiO_2, ammonia and ammonium salts get converted to nitrites upon exposure to sunlight. This could be explained by a photo-oxidization process. In 1938, the ability of TiO_2 to photobleach dyes was reported. The photobleaching was assigned to the formation of active oxygen species on the surface of TiO_2 due to the absorption of ultraviolet (UV) radiation. The term TiO_2 photocatalyst was first coined in 1956. The oxidation of organic solvents and the formation of H_2O_2 by TiO_2 were observed under UV irradiation. Fujishima and Honda first reported water photolysis by TiO_2 in presence of sunlight in the 1970s. Hydrogen was produced by the photocatalytic mechanism using TiO_2 in the 1980s, super hydrophilicity and the photocatalytic properties of the TiO_2 films were demonstrated in the 1990s. The application of TiO_2 catalysts in the fields of Nano engineering and nanotechnology was demonstrated in the early part of the 21st century. All these pioneering research works brought the magical material TiO_2 into limelight and paved the way for numerous industrial applications of TiO_2.

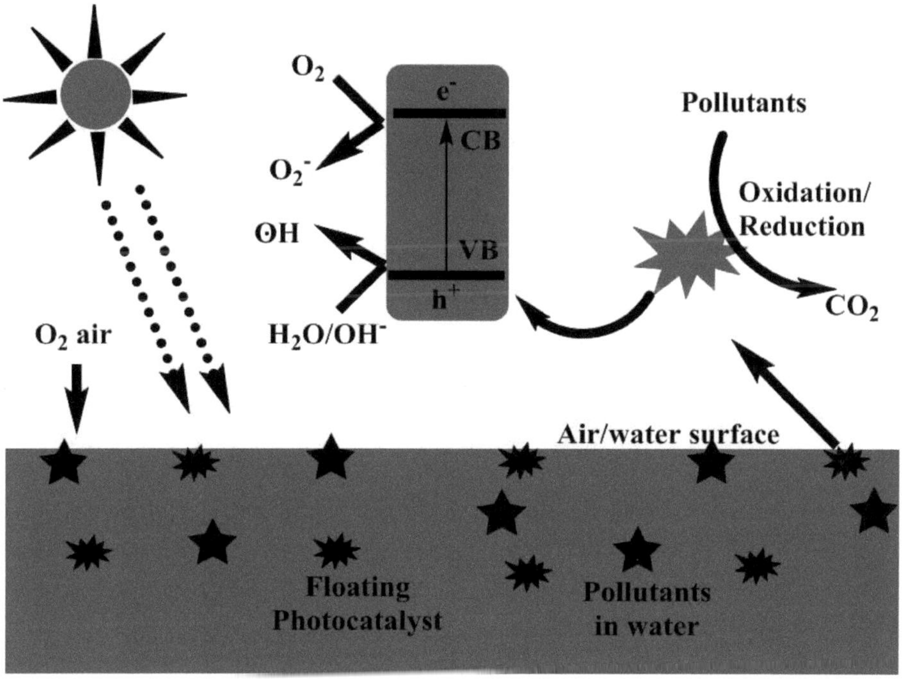

Figure 16.1 Scheme showing the mode of action of the floating photocatalyst.

The field of TiO_2 nanomaterials developed much later and garnered immense interest, owing to the unique properties such as high *surface-to-volume* ratio and quantum confinement. The catalytic activity of nanomaterials mainly arises due to their higher surface area, which leads to increased light absorption. This in turn leads to the increase in photo-induced charge carrier density for semiconductor nanomaterials like TiO_2 (Riu et al. 2006, Magaye et al. 2012).

16.3 DIFFERENT POLYMORPHIC FORMS OF TiO_2

In nature, TiO_2 is known to exist as four common polymorphic forms: (i) anatase (tetragonal), (ii) brookite (orthorhombic), (iii) rutile (tetragonal) and (iv) TiO_2 (B) (monoclinic) (Chen et al. 2015).

16.3.1 Anatase

Anatase TiO_2 has a tetragonal structure comprising TiO_6 octahedra (Figure 16.2). There is significant distortion of the constituent TiO_6 octahedra in anatase, (Linsebigler et al. 1995,

Muscat et al. 2002). The anatase phase is more stable than rutile at low temperature. Selloni, (2008) reported that the reactivity of {001} facet is greater than that of the {101} facet in an anatase crystal.

16.3.2 Rutile

Rutile TiO_2 has a tetragonal structure and contains six atoms per unit cell (Figure 16.2). The TiO_6 octahedra in rutile are only slightly distorted. The rutile phase is usually stable at high temperatures. It was found that anatase and brookite structures got transformed to the rutile phase after reaching a certain particle size, with the rutile phase becoming more stable than anatase for particle sizes greater than 14 nm, (Zhang et al. 2000). However, the catalytic ability of the rutile phase is low.

16.3.3 Brookite

For Brookite, the crystal system is the orthorhombic type. Its unit cell is composed of eight TiO_2 units with edge-sharing of the TiO_6 octahedra (Figure 16.2). Brookite has a larger cell volume than the other polymorphic forms and is also the least dense of the three forms. Table 16.1 summarizes the differences between the anatase, rutile and brookite phase of TiO_2.

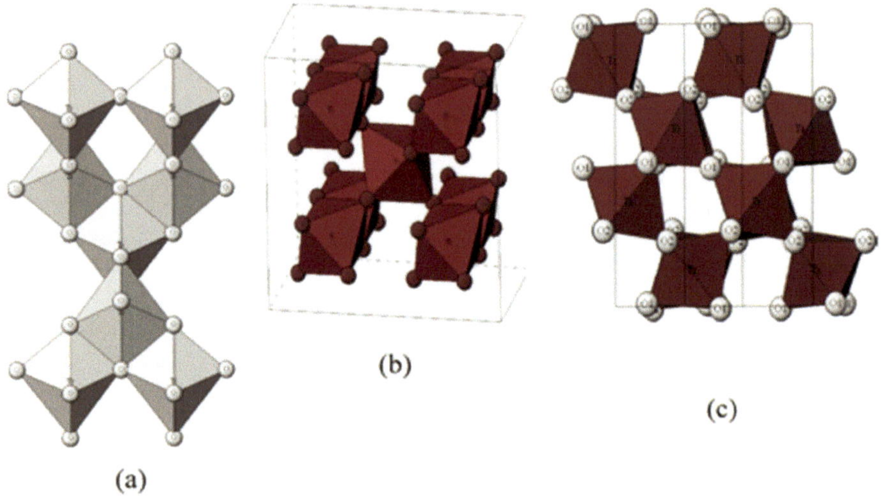

Figure 16.2 Crystalline structures of various polymorphs of TiO_2 : (a) anatase, (b) rutile, (c) brookite. [Reprinted from Applied Catalysis B: Environmental 125 (2012), 331-349., Copyright (2022), with permission from Elsevier]

16.4 SEMICONDUCTING BEHAVIOR OF TiO_2

TiO_2 is a semiconductor popularly used in heterogeneous photocatalysis (Emsley 2011, Goresy et al. 2001). The valence band of TiO_2 comprises the 2p orbitals of oxygen hybridized with the 3d orbitals of titanium, while the conduction band is composed of only the 3d orbitals of titanium (Paxton and Thien 1998). When UV light is incident on TiO_2, valence band electrons get excited to the conduction band leaving behind holes (h^+). The excited electrons (e^-) in the conduction band are now in a purely 3d state and due to dissimilar parity, the probability of the e^- returning to the valence band decreases. This results in reduction in the probability of e^-/h^+ recombination (Banerjee et al. 2006). The band gaps (E_g) are 3.2, 3.02, and 2.96 eV for the anatase, rutile and brookite phases of TiO_2 respectively (Wunderlich et al. 2004). Anatase is the most photoactive phase of TiO_2 followed by rutile. However, the large E_g value of both the anatase and rutile phases (3.0–3.2 eV) means that these phases get activated only by UV light. On the other hand, TiO_2 based photocatalysis will only be a sustainable technology, if photocatalyst materials with high photoresponse under sunlight can be developed. Given that sunlight comprises only 4–5% of UV light, the need of the hour is to develop photocatalyst materials capable of absorbing visible light which contributes 50% to the solar spectrum.

Table 16.1 Physical and structural properties of anatase, rutile and brookite phases of TiO_2

Properties	**Anatase**	**Rutile**	**Brookite**
Molecular weight (g)	79.88[a]	79.88[a]	79.88[b]
Melting point (°C)	1825[a]	1825[a]	1843[b]
Boiling point (°C)	2500–3000[a]	2500–3000[a]	2500–3000[b]
Light absorption (nm)	< 390[a]	< 415[a]	—
Refractive index	2.55[a]	2.77[a]	2.59[b]
Dielectric constant	31[a]	104[a]	—
Crystal structure	Tetragonal[a]	Tetragonal[a]	Orthorhombic[b]
Lattice constant (A°)	a = 3.78[a] c = 9.52[a]	a = 4.59[a] c = 2.96[a]	a = 9.18[b] b = 5.45[b] c = 5.15[b]
Density (g/cm^3)	3.79[a]	4.13[a]	3.99[b]
Ti–O bond length(A°)	1.94[a]	1.95[a]	1.87 – 2.04[b]

[a][These data have been adapted from Pelaez et al. 2012, Appl. Catal. B: Environ. 125: 331–349].
[b][These data have been adapted from Open Google Source].

Thus contemporary research is focused on strategies to modify TiO_2 materials to increase their visible light response. Doping with other materials specially nonmetal atoms has given positive results. The development of heterostructures with TiO_2 coupled to low band gap semiconductors is also an effective strategy (Latroche et al. 1989).

16.5 NEED FOR DEVELOPING MATERIALS FOR ENVIRONMENTAL REMEDIATION

The Brundtland Commission report in 1987 introduced the concept of sustainable development (Wood and Jones 1997). In this century, the world is facing the dire consequences arising from the abuse of nature mainly due to anthropogenic activities. The depletion of the ozone layer, global warming, loss of biodiversity, the health hazards posed by radioactive and hazardous wastes have created a problem not only for one generation but also for future generations. At the United Nations Earth Summit in Rio de Janeiro in 1992, a declaration on environment and development was issued. At the 2012 Rio Summit, four main issues were addressed: (1) generation of toxic and radioactive waste; (2) replacement of fossil fuels by renewable energy; (3) addressing health hazards arising due to vehicle emissions and (4) scarcity of drinking water.

The World Health Organization (WHO) has reported that by 2025, 50% of the human population will be residing in water-stressed zones. Chemical contaminants of water bodies include salts, pesticides, metals, toxins produced by bacteria and drugs. Chemical contaminants are of two major types—organic and inorganic. The main organic pollutants are pharmaceuticals, pesticides, personal care products and manufacturing additives (Lamastra et al. 2016). Thus the need of the hour is to search for viable methods for degradation of these toxic chemicals in water. Efforts are also directed towards mitigating the effects of vehicle exhaust on the environment and hence the search for clean energy sources. In this context, TiO_2 based materials serve to degrade chemical contaminants of water and also help in production of clean energy.

16.6 TiO_2 NANOMATERIALS

Nanomaterials are those materials having at least one dimension lying between 1 to100 nm. As with other materials, for TiO_2 too the activity gets enhanced at the nanoscale. This arises due to the unique and special behavior exhibited by the nanoscale materials *viz.* high *surface-to-volume* ratio, surface plasmon resonance for metal nanoparticles, size-dependent redox potential and quantum confinement effects. Over the past decades, nanostructured materials derived from TiO_2 have been extensively studied for their promising applications *viz.* in solar cells/batteries, self-cleaning coatings, electroluminescent devices and above all in photocatalysis. It is no wonder that contemporary research on materials has witnessed a spurt in research focused on newer methods of synthesis of TiO_2 nanomaterials.

16.7 SYNTHETIC STRATEGIES FOR TiO_2-BASED NANOMATERIALS

16.7.1 Sol–gel Method

The sol–gel method is simple, cost effective and possible under ambient conditions. The sol–gel procedure includes two steps (i) hydrolysis and (ii) polycondensation. During the second step, M–OH–M or M–O–M bridges are established between the metal (M) atoms of the precursor molecules, finally resulting in oxides or hydroxides of the metal. For TiO_2 synthesis by the sol gel process, titanium alkoxides (such as titanium isopropoxide, titanium n-butoxide), alcohol and acid/water mixture are introduced into the reaction system. After continued stirring, densely cross-linked three-dimensional structures are formed which on standing lead to formation of a TiO_2 gel. The reaction mechanism involves the following sequence of reactions:

$$Ti(OR)_n + H_2O \longrightarrow Ti(OH)(OR)_{n-1} + ROH \quad (1)$$

$$Ti(OH)(OR)_{n-1} + H_2O \longrightarrow Ti(OH)_2(OR)_{n-2} + ROH \quad (2)$$

$$-Ti-OH + HO-Ti- \longrightarrow -Ti-O-Ti- + H_2O \quad (3)$$

$$-Ti-OR + RO-Ti- \longrightarrow -Ti-O-Ti- + ROR \quad (4)$$

Ethanol and acid are added as reaction modifiers. The final morphology of the product depends on the synthesis conditions e.g. molar ratios of the reactants, pH, reaction temperature and the reaction time (Tseng et al. 2010). The fabricated TiO_2 gels have low densities, large specific surface areas and normally are amorphous. The gels are subjected to aging and drying, finally annealing results in the formation of the crystalline TiO_2 nanomaterials (Xie et al. 2007).

Additives and shape controllers are often added during hydrolysis to get size and shape tailored TiO_2 nanoparticles (Eiden-Assmann et al. 2004, Lee et al. 2010). In the process used by Eiden-Assmann et al. (2004) monodisperse spherical TiO_2 nanoparticles were produced from $Ti(OEt)_4$ in ethanol by using salts and a polymer. The salts used were alkali halides and nitrates. The polymers employed were diblock-copolymers e.g. Lutensol [$RO(CH_2CH_2O)_xH$] and triblock-copolymers (PEOn–PPOm–PEOn).

Eiden-Assmann et al. (2004) synthesized crystalline TiO_2 nanoparticles with homogeneous size distribution and minimum surface defects by controlling the pH and adding ethylenediamine as a shape controller. The TiO_2 precursor was initially formed by mixing titanium isopropoxide (TTIP) and triethanolamine (TEOA) in a molar ratio of TTIP: TEOA = 1:2, followed by the addition of water. $HClO_4$ or NaOH solution was then added to the prepared precursor solution to adjust the pH from 8.5 to 10.5. Then ethylenediamine was added and the resultant solution heated in an autoclave. The solution was then aged stage wise. During the first aging step, a hydrolyzed gel network was formed and during the second aging step the nucleation and growth of TiO_2 seeds occurred (Figure 16.3). Spherical TiO_2 nanoparticles with an average diameter of 20 nm were obtained at pH 8.5 without ethylenediamine. Increasing the pH to 10.5 changed the shape to rod-like with an average size of 20 nm × 100 nm. The addition of ethylenediamine later doubled the aspect ratio, and the dimension of the final wire-shaped TiO_2 materials was 20 nm × 200 nm.

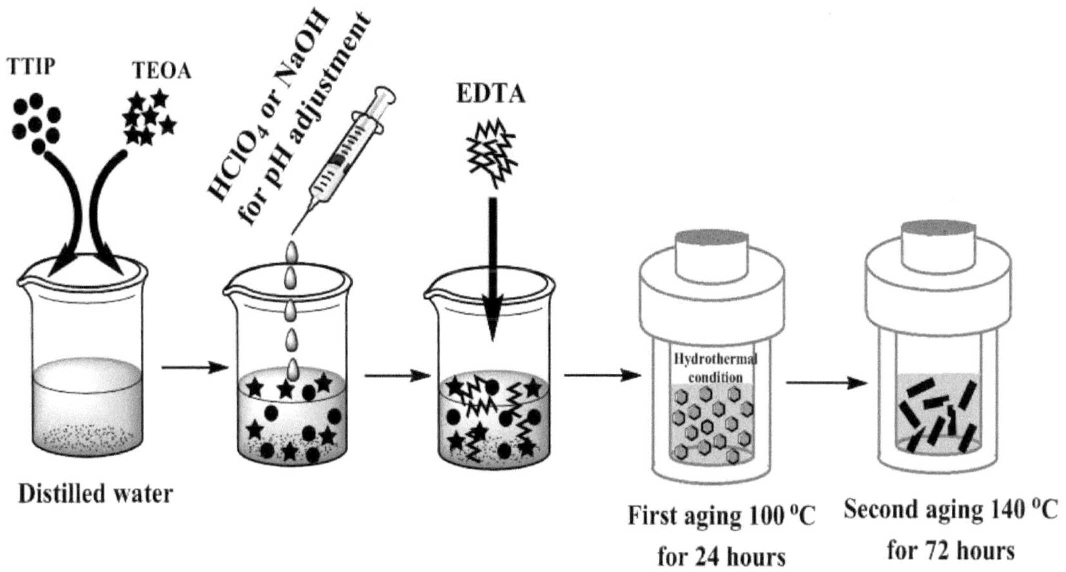

Figure 16.3 Synthesis of TiO_2 nanomaterials by two-step sol-gel method.

16.7.2 Hydrothermal Method

Hydrothermal synthesis is performed in stainless steel autoclaves using water as a solvent, under controlled temperature and pressure. The inner reaction chamber is a Teflon container. The temperature can be elevated above the boiling point of water, this is regulated by the saturation vapor pressure (Chen and Mao 2007). The hydrothermal method was first introduced by Kasuga et al. (1998). By this method, TiO_2 nanomaterials with uniform size distribution and long term stability can be produced.

In a typical synthesis, titanium alkoxide was thoroughly mixed with aqueous alkali and the resultant mixture was heated at 110°C–180°C in a Teflon-lined autoclave. After the hydrothermal treatment, the obtained precipitate was washed and dispersed in aqueous HCl. The resultant precipitate was then calcined to obtain TiO_2 nanomaterials. TiO_2 nanoparticles, nanotubes, nanoribbons and nanowires could be formed depending on reaction conditions (Liu et al. 2007, Wang et al. 2007).

Liu et al. (2007) proposed a method for the preparation of porous TiO_2 nanospheres by the hydrothermal. Briefly, $Ti(SO_4)_2$ and NH_4F were dissolved in water and the mixture was added to a Teflon-lined autoclave (Figure 16.4). The hydrothermal synthesis was performed at 160°C. The formation of porous TiO_2 aggregates was assigned to the HF formed from NH_4F during the reaction. As is well known, HF is a corrosive etching agent which erodes the inside of the TiO_2 nanomaterials to form hollow nanospheres.

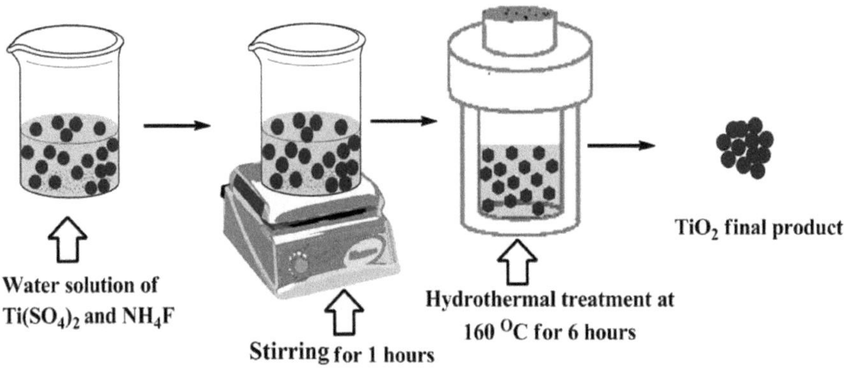

Figure 16.4 Hydrothermal method for the synthesis of TiO_2 nanomaterials.

TiO_2 nanotubes with diameters varying from nanometers to micrometers were prepared by the hydrothermal method (Geng et al. 2009, Liu et al. 2011a, Perera et al. 2012). For example, Geng et al. (2009) performed prolonged hydrothermal treatment at 150°C on pure rutile powder in presence of NaOH to synthesize titanate nanotubes. The titanate nanotubes were subsequently dispersed into a guanidine carbonate solution, dried overnight and calcined to obtain N-doped TiO_2 nanotubes. The N-doped TiO_2 nanotubes were 200–400 nm in length and 10 nm in diameter. However the calcination temperature must be optimum otherwise the TiO_2 nanotubes will sinter together to form TiO_2 nanoparticles at very high temperature.

The formation mechanism of TiO_2 nanotubes could be explained by the following sequential steps:

(i) The TiO_2 nanoparticles are dissolved by NaOH upon hydrothermal treatment. NaOH also reacts with TiO_2 to form layered alkali titanate nanosheets.

$$TiO_2 + 2NaOH \longrightarrow Na_2TiO_3 + H_2O \quad (5)$$

(ii) Gradually, the lamellar nanosheets begin to roll, forming nanotubes. The edges of the lamellar products have many atoms with dangling bonds and thus they get destabilized and subsequently roll to lower the excess energy. The entire process is called the "*dissolution–precipitation*" process.

(iii) After the hydrothermal treatment, the obtained precipitate should be washed with aqueous HCl, thus H^+ will exchange with Na^+ as shown below (Xu and Gao 2004)

$$Na_2TiO_3 + 2H^+ \longrightarrow H_2TiO_3 + 2Na^+ \quad (6)$$

After calcination, TiO_2 nanotubes are finally formed

$$H_2TiO_3 \longrightarrow TiO_2 + H_2O \quad (7)$$

16.7.3 Solvothermal Method

The solvothermal method is another variant of the hydrothermal method where organic solvents are used in lieu of water (Qin et al. 2011, Yang et al. 2013). With organic solvents, the main advantage is that much higher temperature and pressure can be achieved than in a hydrothermal setup. An additional benefit is that by the solvothermal method one can achieve better control over the size and crystallinity (Chen and Mao 2007). The reason for this being that organic solvents of low dielectric constant decrease the solubility of TiO_2, which results in the formation of smaller nanoparticles, also giving rise to better crystallinity (Liu et al. 2011b, Nam et al. 2013). Generally four kinds of organic solvents are used i.e. alcohols, carboxylic acids, toluene and acetone. The precipitates obtained must be washed and dried, but calcination is not necessarily due to the inherently high crystallinity of the TiO_2 nanomaterials prepared by the solvothermal method. The morphology of the final TiO_2 structures depend largely on the organic solvent used, the reaction temperature and time, pH and post-processing calcination (Lü et al. 2011, Nguyen et al. 2010, Yang et al. 2009b).

Yang et al. (2012) prepared anatase TiO_2 nanosheets with dominant {001} facets via alcohol based solvothermal method. Typically, TiF_4 was added to dilute aqueous acid to obtain TiF_4 solution at pH 1.8. This acidified TiF_4 solution, iso-propanol, and HF were added into a Teflon-lined stainless steel autoclave and heated at 180 °C in an electric oven. The anatase TiO_2 nanosheets thus obtained were dried in vacuum. Defluorination was done by heating at 600°C. Iso-propanol and HF promoted the uniform growth of the TiO_2 nanosheets. These TiO_2 nanosheets show promise as photocatalysts for the degradation of organic pollutants (Yang et al. 2009a).

TiO_2 nanorods were synthesized by Li et al. (2006) in presence of Linoleic acid (LA) as the nucleophilic reagent with $Ti(OBu)_4$ as precursor. Titanium alkoxides react with carboxylic acids via esterification reaction, as shown below.

$$NH_4HCO_3 \longrightarrow NH_3 + H_2O + CO_2 \quad (8)$$

$$Ti\,(OBu)_4 + xC_{17}H_{31}CO_2H \longrightarrow Ti\,(OBu)_{4-x}\,(C_{17}H_{31}CO_2)_x + BuOH \quad (9)$$

LA also acts as a coordinating surfactant promoting anisotropic crystal growth.

16.7.4 Anodic Oxidation Method

The anodic oxidation method has been effectively used to synthesize vertical arrays of TiO_2 nanotubes. Briefly, pre-cleaned Ti foils acted as anodes, HF, NaF or KF could be used as electrolytes, the cathode was a Pt foil (Wender et al. 2011). pH was maintained at ≤ 5. The TiO_2 nanotube parameters i.e. shape, pore size and length could be controlled by the anodization parameters (Su and Zhou 2011). Longer TiO_2 nanotubes were formed in presence of polar organic electrolytes and at higher pH in aqueous medium (Yang et al. 2010). If voltage is maintained constant, then pipe-like nanotubes are formed but if voltage is increased tapered nanotubes are formed (Mor et al. 2004).

The dual benefits of increased visible light absorption and enhanced charge carrier separation were obtained by depositing Ag–CdS core-shell nanoparticles on the surface of TiO_2 nanotube arrays (Xie et al. 2011) and ZnTe deposited on TiO_2 nanotube arrays Yang et al. (2010).

N-doped TiO_2 nanotube arrays could be prepared from Ti alloys containing nitrogen. These contained the Ti–O–N structure as seen from X-ray photoelectron spectroscopy (Tsuchiya et al. 2005).

The mechanism of formation of TiO_2 nanotube arrays was explained in detail by these workers. As explained in the initial steps during anodization, electric field-driven dissolution dominates over chemical dissolution as a large electric field exists across the oxide layer. However, the chemical dissolution of the metal oxide leads to the formation of tiny pits which serve as the base for pore formation. Subsequently, two processes occur simultaneously: (i) conversion of small pits to larger pits and (ii) number of pores increase. This is followed by deepening of the pore due to the inward movement of the metal/oxide layer towards the base of the pore (Prakasam et al. 2007). The nanotube increases in length until an equilibrium is reached between the rate of electrochemical etching and the chemical dissolution at the top surface of the nanotubes (Mor et al. 2006). The overall reactions are:

$$2H_2O \longrightarrow O_2 + 4H^+ + 4e^- \quad (10)$$

$$Ti + O_2 \longrightarrow TiO_2 \quad (11)$$

16.7.5 Microwave-assisted Method

The microwave method is an assisted method i.e. it must be used in conjunction with the sol–gel process, hydrothermal process or solvothermal process. The combined method is a clean, cheap and eco-friendly method of synthesis (Hoseinzadeh Hesas et al. 2013). The advantage of the microwave method is that uniform heating occurs throughout the sample unlike the traditional methods of heating, leading to less by-products. Interesting TiO_2 nanomaterials have been prepared by this method (Shojaie and Loghmani 2010, Wilson et al. 2002, Zhu et al. 2011). Chung et al. (2008) prepared TiO_2 nanowires using commercial TiO_2 powder and NaOH via the microwave-assisted hydrothermal method.

16.7.6 Hard Template Method

In this method, tubular nanostructures are synthesized within the pores of a membrane or nanoporous solid. Thus the membrane /nanoporous solid acts as a mould whereby its morphology and porosity get duplicated onto the synthesized nanostructures. The templates used commonly are anodic alumina membrane (AAM), polymeric membranes, activated carbon and silica. These templates are finally removed by combustion or chemical etching (Gong et al. 2012, Hu et al. 2008).

Activated carbon fibers (ACF) have been used as templates to prepare hollow TiO_2 nanofibers (Yuan et al. 2006). The method consisted of dipping ACF into a mixture of titanium isopropoxide and anhydrous ethanol. Then the solution filled ACF was placed in a steel autoclave and heated. Calcining the composite fibers at 600°C in air removes the ACF template resulting in pure TiO_2 nanofibres.

16.7.7 Reverse Micro Emulsion Method

Thermodynamically stable, isotropic, clear liquid mixtures of oil, water and surfactant are called micro emulsions. Often a co-surfactant is also used. The co-surfactants used are generally long chain alcohols (Gupta and Tripathi 2012, Wang et al. 2008). Zhang et al. (2010a) synthesized TiO_2 nanomaterials via a reverse micro emulsion method using urea. $TiCl_3$ was used as the Ti precursor, Triton X-100 as the surfactant, n-hexanol as the co-surfactant and cyclohexane as the oil phase. They could achieve good control over the proportion of anatase and rutile phase in the synthesized nanomaterials.

16.8 GREEN SYNTHESIS OF TiO_2 NANOPARTICLES

In a novel green synthetic approach, leaves of the *Jatropha curcas* plant were used to synthesize TiO_2 nanoparticles (NPs). The leaf extract was used for further synthesis of TiO_2 nanomaterials. Briefly, $TiCl_4$ was added to the filtered leaf extract at room temperature. Appearance of whitish-brown coloration indicated the formation of TiO_2 NPs (Goutam et al. 2018). The solution of TiO_2 NPs was made ammoniacal. The resulting precipitate was filtered and washed with ethyl alcohol to remove the ionic impurities. The precipitate was then air dried, calcined and finally ground in a mortar pestle to obtain the powdered TiO_2 NPs. The reaction mechanism proposed is illustrated in Figure 16.5.

Aloe Vera leaf extract was also used to prepare TiO_2 nanoparticles. $TiCl_4$ was added to the leaf extract at pH 7. The mixture was stirred continuously. The TiO_2 nanoparticles thus formed were dried overnight and calcined at 500°C. The nanoparticle size varied from 60–80 nm depending on the reaction conditions (Ajitha et al. 2015).

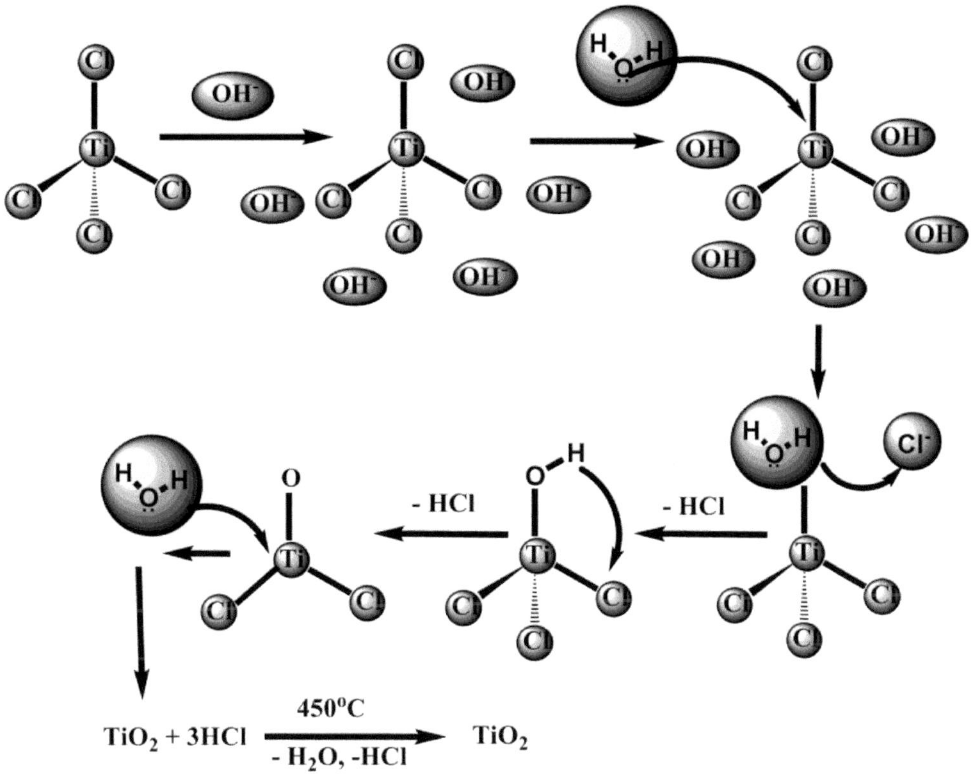

Figure 16.5 Probable mechanism for the formation of TiO_2 NPs in the leaf extract of *Jatropha curcas* L.

16.9 SYNTHESIS OF MAGNELI PHASE TiO_2

The Magneli phases of titanium oxides are a rare series comprising nonstoichiometric TiO_2 with the formula Ti_nO_{2n-1}, n varies from 4–10 (Andersson et al. 1957). These titanium sub oxides have generated much interest due to their high electrical conductivity and long term chemical stability even under harsh conditions such as strongly acidic and alkaline media (Luo et al. 2012).

Some titanium sub oxides under the commercial trade name Ebonex® are used as electrodes and also electrocatalystic supports (Couper et al. 1990). Ellis et al. (2004) used the Magneli phase titanium sub oxides in bipolar lead-acid batteries. Rashkova et al. (2005) used Ebonex® in gas-diffusion oxygen electrodes. Ebonex® materials fared better than carbon black support as electrocatalysts. Titanium sub oxides also function well in photocatalysis. Carbon-coated Magneli phases showed good photocatalytic action under visible light (Toyoda et al. 2009). At very high temperature i.e. 1223 K, rutile could be transformed to the magneli phases Ti_9O_{17} and Ti_6O_{11} in NH_3 atmosphere. However, at even higher temperature i.e., 1373 K, these Magneli phases got converted to the non-Magneli phase Ti_3O_5.

16.10 DEVELOPMENT OF VISIBLE LIGHT ACTIVE (VLA) TITANIA BASED PHOTOCATALYSTS

A big disadvantage of the TiO_2 based photocatalysts is their wide band gap and low absorption in the visible spectral region. A large portion i.e., 40% of the solar spectrum covers the visible region and a meagre 4–5% spans the UV region. Thus, pure TiO_2 is activated mainly by UV photons

(Morikawa et al. 2001). TiO_2 must therefore be modified to enhance its visible light absorption in order to increase its solar light dependent applications. The idea is to prepare visible light active (VLA) TiO_2 nanomaterials. Kamat (2012) has written an Editorial article on the recent advances in the physical chemistry of TiO_2 nanomaterials that will be useful for photocatalysis and in solar cell applications.

16.11 NITROGEN DOPING

Nitrogen is by far the most promising dopant for producing VLA photocatalyst (Fujishima et al. 2008). VLA TiO_2 was formed for the first time by exposing sputtered TiO_2 to N_2 gas (Asahi et al. 2001). This was followed by a series of works targeted towards N-doping of TiO_2 (Di Valentin et al. 2007, Izumi et al. 2009, Zhang et al. 2010b).

TiO_2-based VLA photocatalysts have been useful in degrading organic pollutants *viz.* phenols, dyes and several gaseous pollutants [e.g., volatile organic compounds (VOCs), nitrogen oxides (NO_x)]. N–TiO_2 has been prepared by sputtering (Abadias et al. 2010); ion implantation (Jinlong et al. 2010); gas phase reaction (Sarantopoulos et al. 2009); atomic layer deposition (Pore et al. 2006) and pulsed laser deposition (Zhao et al. 2008). However, the sol-gel method is by far the most popular method for the synthesis of N–TiO_2. In the sol-gel method, simultaneous formation of TiO_2 and N-doping occurs upon hydrolysis of titanium alkoxide precursors in the presence of nitrogen sources. Nitrogen sources include amines, ammonium salts, nitrates, ammonia and urea (Mitoraj and Kisch 2010, Jagadale et al. 2008). An alternative method of increasing the nitrogen content in the TiO_2 lattice is combining titanium precursors with a N-containing ligand, such as complexes of Ti^{4+}-bipyridine or Ti^{4+}-amine (Belver et al. 2006, Sano et al. 2004). One example is the synthesis of N-doped TiO_2 VLA photocatalyst using dodecylammonium chloride (DDAC) as surfactant (Choi et al. 2007). DDAC performs a dual role i.e., (i) as a structure directing agent as well as (ii) a nitrogen dopant.

16.11.1 Doping Other Nonmetals (F, C, S) into TiO_2

Fluorine doping improves the surface acidity and forms reduced Ti^{3+} ions thus the efficiency of charge separation is enhanced. This in turn enhances the photocatalytic performance (Czoska et al. 2008). The photocatalyst prepared by Padmanabhan et al. (2007) by modifying titanium isopropoxide with trifluoroacetic acid was found to be more photocatalytically active than the commercial photocatalyst P25. Doping by carbon, phosphorous and sulphur have also resulted in VLA photocatalysts based on TiO_2 (Irie et al. 2003, Sakthivel and Kisch 2003). Non-metals lead to narrowing of the band gap of TiO_2 by changing lattice parameters and introduction of trap states (Morikawa et al. 2001, Nagaveni et al. 2004). VLA sulfur doped TiO_2 films were prepared using a nonionic surfactant for morphology control and H_2SO_4 as the sulfur source (Han et al. 2011).

16.11.2 Co-doping of Nonmetals into TiO_2

N–F co-doped TiO_2 is a good photocatalyst (Liu et al. 2010). Co-doping leads to double advantage i.e., enhancing the visible light response due to N-doping while F-doping leads to better charge separation. Additionally, it was shown that surface fluorination also prevents conversion from anatase to rutile and protects the N-dopants during annealing (Wu et al. 2010). The synergistic effect of the N–F co-doping has been further observed using a modified sol gel technique employing a nitrogen precursor and a Zonyl FS-300 nonionic fluorosurfactant. The latter functions as both source of fluorine and a pore template (Pelaez et al. 2009). In recent times the N–F co-doped TiO_2 nanomaterials were immobilized on glass substrates via dip-coating (Pelaez et al. 2010).

16.12 OXYGEN RICH TiO_2 MODIFICATION

In another approach, novel VLA photocatalysts were synthesized by *in situ* oxygen generation along with TiO_2 by the thermal decomposition of peroxo-titania complex (Etacheri et al. 2011). Increased Ti–O–Ti bond strength and upward shifting of the valence band (VB) maximum occurred and these effects were responsible for the enhanced visible light activity.

16.12.1 Metal-doped TiO_2

VLA activity as well as increased photocatalytic efficiency was observed by doping transition metals such as Cr, Co, V and Fe onto TiO_2 (Czoska et al. 2008, Murakami et al. 2010). The doping of transition metals on TiO_2 may result in the formation of new energy levels between the VB and CB of TiO_2, thus causing a shift of light absorption towards the visible region. However, transition metal doping suffers from a disadvantage as these centers may act as recombination sites for the photo induced charge carriers thus, lowering the efficiency. In this respect, doping with noble metals like Ag, Au, Pt and Pd on the surface of TiO_2 offers a distinct advantage as it not only enhances VLA action but also delays recombination as the noble metals act as electron traps (Zeng et al. 2007). The enhanced VLA of TiO_2 in presence of Ag was assigned to the surface plasmon resonance of the silver nanoparticles (Nolan et al. 2010). Seery et al. (2007) obtained enhanced VLA photocatalysis with Ag-doped TiO_2. Gunawan et al. (2009) reported reversible photo switching of silver nanoparticles on TiO_2 surface. When the reduced silver clusters on the TiO_2 support were illuminated with visible light (> 450 nm), reverse electron flow from silver to the TiO_2 support was observed resulting in oxidized silver ions ($Ag^{\circ} \rightarrow Ag^{+}$).

16.13 CHARACTERIZATION OF TiO_2 NANO MATERIALS

To confirm that the synthesized TiO_2 nanomaterials actually lie in the nano size regime and to get information about their morphology, various characterization techniques like electron microscopy, powder X-ray diffraction (XRD) pattern, optical absorption and luminescence are used.

16.14 SCANNING ELECTRON MICROSCOPE (SEM)

SEM is used to observe the surface morphology by generating a 3D view of the nanoparticle containing surface. For TiO_2 nanomaterials, various interesting shapes have been observed e.g. nanospheres (Al-Attafi et al. 2017) and nanoflowers (Fang et al. 2011) among many others.

16.15 TRANSMISSION ELECTRON MICROSCOPY (TEM)

TEM yields the size and shape of nanomaterials (up to 1 nm resolution). The histogram of number distribution of particles can also be obtained from TEM. The d spacing value of the crystalline lattice can be directly obtained from high resolution TEM (HRTEM) image, d spacing value of 0.35 nm corresponds to the {101} plane of anatase (Yang et al. 2020).

The selected area electron diffraction (SAED) pattern can give information about the crystal plane. Another useful tool is elemental mapping in TEM. This is used to confirm chemical composition of a nanoparticle.

16.16 POWDER X-RAY DIFFRACTION (XRD)

Powder XRD is used to determine the crystalline phase, crystallographic orientation and size of nanomaterials. Powder XRD is based on Bragg's law:

$$2d \sin\theta = n\lambda \tag{12}$$

where d, 2θ, n and λ are spacing between two parallel planes of the crystal, angle between incident and reflected beam of the X rays, order of diffraction and wavelength of incident X ray respectively. The final XRD plot of reflected intensity versus 2θ gives several peaks, each peak is characteristic of a definite crystal plane of the material. Assignment is done by comparing to the literature XRD pattern obtained from JCPDS files.

TiO_2 nanoparticles exist mainly as three crystalline phases (i) anatase, (ii) rutile and (iii) brookite. The XRD patterns of anatase, rutile and brookite phases of TiO_2 nanomaterials are different (Reyes-Coronado et al. 2008).

16.17 ENERGY DISPERSIVE X-RAY SPECTROSCOPY (EDX)

Energy Dispersive X-ray Spectroscopy (EDAX or EDX) is used for qualitative and quantitative estimation of constituent elements. It is based on the characteristic X-rays emitted by different elements when electrons (emitted in SEM or TEM) interact with the specimen. The detector detects these different X-rays and produces peaks at different positions along the X axis in an intensity versus energy plot.

16.18 X-RAY PHOTOELECTRON SPECTROSCOPY (XPS)

In XPS, a monochromatic X ray beam falls on the surface of a nanomaterial leading to emission of photoelectrons having kinetic energy ranging from 10–1000 eV. From the kinetic energy of emitted photoelectrons and energy of the incident X-rays, binding energy of the emitted electrons can be obtained. By plotting number of emitted photoelectrons versus corresponding binding energy, the XPS spectrum is obtained.

16.19 ULTRA VIOLET (UV) – VISIBLE (VIS) ABSORPTION

The band gap of the TiO_2 nanomaterials can be obtained from the Tauc plot via the following equation using data from the absorption spectra:

$$\alpha h\gamma = E_d (h\gamma - E_g)^{1/2} \tag{13}$$

By plotting $(\alpha h\gamma)^2$ versus $h\gamma$, the band gap E_g can be obtained.

α = optical absorption coefficient
$h\gamma$ = photon energy
E_g = band gap
E_d = constant

16.20 PHOTO LUMINESCENCE (PL) SPECTRA

For semiconductor nanoparticles excitation of electrons from valence to conduction band and their subsequent recombination generates photoluminescence. TiO_2 nanoparticles when excited at 275 nm emit photoluminescence at around 387 nm. In addition to the normal band edge photoluminescence, luminescence due to surface defect states like oxygen vacancies are also observed in the visible region.

16.21 BRUNAUER, EMMETT AND TELLER (BET) STUDY FOR SURFACE CHARACTERIZATION

In BET surface analysis, the adsorption and desorption of a gas e.g., N_2 on the nanomaterial surface gives an adsorption isotherm as per the BET equation. From this, the specific surface area is obtained as also the pore diameter.

16.22 APPLICATIONS OF TiO_2 NANOMATERIALS

New TiO_2 materials have been developed with special focus on environmental applications. Some of the environmental applications of TiO_2 nanomaterials have been summarized below.

Nanosized TiO_2, particularly in the anatase form, exhibits photocatalytic activity under ultraviolet (UV) irradiation. The strong oxidative potential of the positive holes oxidizes water to create hydroxyl radicals. Oxygen or organic materials can be directly oxidized too. Hence, in addition to its use as a pigment, titanium dioxide can be added to paints, cements, windows, tiles or other products for its sterilizing and anti-fouling action. It is also used in dye-sensitized solar cells or Graetzel solar cells.

Titanium dioxide, in thin film and nanoparticle form has great potential for use in energy production; since as a photocatalyst, it can split water into hydrogen and oxygen. The hydrogen thus collected can be used as a fuel. The efficiency of this process can be greatly improved by doping the oxide with carbon (Teh and Mohamed, 2011).

Fujishima and Rao (1997) discovered that when TiO_2 coated glass is exposed to sunlight it exhibits superhydrophilicity. This resulted in the development of self-cleaning glass and anti-fogging coatings.

Nanosized TiO_2 incorporated into outdoor building materials can substantially reduce concentrations of airborne pollutants such as volatile organic compounds (VOCs) and nitrogen oxides (Hogan 2004). The Italcementi Group produced cement containing TiO_2 as a photocatalytic component. This was included in Time Magazine's Top 50 Inventions of 2008.

TiO_2 is useful for detoxification or remediation of wastewater due to several factors (Konstantinou and Albanis, 2004):

1. The process uses natural oxygen and sunlight and occurs under ambient conditions.
2. TiO_2 photocatalyst is cheap, easily available, non-toxic and has a high turnover number.
3. Photocyclized intermediates are not formed unlike direct photolysis processes.
4. Complete oxidation of the materials to CO_2 occurs.
5. TiO_2 can be deposited easily as thin films on suitable reactor substrates, which can be readily separated from treated water (Hanaor and Sorrell, 2014).

The photocatalytic degradation of organic molecules by TiO_2 is useful in photocatalytic antimicrobial coatings (Ramsden 2015). These are thin films applied to hospital furniture and other such surfaces susceptible to contamination by bacteria, fungi, and viruses.

16.22.1 Photoelectrochemical Splitting of Water on the Surface of TiO_2 for the Generation of Clean Energy

The basic principle that guides the photovoltaic chemistry of TiO_2 nanoparticles is the mechanism of the formation of hole and electron pairs upon irradiation by light of suitable wavelengths. When UV light of wavelength in the range 280–400 nm is incident upon TiO_2 surface, one electron is promoted from the valence band to the conduction band as shown in Figure 16.6, thus generating a hole in the valence band. This electron-hole pair helps in the conduction process and is responsible for the photovoltaic applications of TiO_2 nanoparticles.

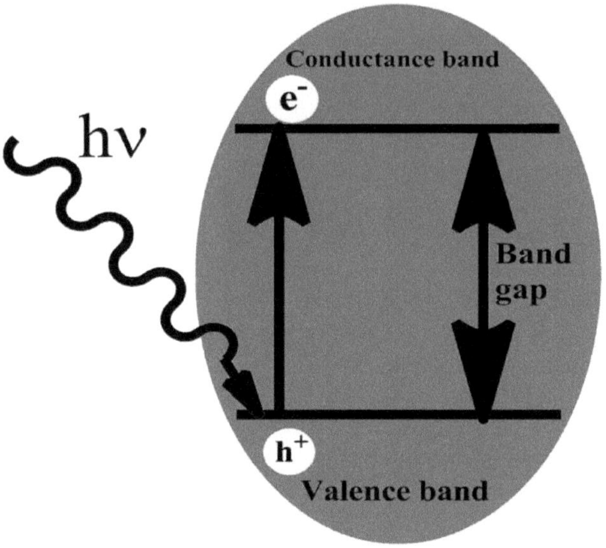

Figure 16.6 Scheme showing generation of an electron-hole pair in a TiO_2 nanoparticle subjected to UV light.

The following equation represents this:

$$TiO_2 + h\nu \longrightarrow e^- (CB) + h^+ (VB) \tag{14}$$

Fujishima and Honda (1972) pioneered the work on photoelectrochemical cell which consisted of rutile as photoanode and a Platinum counter electrode. The cell was irradiated by a xenon lamp. Upon irradiation, electrons were generated and these could flow into the external circuit culminating in photocurrent.

The photoelectrochemical cell could also be employed for water splitting. The steps of the reaction are as follows:

$$H_2O + 2H^+ \longrightarrow \tfrac{1}{2}O_2 + 4H^+ \tag{15}$$

At the TiO_2 electrode and

$$2H^+ + 2e^- \longrightarrow H_2\,(g) \tag{16}$$

At the Pt electrode,

Thus, the overall reaction is:

$$H_2O + h\nu \longrightarrow {}^{1}/_{2}O_2 + H_2\,(g) \tag{17}$$

The water splitting process can be shown by the following schematic diagram (Figure 16.7):

Later, more advanced systems were developed that could split water into the two constituent elements using visible light and solar light too.

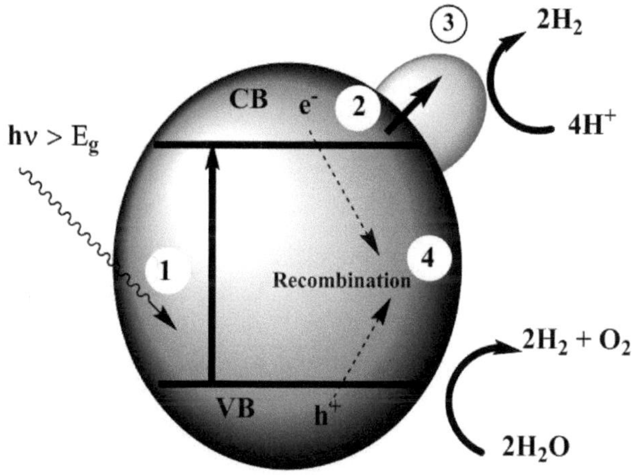

Total reaction: $2H_2O \xrightarrow{h\nu} 2H_2 + O_2$

1. Light absorption
2. e^- transferred from CB of TiO_2 to water
3. H^+ ion is reduced to H_2 gas
4. e^- and h+ recombination

Figure 16.7 Schematic diagram of water splitting reaction on the surface of a TiO_2 nanoparticle.

Jamali et al. (2019) reported that Ag-doped TiO_2 nanoparticles show enhanced photoelectrochemical water splitting efficiency and enhancement in generated photocurrent density due to effective separation of the photogenerated electron-hole pairs.

Khan and Qurashi, (2018) prepared Ag/α–Fe_2O_3/TiO_2 heteronanostructures via electrochemical anodization coupled with sonication. The photocurrent density of Ag/α–Fe_2O_3/TiO_2 (2.59 mA/cm^2) was much higher than that of α–Fe_2O_3/TiO_2 (1.05 mA/cm^2) or pristine TiO_2 (0.54 mA/cm^2), respectively. This enhancement in photocurrent density could be explained by the plasmonic behavior of Ag and lower band gap of α–Fe_2O_3.

Marchal et al. (2016) showed that water splitting in presence of artificial solar light was more enhanced on Au-doped TiO_2 than bare TiO_2.

Chen et al. (2019) prepared CdS-sensitized TiO_2 nanorod array films which showed a broad absorption in the range 350 nm to 570 nm. This led to high photocurrent density (2.03 mA/cm^2) with photoconversion efficiency of ~88%.

CuS/TiO_2 heterostructured nanocomposites with varying TiO_2 content were prepared by Chandra et al. (2018). The heterostructure with optimum ratio leads to very high H_2 production i.e., about ten times higher than that for pristine TiO_2 nanospindles and pure CuS nanoflakes under same irradiation conditions. The increased light absorption and efficient charge separation leads to enhancement in the H_2 evolution rate.

Hematite (α– Fe_2O_3) is a prospective dopant for TiO_2 owing to the high stability, low cost, non-toxicity, narrow band gap (2.0 eV ~ 2.2 eV) and natural abundance of the former. Luan et al. (2015) prepared phosphate bridged TiO_2–Fe_2O_3 nanocomposites that resulted in enhanced photocurrent generation.

Han et al. (2017) synthesized 3D hierarchical hetero-nanostructures consisting of thin α–Fe_2O_3 nanoflakes deposited on TiO_2 nanotubes that had been pre-deposited on FTO glass. These hetero-nanostructures showed efficient photoelectrochemical water splitting due to better charge separation. This occurred as electrons photogenerated in TiO_2 were efficiently collected on FTO, while holes were transferred to the α–Fe_2O_3.

Sohail et al. (2020) synthesized non-stoichiometric Sn–TiO_2 nanoflakes (Figure 16.8) which were further deposited on FTO glass. Chronoamperometric study revealed higher photocurrent for the oxygen evolution reaction (OER) compared to bare TiO_2.

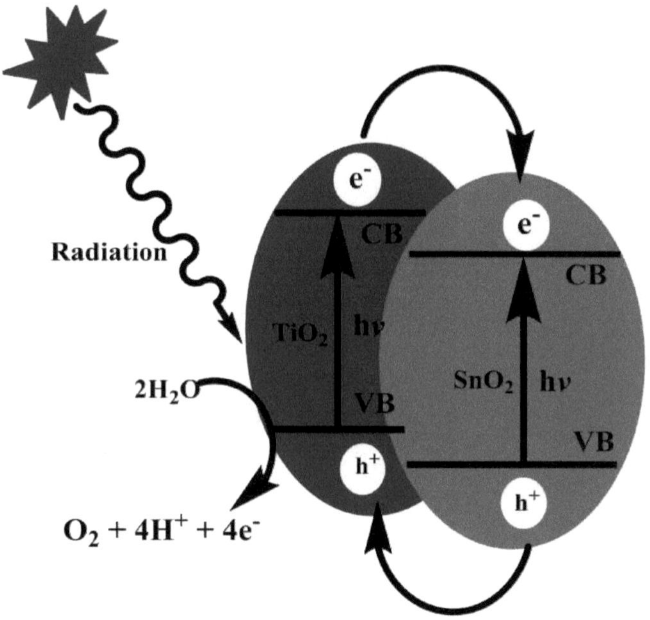

Figure 16.8 Scheme depicting OER on the surface of Sn–TiO_2 nanoflakes.

Cho et al. (2021) and Ahmed et al. (2017) reported the doping of lanthanides e.g., Erbium (Er) and Gadolinium (Gd) on TiO_2 nanotubes and nanorods respectively leading to enhanced photoelectrochemical water splitting compared to pristine TiO_2.

Charge carrier separation and hence long charge carrier lifetimes were obtained upon combining TiO_2 with carbon nanotubes (CNTs) (Dai et al. 2009, Lam et al. 2014). Zhou et al. (2013) used porous CNT@TiO_2 composite nanocables for improving the electrochemical behavior of lithium ions. Wang et al. (2006) used nanowires to design super capacitors based on CNTs@TiO_2. The CNT@TiO_2 nanocomposites were used as photoelectrodes for H_2 production (Dai et al. 2014). Ahmed et al. (2017) prepared CNT–TiO_2 nanoribbons and obtained enhanced photoelectrochemical (PEC) water splitting under visible light. The incident photon-to-current conversion efficiency (IPCE) was 97% at 510 nm i.e., much higher than that reported for pristine TiO_2 photoanodes. Thus the carbon nanotube-TiO_2 nanoribbons could serve as novel photocatalysts for water splitting under solar irradiation for hydrogen production, contributing effectively to generation of a green fuel.

Ahmed et al. (2017) synthesized TiO_2 hollow porous sphere-graphene hybrids which showed enhanced photoelectrochemical water splitting compared to the bare TiO_2 hollow spheres. The enhancement in photocurrent generation upon doping by graphene was assigned to the better electron accepting and transporting behavior of the graphene sheets.

Bellamkonda et al. (2019) developed a novel multiwalled carbon nanotube-graphene-TiO_2 (CNT–GR–TiO_2) composite material for photocatalytic decomposition of water using solar light. The rate of H_2 production was eight-fold higher than the commercial TiO_2 catalyst (Degussa P25). In this case, Graphene acts as an electron reservoir for the water splitting reaction. CNTs prevent the restacking of graphene nanosheets and provide additional electron transport channels thus reducing recombination.

Atrees et al. (2021) reported the Cobalt (Co) and Nickel (Ni)–doped Graphene oxide (GO) supported TiO_2 nanomaterial with significant VLA. H_2 evolution rate was also markedly enhanced.

16.22.2 Photocatalytic Degradation of Hydrocarbons and Halogenated Hydrocarbons

Chlorophenol and its derivatives are refractory xenobiotics and are known to be precursors of the highly toxic polychlorodibenzo-p-dioxins (PCDDs) and polychlorodibenzofurans (PCDFs) during the incineration of industrial waste. Phenols and Chlorophenols are widely used in the manufacturing of pharmaceuticals, petrochemicals, paints, textiles, leather products and wood preservatives. The release of these harmful chemicals cause adverse ecological effects in the biosphere. Due to high water solubility, these chloro-derivatives of phenol contaminate the water. They also induce genotoxicity, mutagenicity and carcinogenicity in humans and animals. Their removal depends on: (i) adsorption; (ii) membrane-based techniques and (iii) use of advanced oxidation processes. TiO_2 is a popular photocatalyst for such removal.

Khraisheh et al. (2012) made a comparison of the effect of bare and Cu-doped TiO_2 nanomaterials on phenol. Cu–TiO_2 showed faster rate of photocatalytic degradation of phenol compared to bare TiO_2. Chun-Te Lin et al. (2018) studied the degradation of 2-chlorophenol under visible light irradiation in presence of Cu-doped TiO_2. They found that Cu^{2+} ion acts as an electron scavenger and prevents the electron-hole recombination in TiO_2, thus facilitating the photodegradation process.

He et al. (2011) studied the photocatalytic degradation of 2-chlorophenol in presence of visible light using iodine-doped TiO_2/SnO_2 nanocomposites. The efficiency of the degradation process was attributed to: (i) migration of the photogenerated carriers to the catalyst interface and (ii) increased absorption in the visible region. It was inferred that the hole in the valence band plays an important role for the photocatalytic removal of 2-Chlorophenol.

Ranjit et al. (2001) carried out a systematic study on the photocatalytic degradation of p-chlorophenoxyacetic acid in presence of lanthanide oxide-doped TiO_2 photocatalysts. The enhanced degradation rate was explained by the formation of Lewis acid-base complex between the lanthanide ion and the TiO_2 substrate via the f-orbitals of the lanthanide ions.

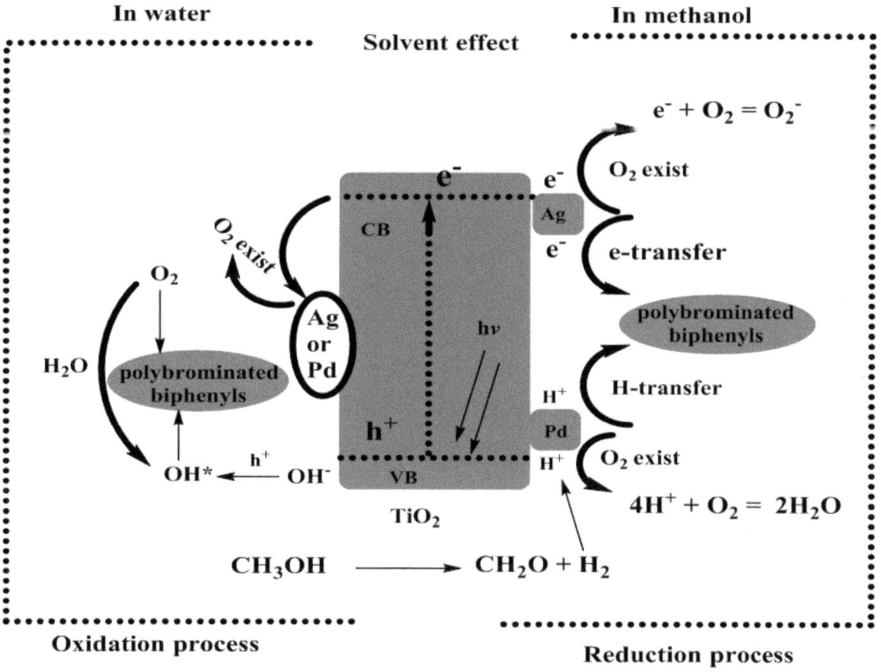

Figure 16.9 Mechanistic pathway for the photodegradation of polybrominated biphenyl pollutants using Pd-TiO_2 and Ag-TiO_2 in various solvents.

Wang et al. (2019) studied the photodegradation of polybrominated biphenyls by Pd and Ag-doped TiO_2 nanomaterials (Figure 16.9). They reported that the Pd–TiO_2 photocatalyst preferentially debrominates the *para*-bromine substituent while the Ag–TiO_2 system preferentially debrominates the *ortho*-bromine substituent. Thus, the Pd-doped TiO_2 and Ag-doped TiO_2 function via different mechanistic pathways. The solvent polarity was found to play a critical role in the photo-degradation due to modulation of the HOMO and LUMO of the pollutant.

16.22.3 TiO_2 Based Nanocatalysts for NO_x, SO_x, H_2S and Other Pollutant Removal

Noble metal doped TiO_2 (M–TiO_2, M = Pd, Ag, Au, Pt) shows enhanced photocatalytic effect (Xu et al. 2017, Hernández Rodríguez et al. 2017). This was manifested in the facile NO_x removal efficiency (55–100%) compared to that of pure TiO_2 (40–90%). The M-doped TiO_2 photocatalyst showed higher stability and selectivity compared to pure TiO_2 (Xu et al. 2017, Tan et al. 2017). This was assigned to: (i) narrowing of the bandgap upon metal doping of TiO_2 (Tan et al. 2017), (ii) plasmon effect of the metal dopants (Xu et al. 2017, Hernández Rodriguez et al. 2017).

For TiO_2 loaded graphene oxide (GO) and reduced graphene oxide (rGO), the NO_x degradation efficiency increased to 31–43% for TiO_2/GO and TiO_2/rGO (compared to 16–24% in pure undoped TiO_2) (Trapalis et al. 2016). This enhancement arises due to effective charge carrier separation by the GO support (Trapalis et al. 2016).

TiO_2 photocatalyst has been effectively used for SO_2 and H_2S degradation using UV light as TiO_2 is an effective photocatalyst for the desulphurization of flue gas (Wang et al. 2016). For visible light irradiation, the catalyst used was multiple element doped TiO_2 (Han et al. 2016) e.g., N–Zr–Ni doped TiO_2, Ce–TiO_2 nanoparticles and Cu–TiO_2 nanoparticle-based photocatalysts. This causes narrowing of the bandgap upon doping (Wang et al. 2017a, Han et al. 2016). Thus a visible-ultraviolet (VUV) lamp with TiO_2 coated titanium mesh filter was designed (Ochiai et al. 2013).

TiO_2 based photocatalysts have also been used for the simultaneous removal of multiple air pollutants in a single operation (Yuan et al. 2012). An example is the TiO_2-aluminum silicate fiber (TAS) developed for simultaneous removal of NO, SO_2 and mercury ion (Hg^{2+}) in the flue gas sample. Thus, TiO_2 based photocatalysts can be employed for simultaneous removal of multiple air pollutants in the future (Yuan et al. 2012). Huang et al. (2011) showed that toluene degradation and air disinfection can both be achieved in a honeycomb type reactor with $TiO_2/\gamma\text{-}Al_2O_3$/NF. High toluene removal efficiency of 82% was obtained and *E. coli* were destroyed within five minutes (Huang et al. 2011).

16.23 DEGRADATION OF PESTICIDES AND PHARMACEUTICAL COMPOUND (PPCs)

16.23.1 Photocatalytic Degradation of PPCs under UV Light Source

UV light mediated photocatalytic degradation of the PPCs, e.g., chloramphenicol (Shokri et al. 2013), metronidazole (Malakootian et al. 2019), amoxicillin trihydrate (Mohammadi et al. 2012), bisphenol A (Gao et al. 2010), naproxen sodium (Hinojosa-Reyes et al. 2019) and the pesticide malathion (Kadam et al. 2014) have been probed using metal/non-metal doped TiO_2 nanomaterials.

Shokri et al. (2013) prepared Ag-doped TiO_2 nanomaterials via photodeposition method where silver ion was doped onto commercial Degussa P25 catalyst. This photocatalyst led to complete degradation of chloramphenicol. Malakootian et al. (2019) used Fe-doped TiO_2 nanomaterials for degradation of metronidazole and industrial waste water. 69.85% degradation of industrial waste water was achieved using Fe–TiO_2 photocatalyst. Gao et al. (2010) prepared undoped TiO_2 and

Zr doped TiO_2 nanomaterials for photodegradation of bisphenol A under UV light. The Zr–TiO_2 nanomaterial showed 90% removal of total organic carbon (TOC), whereas for bare TiO_2 nanomaterial only 60% TOC was removed. Hinojosa-Reyes et al. (2019) used Ni–TiO_2, Cu–TiO_2 and Fe–TiO_2 nanomaterials for photodegradation of naproxen sodium. Kadam et al. (2014) synthesized N–TiO_2 nanomaterials via microwave-mediated method. This photocatalyst was used for the degradation of the pesticide Malathion.

16.23.2 Photocatalytic Degradation of PPCs Under Visible Light Source

Sood et al. (2015) synthesized VLA Fe-doped TiO_2 nanomaterials and observed photocatalytic degradation of p- nitrophenol under 150 W CFL (compact fluorescent lamp) as the visible light source. It was observed that photoluminescence (PL) intensity decreased with increase in Fe dopant concentration due to decreased e–h recombination and enhanced charge carrier separation. This was reflected in the enhanced photocatalytic degradation as Fe concentration increased.

Boxi and Paria (2015) synthesized Ag-doped hollow TiO_2 nanoparticles using AgBr as the sacrificial core. Photocatalytic degradation of metronidazole was studied using a mercury lamp as visible light source. Due to Ag doping, the absorption shifted towards the higher wavelength and thus visible light could be used. Ag-doped hollow TiO_2 nanoparticles showed higher photocatalytic performance compared to undoped TiO_2 nanomaterials. Leong et al. (2014) prepared Ag-doped TiO_2 nanomaterials for degradation of amoxicillin. The bare TiO_2 nanomaterials show absorption in the UV region whereas the doped nanomaterials show significant absorption in the visible region due to surface plasmon resonance arising due to the Ag-dopant. The (VLA) Ag-doped TiO_2 nanomaterials showed higher photocatalytic degradation of amoxicillin compared to bare TiO_2 nanomaterials.

Blanco-Vega et al. (2017) synthesized undoped and Ni-doped TiO_2 nanomaterials with different Ni dopant content. These were employed for the photocatalysed degradation of Bisphenol A using a visible light source. The doped materials performed better.

Yi et al. (2019) synthesized sulphur-doped TiO_2 (S–TiO_2) nanomaterials at three different calcination temperatures (300, 350 and 400°C). They compared the photocatalytic performance against the pollutant diclofenac under white light emitting diode (LED) illumination. Highest photocatalytic degradation efficiency was observed for the S–TiO_2 nanomaterials which were calcined at 300°C mainly due to lower band gap energy.

Abdelhaleem and Chu (2017) synthesized N-doped TiO_2 nanomaterials and used them for degradation of 4-chlorophenoxyacetic acid.

B Lavand and S Malghe (2015) prepared carbon-doped TiO_2 (C-doped TiO_2) nanomaterials by using reverse microemulsion method and studied the photocatalytic degradation of 2, 4, 6-trichlorophenol using these materials. Carbon doping was found to decrease the band gap of the materials.

16.23.3 Photocatalytic Degradation of Pharmaceuticals Under Natural Solar Light Source

The photocatalytic degradation of the pharmaceutical compounds *viz.* ofloxacin, ibuprofen, carbendazim, propiconazole and acephate under natural solar light irradiation was studied using various metal/non-metal doped TiO_2 nanomaterials. Kundu et al. (2014) synthesized Ni-doped TiO_2 nanomaterials and showed that considerable photodegradation of ofloxacin occurs in presence of Ni doped TiO_2 nanomaterials. Bhatia and Dhir (2016) synthesized Bi–TiO_2 and Ni–TiO_2 nanomaterials at different dopant concentrations. Under natural solar light illumination, significant photocatalytic degradation efficiency against ibuprofen was observed. Kaur et al. (2016) synthesized Fe-doped TiO_2 nanomaterials by the impregnation method. Photocatalytic performance was checked against the fungicides carbendazim and propiconazole.

Zhang et al. (2015) prepared N-doped TiO_2 nanotube arrays and utilized them for the photocatalytic degradation of the insecticide acephate under natural solar light illumination. The N–TiO_2 nanotube arrays showed higher photocatalytic performance compared to the undoped TiO_2 nanotube arrays.

16.24 PHOTOCATALYTIC DEGRADATION OF ORGANIC DYES

The wastewater from the printing and dyeing industries contains high amounts of dye contaminants. Several groups studied the dye degradation ability of unmodified and modified TiO_2 nanoparticles against Methylene Blue (MB) (Saranya et al. 2018) and Sudan III (Dodoo-Arhin et al. 2018).

Singh and Mehta, (2020) prepared transition metal ion (Ag^+, Cu^{2+} and Ni^{2+}) doped anatase nanoparticles and showed that Cu^{2+}-doped TiO_2 NPs showed superior photocatalytic activity. Zhao et al. (1998) combined TiO_2 nanoparticles with silver-coated carbon nanotubes (CNTs) to form CNT–Ag–TiO_2 (CAT) ternary nanocomposite structures. The results showed that the photocatalytic activity of TiO_2 under both visible light and UV light was significantly enhanced in presence of CNTs. Wang et al. (2020a) synthesized TiO_2 nanoparticles in presence of Argon plasma, these nanoparticles contained a large number of oxygen vacancies and Ti^{3+} defects. The *as-obtained* Ar plasma-treated TiO_2 (Ar–TiO_2) nanoparticles showed decreased band gap of 3.17 eV compared to pure anatase. The Ar–TiO_2 nanoparticles showed enhanced photocatalytic degradation of methyl orange (MO) dye under sunlight irradiation.

Saqib et al. (2019) synthesized mixed phase TiO_2 NPs immobilized on crystalline zeolites using liquid impregnation (LI) method. This could effectively photocatalyze the degradation of Methylene Blue (MB). Thus mixed-phase TiO_2 and the zeolite support led to a major enhancement in the rate of photocatalytic degradation of MB as compared to the pure anatase.

Liu and Wu, (2019) synthesized a p-n junction hybrid nanocatalyst, $BiFeO_3/TiO_2$ core-shell nanocomposite which showed superior photocatalytic activity, due to the combination of the piezo- and photo-catalytic activities (referred to as the *piezophototronic effect*). With the combined application of ultrasound and light irradiation, the dye degradation ability of the $BiFeO_3/TiO_2$ core-shell nanocomposites increases. The rate constant in presence of the *piezophototronic effect* is higher than that in presence of the piezocatalytic and the photocatalytic processes alone. Thus, hybrid catalysts containing a ferroelectric material, provide a new catalyst design strategy.

Mohammed, (2020) prepared a novel VLA photocatalyst, TiO_2/Au/single-walled carbon nanotube (TiO_2/Au/SWCNT) nanohybrids and showed that these nanohybrids act as efficient photocatalysts for the degradation of Methylene Blue (MB). The improved photocatalytic performance of the nanohybrids arises due to their large surface area, VLA and the incorporation of Au and SWCNTs, which leads to enhanced charge-carrier separation.

16.25 PHOTOCATALYTIC REDUCTION OF CO_2

Enhanced photocatalytic activity was observed upon combining anatase TiO_2 and VLA graphitic carbon nitride (g–C_3N_4), (Zhang et al. 2020, Sun et al. 2013). The conduction band minimum of TiO_2 is around –0.2 V (vs. NHE at pH = 7), which is not negative enough to achieve efficient CO_2 photoreduction. However, the conduction band minimum of g–C_3N_4 (–1.20 V vs. NHE at pH = 7) is adequate for CO_2 photoreduction (Shi et al. 2019, Li et al. 2018, Wang et al. 2020b). By combining the advantages of both g–C_3N_4 and TiO_2, TiO_2/g–C_3N_4 and TiO_2/ g–C_3N_4-based metal/polymer heterostructures, larger extent of CO_2 photoreduction could be achieved. The reason for enhanced CO_2 reduction ability being: (i) suitable band gap energy; (ii) extended range of light absorption; (iii) suppression of photogenerated charge carrier recombination; (iv) improved charge transfer efficiency and (v) high-resistance to photocorrosion (Shi et al. 2019,

Pham and Lee 2017, Wang et al. 2020b). For instance, Adekoya et al. (2017) developed Cu incorporated TiO_2/g–C_3N_4 composites (with a g–C_3N_4 to Cu/TiO_2 ratio of 30:70) showing CH_3OH evolution rate of 2574 μmol/g/h and an HCOOH generation rate of 5069 μmol/g/h under visible light irradiation.

As shown by Wang et al. (2020b), 3D hierarchical TiO_2 micro flowers (MFs) immobilized on porous graphitic carbon nitride (p–g–C_3N_4) heterostructures (Figure 16.10) could be formed.

CO_2 photoreduction on pristine g–C_3N_4 and the p–g–C_3N_4/TiO_2 (porous–g–C_3N_4/TiO_2) composites in presence of visible light was assessed. CO and CH_4 were the main products of the CO_2 photoreduction (Wang et al. 2020b). The primary carbon source for the CO and CH_4 was CO_2 (not g–C_3N_4). Hydrothermal treatment in presence of HNO_3 introduces carbon vacancies and defects on g–C_3N_4.

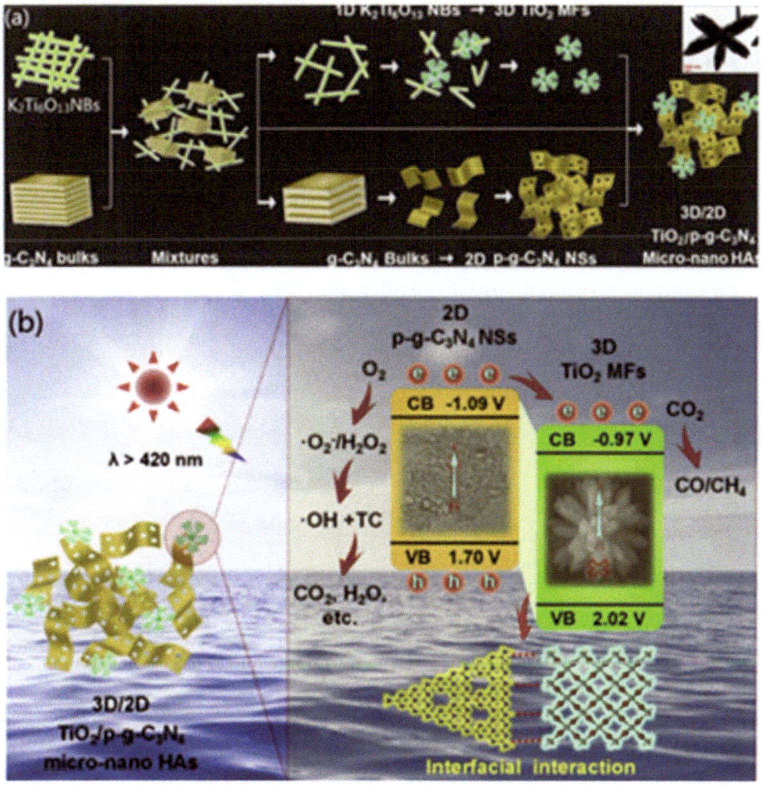

Figure 16.10 (a) Formation of the porous g–C_3N_4 heteronanostructures incorporating TiO_2 microflowers. (b) Proposed mechanism for tetracycline photodegradation and CO_2 photoreduction by these heterostructures. [Reprinted from Chemical Engineering Journal 396 (2020)125347., Copyright (2022), with permission from Elsevier]

16.26 SELF-CLEANING APPLICATIONS OF TiO_2 NANOMATERIALS

Self-cleaning materials have important applications in energy and environment related appliances. Recent examples of many such TiO_2-derived materials have been discussed here in order to gain insight into the fundamentals of self-cleaning hydrophilic and hydrophobic surfaces. Additionally, there are some examples of TiO_2 materials manifesting both superhydrophilic and superhydrophobic behavior. Such unique surfaces find applications in anti-bacterial surfaces, biomedical devices, water purifiers, printing devices, microfluidics and photovoltaic materials. Strategies to promote photocatalytic self-cleaning include doping, heterojunction formation between TiO_2 and low band gap semiconductors and fabrication of graphene-TiO_2 nanocomposites. It has been observed that doped TiO_2 photocatalysts show enhanced self-cleaning behavior which has been assigned to band gap decrease, formation of new energy levels within the band gap and formation of defects i.e., oxygen vacancies. Smart synthetic strategies for making superhydrophilic–superhydrophobic patterns have come up. Surfaces with tuneable and/or reversible hydrophilic/hydrophobic behavior have been devised. In addition, antireflective surfaces with inherent self-cleaning properties have been developed for flat panel displays and solar cells. Another category of TiO_2 based self-cleaning materials show underwater superoleophobic behavior. Such materials find applications in water management, antifouling and separation of *oil-in-water* emulsions. The principle of photo-induced hydrophilic conversion of TiO_2 surfaces finds applications in commercial products such as cement, paint, tiles, glass and fabrics.

16.27 PHOTOCATALYTIC HYDROPHILIC SURFACES

The photocatalytic hydrophilic surfaces can decompose dirt and other impurities in presence of sunlight and even indoor light (Pelaez et al. 2012, Hernandez-Alonso et al. 2009, Zheng et al. 2012, Kisch 2013). The photoexcited electrons on illuminated TiO_2 surface can reduce atmospheric oxygen to give superoxide radicals ($^{\bullet}O_2^-$) and hydroperoxyl radicals ($HO_2^{\bullet}$). The hole left behind in the valence band can oxidize adsorbed water to generate $^{\bullet}OH$. These reactive oxygen species (ROS) decompose organic pollutants to CO_2 and water thus cleansing the surface.

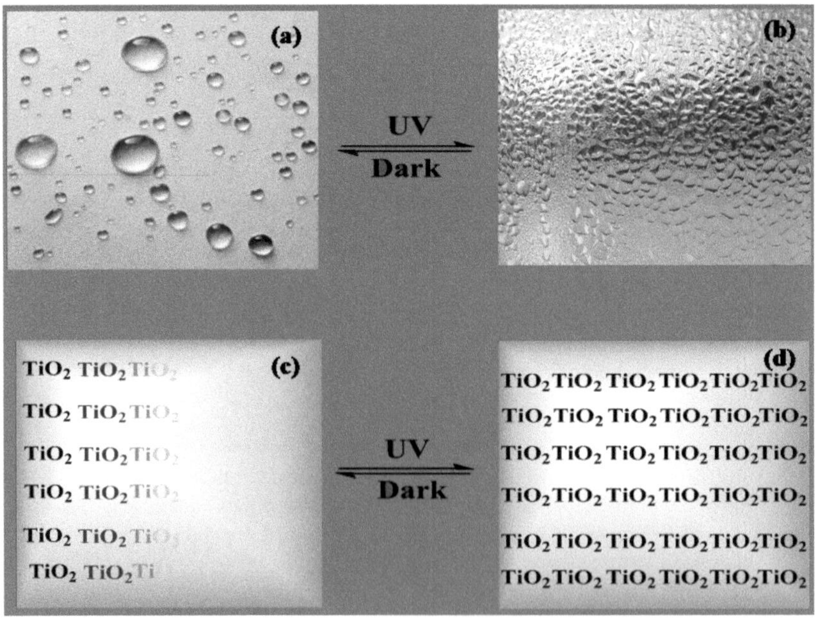

Figure 16.11 Photoinduced switching of TiO_2 surface; (a) a hydrophobic TiO_2 surface; (b) the hydrophobic surface is converted into a superhydrophilic surface upon UV–irradiation; (c) when a hydrophobic TiO_2 coated glass gets exposed to water vapor, fog (small water droplets) is formed, (d) antifogging effect induced by UV-illumination.

The contact angle for water on a thin anatase film is about 72° in absence of light (Wang et al. 1997). However, if the TiO_2 film is illuminated by UV light, the water contact angle (WCA) reduces to 0° implying that the water droplets spread evenly on the surface, Fujishima and Rao (1997). Similar behavior was also observed for oily liquids such as glycerol trioleate and hexadecane (Wang et al. 1997).

Thin films of anatase nanowire arrays were synthesized, these have exposed highly reactive {001} facets (Zhao et al. 2013). When these TiO_2 nanowire arrays were illuminated by UV light, they demonstrated high transparency, photocatalytic activity and superhydrophilicity, implying the potential for use as self-cleaning layers. Self-cleaning TiO_2 nanosheets were deposited on quartz slides by layer-by-layer *(l-b-l)* method (Shibata et al. 2007). A cheap electrospinning technique has been developed to produce photocatalytic, superhydrophilic and optically transparent TiO_2 films which can be used on windows and photovoltaic cells (Anand Ganesh et al. 2012). Emeline et al. (2013) demonstrated that the photoinduced superhydrophilic conversion of TiO_2 surface bears a strong dependence on the intensity and wavelength of the incident light.

Many theories have been put forward to explain the photoinduced hydrophilicity of TiO_2: (i) formation of surface vacancies upon irradiation (Wang et al. 1997); (ii) reconstruction of surface hydroxyl groups induced by photons (Sakai et al. 2003) and (iii) photoinduced removal of the carbonaceous layer formed on TiO_2 surfaces in presence of air (Zubkov et al. 2005). Figure 16.11 below shows how a hydrophobic TiO_2 surface can be converted into a superhydrophilic surface upon UV irradiation.

16.28 GENERATION OF LIGHT INDUCED SURFACE VACANCIES

The first accepted mechanism for photoinduced hydrophilicity arising due to formation of surface defects upon UV irradiation was put forward by Wang et al. (1997). Friction force microscopy shows that UV irradiation causes a structural change on the TiO_2 surface, which in turn affects the interfacial force thus changing the contact angle (Wang et al. 1998). There are five-coordinated Ti atoms with the sixth position occupied by H_2O or OH^- on the surface of TiO_2. UV irradiation gives rise to oxygen vacancies at the position of the two coordinated oxygen bridging sites thus converting Ti^{4+} ions to Ti^{3+} (Wang et al. 1998). Such defects increase the affinity for OH^- which is formed by dissociation of chemisorbed water molecules. This creates hydrophilic domains on the TiO_2 surface.

The {110} and {100} crystal planes of rutile with bridging oxygen sites show greater hydrophilic conversion compared to the {001} planes without bridging oxygen sites (Nakajima et al. 2001). Atomic force microscopic study of UV-irradiated rutile showed that the surface comprises microscopic hydrophilic and oleophilic domains. These can generate capillary flow channels for water and oil respectively (Wang et al. 1998).

However, it was found that if the hydrophilic TiO_2 material is kept in the dark for few days, the hydrophilic character decreases as O_2 molecules from air replace the chemisorbed OH^- ions and water. However, the hydrophilic character can be restored upon further UV illumination. Some groups have also reported photoinduced amphiphilic surface formation for anatase thin films (Nakajima et al. 2000). The hydrophilic conversion of the surface of rutile was accompanied by photoinduced surface hardness (Shibata et al. 2009). Such photoinduced surface hardness was assigned to expansion in surface volume arising due to increase in distance between the adjacent Ti atoms due to adsorption of water molecules upon UV light exposure.

16.29 PHOTOINDUCED RECONSTRUCTION OF SURFACE HYDROXYL GROUPS

It was shown that UV light induced hydrophilic conversion rate of TiO_2 electrodes was enhanced at high positive values of the electrode potentials and diminished in presence of hole scavenging agents (Sakai et al. 2001). Thus, it was inferred that hole diffusion is crucial for the hydrophilic conversion. Sakai et al. (2003) proposed that UV illumination causes reconstruction of surface

–OH groups. The density of surface hydroxyl groups can be correlated to the reciprocal of the contact angle. Their model proposed that the photogenerated holes diffuse to the surface and get trapped at lattice oxygen sites. Hence the binding energy between the Ti and lattice oxygen weakens causing rupturing of water molecules and forming new hydroxyl bonds. The process gets reversed in the absence of light, when the hydroxyl groups undergo surface desorption in the form of H_2O_2 or ($H_2O + O_2$) and the surface goes back to the original less hydrophilic form.

16.30 PHOTO INDUCED REMOVAL OF THE CARBONACEOUS LAYER PRESENT ON TiO_2 SURFACE

Zubkov et al. (2005) proposed that the organic pollutant is photocatalytically degraded when UV light is incident upon a water droplet and this effect increases with UV irradiation time. When the critical point is reached, surface coverage by the organic contaminant approaches zero thus permitting rapid spread of the water droplet.

The WCA for pure TiO_2 coatings is about 10° and it reaches below 5° when 40 mol% SiO_2 is addedto TiO_2. The reduced WCA of TiO_2 in presence of SiO_2 is ascribed to two effects, i.e., (i) the high surface acidity in presence of Si cations and (ii) the ability to adsorb more hydroxyl groups.

16.31 RECENT DEVELOPMENTS IN THE COMBINED SELF-CLEANING AND PHOTOCATALYTIC ACTIVITY OF TiO_2

Some recent developments show that improvement in the photocatalytic and self-cleaning activity of TiO_2 nanomaterials can be achieved by various strategies. Some of these have been discussed in the following paragraphs.

16.32 MODULATING SURFACE ROUGHNESS

Some workers studied the relationship between surface roughness and photoinduced hydrophilicity. They demonstrated that the contact angle of rough Titania surface was 15°, while the smooth Titania surfaces had a contact angle of 70°. They also showed that the hydrophilicity recovery rate of TiO_2 surface under UV irradiation had inverse dependence on the surface roughness. The rough Titania surface was found to possess high numbers of Ti^{3+} ions produced in presence of UV light. These non-stoichiometric species played an important role in adsorption of –OH formed from water molecules (Sakai et al. 1998, Lee et al. 2000). Thus, increase of surface roughness is directly related to enhanced hydrophilicity. Hence, an alternative approach towards achieving superhydrophilic surface is forming a highly porous structure to increase the surface roughness. This happens in reality as observed for a superhydrophilic TiO_2/SiO_2 composite film containing polyethylene glycol (PEG). This composite film showed excellent antifogging behavior due to increased surface roughness and corresponding decrease of WCA from 15° to 3° (Wang et al. 2011). WCA measurement on the porous TiO_2 coating presented good correlation between superhydrophilicity and number of surface OH groups and also surface roughness. Porous TiO_2 meso-channels were formed by chemical etching (Murakami et al. 2008) which showed WCA below 5° upon UV irradiation.

16.33 DOPING TiO_2 WITH NONMETALS

TiO_2 thin films doped with carbon were prepared. They were found to undergo conversion to hydrophilic surface (with WCA < 5°) under both UV and visible light irradiation (Irie et al. 2006). It has been found that surface fluorination of TiO_2 enhances hydrophilicity under UV light. The reason has been assigned to the enhanced acidity. Additionally, fluorination also promotes formation of Ti^{3+} via charge compensation between Ti^{4+} and F^{and}. This promotes charge carrier separation (Czoska et al. 2008).

N–F co-doped TiO_2 synthesized in presence of a fluorosurfactant showed WCA ~8° under visible light (Kontos et al. 2011). The photocatalytic action of the co-doped films showed good correspondence with their hydrophilic conversion properties.

C−N−F-codoped TiO_2 films with self-cleaning property were synthesized by a layer-by-layer (*l-b-l*) dip-coating method using TiO_2 and NH_4F as precursors (Xu et al. 2010). These C−N−F-codoped TiO_2 films also showed enhanced photocatalytic degradation of stearic acid under visible light.

16.34 DOPING TiO_2 WITH METALS

Y_2O_3 doped TiO_2 nanocomposite film deposited on indium tin oxide (ITO) got converted into a hydrophilic surface (with WCA ~8°) under a daylight lamp (Zhang et al. 2008). Y_2O_3 addition created large number of oxygen vacancies on TiO_2 during the UV irradiation (Zhang et al. 2008, Yang et al. 2009). These oxygen vacancies account for the increased hydrophilicity of the Y_2O_3–TiO_2 nanocomposite film.

Zn^{2+}-doped anatase films deposited on ITO glass presented enhanced hydrophilicity compared to the undoped TiO_2 films also due to surface oxygen vacancies in the doped films (Qu et al. 2010). Upon addition of Zn^{2+} and an anionic surfactant like sodium dodecylbenzenesulfonate (DBS), the surface roughness of TiO_2 films increased yielding a superhydrophilic surface (WCA 3°) which showed long temporal stability (Qu et al. 2010).

Wang et al. (2014) showed that stable ZnO@TiO_2 core/shell nanotube arrays could be prepared (Figure 16.12) and these serve as self-cleaning coatings.

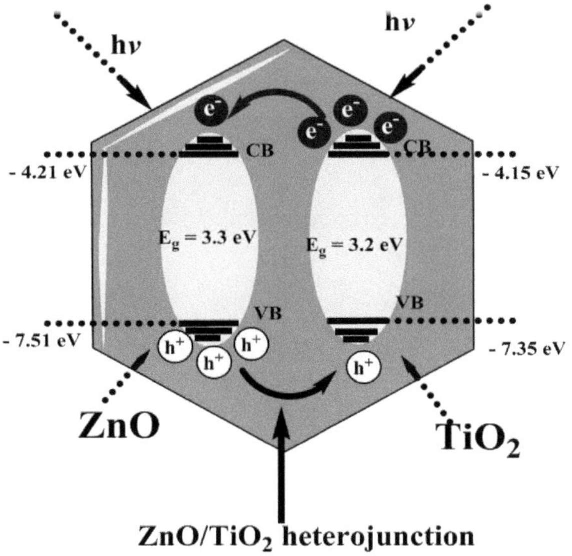

Figure 16.12 Schematic diagram demonstrating the charge transfer between ZnO and TiO_2 nanotube arrays.

16.35 HETEROJUNCTION FORMATION

Hierarchical Bi_2MoO_6/TiO_2 bilayer films showed enhancement in both photocatalytic activity and self-cleaning ability under visible light (Tian et al. 2013).

VLA TiO_2/Multiwalled carbon nanotube (CNT) heterostructures were synthesized by chemical vapor deposition technique (Abdi et al. 2011). The TiO_2/CNT heterostructures were superhydrophilic in presence of UV light.

16.36 GRAPHENE BASED TiO_2 NANOCOMPOSITES

Highly conductive and optically transparent thin films comprising TiO_2 loaded with graphene showed appreciable photocatalytic behavior and superhydrophilicity, much more than pure TiO_2 under a white fluorescent lamp (Anandan et al. 2013). The enhanced photocatalytic activity of graphene/TiO_2 arises due to efficient electron injection from the conduction band of TiO_2 to graphene. WCA for the graphene loaded TiO_2 film decreased with increasing illumination time. This film did not show hydrophilic conversion under visible light.

TiO_2/reduced graphene oxide (RGO) hybrid films converted into a superhydrophilic film with a WCA 4.2° upon illumination by broadband light and also demonstrated antifogging behavior. These hybrid films thus have the potential to be used in self-cleaning optoelectronic devices developed by Zhu et al. (2014).

16.37 ORGANIC POLYMER COATING

Polydimethylsiloxane (PDMS)/TiO_2 nanoparticle coatings on glass form "smart glass" showing self-cleaning behavior (Syafiq et al. 2018) as demonstrated by water contact angle (WCA) analysis. Normal glass has a WCA of 30° and is relatively hydrophilic in nature as shown in Figure 16.13a.

When coated with only PDMS, the hydrophilicity of normal glass changes and it becomes hydrophobic with WCA 92.6° (Figure 16.13b). Again when PDMS is modified by TiO_2 nanoparticles, the hydrophobic nature is lost and the coated glass becomes superhydrophilic and WCA decreases markedly (Figure 16.13c).

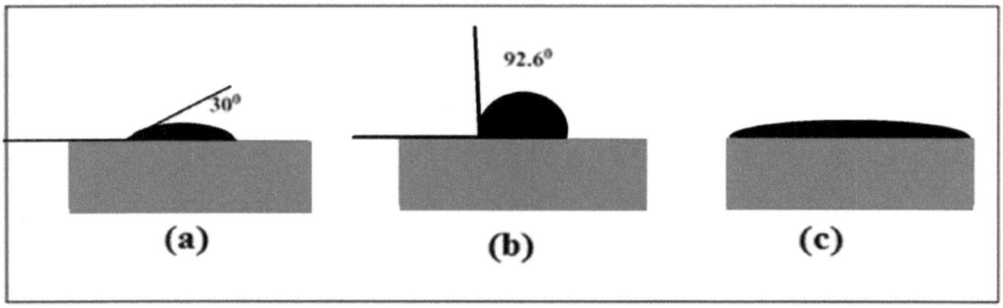

Figure 16.13 (a) WCA of bare glass (WCA= 30°), (b) WCA of glass coated with PDMS (WCA = 92.6°), (c) WCA of glass coated with PDMS/TiO_2 forming a thin layer of water (WCA = 4.2°).

16.38 ANTI-FOGGING APPLICATIONS

It is known that lotus leaves do not allow water to spread and wet their surface. TiO_2 also shows similar behaviour. Due to the similarity between the natural phenomenon shown by lotus leaves and the behavior shown by TiO_2 nanoparticles, it is said that TiO_2 shows the *Lotus Effect*. The *Lotus Effect* refers to self-cleaning and water-repelling behavior arising due to superhydrophobicity. Thus, superhydrophobicity causes the TiO_2 coated glass surface to repel water and the superhydrophilicity causes the water droplets to drain away. For this reason, TiO_2 coated glasses find applications in windshields of automobiles, window panes in buildings and in spectacle lenses. The anti-fogging lens is not only useful in daily lives but is also useful in aquatic sports where special goggles whose surfaces cannot be wetted easily, are used. TiO_2 polycarbonate coated glass repels water from its surface (Tricoli et al. 2009).

16.39 TiO_2/SiO_2 HETEROSTRUCTURES

Modified self-cleaning TiO_2 coatings with low anti-reflection property and high light transmission ability are useful in flat panel displays, solar energy collectors and in greenhouses (Faustini et al. 2010, Yao and He 2014). For practical applications, the long-term stability of self-cleaning coatings even in the absence of light is desired. However, photoinduced superhydrophilic character of pure TiO_2 films gets gradually lost in absence of light. Modified TiO_2/SiO_2 composite films show long-term hydrophilicity which gets retained even in the dark [(Xu et al. 2013(a)]. Increased Lewis acidity of the TiO_2/SiO_2 composite films causes increase in the number of surface hydroxyl groups, resulting in enhanced hydrophilicity (Guan et al. 2003). Additionally, SiO_2 also causes decrease in the refractive index of the TiO_2/SiO_2 composite film and hence increases the extent of light absorption (Kapridaki et al. 2014).

TiO_2/SiO_2 composite films doped with 3*d* metals demonstrated photoinduced superhydrophilicity (Mokhtarimehr et al. 2013). The metal ions act as traps for the holes, thus charge carrier separation increases. Several groups have reported fabrication of superhydrophilic wool and cotton using coatings of TiO_2/SiO_2 composites. These fabrics also showed photocatalytic self-cleaning activity against bacterial adhesion (Erdural et al. 2014). Such cotton fibers have also been used for the removal of food, coffee and wine strains (Pakdel et al. 2013).

Anti-reflective, self-cleaning coatings using layer-by-layer (*l-b-l*) assembly of SiO_2–TiO_2 core-shell nanoparticles on glass supports have been prepared. These show superhydrophilic behavior both in the presence and absence of UV illumination (Li et al. 2013). The anti-reflection behavior arises from a layer of sub micrometer-sized SiO_2 particles. This layer also imparts porosity to the composite. The superhydrophilic behavior arises from higher surface roughness of the coating.

Anti-reflective coatings were produced by coating poly(ethylene terephthalate) with TiO_2 nanoparticles (Nakata et al. 2011) and poly(methyl methacrylate) with TiO_2 nanoparticles (Nakata et al. 2014). The anti-reflection and self-cleaning properties of such coatings can be used in solar cells and flat panel displays. Dust and other organic pollutants drastically reduce the conversion efficiency of solar cells and thus such self-cleaning coatings can improve the efficiency of solar cells.

16.40 PHOTOCATALYTIC HYDROPHOBIC SURFACES

Superhydrophobic surfaces possess a WCA greater than 150°. In distinct contrast to superhydrophilic photocatalytic materials, there are only few examples of superhydrophobic photocatalytic surfaces (Nakajima et al 2000, Macias-Montero et al. 2011, Ragesh et al. 2014). However, these surfaces offer several additional benefits as self-cleaning surfaces over superhydrophilic surfaces. Some of these are distinct decrease in bacterial adhesion (Crick et al. 2011), enhanced self-cleaning action due to "*lotus effect*" (Wooh et al. 2014) and anti-misting behavior (Crick and Parkin 2010).

Transparent superhydrophobic TiO_2 films could be prepared by strong heating of aluminum acetylacetonate and titanium acetylacetonate mixture and then coating by a fluoroalkylsilane (Watanabe and Yoshida 2008, Nakajima et al. 2000). The resultant film comprising 2 wt% TiO_2 had WCA 140° upon UV illumination. The film could retain its hydrophobicity and also demonstrated stain removal via photocatalytic degradation in presence of outdoor light. The self-cleaning activity by superhydrophobic films was assigned to photocatalytic action of TiO_2 and its photoinduced amphiphilic property *i.e.* both hydrophilicity and lipophilicity. By partially substituting Ca^{2+} ions by Ti^{4+} ions, Ti-doped calcium hydroxyapatite (HAP) photocatalysts have been prepared (Wakamura et al. 2003). Ti-doped HAP particles photocatalyzed decomposition of acetaldehyde upon illumination by UV light. When Ti-doped HAP particles were coated with a fluoroalkyl silane (FAS) and a methacrylate based polymer, a superhydrophobic film with WCA 155° was obtained (Yoshida et al. 2006). These films retained their hydrophobicity for a long period.

Kamegawa et al. (2012) prepared a superhydrophobic coating of TiO_2 and polytetrafluoroethylene (PTFE) via radio frequency-magnetron sputtering (RF–MS) method. Photocatalytic degradation of oleic acid in presence of UV light was obtained using this film. Its superhydrophobicity was maintained even after five cycles of oleic acid degradation. TiO_2 nanoparticles in polydimethylsiloxane (PDMS) were used to fabricate superhydrophobic thin films by chemical vapor deposition technique (Crick et al. 2012). TiO_2-high-density polyethylene (HDPE) nanocomposite films were prepared by template-based lamination (Xu et al. 2013b). The nanocomposite surface was rough and it showed superhydrophobicity (WCA 158°) and self-cleaning behavior. Upon UV illumination this superhydrophobic surface transformed to a hydrophilic surface due to the creation of oxygen vacancies. However, the hydrophobicity could be restored upon heating due to conversion of the UV light induced Ti–O–H bonds to more hydrophobic Ti–O bonds.

Superhydrophobic films with WCA 155.5° and long-term retention of hydrophobicity were prepared using TiO_2 and dodecylamine (Kartini et al. 2014). The enhanced hydrophobicity arises due to outward orientation of hydrocarbon chains in the film.

16.41 SMART COATINGS WITH PHOTO-REVERSIBLE HYDROPHOBIC-HYDROPHILIC NATURE

The fabrication of TiO_2 based surfaces with controllable wetting behavior and photo-reversible hydrophobic-hydrophilic behavior have revolutionized this technology (Wan et al. 2008, Uyama et al. 2011, Li et al. 2013). These so-called *'smart coatings'* find useful applications in sensors, drug delivery and coatings (Koch et al. 2009, Li et al. 2013). Surface roughness, morphology and the polarity of the surface can control the photo-switching behavior (Yan et al. 2008, Uyama et al. 2011).

An example of such a "*smart coating*" was obtained using a titania/silica composite film by using SiO_2 microspheres in combination with TiO_2 nanospheres and a commercial water-resistant material i.e. Aquapel, (Li et al. 2013). The WCA of these is dependent on the microscopic surface roughness. Yan et al. (2008) have proposed the novel concept that photo-induced superhydrophilicity in TiO_2 originates from the surface restructuring undergone upon light illumination by an intrinsically hydrophobic Titania surface.

16.42 APPLICATION OF SUPEROLEOPHOBIC TiO_2 NANO MATERIALS IN OIL RECOVERY

Contemporary research has seen a shift in focus to materials showing underwater superoleophobicity owing to their applications in under-water management, marine antifouling and cleaning up of oil spills (Liu et al. 2013, Zhang et al. 2013(b)). Underwater superhydrophobicity of TiO_2 has been reported for TiO_2 thin films subjected to light irradiation (Sawai et al. 2013). Earlier it had been mentioned that for special type of modified TiO_2 films, UV illumination creates amphiphilic TiO_2 surface. Such surfaces show underwater superoleophobicity with an oil contact angle (OCA) higher than 160°.

(Wang and Guo 2014) prepared superoleophobic coatings using flower-like rutile TiO_2 grown on FTO substrate for underwater use. The coatings exhibited superamphiphilicity with both WCA and OCA values of 0°. The oil-repellent behavior exhibited by these coatings resembled that of fish scales with an underwater OCA ~ 155°. This arises due to water molecules trapped in the cavities of the rough TiO_2 surface (with flower like morphology), these trapped water molecules repel oil molecules.

Effective separation was achieved in an *oil-in-water* emulsion using superhydrophobic and superoleophobic (underwater) membranes made by using sulfonated graphene oxide (SGO) nanosheets in combination with TiO_2 nanospheres (Gao et al. 2014a). The cross-linked network of the SGO/TiO_2 nanostructured membrane provides mechanical flexibility and produces high oil rejection rate combined with low membrane fouling. A multipurpose ultrathin film was prepared using single-walled carbon nanotubes (SWCNTs) and TiO_2. This film showed superhydrophilicity, superoleophilicity and underwater superoleophobicity upon UV-illumination (Gao et al. 2014b).

16.43 COMMERCIAL APPLICATIONS OF TiO_2 NANOMATERIALS AS PHOTOCATALYTIC SELF-CLEANING AGENTS

The photoinduced hydrophilic conversion of TiO_2 surface manifested by self-cleaning and anti-fogging behavior finds commercial applications in glass, tiles, cement, paints and fabrics. Transparent TiO_2 photocatalyst based thin films with self-cleaning ability in presence of sunlight have been used to construct building materials such as tiles and glass windows (Midtdal and Jelle 2013). Activ™ developed by Pilkington Glass is the first example of self-cleaning glass, comprising a 15 nm thick nanocrystalline TiO_2 film deposited on glass surface (Wardak et al. 2006). Activ™ is one of the most successful self-cleaning products and is widely used for numerous commercial applications all over the world. The photocatalytic activity of Activ™ is assessed in terms of its ability to degrade stearic acid (Mills et al. 2003). Activ™ possesses many attributes that make it a superior photocatalytic and self-cleaning material: (i) high mechanical stability and good photocatalytic recyclability; (ii) low visible reflectance (~7%); (iii) adequate absorption of solar radiation and (iv) high transmittance. Other self-cleaning glasses *viz.* Radiance Ti™, Sunclean™ and Bioclean™ are also picking up momentum.

Hydrotech™ introduced by the Japanese company TOTO Limited uses sunlight to break down pollutants and subsequently these are washed away by rain/water. Hydrotech™ has been manufactured by spraying a liquid suspension of TiO_2 on the surface followed by sintering at 600–800°C to ensure strong attachment of the TiO_2 layer on the targeted surface (Shimohigoshi and Saeki, 2007). As high processing temperatures are required for commercial products based on TiO_2, self-cleaning TiO_2 materials need to have high thermal stability. Non-metallic doping can increase the thermal stability of anatase (Periyat et al. 2008, Yan et al. 2005), however, this also promotes electron-hole recombination and thus reduces photocatalytic activity. VLA oxygen-rich TiO_2 with anatase phase is stable up to 900°C (Etacheri et al. 2011). Superhydrophobic paint prepared from an ethanolic suspension of perfluorooctyltriethoxysilane ($C_{14}H_{19}F_{13}O_3Si$) coated anatase nanoparticles can be applied on stainless steel, glass, clothes and paper.

Water droplets can bounce off from the superhydrophobic surface without wetting the material (Lu et al. 2015).

Self-cleaning glass coated with TiO_2 nanoparticles has been applied on the building surface of the National Opera Hall, China (Bai 2005). TiO_2 nanoparticles containing white cement have also been used in *Dives in Misericordia* Church in Rome (Altan 2010) and Roof of Dubai Sports City's Cricket Stadium. The Italian multinational company, Italcementi has developed a host of photocatalytic cements that are commercially sold as TX Aria^{TM}, TX Active^{TM} and TX Arca^{TM} (Altan 2010). TioCem^{TM} produced by Heidelberg Cement Technology Centre, GmbH, is another commercial self-cleaning photocatalytic cement (Spasiano et al. 2015). Self-cleaning windows and roof tiles are widely used in Japan. Clean and glare free windows, headlights and automobile mirrors employing self-cleaning coatings with antifogging properties (Zhang et al. 2012) are used in the automobile industry. Usually water droplets form a continuous film on a superhydrophilic surface, which prevents light scattering thus providing a clear and unhindered view.

TiO_2 thin films containing Ni^{2+} and Fe^{3+} dopants were prepared so as to obtain VLA sell-cleaning materials (Murugan et al. 2013). The dopants lead to oxygen vacancies or Ti interstitial vacancies in TiO_2 depending on the valence state of the dopant, this leads to VLA. VLA glazed ceramic tiles were manufactured by coating with Degussa P25 modified with tetraethyl orthosilicate (TEOS) (Zhang et al. 2013a). The modified tiles exhibited enhanced photocatalytic activity and improved hydrophilicity under visible light irradiation.

Photoinduced antimicrobial action of TiO_2 is useful for building materials for both indoor and outdoor applications (Amézaga-Madrid et al. 2002). Antimicrobial activity of TiO_2 is in fact most important for applications in the medical field and in the food industry. Companies such as TOTO, Karpery and Biocera manufactured ceramics with a thin film of photocatalyst deposited to function as antimicrobial agent (Radetić 2013). Galkina et al. (2014) have provided a comprehensive review on TiO_2 functionalized textile and wool, thus only a brief summary of the recent developments is discussed here.

Self-cleaning textile fibers based on cotton, polyester, polyamide and cellulose coated with TiO_2 NPs have been manufactured (Galkina et al. 2014, Pasqui and Barbucci 2014). Usually such functionalization is done using carboxylic acid as the anchoring group as it can coordinate well to the Ti atom and also binds to TiO_2 via H-bonding (Radetić 2013). Nanocrystalline TiO_2 also contributes to the UV protection factor of the fabric (Daoud and Xin 2004, Daoud et al. 2005). TiO_2 coating can also improve the mechanical strength of cotton fibers (Qi et al. 2011). Such coated fibers show high stain removal and antibacterial properties (Daoud et al. 2005). Self-cleaning cotton fabrics have been manufactured by treating the cotton fibers with TiO_2 nanoparticles and multi-walled carbon nanotubes (MWCNTs) using succinic acid as cross linker (Karimi et al. 2014). The TiO_2–MWCNT coating also improved UV blocking capacity of the cotton fibers.

Self-cleaning textiles coated with hydrophilic TiO_2 surfaces have been fabricated by Kiwi and Pulgarin (2010) and Yuranova et al. 2006. Polar functional groups aid effective adhesion of metal oxides to the fabric surface. The self-cleaning ability of the TiO_2–SiO_2-coated textiles was understood by the removal of red wine stains (Figure 16.14) (Yuranova et al. 2006).

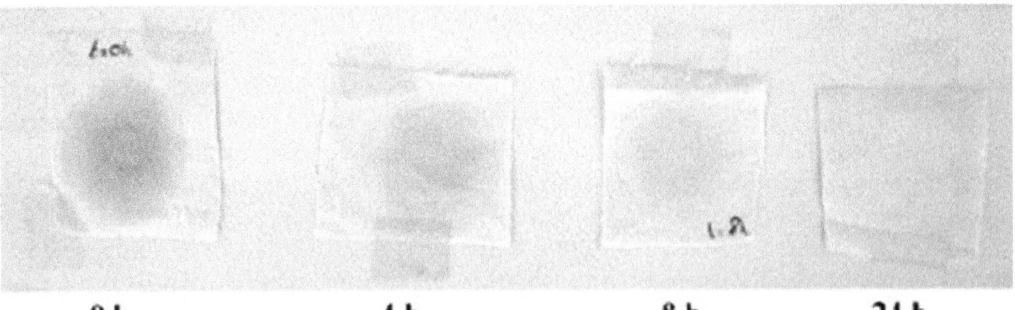

Figure 16.14 Removal of wine stains from cotton textiles coated with TiO_2–SiO_2 after irradiating for 0, 4, 8 and 24 hours. [Reprinted from Journal of Molecular Catalysis A: Chemical244 (2006)160-167., Copyright (2022), with permission from Elsevier]

VLA self-cleaning cotton fabric has been prepared by coating cotton fibers with TiO_2-noble metal composites (Wu and Long, 2011). Wool coated with TiO_2–Ag nanocomposites using citric acid as cross linker could effectively erase methylene blue and saffron stains (Montazer et al. 2012). Polyester coated with TiO_2 and Ag nanoparticles showed improved UV protection and improved antimicrobial and self-cleaning action (Mihailović et al. 2011). Polyester fabrics coated with AgI/AgCl/TiO_2 nanocomposites demonstrated efficient degradation of methylene blue and also showed antimicrobial action against *E. coli* (Rehan et al. 2013). Afzal et al. 2014 demonstrated the superior self-cleaning activity of TiO_2/TCPP functionalized cotton fabrics coated with trimethoxy (octadecyl) silane (OTMS), TCPP being tetra (4-carboxyphenyl) porphyrin. The functionalized cotton fabrics displayed a superhydrophobic surface with WCA of 156°.

Self-cleaning keratin wool fibers were prepared by depositing anatase nanoparticles on wool (Daoud et al. 2008).

16.44 USE OF TiO_2 AND SURFACE MODIFIED TiO_2 NANOMATERIALS IN DYE SENSITIZED SOLAR CELLS

Solar cells are promising renewable energy sources. A solar cell is a photonic device that converts photons with specific wavelengths to electricity. The dye-sensitized solar cells (DSSC) are based on thin film technology. Grätzel (2003) developed the "*Gratzel Cell*" whereby a nanocrystalline TiO_2 film was sensitized by a Ru bipyridyl complex (Saadoun et al. 2000). Mesoporous dye-sensitized TiO_2 forms the photo-anode of the solar cell (O'Regan et al. 1991) as shown in Figure 16.15. The mesoporous nanoparticle network provides enhanced adsorption of the dye molecules. The adsorbed dye molecules can be excited by solar energy to generate electron-hole pairs that are subsequently separated and transported into the TiO_2 network (Nuran et al. 2015). The absorption spectrum of the dye and the binding of the dye to TiO_2 surface are important parameters that determine the efficiency of the DSSC (Hao et al. 2006). A lot of research has been focused on finding effective sensitizer dyes (Kim et al. 2014).

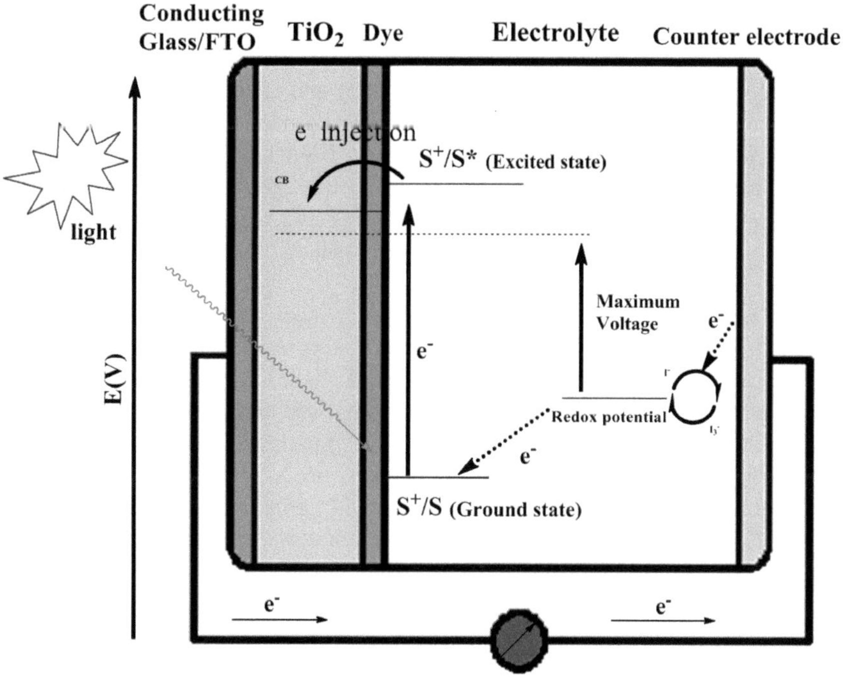

Figure 16.15 Schematic diagram of a dye-sensitized solar cell.

The layout of the DSSC (Figure 16.15) comprises a nanocrystalline porous TiO_2 electrode with adsorbed dye, a counter electrode and an electrolyte couple comprising iodide-triiodide ions. The working principle of the DSSC consists of the following steps:

1. Excitation of the Dye (D) by sunlight,
2. Electron injection from the excited dye (D*) to the semiconductor (TiO_2) network and iodine reduction at the counter electrode,
3. Movement of the photogenerated electron through the TiO_2 network to the electrode.
4. Diffusion in the electrolyte,
5. Regeneration of the oxidized dye to continue the next cycle of photoexcitation.

These above steps can be summarized by the following sequence of reactions.

Reaction

$$D + hv \longrightarrow D^* \quad (18)$$

$$D^* \longrightarrow D^+ + e^- \quad (19)$$

$$2D^+ + 2I^- \longrightarrow 2D + I_2 \quad (20)$$

$$I^- + I_2 \longrightarrow I_3^- \quad (21)$$

$$I_3^- + 2e^- \longrightarrow 3I^- \quad (22)$$

Guo et al. (2017) synthesized hierarchical TiO_2 submicrorods (HTRs) assembled by using tiny nanoparticles and nanorods (as a self-template). These submicrorods possessed higher surface area than P25. Composite photoanodes in dye-sensitized solar cells (DSSCs) were prepared by integrating the prepared HTRs and P25 NPs, and a photovoltaic conversion efficiency of 8.09% was obtained, which was higher than that obtained with the pristine P25 based photoanode (5.37%). The hierarchical TiO_2 submicrorods (i) increase the specific surface area for effective dye adsorption (ii) enhance the light harvesting efficiency and (iii) facilitate electron transport.

Grätzel (2003) prepared TiO_2 nanotube arrays by electrochemical anodization of Ti foils. The length of the TiO_2 nanotube arrays (5–14 μm) could be controlled by varying the anodization time. The influence of nanotube length on the photovoltaic performance of DSSCs was studied (Tan and Wu 2006). A 3.6% conversion efficiency was obtained.

Ran et al. (2018) prepared a novel double-layer film and used it in DSSCs using commercial TiO_2 nanoparticles as an underlayer and Au and Ag nanoparticles (NP) and nanowires (NW) doped onto TiO_2. The influence of Au NPs, Ag NPs, Au NWs and Ag NWs on the thin film, TiO_2 electrodes and the photovoltaic (PV) performances of the DSSCs was evaluated. The sunlight absorption and the photoelectric conversion efficiency (PCE) of the cells were greatly enhanced due to the plasmon effect of the noble metals. This resulted in higher photoelectric conversion efficiencies (PCE) than the pure TiO_2 based solar cells.

Graphene-TiO_2 based photoanodes (g–TiO_2) were also fabricated (Low and Lai 2018). Graphene causes better dye adsorption and enhanced charge separation, it also facilitates electron transport and hence minimizes charge recombination (Eshaghi and Aghaei 2015). This enhances the photoconversion efficiency (PCE) of the DSSCs (Eshaghi and Aghaei 2015).

Kazmi et al. (2017) prepared graphene-TiO_2 (g–TiO_2) nanocomposites with different graphene concentrations. The band gap was reduced from 3.16 eV to 2.2 eV with the increase in graphene concentration. Additionally, conductivity increases with graphene concentration. The PCE of the DSSCs increased from 4.98 % to 7.68 % in presence of graphene. Graphene being a good conductor serves as a bridge in promoting electron transfer from TiO_2 to the electrode thus reducing electron–hole recombination. Hence the PCE gets enhanced.

16.45 CONCLUSIONS AND FUTURE PERSPECTIVES

This chapter takes the reader on a journey through the initial advent of the semiconductor Titanium oxide (TiO_2) as a purely commercial paint product in the early 20th century to its present day applications as a "*magical nanomaterial*" in multivaried domains. It has indeed been a fascinating journey through a century whereby the readily available cost-effective TiO_2 paint material has evolved into the main building block of "smart materials" that are almost indispensable in the modern world.

The chapter begins with the history of development of the field of bulk TiO_2 materials, followed by a discussion on the basic structure, morphology and semiconducting behavior of the bulk materials. Then the reader is explained the objective of this review i.e. the need for developing materials for environmental remediation with special emphasis on the matter of sustainability.

Subsequently, the reader is given a detailed overview of existing literature on TiO_2 nanomaterials – the various synthetic strategies, characterization of these nanomaterials and finally the applications of TiO_2 based nanomaterials. The applications discussed in this chapter include photoelectrochemical water splitting and H_2 production; photocatalytic degradation of organic pollutants, pesticides, pharmaceuticals, removal of NO_x and SO_x, CO_2 reduction; visible light active (VLA) materials; self-cleaning behavior; superhydrophilicity and superhydrophobicity; superoleophobic surfaces for oil recovery and dye sensitized solar cells.

The main aim of this review is on applications of TiO_2 nanomaterials for protecting our environment and sustainable living. To this end, this review provides a comprehensive account on the use of TiO_2 nanomaterials in sustainable practices.

The interesting field of TiO_2 nanomaterials will continue to attract researchers to introduce new modifications to basic TiO_2 nanomaterials to augment their environmental applications. There will be further developments in discovery of novel TiO_2 based nanomaterials for use in sustainable technologies and for environmental applications.

ACKNOWLEDGEMENTS

S. De thanks University of Kalyani for generous grant of the RUSA Scheme IP/RUSA(C-10)/11/2021 and DST Fist Level (II) grant of the Department of Chemistry, University of Kalyani.

T. Dagar acknowledges University of Kalyani for the Research Fellowship. S. Sarkar thanks DST, Government of India for the research fellowship [No. DST/INSPIRE Fellowship/2016/IF160188]. B. Pal Choudhury thanks the Govt. of West Bengal for research scholarship in the form of the Swami Vivekanada Merit cum Means Fellowship. Above all, the authors thank the Editors, Professor Jayanta Kumar Biswas and Professor Mahendra Rai for the kind invitation to submit a Book Chapter.

References

Abadias, G., Paumier, F., Eyidi, D., Guérin, P. and Girardeau, T. 2010. Structure and properties of nitrogen-doped titanium dioxide thin films produced by reactive magnetron sputtering. Surf. Interface Anal. 42: 970–973.

Abdelhaleem, A. and Chu, W. 2017. Photodegradation of 4-chlorophenoxyacetic acid under visible LED activated N-doped TiO_2 and the mechanism of stepwise rate increment of the reused catalyst. J. Hazard. Mater. 338: 491–501.

Abdi, Y., Khalilian, M. and Arzi, E. 2011. Enhancement in photo-induced hydrophilicity of TiO_2/CNT nanostructures by applying voltage. J. Phys. D: Appl. Phys. 44: 255405.

Adekoya, D.O., Tahir, M. and Amin, N.A.S. 2017. g–C_3N_4/(Cu/TiO_2) nanocomposite for enhanced photoreduction of CO_2 to CH_3OH and HCOOH under UV/visible light. J. CO_2 Util. 18: 261–274.

Afzal, S., Daoud, W.A. and Langford, S.J. 2014. Superhydrophobic and photocatalytic self-cleaning cotton. J. Mater. Chem. A. 2: 18005–18011.

Ahmed, N., Farghali, A.A., El Rouby, W.M.A. and Allam, N.K. 2017. Enhanced photoelectrochemical water splitting characteristics of TiO_2 hollow porous spheres by embedding graphene as an electron transfer channel. Int. J. Hydrog. Energy. 42: 29131–29139.

Ajitha, B., Reddy, Y.A.K. and Reddy, P.S. 2015. Biosynthesis of silver nanoparticles using Momordica charantia leaf broth: Evaluation of their innate antimicrobial and catalytic activities. J. Photochem. Photobiol. B. Biol. 146: 1–9.

Al-Attafi, K., Nattestad, A., Yamauchi, Y., Dou, S.X. and Kim, J.H. 2017. Aggregated mesoporous nanoparticles for high surface area light scattering layer TiO_2 photoanodes in dye-sensitized solar cells. Sci. Rep. 7: 10341.

Altan, O. 2010. Photocatalytic concrete/cement—how does it function and possible product areas. Independent Bachelor Thesis, University of Linkoping, Sweden.

Amézaga-Madrid, P., Nevárez-Moorillón, G.V., Orrantia-Borunda, E. and Miki-Yoshida, M. 2002. Photo-induced bactericidal activity against Pseudomonas aeruginosa by TiO_2 based thin films. FEMS Microbiol. Lett. 211: 183–188.

Anandan, S., Narasinga Rao, T., Sathish, M., Rangappa, D., Honma, I. and Miyauchi, M. 2013. Super-hydrophilic graphene-loaded TiO_2 thin film for self-cleaning applications. ACS Appl. Mater. Interfaces 5: 207–212.

Anand Ganesh, V., Sreekumaran Nair, A., Kumar Raut, H., Michael Walsh, T. and Ramakrishna, S. 2012. Photocatalytic superhydrophilic TiO_2 coating on glass by electrospinning. RSC Adv. 2: 2067–2072.

Andersson, S., Collén, B., Kuylenstierna, U. and Magnéli, A. 1957. Phase analysis studies on the titanium-oxygen system. Acta Chem. Scand. 11: 1641–1652.

Asahi, R., Morikawa, T., Ohwaki, T., Aoki, K. and Taga, Y. 2001. Visible-light photocatalysis in nitrogen-doped titanium oxides. Science. 293, 269–271.

Atrees, M.S., Ebraheim, E.E., Ali, M.E.M., Khawassek, Y.M., Mahmoud, M.S. and Almutairi, M.M. 2021. Synergetic effect of metal-doped GO and TiO_2 on enhancing visible-light-driven photocatalytic hydrogen production from water splitting. Energy Sources A: Recovery Util. Environ. Eff. 43: 484–494.

Bai, C. 2005. Ascent of nanoscience in China. Science. 309: 61–63.

Banerjee, S., Gopal, J., Muraleedharan, P., Tyagi, A.K. and Raj, B. 2006. Physics and chemistry of photocatalytic titanium dioxide: visualization of bactericidal activity using atomic force microscopy. Curr. Sci. 90: 1378–1383.

Bellamkonda, S., Thangavel, N., Hafeez, H.Y., Neppolian, B. and Ranga Rao, G. 2019. Highly active and stable multi-walled carbon nanotubes-graphene-TiO_2 nanohybrid: an efficient non-noble metal photocatalyst for water splitting. Catal. Today, SI: Photocatalytic Materials. 321: 120– 127.

Belver, C., Bellod, R., Fuerte, A. and Fernández-García, M. 2006. Nitrogen-containing TiO_2 photocatalysts: Part 1. Synthesis and solid characterization. Appl. Catal. B: Environ. 65: 301–308.

Bhatia, V. and Dhir, A. 2016. Transition metal doped TiO_2 mediated photocatalytic degradation of anti-inflammatory drug under solar irradiations. J. Environ. Chem. Eng. 4: 1267–1273.

Blanco-Vega, M.P., Guzmán-Mar, J.L., Villanueva-Rodríguez, M., Maya-Treviño, L., Garza-Tovar, L.L., Hernández-Ramírez, A. and Hinojosa-Reyes, L. 2017. Photocatalytic elimination of bisphenol A under visible light using Ni-doped TiO_2 synthesized by microwave assisted sol-gel method. Mater. Sci. Semicond. Process. 71: 275–282.

Boxi, S.S. and Paria, S. 2015. Visible light induced enhanced photocatalytic degradation of organic pollutants in aqueous media using Ag doped hollow TiO_2 nanospheres. RSC Adv. 5: 37657–37668.

Chandra, M., Bhunia, K. and Pradhan, D. 2018. Controlled synthesis of CuS/TiO_2 heterostructured nanocomposites for enhanced photocatalytic hydrogen generation through water splitting. Inorg. Chem. 57: 4524–4533.

Chen, X. and Mao, S.S. 2007. Titanium dioxide nanomaterials: synthesis, properties, modifications, and applications. Chem. Rev. 107: 2891–2959.

Chen, X., Liu, L. and Huang, F. 2015. Black titanium dioxide (TiO_2) nanomaterials. Chem. Soc. Rev. 44(7): 1861–1885.

Chen, S., Li, C. and Hou, Z. 2019. A novel *in situ* synthesis of TiO_2/CdS heterojunction for improving photoelectrochemical water splitting. Int. J. Hydrog. Energy. 44: 25473–25485.

Cho, H., Joo, H., Kim, H., Kim, J.E., Kang, K.S. and Yoon, J. 2021. Improved photoelectrochemical properties of TiO_2 nanotubes doped with Er and effects on hydrogen production from water splitting. Chemosphere. 267: 129289.

Choi, H., Antoniou, M.G., Pelaez, M., de la Cruz, A.A., Shoemaker, J.A. and Dionysiou, D.D. 2007. Mesoporous nitrogen-doped TiO_2 for the photocatalytic destruction of the cyanobacterial toxin microcystin-lr under visible light irradiation. Environ. Sci. Technol. 41: 7530–7535.

Chun, H.Y., Park, S.S., You, S.H., Gi Hyeon, Bae, W.T., Kim, K.W., Park, J., Ozturk, A., et al. 2009. Preparation of a transparent hydrophilic TiO_2 thin film photocatalyst. J. Ceram. Process. Res. 10: 219–223.

Chung, C.C., Chung, T.W. and Yang, T.C.K. 2008. Rapid synthesis of titania nanowires by microwave-assisted hydrothermal treatments. Ind. Eng. Chem. Res. 47: 2301–2307.

Chun-Te Lin, J., Sopajaree, K., Jitjanesuwan, T. and Lu, M.C. 2018. Application of visible light on copper-doped titanium dioxide catalyzing degradation of chlorophenols. Sep. Purif. Technol. 191: 233–243.

Corredor, J., Rivero, M.J., Rangel, C.M., Gloaguen, F. and Ortiz, I. 2019. Comprehensive review and future perspectives on the photocatalytic hydrogen production. J. Chem. Technol. Biotechnol. 94: 3049–3063.

Couper, A.M., Pletcher, D. and Walsh, F.C. 1990. Electrode materials for electrosynthesis. Chem. Rev. 90(5): 837–865.

Crick, C.R. and Parkin, I.P. 2010. Preparation and characterisation of super-hydrophobic surfaces. Chem. Eur. J. 16: 3568–3588.

Crick, C.R., Ismail, S., Pratten, J. and Parkin, I.P. 2011. An investigation into bacterial attachment to an elastomeric superhydrophobic surface prepared via aerosol assisted deposition. Thin Solid Films. 519: 3722–3727.

Crick, C.R., Bear, J.C., Kafizas, A. and Parkin, I.P. 2012. Superhydrophobic photocatalytic surfaces through direct incorporation of titania nanoparticles into a polymer matrix by aerosol assisted chemical vapor deposition. Adv. Mater. 24: 3505–3508.

Czoska, A.M., Livraghi, S., Chiesa, M., Giamello, E., Agnoli, S., Granozzi, G., Finazzi, E., et al. 2008. The nature of defects in fluorine-doped TiO_2. J. Phys. Chem. C. 112: 8951– 8956.

Dai, K., Peng, T., Ke, D. and Wei, B. 2009. Photocatalytic hydrogen generation using a nanocomposite of multi-walled carbon nanotubes and TiO_2 nanoparticles under visible light irradiation. Nanotechnology. 20: 125603.

Dai, K., Zhang, X., Fan, K., Zeng, P. and Peng, T. 2014. Multiwalled carbon nanotube-TiO_2 nanocomposite for visible-light-induced photocatalytic hydrogen evolution. J. Nanomaterials. 2014: 1–8.

Daoud, W.A. and Xin, J.H. 2004. Low temperature sol-gel processed photocatalytic titania coating. J. Solgel. Sci. Technol. 29: 25–29.

Daoud, W.A., Xin, J.H. and Zhang, Y.H. 2005. Surface functionalization of cellulose fibers with titanium dioxide nanoparticles and their combined bactericidal activities. Surf. Sci. 599: 69–75.

Daoud, W.A., Leung, S.K., Tung, W.S., Xin, J.H., Cheuk, K. and Qi, K. 2008. Self-cleaning keratins. Chem. Mater. 20: 1242–1244.

Di Valentin, C., Finazzi, E., Pacchioni, G., Selloni, A., Livraghi, S., Paganini, M.C. and Giamello, E. 2007. N-doped TiO_2: theory and experiment. Chem. Phys. 339: 44–56.

Dodoo-Arhin, D., Buabeng, F.P., Mwabora, J.M., Amaniampong, P.N., Agbe, H., Nyankson, E., Obada, D.O., et al. 2018. The effect of titanium dioxide synthesis technique and its photocatalytic degradation of organic dye pollutants. Heliyon. 4(7): 00681.

Dozzi, M.V. 2012. Improving the photocatalytic activity of TiO_2 for environmental applications: effects of doping and of surface modification. Doctoral Thesis, University of Milan.

Eiden-Assmann, S., Widoniak, J. and Maret, G. 2004. Synthesis and characterization of porous and nonporous monodisperse colloidal TiO_2 particles. Chem. Mater. 16: 6–11.

Ellis, K., Hill, A., Hill, J., Loyns, A. and Partington, T. 2004. The performance of Ebonex® electrodes in bipolar lead-acid batteries. J. Power Sources. 136: 366–371.

Emeline, A.V., Rudakova, A.V., Sakai, M., Murakami, T. and Fujishima, A. 2013. Factors affecting UV-induced superhydrophilic conversion of a TiO_2 surface. J. Phys. Chem. C. 117: 12086–12092.

Emsley, J. 2011. Nature's Building Blocks: An A-Z Guide to the Elements. OUP Oxford.

Erdural, B., Bolukbasi, U. and Karakas, G. 2014. Photocatalytic antibacterial activity of TiO_2–SiO_2 thin films: the effect of composition on cell adhesion and antibacterial activity. J. Photochem. Photobiol., A: Chem. 283, 29–37.

Eshaghi, A. and Aghaei, A.A., 2015. Effect of TiO_2–graphene nanocomposite photoanode on dye-sensitized solar cell performance. Bull. Mater. Sci. 38: 1177–1182.

Etacheri, V., Seery, M.K., Hinder, S.J. and Pillai, S.C. 2011. Oxygen rich titania: a dopant free, high temperature stable, and visible-light active anatase photocatalyst. Adv. Funct. Mater. 21: 3744–3752.

Fang, W.Q., Zhou, J.Z., Liu, J., Chen, Z.G., Yang, C., Sun, C.H., Qian, G.R., et al. 2011. Hierarchical structures of single-crystalline anatase TiO_2 nanosheets dominated by 001 facets. Chem. Eur. J. 17, 1423–1427.

Faustini, M., Nicole, L., Boissiere, C., Innocenzi, P., Sanchez, C., Grosso, D. and Hydrophobic, A. 2010. Self-cleaning, and antifogging sol-gel coatings: an example of multifunctional nanostructured materials for photovoltaic cells. Chem. Mater. 22: 4406–4413.

Flannigan, B., Samson, R.A. and Miller, J.D. (eds). 2002. Microorganisms in Home and Indoor Work Environments: Diversity, Health Impacts, Investigation and Control. CRC Press. London.

Fox, M.A. and Dulay, M.T. 1993. Acceleration of secondary dark reactions of intermediates derived from adsorbed dyes on irradiated TiO_2 powders. J. Photochem. Photobiol. A: Chem. 98: 91–101.

Frank, S.N. and Bard, A.J. 1977. Heterogeneous photocatalytic oxidation of cyanide ion in aqueous solutions at titanium dioxide powder. J. Am. Chem. Soc. 99: 303–304.

Fujishima, A. and Honda, K. 1972. Electrochemical photolysis of water at a semiconductor electrode. Nature. 238: 37–38.

Fujishima, A. and Rao, T.N. 1997. Recent advances in heterogeneous TiO_2 photocatalysis. J. Chem. Sci. 109: 471–486.

Fujishima, A., Zhang, X. and Tryk, D.A. 2008. TiO_2 photocatalysis and related surface phenomena. Surf. Sci. Rep. 63: 515–582.

Galkina, O.L., Sycheva, A., Blagodatskiy, A., Kaptay, G., Katanaev, V.L., Seisenbaeva, G.A., Kessler, V.G., et al. 2014. The sol–gel synthesis of cotton/TiO_2 composites and their antibacterial properties. Surf. Coat. Technol. 253: 171–179.

Ganguly, P., Harb, M., Cao, Z., Cavallo, L., Breen, A., Dervin, S., Dionysiou, D.D., et al. 2019. 2D nanomaterials for photocatalytic hydrogen production. ACS Energy Lett. 4: 1687–1709.

Gao, B., Lim, T.M., Subagio, D.P. and Lim, T.T. 2010. Zr-doped TiO_2 for enhanced photocatalytic degradation of bisphenol A. Appl. Catal. A-Gen. 375: 107–115.

Gao, P., Liu, Z., Sun, D.D. and Ng, W.J. 2014a. The efficient separation of surfactant-stabilized oil–water emulsions with a flexible and superhydrophilic graphene–TiO_2 composite membrane. J. Mater. Chem. A. 2: 14082–14088.

Gao, S.J., Shi, Z., Zhang, W.B., Zhang, F. and Jin, J. 2014b. Photoinduced superwetting single-walled carbon nanotube/TiO_2 ultrathin network films for ultrafast separation of oil-in-water emulsions. ACS. Nano. 8: 6344–6352.

Geng, J., Yang, D., Zhu, J., Chen, D. and Jiang, Z. 2009. Nitrogen-doped TiO_2 nanotubes with enhanced photocatalytic activity synthesized by a facile wet chemistry method. Mater. Res. Bull. 44: 146–150.

Gong, L., Liu, X. and Lu, L. 2012. Synthesis of MnO_2 nanorods from a ZnO template and their capacitive performances. Mater. Lett. 67: 226–228.

Goresy, A.E., Chen, M., Dubrovinsky, L., Gillet, P. and Graup, G. 2001. An ultradense polymorph of rutile with seven-coordinated titanium from the ries crater. Science. 293: 1467–1470.

Goutam, S.P., Saxena, G., Singh, V., Yadav, A.K., Bharagava, R.N. and Thapa, K.B. 2018. Green synthesis of TiO_2 nanoparticles using leaf extract of *Jatropha curcas* L. for photocatalytic degradation of tannery wastewater. Chem. Eng. J. 336: 386–396.

Grätzel, M. 2003. Dye-sensitized solar cells. J. Photochem. Photobiol. C: Photochem. Rev. 4: 145–153.

Grimes, C.A. 2007. Synthesis and application of highly ordered arrays of TiO_2 nanotubes. J. Mater. Chem. 17: 1451–1457.

Guan, K., Lu, B. and Yin, Y. 2003. Enhanced effect and mechanism of SiO_2 addition in super-hydrophilic property of TiO_2 films. Surf. Coat. Technol. 173: 219–223.

Gunawan, C., Teoh, W.Y., Marquis, C.P., Lifia, J. and Amal, R. 2009. Reversible antimicrobial photoswitching in nanosilver. Small. 5: 341–344.

Guo, D., Xiao, S., Fan, K. and Yu, J. 2017. Hierarchical TiO_2 submicrorods improve the photovoltaic performance of dye-sensitized solar cells. ACS Sustain. Chem. Eng. 5(2): 1315–1321.

Guo, Q., Zhou, C., Ma, Z. and Yang, X. 2019. Fundamentals of TiO_2 photocatalysis: concepts, mechanisms, and challenges. Adv. Mater. 31(50): 1901997.

Gupta, S. and Tripathi, M. 2012. A review on the synthesis of TiO_2 nanoparticles by solution route. Open Chem. 10: 279–294.

Han, C., Pelaez, M., Likodimos, V., Kontos, A.G., Falaras, P., O'Shea, K. and Dionysiou, D.D. 2011. Innovative visible light-activated sulfur doped TiO_2 films for water treatment. Appl. Catal. B: Environ. 107: 77–87.

Han, Y., Zhang, J. and Zhao, Y. 2016. Visible-light-induced photocatalytic oxidation of nitric oxide and sulfur dioxide: discrete kinetics and mechanism. Energy. 103: 725–734.

Han, H., Riboni, F., Karlicky, F., Kment, S., Goswami, A., Sudhagar, P., Yoo, J., et al. 2017. α–Fe_2O_3/ TiO_2 3D hierarchical nanostructures for enhanced photoelectrochemical water splitting. Nanoscale. 9: 134–142.

Hanaor, D.A.H. and Sorrell, C.C. 2014. Sand supported mixed-phase TiO_2 photocatalysts for water decontamination applications. Adv. Eng. Mater. 16: 248–254.

Hao, S., Wu, J., Huang, Y. and Lin, J. 2006. Natural dyes as photosensitizers for dye-sensitized solar cell. Solar Energy. Sol. Energy Mater. Sol. Cells. 80: 209–214.

He, Z., Wang, C., Wang, H., Hong, F., Xu, X., Chen, J. and Song, S. 2011. Increasing the catalytic activities of iodine doped titanium dioxide by modifying with tin dioxide for the photodegradation of 2- chlorophenol under visible light irradiation. J. Hazard. Mater. 189: 595–602.

Hernández-Alonso, M.D., Fresno, F., Suárez, S. and Coronado, J.M. 2009. Development of alternative photocatalysts to $TiO\ _2$: challenges and opportunities. Energy Environ. Sci. 2(12): 1231–1257.

Hernández Rodríguez, M.J., Pulido Melián, E., García Santiago, D., González Díaz, O., Navío, J.A. and Doña Rodríguez, J.M. 2017. NO photooxidation with TiO_2 photocatalysts modified with gold and platinum. Appl. Catal. B: Environ. 205: 148–157.

Hinojosa–Reyes, M., Camposeco–Solis, R., Ruiz, F., Rodríguez–González, V. and Moctezuma, E. 2019. Promotional effect of metal doping on nanostructured TiO_2 during the photocatalytic degradation of 4-chlorophenol and naproxen sodium as pollutants. Mater. Sci. Semicond. Process. 100: 130–139.

Hoffmann, M.R., Martin, S.T., Choi, W. and Bahnemann, D.W. 1995. Environmental applications of semiconductor photocatalysis. Chem. Rev. 95: 69–96.

Hogan, J. 2004. Smog-busting paint soaks up noxious gases. New Scientist. 4(10).

Hoseinzadeh Hesas, R., Wan Daud, W.M.A., Sahu, J.N. and Arami-Niya, A. 2013. The effects of a microwave heating method on the production of activated carbon from agricultural waste: a review. J. Anal. Appl. Pyrolysis. 100: 1–11.

Hu, J.S., Zhong, L.S., Song, W.G. and Wan, L.J. 2008. Synthesis of hierarchically structured metal oxides and their application in heavy metal ion removal. Adv. Mater. 20: 2977–2982.

Huang, H., Leung, D.Y.C., Li, G., Leung, M.K.H. and Fu, X. 2011. Photocatalytic destruction of air pollutants with Vacuum Ultraviolet (VUV) irradiation. Catal. Today. 175: 310–315.

Irie, H., Watanabe, Y. and Hashimoto, K. 2003. Carbon-doped anatase TiO_2 powders as a visible-light sensitive photocatalyst. Chem. Lett. 32: 772–773.

Irie, H., Washizuka, S. and Hashimoto, K. 2006. Hydrophilicity on carbon-doped TiO_2 thin films under visible light. Thin Solid Films. 510: 21–25.

Izumi, Y., Itoi, T., Peng, S., Oka, K. and Shibata, Y. 2009. Site structure and photocatalytic role of sulfur or nitrogen-doped titanium oxide with uniform mesopores under visible light. J. Phys. Chem. C. 113: 6706–6718.

Jagadale, T.C., Takale, S.P., Sonawane, R.S., Joshi, H.M., Patil, S.I., Kale, B.B. and Ogale, S.B. 2008. N-doped TiO_2 nanoparticle based visible light photocatalyst by modified peroxide sol–gel method. J. Phys. Chem. C. 112: 14595–14602.

Jamali, S., Moshaii, A. and Mohammadi, K. 2019. Optimization of Ag loaded TiO_2 nanotube arrays for plasmonic photoelectrochemical water splitting. Iran. J. Hydrog. Fuel cell. 6: 133–140.

Jinlong, L., Xinxin, M., Mingren, S., Li, X. and Zhenlun, S. 2010. Fabrication of nitrogen-doped mesoporous TiO_2 layer with higher visible photocatalytic activity by plasma-based ion implantation. Thin Solid Films. 519: 101–105.

Kadam, A.N., Dhabbe, R.S., Kokate, M.R., Gaikwad, Y.B. and Garadkar, K.M. 2014. Preparation of N doped TiO_2 via microwave-assisted method and its photocatalytic activity for degradation of malathion. Spectrochim. Acta A: Mol. Biomol. Spectrosc. 133: 669–676.

Kamat, P.V. 2012. TiO_2 nanostructures: recent physical chemistry advances. J. Phys. Chem. C. 116(22): 11849–11851.

Kamegawa, T., Shimizu, Y. and Yamashita, H. 2012. Superhydrophobic surfaces with photocatalytic self-cleaning properties by nanocomposite coating of TiO_2 and polytetrafluoroethylene. Adv. Mater. 24: 3697–3700.

Kapridaki, C., Pinho, L., Mosquera, M.J. and Maravelaki-Kalaitzaki, P. 2014. Producing photoactive, transparent and hydrophobic SiO_2-crystalline TiO_2 nanocomposites at ambient conditions with application as self-cleaning coatings. Appl. Catal. B: Environ. 156: 416–427.

Karimi, L., Zohoori, S. and Amini, A. 2014. Multi-wall carbon nanotubes and nano titanium dioxide coated on cotton fabric for superior self-cleaning and UV blocking. New Carbon Mater. 29: 380–385.

Kartini, I., Santosa, S.J., Febriyanti, E., Nugroho, O.R., Yu, H. and Wang, L. 2014. Hybrid assembly of nanosol titania and dodecylamine for superhydrophobic self-cleaning glass. J. Nanopart. Res. 16: 2514.

Kasuga, T., Hiramatsu, M., Hoson, A., Sekino, T. and Niihara, K. 1998. Formation of titanium oxide nanotube. Langmuir. 14: 3160–3163.

Kaur, T., Sraw, A., Toor, A.P. and Wanchoo, R.K. 2016. Utilization of solar energy for the degradation of carbendazim and propiconazole by Fe doped TiO_2. Sol. Energy. 125: 65–76.

Kazmi, S.A., Hameed, S., Ahmed, A.S., Arshad, M. and Azam, A. 2017. Electrical and optical properties of graphene-TiO_2 nanocomposite and its applications in dye sensitized solar cells (DSSC). J. Alloys Compd. 691: 659–665.

Khan, I. and Qurashi, A. 2018. Sonochemical-assisted in situ electrochemical synthesis of Ag/α–Fe_2O_3/TiO_2 nanoarrays to harness energy from photoelectrochemical water splitting. ACS Sustainable Chem. Eng. 6: 11235–11245.

Khraisheh, M., Wu, L., Al-Muhtaseb, A.H., Albadarin, A.B. and Walker, G.M. 2012. Phenol degradation by powdered metal ion modified titanium dioxide photocatalysts. Chem. Eng. J. 213: 125–134.

Kim, T.Y., Park, K.H., Lee, J.W., Han, S. and Cho, S.Y. 2014. Adsorption equilibrium and kinetics of gardenia blue on TiO_2 photoelectrode for dye-sensitized solar cells. Int. J. Photoenergy.

Kisch, H. 2013. Semiconductor Photocatalysis—Mechanistic and Synthetic Aspects. Angew. Chem. Int. Ed. 52: 812–847.

Kiwi, J. and Pulgarin, C. 2010. Innovative self-cleaning and bactericide textiles. Catal. Today. EAAOP-2. 151(1–2): 2–7.

Koch, K., Bhushan, B. and Barthlott, W. 2009. Multifunctional surface structures of plants: an inspiration for biomimetics. Prog. Mater. Sci. 54: 137–178.

Konstantinou, I.K. and Albanis, T.A. 2004. TiO_2-assisted photocatalytic degradation of azo dyes in aqueous solution: kinetic and mechanistic investigations: a review. Appl. Catal. B: Environ. 49: 1–14.

Kontos, A.G., Pelaez, M., Likodimos, V., Vaenas, N., Dionysiou, D.D. and Falaras, P. 2011. Visible light induced wetting of nanostructured N-F co-doped titania films. Photochem. Photobiol. Sci. 10: 350–354.

Kumaravel, V., Bartlett, J. and Pillai, S.C. 2020. Photoelectrochemical conversion of carbon dioxide (CO_2) into fuels and value-added products. ACS Energy Lett. 5: 486–519.

Kundu, P., Kaur, A., Mehta, S.K. and Kansal, S.K. 2014. Removal of ofloxacin from aqueous phase using Ni-doped TiO_2 nanoparticles under solar irradiation. J. Nanosci. Nanotechnol. 14: 6991–6995.

Lam, S.M., Sin, J.C., Abdullah, A.Z. and Mohamed, A.R. 2014. Photocatalytic TiO_2/carbon nanotube nanocomposites for environmental applications: an overview and recent developments. Fulluerenes, NTs and C-NS. 22: 471–509.

Lamastra, L., Balderacchi, M. and Trevisan, M. 2016. Inclusion of emerging organic contaminants in groundwater monitoring plans. Methods X. 3: 459–476.

Latroche, M., Brohan, L., Marchand, R. and Tournoux, M. 1989. New hollandite oxides: TiO_2(H) and $K_{0.06}TiO_2$. J. Solid State Chem. 81: 78–82.

Lavand, A.B. and Malghe, Y.S. 2015. Nano sized C-doped TiO_2 as a visible-light photocatalyst for the degradation of 2, 4, 6-trichlorophenol. Adv. Mater. Lett. 6(8): 695–700.

Lee, H.Y., Park, Y.H. and Ko, K.H. 2000. Correlation between surface morphology and hydrophilic/hydrophobic conversion of MOCVD−TiO_2 films. Langmuir. 16: 7289–7293.

Lee, S., Cho, I.S., Lee, J.H., Kim, D.H., Kim, D.W., Kim, J.Y., Shin, H., et al. 2010. Two-step sol−gel method-based TiO_2 nanoparticles with uniform morphology and size for efficient photo-energy conversion devices. Chem. Mater. 22: 1958– 1965.

Leong, K.H., Gan, B.L., Ibrahim, S. and Saravanan, P. 2014. Synthesis of surface plasmon resonance (SPR) triggered Ag/TiO_2 photocatalyst for degradation of endocrine disturbing compounds. Appl. Surf. Sci. 319: 128–135.

Li, X.L., Peng, Q., Yi, J.X., Wang, X. and Li, Y. 2006. Near monodisperse TiO_2 nanoparticles and nanorods. Chem. Eur. J. 12: 2383–2391.

Li, W., Guo, T., Meng, T., Huang, Y., Li, X., Yan, W., Wang, S., et al. 2013. Enhanced reversiblewettability conversion of micro-nano hierarchical TiO_2/SiO_2 composite films under UV irradiation. Appl. Surf. Sci. 283: 12–18.

Li, Xin, Yu, J., Low, J., Fang, Y., Xiao, J. and Chen, X. 2015. Engineering heterogeneous semiconductors for solar water splitting. J. Mater. Chem. A. 3: 2485–2534.

Li, X., Shen, R., Ma, S., Chen, X. and Xie, J. 2018. Graphene-based heterojunction photocatalysts. 2nd International Workshop on Graphene and C_3N_4-based Photocatalysts. Appl. Surf. Sci. 430: 53–107.

Linsebigler, A.L., Lu, G. and Yates Jr, J.T. 1995. Photocatalysis on TiO_2 surfaces: principles, mechanisms, and selected results. Chem. Rev. 95(3): 735–758.

Liu, Z., Sun, D.D., Guo, P. and Leckie, J.O. 2007. One-step fabrication and high photocatalytic activity of porous TiO_2 hollow aggregates by using a low-temperature hydrothermal method without templates. Chem. Eur. J. 13(6): 1851–1855.

Liu, S., Yu, J. and Wang, W. 2010. Effects of annealing on the microstructures and photoactivity offluorinated N-doped TiO_2. Phys. Chem. Chem. Phys. 12: 12308–12315.

Liu, Y., Zhang, X., Liu, R., Yang, R., Liu, C. and Cai, Q. 2011a. Fabrication and photocatalytic activity of high-efficiency visible-light-responsive photocatalyst ZnTe/TiO_2 nanotube arrays. J. Solid State Chem. 184: 684–689.

Liu, J., Zhao, Y., Shi, L., Yuan, S., Fang, J., Wang, Z. and Zhang, M. 2011b. Solvothermal synthesis of crystallinephase and shape controlled Sn^{4+}-doped TiO_2 nanocrystals: effects of reaction solvent. ACS Appl. Mater. Interfaces. 3: 1261–1268.

Liu, K., Tian, Y. and Jiang, L. 2013. Bio-inspired superoleophobic and smart materials: design, fabrication, and application. Prog. Mater. Sci. 58: 503–564.

Liu, Y., He, X., Duan, X., Fu, Y., Fatta-Kassinos, D. and Dionysiou, D.D. 2016. Significant role of UV and carbonate radical on the degradation of oxytetracycline in UV-AOPs: kinetics and mechanism. Water Res. 95: 195–204.

Liu, Y.L. and Wu, J.M. 2019. Synergistically catalytic activities of $BiFeO_3$/TiO_2 core-shell nanocomposites for degradation of organic dye molecule through piezophototronic effect. Nano Energy. 56: 74–81.

Low, F.W. and Lai, C.W. 2018. Recent developments of graphene-TiO_2 composite nanomaterials as efficient photoelectrodes in dye-sensitized solar cells: a review. Renew. Sustain. Energy Rev. 82: 103–125.

Lü, X., Ding, S., Xie, Y. and Huang, F. 2011. Non-aqueous preparation of high-crystallinity hierarchical TiO_2 hollow spheres with excellent photocatalytic efficiency. Eur. J. Inorg. Chem. 2879–2883.

Lu, Y., Sathasivam, S., Song, J., Crick, C.R., Carmalt, C.J. and Parkin, I.P. 2015. Robust self-cleaning surfacesthat function when exposed to either air or oil. Science. 347: 1132–1135.

Luan, P., Xie, M., Fu, X., Qu, Y., Sun, X. and Jing, L. 2015. Improved photoactivity of TiO_2–Fe_2O_3 nanocomposites for visible-light water splitting after phosphate bridging and its mechanism. Phys. Chem. Chem. Phys. 17: 5043–5050.

Luo, Z., Sang, S., Wu, Q. and Liu, S. 2012. A Conductive additive for Zn electrodes in secondary Ni/Zn batteries: the magneli phase titanium sub-oxides conductive ceramic $TinO_{2n-1}$. ECS Electrochem. Lett. 2: A21.

Luo, J., Zhang, S., Sun, M., Yang, L., Luo, S. and Crittenden, J.C. 2019. A Critical review on energy conversion and environmental remediation of photocatalysts with remodeling crystal lattice, surface, and interface. ACS Nano. 13: 9811–9840.

Macias-Montero, M., Borras, A., Saghi, Z., Romero-Gomez, P., Sanchez-Valencia, J.R., Gonzalez, J.C., Barranco, A., et al. 2011. Superhydrophobic supported Ag-NPs@ZnO-nanorods with photoactivity in the visible range. J. Mater. Chem. 22: 1341–1346.

Magaye, R., Zhao, J., Bowman, L. and Ding, M.I.N. 2012. Genotoxicity and carcinogenicity of cobalt-, nickel-and copper-based nanoparticles. Exp. Ther. Med. 4(4): 551–561.

Malakootian, M., Olama, N., Malakootian, M. and Nasiri, A. 2019. Photocatalytic degradation of metronidazole from aquatic solution by TiO_2-doped Fe^{3+} nano-photocatalyst. Int. J. Environ. Sci. Technol. 16: 4275–4284.

Marchal, C., Behr, M., Vigneron, F., Caps, V. and Keller, V. 2016. Au/TiO_2 photocatalysts prepared by solid grinding for artificial solar-light water splitting. New J. Chem. 40: 4428–4435.

Midtdal, K. and Jelle, B.P. 2013. Self-cleaning glazing products: a state-of-the-art review and future research pathways. Sol. Energy Mater. Sol. Cells. 109: 126–141.

Mihailović, D., Šaponjić, Z., Vodnik, V., Potkonjak, B., Jovančić, P., Nedeljković, J.M. and Radetić, M. 2011. Multifunctional PES fabrics modified with colloidal Ag and TiO_2 nanoparticles. Polym. Adv. Technol. 22: 2244–2249.

Mills, A., Lepre, A., Elliott, N., Bhopal, S., Parkin, I.P. and O'Neill, S.A. 2003. Characterisation of the photocatalyst Pilkington Activ^{TM}: a reference film photocatalyst? J. Photochem. Photobiol. A: Chem. 160: 213–224.

Mitoraj, D. and Kisch, H. 2010. On the mechanism of urea-induced titania modification. Eur. J. Chem. 16: 261–269.

Mohammadi, R., Massoumi, B. and Rabani, M. 2012. Photocatalytic decomposition of amoxicillin trihydrate antibiotic in aqueous solutions under UV irradiation using Sn/TiO_2 nanoparticles. Int. J. Photoenergy. 51: 1–11.

Mohammed, M.K.A. 2020. Sol-gel synthesis of Au-doped TiO_2 supported SWCNT nanohybrid with visible-light-driven photocatalytic for high degradation performance toward methylene blue dye. Optik. 223: 165607.

Mokhtarimehr, M., Pakshir, M., Eshaghi, A. and Shariat, M.H. 2013. Super-hydrophilic property of vanadium doped TiO_2–SiO_2 sol–gel derived thin film. Thin Solid Films. 532: 123–126.

Montazer, M., Behzadnia, A. and Moghadam, M.B. 2012. Superior self-cleaning features on wool fabric using TiO_2/Ag nanocomposite optimized by response surface methodology. J. Appl. Polym. Sci. 125: E356–E363.

Mor, G.K., Carvalho, M.A., Varghese, O.K., Pishko, M.V. and Grimes, C.A. 2004. A room-temperature TiO_2-nanotube hydrogen sensor able to self-clean photoactively from environmental contamination. J. Mater. Res. 19: 628–634.

Mor, G.K., Varghese, O.K., Paulose, M., Shankar, K. and Grimes, C.A. 2006. A review on highly ordered, vertically oriented TiO_2 nanotube arrays: Fabrication, material properties, and solar energy applications. Sol. Energy Mater. Sol. Cells. 90: 2011–2075.

Morikawa, T., Asahi, R., Ohwaki, T., Aoki, K. and Taga, Y. 2001. Band-gap narrowing of titanium dioxide by nitrogen doping. Jpn. J. Appl. Phys. 40: L561.

Murakami, A., Yamaguchi, T., Hirano, S. and Kikuta, K. 2008. Synthesis of porous titania thin films using carbonatation reaction and its hydrophilic property. Thin Solid Films. 516: 3888–3892.

Murakami, N., Ono, A., Nakamura, M., Tsubota, T. and Ohno, T. 2010. Development of a visible-light-responsive rutile rod by site-selective modification of iron(III) ion on {111} exposed crystal faces. Appl. Catal. B: Environ. 97: 115–119.

Murugan, K., Subasri, R., Rao, T.N., Gandhi, A.S. and Murty, B.S. 2013. Synthesis, characterization and demonstration of self-cleaning TiO_2 coatings on glass and glazed ceramic tiles. Prog. Org. Coat. 76: 1756–1760.

Muscat, J., Swamy, V. and Harrison, N.M. 2002. First-principles calculations of the phase stability of TiO_2. Phys. Rev. B. 65(22): 224112.

Nagaveni, K., Hegde, M.S., Ravishankar, N., Subbanna, G.N. and Madras, G. 2004. Synthesis and structure of nanocrystalline TiO_2 with lower band gap showing high photocatalytic activity. Langmuir. 20: 2900–2907.

Nakajima, A., Koizumi, S., Watanabe, T. and Hashimoto, K. 2000. Photoinduced amphiphilic surface on polycrystalline anatase TiO_2 thin films. Langmuir. 16: 7048–7050.

Nakajima, A., Koizumi, S., Watanabe, T. and Hashimoto, K. 2001. Effect of repeated photo-illumination on the wettability conversion of titanium dioxide. J. Photochem. Photobiol. A: Chem. 146: 129–132.

Nakata, K., Sakai, M., Ochiai, T., Murakami, T., Takagi, K. and Fujishima, A. 2011. Antireflection and self-cleaning properties of a moth-eye-like surface coated with TiO_2 particles. Langmuir. 27: 3275–3278.

Nakata, K., Terashima, C. and Fujishima, A. 2014. Moth-eye-structured antireflective poly (methyl methacrylate) film with visible-light-responsive self-cleaning ability. Chem. Lett. 43: 1511–1513.

Nam, C.T., Yang, W.-D. and Duc, L.M. 2013. Solvothermal synthesis of TiO_2 photocatalysts in ketone solvents with low boiling points. J. Nanomater. 36: 36.

Nguyen, T.D., Dinh, C.T. and Do, T.O. 2010. Shape- and size-controlled synthesis of monoclinic ErOOH and cubic Er_2O_3 from micro- to nanostructures and their upconversion luminescence. ACS Nano. 4: 2263–2273.

Nolan, N.T., Seery, M.K., Hinder, S.J., Healy, L.F. and Pillai, S.C. 2010. A Systematic study of the effect of silver on the chelation of formic acid to a titanium precursor and the resulting effect on the anatase to rutile transformation of TiO_2. J. Phys. Chem. C. 114: 13026–13034.

Nuran, A., Kento, T., Daiki, M. and Fujiki, A. 2015. The basic research on the dye-sensitized solar cells (DSSC). J. Clean Energy Technol. 3: 382–387.

Ochiai, T., Masuko, K., Tago, S., Nakano, R., Niitsu, Y., Kobayashi, G., Horio, K., et al. 2013. Development of a hybrid environmental purification unit by using of excimer VUV lamps with TiO_2 coated titanium mesh filter. Chem. Eng. J. 218: 327–332.

O'Regan, B., Gratzel, M., Regan, B.O., Gratzel, M., Regan, B.O., Gratzel, M., Regan, B.O., et al. 1991. Synthesise of nanocrystalline titanium dioxide dye sensitised solar cell. Nature. 353. 737–781.

Padmanabhan, S.C., Pillai, S.C., Colreavy, J., Balakrishnan, S., McCormack, D.E., Perova, T.S., Gun'ko, Y., et al. 2007. A simple sol−gel processing for the development of high-temperature stable photoactive anatase titania. Chem. Mater. 19(18): 4474–4481.

Pakdel, E., Daoud, W.A. and Wang, X. 2013. Self-cleaning and superhydrophilic wool by TiO_2/SiO_2 nanocomposite. Appl. Surf. Sci. 275: 397–402.

Parkin, I.P. and Palgrave, R.G. 2005. Self-cleaning coatings. J. Mater. Chem. 15: 1689–1695.

Pasqui, D. and Barbucci, R. 2014. Synthesis, characterization and self cleaning properties of titania nanoparticles grafted on polyester fabrics. J. Photochem. Photobiol. A: Chem. 274: 1–6.

Paxton, A.T. and Thiên-Nga, L. 1998. Electronic structure of reduced titanium dioxide. Phys. Rev. B. 57: 1579–1584.

Pelaez, M., de la Cruz, A.A., Stathatos, E., Falaras, P. and Dionysiou, D.D. 2009. Visible light-activated N-F-codoped TiO_2 nanoparticles for the photocatalytic degradation of microcystin-LR in water. Catalysis Today, Selected papers presented at the 5th European Conference on Solar Chemistry and Photocatalysis: Environmental Applications (SPEA 5). 4–8th October 2008. 144: 19–25.

Pelaez, M., Falaras, P., Likodimos, V., Kontos, A.G., de la Cruz, A.A., O'shea, K. and Dionysiou, D.D. 2010. Synthesis, structural characterization and evaluation of sol–gel-based NF-TiO_2 films with visible light-photoactivation for the removal of microcystin-LR. Applied Catalysis B: Environmental, TiO_2 photocatalysis-25 years. 99: 378–387.

Pelaez, M., Nolan, N.T., Pillai, S.C., Seery, M.K., Falaras, P., Kontos, A.G., Dunlop, P.S.M., et al. 2012. A review on the visible light active titanium dioxide photocatalysts for environmental applications. Appl. Catal. B: Environ. 125: 331–349.

Perera, S.D., Mariano, R.G., Vu, K., Nour, N., Seitz, O., Chabal, Y. and Balkus, K.J. 2012. Hydrothermal synthesis of graphene-TiO_2 nanotube composites with enhanced photocatalytic activity. ACS Catal. 2: 949–956.

Periyat, P., Pillai, S.C., McCormack, D.E., Colreavy, J. and Hinder, S.J. 2008. Improved high-temperature stability and sun-light-driven photocatalytic activity of sulfur-doped anatase TiO_2. J. Phys. Chem. C. 112: 7644–7652.

Pham, T.D. and Lee, B.K. 2017. Novel capture and photocatalytic conversion of CO_2 into solar fuels by metals co-doped TiO_2 deposited on PU under visible light. Appl. Catal. A: Gen. 529: 40–48.

Pore, V., Heikkilä, M., Ritala, M., Leskelä, M. and Areva, S. 2006. Atomic layer deposition of TiO_2–xN_x thin films for photocatalytic applications. J. Photochem. Photobiol. A: Chem. 177: 68–75.

Prakasam, H.E., Shankar, K., Paulose, M., Varghese, O.K. and Grimes, C.A. 2007. A new benchmark for tio_2nanotube array growth by anodization. J. Phys. Chem. C. 111: 7235–7241.

Qi, K., Wang, X. and Xin, J.H., 2011. Photocatalytic self-cleaning textiles based on nanocrystalline titanium dioxide. Text. Res. J. 81: 101–110.

Qin, D.-D., Tao, C.-L., In, S.-il, Yang, Z.-Y., Mallouk, T.E., Bao, N. and Grimes, C.A. 2011. Facile solvothermal method for fabricating arrays of vertically oriented α-Fe_2O_3 nanowires and their application in photoelectrochemical water oxidation. Energy Fuels. 25: 5257–5263.

Qu, Y., Song, S., Jing, L., Luan, Y. and Fu, H. 2010. Effects of the co-addition of Zn^{2+} and sodium dodecyl benzene sulfonate on photocatalytic activity and wetting performance of anatase TiO_2 nanoparticle films. Thin Solid Films. 518: 3177–3181.

Radetić, M. 2013. Functionalization of textile materials with TiO_2 nanoparticles. J. Photochem. Photobiol. C. 16: 62–76.

Ragesh, P., Nair, S.V. and Nair, A.S. 2014. An attempt to fabricate a photocatalytic and hydrophobic self-cleaning coating via electrospinning. RSC Adv. 4: 38498–38504.

Ramsden, J. 2015. Photocatalytic antimicrobial coatings. Nanotechnol. Percept. 11: 146–168.

Ran, H., Fan, J., Zhang, X., Mao, J. and Shao, G., 2018. Enhanced performances of dye-sensitized solar cells based on Au-TiO_2 and Ag-TiO_2 plasmonic hybrid nanocomposites. Appl. Surf. Sci. 430: 415–423.

Ranjit, K.T., Willner, I., Bossmann, S.H. and Braun, A.M. 2001. Lanthanide oxide-doped titanium dioxide photocatalysts: novel photocatalysts for the enhanced degradation of p-chlorophenoxyacetic acid. Environ. Sci. Technol. 35: 1544–1549.

Rashkova, V., Kitova, S. and Vitanov, T. 2005. Influence of the support type on the electrocatalytic properties of thin vacuum deposited films of cobalt oxides. J. Optoelectron. Adv. Mater. 7: 1341–1346.

Rehan, M., Hartwig, A., Ott, M., Gätjen, L. and Wilken, R. 2013. Enhancement of photocatalytic self-cleaning activity and antimicrobial properties of poly(ethylene terephthalate) fabrics. Surf. Coat. Technol. 219: 50–58.

Reyes-Coronado, D., Rodríguez-Gattorno, G., Espinosa-Pesqueira, M.E., Cab, C., Coss, R. and de, Oskam, G. 2008. Phase-pure TiO_2 nanoparticles: anatase, brookite and rutile. Nanotechnology 19(14): 145605.

Riu, J., Maroto, A. and Rius, F.X. 2006. Nanosensors in environmental analysis. Talanta. 69: 288–301.

Saadoun, L., Ayllón, J.A., Jiménez-Becerril, J., Peral, J., Domènech, X. and Rodrıǵ uez-Clemente, R. 2000.

Synthesis and photocatalytic activity of mesoporous anatase prepared from tetrabutylammonium-titania composites. Mater. Res. Bull. 35: 193–202.

Sakai, N., Wang, R., Fujishima, A., Watanabe, T. and Hashimoto, K. 1998. Effect of ultrasonic treatment on highly hydrophilic TiO_2 surfaces. Langmuir. 14: 5918–5920.

Sakai, N., Fujishima, A., Watanabe, T. and Hashimoto, K. 2001. Enhancement of the photoinduced hydrophilic conversion rate of TiO_2 film electrode surfaces by anodic polarization. J. Phys. Chem. B. 105: 3023–3026.

Sakai, N., Fujishima, A., Watanabe, T. and Hashimoto, K. 2003. Quantitative evaluation of the photoinduced hydrophilic conversion properties of TiO_2 thin film surfaces by the reciprocal of contact angle. J. Phys. Chem. B. 107: 1028–1035.

Sakthivel, S. and Kisch, H. 2003. Daylight Photocatalysis by Carbon-Modified Titanium Dioxide. Angew. Chem. Int. Ed. 42: 4908–4911.

Sano, T., Negishi, N., Koike, K., Takeuchi, K. and Matsuzawa, S. 2004. Preparation of a visible light-responsive photocatalyst from a complex of Ti^{4+} with a nitrogen-containing ligand. J. Mater. Chem. 14: 380–384.

Saqib, N.U., Adnan, R. and Shah, I. 2019. Zeolite supported TiO_2 with enhanced degradation efficiency for organic dye under household compact fluorescent light. Mater. Res. Express. 6(9): 095506.

Sarantopoulos, C., Gleizes, A.N. and Maury, F. 2009. Chemical vapor deposition and characterization of nitrogen doped TiO_2 thin films on glass substrates. Thin Solid Films. 518: 1299–1303.

Saranya, K.S., Vellora Thekkae Padil, V., Senan, C., Pilankatta, R., Saranya, K., George, B., Wacławek, S., et al. 2018. Green synthesis of high temperature stable anatase titanium dioxide nanoparticles using gum kondagogu: characterization and solar driven photocatalytic degradation of organic dye. Nanomat. 8: 1002.

Sawai, Y., Nishimoto, S., Kameshima, Y., Fujii, E. and Miyake, M. 2013. Photoinduced underwater superoleophobicity of TiO_2 thin films. Langmuir. 29: 6784–6789.

Seery, M.K., George, R., Floris, P. and Pillai, S.C. 2007. Silver doped titanium dioxide nanomaterials for enhanced visible light photocatalysis. J. Photochem. Photobiol. A: Chem. 189: 258–263.

Selloni, A. 2008. Anatase shows its reactive side. Nat. Mater. 7(8): 613–615.

Shi, H., Long, S., Hu, S., Hou, J., Ni, W., Song, C., Li, K., et al. 2019. Interfacial charge transfer in 0D/2D defect-rich heterostructures for efficient solar-driven CO_2 reduction. Appl. Catal. B: Environ. 245: 760–769.

Shibata, T., Irie, H. and Hashimoto, K. 2009. Photoinduced hardness change on TiO_2 single crystal surfaces. Chem. Commun. 25: 3735–3737.

Shibata, T., Sakai, N., Fukuda, K., Ebina, Y. and Sasaki, T. 2007. Photocatalytic properties of titania nanostructured films fabricated from titania nanosheets. Phys. Chem. Chem. Phys. 9: 2413–2420.

Shimohigoshi, M. and Saeki, Y. 2007. Research and applications of photocatalyst tiles. pp. 291–297 *In*: Baglioni, P. and Casssar, L. (eds). International RILEM Symposium on Photocatalysis, Environment and Construction Materials—TDP. RILEM Publications SARL.

Shojaie, A.F. and Loghmani, M.H. 2010. La^{3+} and Zr^{4+} co-doped anatase nano TiO_2 by sol-microwave method. Chem. Eng. J. 157: 263–269.

Shokri, M., Jodat, A., Modirshahla, N. and Behnajady, M.A. 2013. Photocatalytic degradation of chloramphenicol in an aqueous suspension of silver-doped TiO_2 nanoparticles. Environ. Technol. 34: 1161–1166.

Singh, M.K., Mehata, M.S. 2020. Enhanced photoinduced catalytic activity of transition metal ions incorporated TiO_2 nanoparticles for degradation of organic dye: absorption and photoluminescence spectroscopy. Opt. Mater. 109: 110309.

Sohail, M., Baig, N., Sher, M., Jamil, R., Altaf, M., Akhtar, S. and Sharif, M. 2020. A novel tin-doped titanium oxide nanocomposite for efficient photo-anodic water splitting. ACS Omega. 5: 6405–6413.

Sood, S., Umar, A., Mehta, S.K. and Kansal, S.K. 2015. Highly effective Fe-doped TiO_2 nanoparticles photocatalysts for visible-light driven photocatalytic degradation of toxic organic compounds. J. Colloid Interface Sci. 450: 213–223.

Spasiano, D., Marotta, R., Malato, S., Fernandez-Ibañez, P. and Di Somma, I. 2015. Solar photocatalysis: Materials, reactors, some commercial, and pre-industrialized applications. A comprehensive approach. Appl. Catal. B: Environ. 170–171: 90–123.

Su, C., Tseng, C.-M., Chen, L.-F., You, B.-H., Hsu, B.-C. and Chen, S.-S. 2006. Sol–hydrothermal preparation and photocatalysis of titanium dioxide. Thin Solid Films, Proceedings of the Third Asian Conference on Chemical Vapor Deposition (Third Asian-CVD). 498: 259–265.

Su, Z. and Zhou, W. 2011. Formation, morphology control and applications of anodic TiO_2 nanotube arrays. J. Mater. Chem. 21: 8955–8970.

Sui, Y., Liu, J., Zhang, Y., Tian, X. and Chen, W. 2013. Dispersed conductive polymer nanoparticles on graphitic carbon nitride for enhanced solar-driven hydrogen evolution from pure water. Nanoscale. 5: 9150–9155.

Syafiq, A., Vengadaesvaran, B., Pandey, A.K. and Rahim, N.A. 2018. Superhydrophilic smart coating for self- cleaning application on glass substrate. J. Nanomater. 1–10.

Tan, B. and Wu, Y. 2006. Dye-Sensitized solar cells based on anatase TiO_2 nanoparticle/nanowire composites. J. Phys. Chem. B. 110: 15932–15938.

Tan, B., Zhang, X., Li, Y., Chen, H., Ye, X., Wang, Y. and Ye, J. 2017. Anatase TiO_2 mesocrystals: green synthesis, *in situ* conversion to porous single crystals, and self-doping Ti^{3+} for enhanced visible light driven photocatalytic removal of NO. Chem. Eur. J. 23: 5478–5487.

Teh, C.M. and Mohamed, A.R. 2011. Roles of titanium dioxide and ion-doped titanium dioxide on photocatalytic degradation of organic pollutants (phenolic compounds and dyes) in aqueous solutions: a review. J. Alloys Compd. 509: 1648–1660.

Tian, G., Chen, Y., Zhai, R., Zhou, J., Zhou, W., Wang, R. and Pan, K., et al. 2013. Hierarchical flake- like Bi_2MoO_6/TiO_2 bilayer films for visible-light-induced self-cleaning applications. J. Mater. Chem. A. 1: 6961–6968.

Toyoda, M., Yano, T., Tryba, B., Mozia, S., Tsumura, T. and Inagaki, M. 2009. Preparation of carbon-coated Magneli phases Ti_nO_{2n-1} and their photocatalytic activity under visible light. Appl. Catal. B: Environ. 88: 160–164.

Trapalis, A., Todorova, N., Giannakopoulou, T., Boukos, N., Speliotis, T., Dimotikali, D. and Yu, J. 2016. TiO_2/graphene composite photocatalysts for NO_x removal: a comparison of surfactant-stabilized graphene and reduced graphene oxide. Appl. Catal. B: Environ.180: 637– 647.

Tricoli, A., Righettoni, M. and Pratsinis, S.E. 2009. Anti-fogging nanofibrous SiO_2 and nanostructured SiO_2–TiO_2 films made by rapid flame deposition and *in situ* annealing. Langmuir. 25: 12578– 12584.

Tseng, T.K., Lin, Y.S., Chen, Y.J. and Chu, H. 2010. A review of photocatalysts prepared by sol-gel method for VOCs removal. Int. J. Mol. Sci. 11: 2336–2361.

Tsuchiya, H., Macak, J.M., Taveira, L. and Schmuki, P. 2005. Fabrication and characterization of smooth high aspect ratio zirconia nanotubes. Chem. Phys. Lett. 410: 188–191.

Uyama, A., Yamazoe, S., Shigematsu, S., Morimoto, M., Yokojima, S., Mayama, H., Kojima, Y., et al. 2011. Reversible photocontrol of surface wettability between hydrophilic and superhydrophobic surfaces on an asymmetric diarylethene solid surface. Langmuir. 27: 6395– 6400.

Wakamura, M., Hashimoto, K. and Watanabe, T. 2003. photocatalysis by calcium hydroxyapatite modified with Ti(IV): albumin decomposition and bactericidal effect. Langmuir. 19: 3428–3431.

Wan, P., Jiang, Y., Wang, Y., Wang, Z. and Zhang, X. 2008. Tuning surface wettability through photocontrolled nreversible molecular shuttle. Chem. Commun. 44: 5710–5712.

Wang, R., Hashimoto, K., Fujishima, A., Chikuni, M., Kojima, E., Kitamura, A., Shimohigoshi, M., et al. 1997. Light-induced amphiphilic surfaces. Nature. 388: 431–432.

Wang, R., Hashimoto, K., Fujishima, A., Chikuni, M., Kojima, E., Kitamura, A., Shimohigoshi, M., et al. 1998. Photogeneration of highly amphiphilic TiO_2 surfaces. Adv. Mater. 10: 135–138.

Wang, Q., Wen, Z.H. and Li, J.H. 2006. A hybrid supercapacitor fabricated with a carbon nanotube cathode and a TiO_2–B nanowire anode. Adv. Funct. Mater. 16(16): 2141–2146.

Wang, F., Shi, Z., Gong, F., Jiu, J. and Adachi, M. 2007. Morphology control of anatase TiO_2 by surfactant-assisted hydrothermal method. Chin. J. Chem. Eng. 15: 754–759.

Wang, Y., Zhou, A. and Yang, Z. 2008. Preparation of hollow TiO_2 microspheres by the reverse microemulsions. Mater. Lett. 62: 1930–1932.

Wang, Y., Huang, Y., Ho, W., Zhang, L., Zou, Z. and Lee, S. 2009. Biomolecule-controlled hydrothermal synthesis of C–N–S-tridoped TiO_2 nanocrystalline photocatalysts for NO removal under simulated solar light irradiation. J. Hazard. Mater. 169: 77–87.

Wang, J.J., Wang, D.S., Wang, J., Zhao, W.L. and Wang, C.W. 2011. High transmittance and superhydrophilicity of porous TiO_2/SiO_2 bi-layer films without UV irradiation. Surf. Coat. Technol. 205: 3596–3599.

Wang, H. and Guo, Z. 2014. Design of underwater superoleophobic TiO_2 coatings with additional photo-induced self-cleaning properties by one-step route bio-inspired from fish scales. Appl. Phys. Lett. 104: 183703.

Wang, R., Tan, H., Zhao, Z., Zhang, G., Song, L., Dong, W. and Sun, Z. 2014. Stable ZnO@TiO_2 core/shell nanorod arrays with exposed high energy facets for self-cleaning coatings with anti-reflective properties. J. Mater. Chem. A. 2: 7313–7318.

Wang, H., Xie, D., Chen, Q. and You, C. 2016. Kinetic modeling for the deactivation of TiO_2 during the photocatalytic removal of low concentration SO_2. Chem. Eng. J. 303: 425–432.

Wang, X., Wang, X., Zhao, J., Song, J., Zhou, L., Wang, J., Tong, X., et al. 2017a. An alternative to in situ photocatalytic degradation of microcystin-LR by worm-like N, P co-doped TiO_2/expanded graphite by carbon layer (NPT-EGC) floating composites. Appl. Catal. B: Environ. 206: 479–489.

Wang, L., Zhao, Y. and Zhang, J. 2017b. Photochemical removal of SO_2 over TiO_2-based nanofibers by a dry photocatalytic oxidation process. Energy Fuels. 31: 9905–9914.

Wang, R., Tang, T., Lu, G., Huang, K., Feng, S., Zhang, X., Tao, X., et al. 2019. Photocatalytic degradation of polybrominated biphenyls (PBBs) on metal doped TiO_2 nanocomposites in aqueous environments: mechanisms and solution effects. Environ. Sci.: Nano. 6: 1111–1120.

Wang, C., Zhao, Y., Xu, H., Li, Y., Wei, Y., Liu, J. and Zhao, Z. 2020a. Efficient Z-scheme photocatalysts of ultrathin g–C_3N_4-wrapped Au/TiO_2-nanocrystals for enhanced visible-light-driven conversion of CO_2 with H_2O. Appl. Catal. B: Environ. 263: 118314.

Wang, Q., Zhang, L., Guo, Y., Shen, M., Wang, M., Li, B. and Shi, J. 2020b. Multifunctional 2D porous g–C_3N_4 nanosheets hybridized with 3D hierarchical TiO_2 microflowers for selective dye adsorption, antibiotic degradation and CO_2 reduction. Chem. Eng. J. 396: 125347.

Wardak, A., Swami, N. and Gorman, M.E. 2006. The product life cycle and challenges to nanotechnology regulation. Nanotech. L. Bus. 3: 507.

Watanabe, T. and Yoshida, N. 2008. Wettability control of a solid surface by utilizing photocatalysis. Chem. Rec. 8: 279–290.

Wender, H., Feil, A.F., Diaz, L.B., Ribeiro, C.S., Machado, G.J., Migowski, P., Weibel, D.E., et al. 2011. Self-organized TiO_2 nanotube arrays: synthesis by anodization in an ionic liquid and assessment of photocatalytic properties. ACS Appl. Mater. Interfaces. 3: 1359–1365.

Wilson, G.J., Will, G.D., Frost, R.L. and Montgomery, S.A. 2002. Efficient microwave hydrothermal preparation of nanocrystalline anatase TiO_2 colloids. J. Mater. Chem. 12: 1787–1791.

Wood, C. and Jones, C.E. 1997. The effect of environmental assessment on UK local planning authority decisions. Urban Studies. 34(8): 1237–1257.

Wooh, S., Koh, J.H., Lee, S., Yoon, H. and Char, K. 2014. Trilevel-structured superhydrophobic pillar arrays with tunable optical functions. Adv. Funct. Mater. 24: 5550–5556.

Wu, G., Wen, J., Nigro, S. and Chen, A. 2010. One-step synthesis of N-and F-codoped mesoporous TiO_2 photocatalysts with high visible light activity. Nanotechnology. 21: 085701.

Wu, D. and Long, M. 2011. Realizing visible-light-induced self-cleaning property of cotton through coating N-TiO_2 film and loading AgI particles. ACS Appl. Mater. Interfaces. 3: 4770–4774.

Wunderlich, W., Oekermann, T., Miao, L., Hue, N., Tanemura, S. and Tanemura, M. 2004. Electronic properties of Nano-porous TiO_2- and ZnO-Thin films—comparison of simulations and experiments. J. Ceram. Process. Res. 5: 343–354.

Xie, Y., Zhao, Q., Zhao, X.J. and Li, Y. 2007. Low temperature preparation and characterization of N-doped and NS-codoped TiO_2 by sol–gel route. Catal. Lett. 118: 231–237.

Xie, K., Wu, Q., Wang, Y., Guo, W., Wang, M., Sun, L. and Lin, C. 2011. Electrochemical construction of Z-scheme type CdS–Ag–TiO_2 nanotube arrays with enhanced photocatalytic activity Electrochem. Commun. 13(12): 1469–1472.

Xing, Z., Zhang, J., Cui, J., Yin, J., Zhao, T., Kuang, J., Xiu, Z., et al. 2018. Recent advances in floating TiO_2-based photocatalysts for environmental application. Appl. Catal. B: Environ. 225: 452–467.

Xu, H. and Gao, L. 2004. Hydrothermal synthesis of high-purity $BaTiO_3$ powders: control of powder phase and size, sintering density, and dielectric properties. Mater. Lett. 58: 1582–1586.

Xu, Q., Wellia, D.V., Sk, M.A., Lim, K.H., Loo, J.S.C., Liao, D.W., Amal, R., et al. 2010. Transparent visible light activated C–N–F-codoped TiO_2 films for self-cleaning applications. J. Photochem. Photobiol. A. 210: 181–187.

Xu, Y.J., Liao, J.X., Cai, Q.W. and Yang, X.X. 2013a. Preparation of a highly-reflective $TiO_2/SiO_2/Ag$ thin film with self-cleaning properties by magnetron sputtering for solar front reflectors. Sol. Energy Mater. Sol. Cells. 113: 7–12.

Xu, Q.F., Liu, Y., Lin, F.J., Mondal, B. and Lyons, A.M. 2013b. Superhydrophobic TiO_2–polymer nanocomposite surface with UV-induced reversible wettability and self-cleaning properties. ACS Appl. Mater. Interfaces. 5: 8915–8924.

Xu, M., Wang, Y., Geng, J. and Jing, D. 2017. Photodecomposition of NO_x on Ag/TiO_2 composite catalysts in a gas phase reactor. Chem. Eng. J. 307: 181–188.

Yan, M., Chen, F., Zhang, J. and Anpo, M. 2005. Preparation of controllable crystalline titania and study on the photocatalytic properties. J. Phys. Chem. B. 109: 8673–8678.

Yan, B., Tao, J., Pang, C., Zheng, Z., Shen, Z., Huan, C.H.A. and Yu, T. 2008. Reversible UV-light-induced ultrahydrophobic-to-ultrahydrophilic transition in an α–Fe_2O_3 nanoflakes film. Langmuir. 24: 10569–10571.

Yang, S., Halliburton, L.E., Manivannan, A., Bunton, P.H., Baker, D.B., Klemm, M., Horn, S., et al. 2009a. Photoinduced electron paramagnetic resonance study of electron traps in TiO_2 crystals: oxygen vacancies and Ti^{3+} ions. Appl. Phys. Lett. 94: 162114.

Yang, H.G., Liu, G., Qiao, S.Z., Sun, C.H., Jin, Y.G., Smith, S.C., Zou, J., et al. 2009b. Solvothermal synthesis and photoreactivity of anatase TiO_2 nanosheets with dominant {001} facets. J. Am. Chem. Soc. 131: 4078–4083.

Yang, L., Luo, S., Cai, Q. and Yao, S. 2010. A review on TiO_2 nanotube arrays: fabrication, properties, and sensing applications. Chin. Sci. Bull. 55: 331–338.

Yang, L., Viste, M., Hossick-Schott, J. and Sheldon, B.W. 2012. Internal stress evolution during field-induced crystallization of anodic tantalum oxide. Electrochim. Acta. 81: 90–97.

Yang, H.K., Moon, B.K., Choi, B.C., Jeong, J.H. and Kim, K.H. 2013. Crystal growth and photoluminescence properties of Sm^{3+} doped CeO_2 nanophosphors by solvothermal method. J. Nanosci. Nanotechnol. 13: 6060–6063.

Yang, W., Shen, H., Min, H. and Ge, J. 2020. Enhanced visible light-driven photodegradation of rhodamine B by Ti^{3+} self-doped TiO_2@Ag nanoparticles prepared using Ti vapor annealing. J. Mater. Sci. 55: 701–712.

Yao, L. and He, J. 2014. Recent progress in antireflection and self-cleaning technology—from surface engineering to functional surfaces. Prog. Mater. Sci. 61: 94–143.

Yi, C., Liao, Q., Deng, W., Huang, Y., Mao, J., Zhang, B. and Wu, G. 2019. The preparation of amorphous TiO_2 doped with cationic S and its application to the degradation of DCFs under visible light irradiation. Sci. Total Environ. 684: 527–536.

Yoshida, N., Takeuchi, M., Okura, T., Monma, H., Wakamura, M., Ohsaki, H. and Watanabe, T. 2006. Super-hydrophobic photocatalytic coatings utilizing apatite-based photocatalyst. Thin Solid Films. 502: 108–111.

Yuan, R., Fu, X., Wang, X., Liu, P., Wu, L., Xu, Y., Wang and Xuxu, Wang, Z. 2006. Template synthesis of hollow metal oxide fibers with hierarchical architecture. Chem. Mater. 18: 4700–4705.

Yuan, Y., Zhang, J., Li, H., Li, Y., Zhao, Y. and Zheng, C. 2012. Simultaneous removal of SO_2, NO and mercury using TiO_2-aluminum silicate fiber by photocatalysis. Chem. Eng. J. 192: 21–28.

Yuranova, T., Mosteo, R., Bandara, J., Laub, D. and Kiwi, J. 2006. Self-cleaning cotton textiles surfaces modified by photoactive SiO_2/TiO_2 coating. J. Mol. Catal. A. Chem. 244: 160–167.

Zeng, Y., Wu, W., Lee, S. and Gao, J. 2007. Photocatalytic performance of plasma sprayed Pt-modified TiO_2 coatings under visible light irradiation. Catal. Commun. 8: 906–912.

Zhang, Q., Gao, L. and Guo, J. 2000. Effects of calcination on the photocatalytic properties of nanosized TiO_2 powders prepared by $TiCl_4$ hydrolysis. Appl. Catal. B: Environ. 26: 207–215.

Zhang, X., Yang, H. and Tang, A. 2008. Optical, electrochemical and hydrophilic properties of Y_2O_3 doped TiO_2 nanocomposite films. J. Phys. Chem. B. 112: 16271–16279.

Zhang, Z., Zuo, F. and Feng, P. 2010a. Hard template synthesis of crystalline mesoporous anatase TiO_2 for photocatalytic hydrogen evolution. J. Mater. Chem. 20: 2206–2212.

Zhang, J., Wu, Y., Xing, M., Leghari, S.A.K. and Sajjad, S. 2010b. Development of modified N doped TiO_2 photocatalyst with metals, nonmetals and metal oxides. Energy Environ. Sci. 3: 715–726.

Zhang, L., Dillert, R., Bahnemann, D. and Vormoor, M. 2012. Photo-induced hydrophilicity and self-cleaning: models and reality. Energy Environ. Sci. 5: 7491–7507.

Zhang, P., Tian, J., Xu, R. and Ma, G. 2013a. Hydrophilicity, photocatalytic activity and stability of tetraethyl orthosilicate modified TiO_2 film on glazed ceramic surface. Appl. Surf. Sci. 266: 141–147.

Zhang, L., Zhong, Y., Cha, D. and Wang, P. 2013b. A self-cleaning underwater superoleophobic mesh for oil-water separation. Sci. Rep. 3: 2326.

Zhang, X., Wang, P., Yang, P. and Jiang, S.P. 2020. Photo-chemical property evolution of superior thin g-C_3N_4 nanosheets with their crystallinity and Pt deposition. Int. J. Hydrog. Energy. 45: 21523–21531.

Zhang, X., Zhou, J., Gu, Y. and Fan, D. 2015. Visible-Light photocatalytic activity of n-doped TiO_2 nanotube arrays on acephate degradation. J. Nanomater. 16: 42.

Zhang, Y. and Xu, Y. 2019. Simultaneous electrochemical dual-electrode exfoliation of graphite toward scalable production of high-quality graphene. Adv. Funct. Mater. 29(37): 1902171.

Zhao, J., Wu, T., Wu, K., Oikawa, K., Hidaka, H. and Serpone, N. 1998. Photoassisted degradation of dye pollutants. 3. Degradation of the cationic dye rhodamine B in aqueous anionic surfactant/TiO_2 dispersions under visible light irradiation: evidence for the need of substrate adsorption on TiO_2 particles. Environ. Sci. Technol. 32: 2394–2400.

Zhao, L., Jiang, Q. and Lian, J. 2008. Visible-light photocatalytic activity of nitrogen-doped TiO_2 thin film prepared by pulsed laser deposition. Appl. Surf. Sci. 254: 4620–4625.

Zhao, Z., Tan, H., Zhao, H., Li, D., Zheng, M., Du, P., Zhang, G., et al. 2013. Orientated anatase TiO_2 nanocrystal array thin films for self-cleaning coating. Chem. Commun. 49: 8958–8960.

Zheng, Y., Liu, J., Liang, J., Jaroniec, M. and Qiao, S.Z. 2012. Graphitic carbon nitride materials: controllable synthesis and applications in fuel cells and photocatalysis. Energy Environ. Sci. 5: 6717–6731.

Zhou, H., Liu, L., Wang, X., Liang, F., Bao, S., Lv, D., Tang, Y., et al. 2013. Multimodal porous CNT@ TiO_2 nanocables with superior performance in lithium-ion batteries. J. Mater. Chem. A. 1: 8525–8528.

Zhu, G., Pan, L., Xu, T. and Sun, Z. 2011. CdS/CdSe-cosensitized TiO_2 photoanode for quantum-dot-sensitized solar cells by a microwave-assisted chemical bath deposition method. ACS Appl. Mater. Interfaces. 3: 3146–3151.

Zhu, J., Cao, Y. and He, J. 2014. Facile fabrication of transparent, broadband photoresponse, self-cleaning multifunctional graphene–TiO_2 hybrid films. J. Photochem. Photobiol. A. 420: 119– 126.

Ziolli, R.L. and Jardim, W.F. 2002. Photocatalytic decomposition of seawater-soluble crude-oil fractions using high surface area colloid nanoparticles of TiO_2. J. Photochem. Photobiol. A. 147: 205–212.

Zubkov, T., Stahl, D., Thompson, T.L., Panayotov, D., Diwald, O. and Yates, J.T. 2005. Ultraviolet light-induced hydrophilicity effect on $TiO_2(110)(1\times1)$. Dominant role of the photooxidation of adsorbed hydrocarbons causing wetting by water droplets. J. Phys. Chem. B. 109: 15454–15462.

Chapter 17

Application of Nanoparticles to Control Microbial Biodeterioration of Concrete

Anirban Chaudhuri[1] and Subarna Bhattacharyya*[1]
[1]School of Environmental Studies, Jadavpur University, Kolkata, India

Punarbasu Chaudhuri[2]
[2]Department of Environmental Science, University of Calcutta, Kolkata, India

17.1 INTRODUCTION

Nanotechnology is the potential way to develop new materials which are being used in the field of construction and building materials. The replication of natural systems through nanotechnology is one of the most promising areas of conservation science. Nanotechnology is concerned with objects between 1 and 100 nm in size. Nanotechnology is the design, characterization, production and application of structures, devices and systems by controlling shape and size at the nanoscale. Nanotechnology needs latest imaging techniques for investigating and discovering the material behaviour for designing and producing very fine powders, liquids or solids materials with particle size between 1 and 100 nm. These materials are known as nanoparticles (NP). The properties of NPs are very different from the properties of the same materials. The other term nano-science is the study of incident and the manipulation of materials at nanoscale and considers an extended part of general sciences into the nanoscale. Nanotechnology is used to create the materials having nano dimension which is actually being used to make devices by controlling of matter at atoms, molecules and nanoscale structures (Roco et al. 1999).

Modern buildings are made by concrete structures, which is composite of cement, sand, iron and steel. Concrete is widely used in high rise commercial and residential buildings, bridge construction, marine environments, sewage, soil and nuclear industries. Concrete is a cost effective, sustainable choice for different structures because of its strength, durability, versatility, reflectivity and long-lasting nature. Physical, chemical and aesthetic impacts are very common which are the consequences of weathering, chloride ion penetration, carbonation, corrosion and deterioration of

*Corresponding Author: barna_kol@yahoo.com

concrete. This increases maintenance costs of building materials, bridges, archaeological heritage sites and reducing the long term capacity of structural materials (Kumar et al. 2009, Sanchez-Silva et al. 2011).

Water availability, pH, climatic exposure, chemical compositions are some of the factors which support microbial growth on construction materials (Verdier et al. 2014, Guerra et al. 2019). Presence of moisture, living microorganisms deteriorate the concrete structures, penetrate into the porous structure of concrete structure; settle down there and dissolve the inorganic materials by secreting organic acids. The whole process is known as biodeterioration. Microorganisms like bacteria, fungi, algae and lichen play major roles in the deterioration of concrete structures because they frequently grow on building materials. They firstly colonize on the concrete surface followed by the production of organic acids like acetic acid, lactic acid etc. leading to the dissolution of calcium containing minerals from the concrete matrices (Cwalina 2008). Sulfur oxidizing bacteria produce sulphuric acid which reacts with the free lime and forms gypsum ($CaSO_4.2H_2O$). This enters the interior of the concrete and deteriorates it (George et al. 2012). Ettringite ($3Ca.Al_2O_3.3CaSO_4.32H_2O$) is also formed by the reaction between gypsum crystals and calcium aluminate present in the concrete. Therefore, the internal pressure of concrete increases and cracks are formed (Aviam et al. 2004). These cracks provide larger area for corrosion and additional sites for acid penetration (Davis et al. 1998). Similarly, nitrifying bacteria produces nitric acid which corrodes the concrete surface (Sand et al. 1991). Fungi have also been found to be involved in the corrosion processes almost in the same way like bacteria. Moreover, their hyphae enter inside the micro-crack and add mechanical force to enhance the process (Chaudhuri et al. 2020).

One of the main factors influencing concrete deterioration in both short- and long-term studies is weight loss (Momber 2000). Biodeterioration could also affect porosity and compressive strength of concrete. During biodeterioration process, a layer of high porosity is produced by biogenic acids leading to greater permeability as well as diffusion within the concrete matrix (Beddoe et al. 2005). For larger porosities the compressive strength of concrete also decreases (Kumar and Bhattacharjee 2003, Chen et al. 2013). Fungi also helps to leach calcium, silica, iron, potassium and sulphur from concrete (Andersen et al. 2017, Chaudhuri et al. 2020).

The rapidly deteriorating modern concrete sewer system in Hamburg, Germany, raised curiosity on the microbiological mechanisms causing concrete pipe degradation (Milde et al. 1993). Deterioration of newly installed sewer system was also reported by US cities (Kulpa and Baker 1990, Mansfeld et al. 1991, Padival et al. 1995). Until now, there is no clear idea of the extent and severity of the damage, the operative technique, and fruitful remedial process for the microbial attack to concrete structures (Nasrazadani and Sudoi 2010). The detailed microbiological characterization studies for concrete with fly ash and superplasticizer and concrete with only superplasticizer was less prone to biodeterioration and biofilm formation compared to normal concrete as fly ash make the concrete impermeable (Viswakarma et al. 2014). Other techniques such as application of fungicide in paints, application of NP during white wash have been developed to deal with this issue (Heaton et al. 1991, Aldosari et al. 2019). Also ultraviolet radiation (van der Molen et al. 1980) and ionizing radiation (Ley 1988) are commonly used for sterilisation of different materials. This review study focuses on comprehensive and practical overview to the biodeterioration as well as conservation of concrete through the application of NP.

17.2 MICRO-ECOLOGY OF CONCRETE

17.2.1 Microorganisms Responsible for Concrete Biodeterioration

Microorganisms are microscopic, minute living organisms, playing a major role in the deterioration of concrete structures (Bastidas-Arteaga et al. 2008). There are sufficient evidences to show that a wide variety of microorganisms are responsible for concrete biodeterioration (Table 17.1).

Table 17.1 Microorganisms involved in the deterioration of concrete

Microorganisms	Sources
Bacteria	
Acidithiobacillus thiooxidans	Lahav et al. 2004, Mustafa 2009, Magniont et al. 2011
Cyanobacterium pleurocapsa	Gaylarde et al. 2005
Thiobacillus intermedius	Giannantonio et al. 2009, Magniont et al. 2011
Thiobacillus neapolitanus	Mustafa 2009, Magniont et al. 2011
Acidiphilium cryptum	Sh et al. 2010
Thiobacillus ferrooxidans	Sh et al. 2010
Fungi	
Mucor sp.	Ghafoori and Mathis 1997
Penicillium oxalicum	Ghafoori and Mathis 1997, Chaudhuri et al. 2020
Fusarium sp.	Gu et al. 1998
Alternaria alternata	Warscheida and Braamsb 2000
Aspergillus niger	Lajili et al. 2008
Aspergillus tamarii	Chaudhuri et al. 2020, Bhattacharyya et al. 2022
Algae	
Chlorella sp.	Miller et al. 2006
Stichococcus sp.	Miller et al. 2006
Ulva fasciata	Javaherdashti and Setareh 2006
Lichen	
Candelariella sp.	Edwards et al. 1999

Parker (1945) observed that the microorganism involved in concrete degradation is acidophilic sulphur-oxidizing bacteria (SOB) *Thiobacillus* sp. *Thiobacillus ferrooxidans*, *Acidiphilium cryptum* and *Leptospirilium ferrooxidans* were also observed in sewer pipes (Sh et al. 2010). *Cyanobacterium pleurocapsa* and *Cyanobacterium chroococcidiopsis* have been found in biofilms of monuments and buildings (Gaylarde and Gaylarde 2005). *Aspergillus tamarii* and *Fusarium* sp. play an important role in the deterioration of heritage building (Chaudhuri et al. 2020, Bhattacharyya et al. 2022) and sewer pipes (Gu et al. 1998). *Chlorella* sp., *Stichococcus* sp. and *Cyanobacterium gloeocapsa* are the most common genera of algae found in the construction materials (Miller et al. 2006).

17.2.2 Microbial Colonization on Concrete

After casting, concrete is highly alkaline (pH $>$ 12), which inhibits the development of microorganisms. At high pH, little microbial activity has occurred (Sand 1987). This high pH is due to the formation of calcium hydroxide, which is a by-product of cement hydration. Usually, the erosive action of water creates roughness on the concrete surface (Ribas-Silva 1995). Additionally, the availability of moisture and nutrients enhance colonization of microbes on concrete surfaces. At first they form a biofilm which reacts with the substrate. Ultimately, hyphae penetrate the material and start to produce pigments (El-Derby et al. 2016). In aquatic environment, colonization by bacteria (sulfur-oxidizing and sulfur-reducing) on concrete has been related with the sulfur cycle in their environment (Satoh et al. 2009). Sulfur is converted into sulfide by anaerobic sulfur reducing bacteria, combines with hydrogen and forms hydrogen sulfide. In the sewer systems, pH of alkaline concrete surface is reduced by the carbonation and neutralization of hydrogen sulfide (Lahav et al. 2004, Nielsen et al. 2005, Zhang et al. 2008).

Neutral pH of concrete surface generates condition for microbial colonization of both aerobic and autotrophic *Thiobacillus* sp. which leads to the production of sulphuric acid (Devereux et al. 1989). Sulfuric acid produced by *Thiobacillus* sp. reacts with $Ca(OH)_2$ to form gypsum and ettringite that leads to internal pressure as well as cracking (Mori et al. 1992). We also conducted similar type of research work with fungi (*Aspergillus tamarii*) in our laboratory and fractured crack surface of concrete sample (Figure 17.1) was observed by scanning electron microscope (SEM). *Thiobacillus* sp. such as *T. thioparus*, *T. thiooxidans*, *T. novellus*, *T. neapolitanus* and *T. intermedius* plays a vital role in these colonization events (Rigdon and Beardsley 1956, Mori et al. 1992). Fungi, algae and lichens also participate in these colonization processes (Gu et al. 1998, Nica et al. 2000, Javaherdashti and Setareh 2006, Chaudhuri et al. 2020).

Figure 17.1 Scanning electron micrographs of crack formation on the concrete cube surface (cited from Chaudhuri et al. 2020).

17.2.3 Microbial Succession on Concrete

Thiobacillus sp. can attach to the concrete surface (pH < 9) and reproduce (Mori et al. 1992) in the presence of sufficient nutrients, moisture and oxygen. *Thiobacillus thioparus* was found to colonize new concrete pipe surfaces and vanishe gradually as deterioration became more severe. As pH continues to fall to weakly acidophilic conditions, *Thiobacillus intermedius, Thiobacillus novellus* and *Thiobacillus neapolitanus* develops on the concrete surface (Milde et al. 1983, Sand 1987). *Thiobacillus thiooxidans* starts to grow at pH levels below 5 and produces high amounts of H_2SO_4, and the pH reaches 1.5 (Sand and Bock 1984). *T. thiooxidans* grows vigorously, when the pH reaches to 3 and also supports the formation of elemental sulfur. It rapidly oxidizes to sulfate ion and at pH equal to 1 it becomes inhibitory even for Acidophilic heterotrophs used the organic waste products released by *T. thiooxidans.* This succession differs in its details in certain cases. Sand and Bock (1984) found no *Thiobacillus thioparus* in studies in Germany, whereas Parker (1951) emphasized *Thiobacillus thioparus* and *Thiobacillus thiooxidans* in their studies in Australia.

17.3 HISTORY OF NPs

Human had initiated to use NPs from fourth century AD, by the Roman people. Probably that was the oldest example of nanotechnology in the ancient world. The Lycurgus cup, exhibited at British Museum, is the oldest achievements of ancient dichroic glass generating from glass industry.

Dichroic glass considers two different types of glass, changing colour in particular lighting conditions. The glass appears green in direct light and red-purple when light shines through the glass. In 1990, the scientists analysed the cup using a transmission electron microscopy (TEM) to explain the phenomenon of dichroism (Barber and Freestone, 1990). The incident is due to the presence of nanoparticles with 50–100 nm in diameter. X-ray analysis showed that these nanoparticles are silver-gold (Ag-Au) alloy, with a ratio of Ag: Au of about 7:3, which contains 10% copper (Cu) dispersed in a glass matrix (Freestone et al. 2007). The Au nanoparticles produce a red colour as result of light absorption (520 nm). The red-purple color is due to the absorption by the bigger particles while the green colour is attributed to the light scattering by colloidal dispersions of Ag nanoparticles with a size > 40 nm. The Lycurgus cup is recognized as one of the oldest synthetic nanomaterials (The New York Times 2019). A similar effect is seen in late medieval church windows, shining a luminous red and yellow colour due to the fusion of Au and Ag nanoparticles into the glass. During the 9th–17th centuries, glowing, glittering "luster" ceramic glazes used in the Islamic world, and later in Europe contained Ag or copper (Cu) or other nanoparticles (Pradell et al. 2007). Renaissance pottery during 16th century was famous for creating Italians nanoparticles. Ottoman techniques were the secret of that technology. "Damascus" saber blades, cementite nanowires and carbon nanotubes were used during the 13th–18th centuries. NP provided strength, resilience, and the ability to hold a keen edge of that material (Poole and Owens 2003). Without knowing the actual science, the medieval artist had manufactured these surprising materials.

Michael Faraday had also studied and tried to prepare the colloidal suspensions of "Ruby" gold. NP present in this material actually developed unique optical and electronic properties. Faraday had also reported the science of how under certain lighting conditions, gold NP was responsible to produce different-coloured solutions (Faraday 1857). The development of NP is one of outcome of nanoscience. At 1981, IBM Zurich Research Laboratory, physicists Gerd Binnig and Heinrich Rohrer invented Scanning Tunneling Microscope. The earlier idea of Feynman was established by this new microscopy and the era of nano-science was developed (Binnig et al. 1982).

17.4 NANOPARTICLES IN CONSTRUCTION INDUSTRY

Concrete is the major component of construction material for many years. Portland cement-based binders, water, and coarse and fine aggregates have been used as primary components of concrete construction. The binders are made from Portland clinker ground, a little calcium sulphate and fine mineral powders, such as pozzolan (usually volcanic ash), limestone, granulated blast furnace slag, and fly ash etc. (Reches 2018). At very small concentration, NP are very potent additives for modification of cement products. The major modifications are reduction of set time (by 1–2 h) and diffusivity (by 4–75%) and increase of strength (by 5–25%), and thermal durability (0–30% increase in residual strength). These help in the unique reactivity of nanoparticles associated with their small size and large surface area (Reches 2018). The durability of concrete depends on the connection interfaces between the voids, the aggregates, and the cement paste (Salemi and Behfarnia 2013). Therefore, NP with properties, such as durability and strength, are of particular interest in the production of concrete. Presently several NP are used to enhance strength of concrete (Table 17.2) in different ways.

Both TiO_2 and ZnO NP have resistance against photo-catalytic activity as they have high surface area. Therefore, when added in concrete, the concrete becomes self-cleaning, self-disinfecting, and environmental pollution cleansing. In the presence of light, TiO_2 breaks down the organic pollutants and dirt on the surface of the concrete into harmless water and CO_2; the products of the catalytic reaction are then easily removed by simple rinsing or rain (Konsta-Gdoutos et al. 2010). Carbon nanotubes (CNT) are being incorporated in concrete as

Table 17.2 Nanoparticles involved in the improvement of concrete properties

NPs used	Mode of action	Sources
Nanosilica	Acts as a nanofiller for the calcium silicate hydrate (Ca–Si–H) particles in the cement and acts as a strong binding agent	Salemi and Behfarnia. 2013
Nanotitania	Increased degree of hydration and self-cleaning capacity	Mohajerani et al. 2019
Carbon nanotube	Mechanical durability and crack prevention	Mohajerani et al. 2019
Zinc oxide	UV radiation protection	Solano et al. 2020
Silicon dioxide	Increased compressive strength of concrete cube and less leaching of calcium ions.	Bhattacharyya et al. 2022

nano-filler as CNTs have high surface area and extraordinary mechanical properties. The added CNTs in concrete can make concrete impervious to salts and water, which in turn significantly enhances the durability of concrete. CNT can both reduce the permeability of concrete to ions and increase the strength of concrete, thereby improving durability. The addition of NP like TiO_2, Al_2O_3, ZrO_2, Fe_2O_3, SiO_2 and metal oxide containing nano-clays improve the concrete and/or cement mortar properties. Metal oxide NP addition accelerates chemical reactions during initial hydration thus strengthening cement composites. In the reaction, $Ca(OH)_2$ present in concrete reacts with the metal oxide nanoparticles, which increase the quantity of calcium silicate hydrate (C–S–H) in the concrete. This phenomena leads to more compaction, decreasing permeability and as a result improving mechanical properties (Hanus and Harris 2013). Nano-silica also reduces the porosity of the concrete, reducing the ability of water and other elements to penetrate the concrete, which avoids the potential for concrete degradation.

17.5 LIMITATIONS

The use of NP in construction industry is restricted due to some reasons. The lack of knowledge and information concerning the suitable NP for construction sector and their behaviour is a major drawback. The standards for design and execution of the construction elements using NP is yet to be standardised. Moreover, high costs and unknown health risks associated with NP are the major limitation of the large scale use of NP.

17.6 APPLICATION OF NPs IN CONSERVATION OF ARCHAEOLOGICAL MONUMENTS

Biodeterioration is a worldwide problem which can be observed even in many areas designated as world cultural heritage sites. Deterioration of cultural heritage built with stone materials, such as exterior archaeological monuments, is a complex process caused by physical, chemical and biological agents. This deterioration is also caused by sunlight, moisture, inorganic pollutants and airborne microorganisms (Basu et al. 2021). The microbial growth on stones, concrete, wood of these monuments causes aesthetic, physical and chemical damage, which is irreversible and results in destruction and permanent loss of stone monuments (Farooq 2015). Different in-vitro conservative and remedial measures have been taken to control biodeterioration of monuments with high archaeological or cultural heritage value (Ditaranto et al. 2011). Eight bacterial and seven fungal genera were isolated from biofilms involved in the biodeterioration of archaeological walls of an ancient city, Mexico State (Carrillo-González et al. 2016). They treated the microbial infection by applying Silver NP in laboratory. Aldosari et al. (2019) proposed a simple material obtained by dispersing commercial ZnO NPs in acrylic polymer as a multifunctional biocidal and consolidating coating. The novel nanocomposite coating was applied to the marble samples collected from various archeological sites in Egypt. The color variation of marble sample subject to various aging procedure prove the efficient protection of the treatment with ZnO nanoparticles against UV exposure, together with minimal or no changes in optical properties following the coating. Goffredo and Munafo (2015) studied the use of titanium dioxide (TiO_2) as a protective coating for stone surfaces of architectural and archaeological interest to impede the formation of potentially harmful soiling. TiO_2 applied on stone is very efficient and able to remove the soiling applied on treated surfaces. On the other hand, the NP seems to lose a relevant part of their photo activity in the long term after an accelerated weathering process. Calcium carbonate NP ($CaCO_3$) was added to acrylic-based polymer in order to improve its physiochemical and mechanical properties and to use it in the consolidation and protection of limestone monuments (Aldosari et al. 2019). The results of experimental study showed that the addition of NP to the acrylic polymer improved its ability to consolidate and protect the limestone samples. Moreover, the addition of ($CaCO_3$) nanoparticles enhanced the mechanical properties of the polymer and improved its interaction with the stone grains. The polymer containing $CaCO_3$ NP could significantly reduce the water absorption rates inside stone bulk in addition to enhancing the stone durability as compared to those treated with polymer without the nanoparticles. Emulsion polymerization method is an appropriate technique for conservation of stone monuments. There is very limited data on the field application of NP used for conservation of archaeological site especially on concrete materials. The aim of restoration of archaeological site is to create hydrophobic, anti-freeze and self-cleaning coatings by using NP. These coatings can also have application in the construction and automotive industries (Eseev et al. 2019). A few works have been informed on conservation studies by using CNTs. CNTs embedded in epoxy resin and are successfully used as coatings for wood (Cestari et al. 2009). This coating consolidates the wood layers of ply board and also increases the mechanical strength (Cestari et al. 2009). CNTs are also efficient and sustainable materials for conservation of cultural heritage made by pentelic marbles from the Basilica Neptuni, Rome, Italy. CNTs can also remove black crust from the surfaces (Valentini et al. 2013). The CNTs have also been tested in laboratory by mixing with cement, stone and silica to make concrete mortar. This study reported that the addition of CNTs in mortar help to improve mechanical strength and stability against microorganisms. Most cultural heritage and monuments are usually made of marble, sandstone, wood and limestone materials. Undoubtedly, there are a few biodeterioration studies of age old concrete material (David et al. 2020).

17.7 CONCLUSION

In the last decade, several innovative techniques and procedures were adopted in the field of conservation of construction materials. This review is intended to compile the advancement of concrete biodeterioration and protection countermeasures by focusing on antimicrobial techniques such as application of nanoparticles that appears to be the most effective and convenient way for the conservation of concrete structures.

In this review, we have shown that:

- To characterise the deterioration or corrosion of concrete in the presence of biological agents such as bacteria, fungi, algae and lichens.
- Acids production can be detrimental to the physical and chemical properties of concrete structures.
- Formation of biofilm will increase the likelihood of attracting more microorganisms that can further deteriorate the concrete.

- The usage of the Ag nanoparticles showed highest inhibition efficiencies against environmental biodeterioration of concrete.

Moreover, continuous monitoring and control of environmental parameters, including pH, temperature, relative humidity, water contents, pollutants and nutrient sources should be a high priority in overall deterioration prevention and conservation activities. This is certainly a matter that will need further research and study in future works.

References

Aldosari, M.A., Darwish, S.S., Adam, M.A., Elmarzugi, N.A. and Ahmed, S.M. 2019. Using Zno nanoparticles in fungal inhibition and self-protection of exposed marble columns in historic sites. Archaeol. Anthropol. Sci. 11: 3407–3422. DOI: 10.1007/s12520-018-0762-z.

Andersen, B., Dosen, I., Lewinska, A.M. and Nielsen, K.F. 2017. Pre-contamination of new gypsum wallboard with potentially harmful fungal species. Indoor Air. 27(1): 6–12. DOI: 10.1111/ina.12298.

Aviam, O., Bar-Nes, G., Zeri, Y. and Sivan, A. 2004. Accelerated biodegradation of cement by sulfur-oxidizing bacteria as a bioassay for evaluating immobilization of low-level radioactive waste. Appl. Environ. Microbiol. 70: 6031–6036.

Barber, D.J. and Freestone, I.C. 1990. An investigation of the origin of the colour of the Lycurgus Cup by analytical transmission electron microscopy. Archaeometry. 32: 33–45. DOI: 10.1111/j.1475-4754.1990.tb01079.x

Bastidas-Arteaga, E., Sánchez-Silva, M., Chateauneuf, A. and Silva, M.R. 2008. Coupled reliability model of biodeterioration, chloride ingress and cracking for reinforced concrete structures. Struct. Saf. 30: 110–129.

Basu, C., Bhattacharyya, S., Chaudhuri, A., Akhtar, S., Chatterjee, A., Thakur, B., Guha, H., et al. 2021. Assessment of potential damage factor: a case study of St. Paul's Cathedral, Kolkata. J. Heritage Manage. 6(1): 53–68. DOI: 10.1177/24559296211108678.

Beddoe, R.E. and Dorner, H.W. 2005. Modelling acid attack on concrete: Part I. The essential mechanisms. Cem. Concr. Res. 35: 2333–2339.

Bhattacharyya, S., Akhtar, S., Chaudhuri, A., Mahanty, S., Chaudhuri P. and Sudarshan, M. 2022. Affirmative nanosilica mediated approach against fungal biodeterioration of concrete materials. Case Stud. Constr. Mater. 17. p.e01258. DOI: https://doi.org/10.1016/j.cscm.2022.e01258.

Binnig, G., Rohrer, H., Gerber, C. and Weibel, E. 1982. Tunneling through a controllable vacuum gap. Appl. Phys. Lett. 40. 178. DOI: 10.1063/1.92999.

Carrillo-González, R., Martínez-Gómez, M.A., González-Chávez, M.D.C.A. and Hernández, J.C.M. 2016. Inhibition of microorganisms involved in deterioration of an archaeological site by silver nanoparticles produced by a green synthesis method. Sci. Total Environ. 565: 872–881.

Cestari, C., Marzi, T. and Invernizzi, S. 2009. Use of nanotechnologies for conservation, reinforcement and maintenance: Timber construction in cultural heritage of the mediterranean basin. In Proceedings of the 4th International Congress on "Science and Technology for the Safeguard of Cultural Heritage in the Mediterranean Basin". pp 6–8.

Chaudhuri, A., Bhattacharyya, S., Chaudhuri, P., Sudarshan, M. and Mukherjee, S. 2020. *In vitro* deterioration study of concrete and marble by *Aspergillus tamarii*. J. Build. Eng. 32: 101774. DOI: 10.1016/j.jobe.2020.101774.

Chen, X., Wu, S. and Zhou, J. 2013. Influence of porosity on compressive and tensile strength of cement mortar. Constr. Build. Mater. 40: 869–874.

Cwalina, B. 2008. Biodeterioration of concrete. Archit. Civ. Eng. Environ. 4: 133–140.

David, M.E., Ion, R.M., Grigorescu, R.M., Iancu, L. and Andrei, E.R. 2020. Nanomaterials used in conservation and restoration of cultural heritage: an up-to-date overview. Materials. 13(9): 2064.

Davis, J.L., Nica, D., Shields, K. and Roberts, D.J. 1998. Analysis of concrete from corroded sewer pipe. Int. Biodeterior. Biodegrad. 42: 75–84.

Devereux, R., Delaney, M., Widdel, F. and Stahl, D.A. 1989. Natural relationships among sulfate-reducing eubacteria. J. of Bacteriol. 171: 6689e6695.

Ditaranto, N., Loperfido, S., van der Werf, I., Mangone, A., Cioffi, N. and Sabbatini, L. 2011. Synthesis and analytical characterisation of copper-based nanocoatings for bioactive stone artworks treatment. Anal. Bioanal. Chem. 399(1): 473–481.

Edwards, K.J., Goebel, B.M., Rodgers, T.M., Schrenk, M.O., Gihring, T.M., Cardona, M.M., Hu, B., et al. 1999. Geomicrobiology of pyrite (FeS_2) dissolution: case study at iron mountain. California. Geomicrobiol. J. 16: 155–179.

El-Derby, A.A., Mansour, M. and Salem, M.Z. 2016. Investigation the microbial deterioration of sandstone from the osirion's sarcophagus chamber as affected by rising ground water level. Mediterran. Archaeol. Archaeom. 16(1): 273–281. DOI: 10.5281/zenodo.46360.

Eseev, M., Goshev, A. and Kapustin, S. 2019. Simple method of creating hydrophobic surfaces based on modified MW-CNTs. Mater. Sci. 1: 1–13.

Faraday, M. 1857. The Bakerian lecture: experimental relations of gold (and other metals) to light. Philos. Trans. R. Soc. Lond. 147: 145–181.

Farooq, M. 2015. Mycobial deterioration of stone monuments of dharmarajika, taxila. J. Microbiol. Exp. 2(1): 29–33.

Freestone, I., Meeks, N., Sax, M. and Higgitt, C. 2007. The lycurgus cup—a roman nanotechnology. Gold Bull. 40: 270–277. DOI: 10.1007/BF03215599.

Gaylarde, C. and Gaylarde, P.A. 2005. A comparative study of the major microbial biomass biofilms on exteriors of buildings in Europe and Latin America. Int. Biodeter. Biodegrad. 55: 131–139.

George, R.P., Vinita, V., Samal, S.S. and Mudali, U.K. 2012. Current understanding and future approaches for controlling microbially influenced concrete corrosion: a review. Concr. Res. Lett. 3(3): 491–506.

Ghafoori, N. and Mathis, R. 1997. Sulfate resistance of concrete pavers. J. Mater. Civil. Eng. 9: 35–40.

Giannantonio, D.J., Kurth, J.C., Kurtis, K.E. and Sobechy, P.A. 2009. Effects of concrete properties and nutrients on fungal colonization and fouling. Int. Biodeterior. Biodegrad. 63: 252–259.

Goffredo, G.B. and Munafò, P. 2015. Preservation of historical stone surfaces by TiO_2 nanocoatings. Coatings. 5(2): 222–231.

Guerra, F.L., Lopes, W., Cazarolli, J.C., Lobato, M., Masuero, A.B., Dal Molin, D.C.C., Bento, F.M., et al. 2019. Biodeterioration of mortar coating in historical buildings: microclimatic characterization, material, and fungal community. Build. Environ. 155: 195–209. DOI: 10.1016/j.buildenv.2019.03.017.

Gu, J.D., Ford, T.E., Berke, N.S. and Mitchell, R. 1998. Biodeterioration of concrete by the fungus Fusarium. Int. Biodeter. Biodegr. 41(2): 101–109. DOI: 10.1016/S0964-8305(98)00034-1.

Gu, J.D. 2003. Microbiological deterioration and degradation of synthetic polymeric materials: recent research advances. Int. Biodeter. Biodegr. 52(2): 69–91. DOI: 10.1016/S0964-8305(02)00177-4.

Hanus, M.J. and Harris, A.T. 2013. Nanotechnology innovations for the construction industry. Prog. Mater. Sci. 58(7): 1056–1102.

Heaton, P.E., Callow, M.E., Butler, G.M. and Milne, A. 1991. Control of mould growth by anti-fungal paints. Int. Biodeterior. 27(2): 163–173. DOI: 10.1016/ 0265-3036(91)90008-F.

Javaherdashti, R. and Setareh, M. 2006. Evaluation of sessile microorganisms in pipelines and cooling towers of some Iranian industries. J. Mater. Eng. Perform. 15: 5–8.

Konsta-Gdoutos, M.S., Metaxa, Z.S. and Shah, S.P. 2010. Multi-scale mechanical and fracture characteristics and early-age strain capacity of high performance carbon nanotube/cement nanocomposites. Cem. Concr. Compos. 32(2): 110–115.

Kulpa, C.F. and Baker, C.J. 1990. Involvement of sulfur-oxidizing bacteria in concrete deterioration. pp. 313–322. *In:* Dowling, N.J., Mittelman, M.W. and Danko, J.C. (eds). Microbially Influenced Corrosion and Biodeterioration National Association of Corrosion Engineers Houston TX 4/7–4/9.

Kumar, R. and Bhattacharjee, B. 2003. Porosity, pore size distribution and in situ strength of concrete. Cem. Concr. Res. 33: 155–164.

Kumar, R., Gardoni, P. and Sanchez-Silva, M. 2009. Effect of cumulative seismic damage and corrosion on the life-cycle cost of reinforced concrete bridges. Earthq. Eng. Struct. Dyn. 38: 887–905.

Lahav, O., Lu, Y., Shavit, U. and Loewenthal, R. 2004. Modeling hydrogen sulphide emission rates in gravity sewage collection systems. J. Environ. Eng. 130(11): 1382e1389.

Lajili, H., Devillers, P. and Grambin-Lapeyre, C. 2008. Alteration of a cement matrix subjected to biolixiviation test. Mater. Struct. 41: 1633–1645.

Ley, F.J. 1988. The control of microorganisms using ionisation radiation. pp. 523–528. *In:* Houghton, D.R., Smith, R.N. and Eggins, H.O.W. (eds). Biodeterioration, Vol. 7. Elsevier Applied Science. London. New York.

Magniont, C., Coutand, M., Bertron, A., Cameleyre, X., Lafforgue, C., Beaufort, S. and Escadeillas, G. 2011. A new test method to assess the bacterial deterioration of cementitious materials. Cem. Concr. Res. 41: 429–438.

Mansfeld, F., Shih, H., Postyn, A., Devinny, J., Islander, R. and Chen, C.L. 1991. Corrosion Monitoring and Control in Concrete Sewer Pipes. Corrosion. 47(5): 369–376. DOI: 10.5006/1.3585267.

Milde, K., Sand, W., Wolff, W. and Bock, E. 1983. Thiobacilli of the corroded concrete walls of the Hamburg sewer system. Microbiology. 129(5): 1327–1333. DOI: 10.1099/00221287-129-5-1327.

Miller, A., Dionisio, A. and Macedo, M. 2006. Primary bioreceptivity: a comparative study of different Portuguese lithotypes. Int. Biodeterior. Biodegrad. 57: 136–142.

Mohajerani, A., Burnett, L., Smith, J.V., Kurmus, H., Milas, J., Arulrajah, A. and Abdul Kadir, A. 2019. Nanoparticles in construction materials and other applications, and implications of nanoparticle use. Materials. 12(19): 3052.

Momber, A.W. 2000. Short-time cavitation erosion of concrete. Wear. 241: 47–52. DOI: 10.1016/S0043-1648(00)00348-3.

Mori, T., Nonaka, T., Tazak, K., Koga, M., Hikosaka, Y. and Nota, S. 1992. Interactions of nutrients, moisture, and pH on microbial corrosion of concrete sewer pipes. Water Res. 26: 29–37.

Mustafa, O.M. 2009. Reinforcement of beams by using carbon fiber reinforced polymer in concrete buildings. Sci. Res. Essays. 4: 1136–1145.

Nasrazadani, S. and Sudoi, E. 2010. A review of biodeterioration of concrete structures. Corrosion. 2010. NACE-10216.

Nica, D., Davis, J.L., Kirby, L., Zuo, G. and Roberts, D.J. 2000. Isolation and characterization of microorganisms involved in the biodeterioration of concrete in sewers. Int. Biodeterior. Biodegrad. 46: 61–68.

Nielsen, A.H., Yongsiri, C., Hvitved-Jacobsen, T. and Vollertsen, J. 2005. Simulation of sulfide buildup in wastewater and atmosphere of sewer networks. Water Sci. Technol. 52: 201e208.

Padival, N.A., Weiss, J.S. and Arnold, R.G. 1995. Control of Thiobacillus by means of microbial competition: implications for corrosion of concrete sewers. Water Environ. Res. 67: 201–205. DOI: 10.2175/106143095X131358.

Parker, C.D. 1945. The corrosion of concrete 1. The isolation of a species of bacterium associated with the corrosion of concrete exposed to atmosphere containing hydrogen sulphide. Aust. J. Exp. Biol. Med. Sci. 23: 81–90.

Parker, C.D. 1951. Mechanics of corrosion of concrete sewers by hydrogen sulfide. Sewage and Industrial Wastes. 23: 1477–1479.

Poole, C.P. and Owens, F.J. 2003. Introduction to Nanotechnology. John Wiley & Sons. New York. NY. USA.

Pradell, T., Climent-Font, A., Molera, J., Zucchiatti, A., Ynsa, M.D., Roura, P. and Crespo, D. 2007. Metallic and nonmetallic shine in luster: an elastic ion backscattering study. J. Appl. Phys. 101: 103518. DOI: 10.1063/1.2734944.

Reches, Y. 2018. Nanoparticles as concrete additives: review and perspectives. Constr. Build. Mater. 175: 483–495.

Ribas-Silva, M. 1995. Study of biological degradation applied to concrete. Proc., Transactions of 13th Int. Conf. on Structural Mechanics in Reactor Technology-SMiRT, Univ. Federal do Rio Grande do Sul, Porto Alegre, Brazil. 327–332.

Rigdon, J.H. and Beardsley, C.W. 1956. Corrosion of concrete by autotrophs. Corrosion. 5: 60–62.

Roco, M.C., Williams, S. and Alivisatos, P. (eds). 1999. Nanotechnology research directions: IWGN workshop report – vision for nanotechnology R&D in the next decade. WTEC, Loyola College in Maryland.

Salemi, N. and Behfarnia, K. 2013. Effect of nano-particles on durability of fiber-reinforced concrete pavement. Constr. Build. Mater. 48: 934–941.

Sanchez-Silva, M., Klutke, G.A. and Rosowsky, D.V. 2011. Life-cycle performance of structures subject to multiple deterioration mechanisms. Struct. Saf. 33: 206–217.

Sand, W. and Bock, E. 1984. Concrete corrosion in the Hamburg sewer system. Environ. Tech. Lett. 5: 517–528.

Sand, W. 1987. Importance of hydrogen sulfide, thiosulfate, and methylmercaptan for growth of *Thiobacilli* during simulation of concrete corrosion. Appl. Environ. Microbiol. 53: 1645–1648.

Sand, W., Ahlers, B. and Bock, E. 1991. The impact of microorganisms - especially nitric acid producing bacteria - on the deterioration of natural stones. pp. 481–484. *In*: Baer, N.S., Sabbioni, C. and Sors. A.I. (eds.). Science, Technology and European Cultural Heritage. Proceedings of the European Symposium, Bologna, Italy, June 13–16, 1989. Butterworth-Heinemann Ltd., Oxford, England.

Satoh, H., Odagiri, M., Ito, T. and Okabe, S. 2009. Microbial community structures and in situ sulfate-reducing and sulfur-oxidizing activities in biofilms developed on mortar specimens in a corroded sewer system. Water Res. 43: 4729–4739.

Sh, W., Sanchez-Silva, M. and Trejo, D. 2010. Microbial mediated deterioration of reinforced concrete structures. Int. Biodeterior. Biodegrad. 64: 748–754.

Solano, R., Patino-Ruiz, D. and Herrera, A. 2020. Preparation of modified paints with nano-structured additives and its potential applications. Nanomater. Nanotechnol. 10, 1847980420909188.

The New York Times. [(accessed on 22 July 2019)]; Available online: www.nytimes.com/imagepages/2005/02/21/science/20050222_NANO1_GRAPHIC.html.

Valentini, F., Carbone, M. and Palleschi, G. 2013. Carbon nanostructured materials for applications in nano-medicine, cultural heritage, and electrochemical biosensors. Anal. Bioanal. Chem. 405: 451–465. DOI: 10.1007/s00216-012-6351-6.

van der Molen, J.M., Garty, J., Aardema, B.W. and Krumbein, W.E. 1980. Growth control of algae and cyanobacteria on historical monuments by a mobile UV unit (MUVU). Stud. Conserv. 25: 71–77.

Verdier, T., Coutand, M., Bertron, A. and Roques, C. 2014. A review of indoor microbial growth across building materials and sampling and analysis methods. Build. Environ. 80: 136–149.

Viswakarma, V., George, R.P., Ramachandran, D., Anandkumar, B. and Madali, U.K. 2014. Studies of detailed biofilm characterization on fly ash concrete in comparison with normal and superplasticizer concrete in seawater environments. Environ. Technol. 35(1): 42–51.

Warscheida, T.H. and Braamsb, J. 2000. Biodeterioration of stone: a review. Inter. Biodeterior. Biodegrad. 46: 343–368.

Zhang, L., Schryver, P.D., Gusseme, B.D., Muynck, W.D., Boon, N. and Verstraete, W. 2008. Chemical and biological technologies for hydrogen sulfide emission control in sewer systems: a review. Water Res. 42: 1e12.

Section VI

Toxicity and Risk

Chapter 18

Assessment of Nanoparticle-Plant Interactions: An Integrative Proteomic Approach

Ghazala Mustafa[1] and Atikah Farooq[1]
[1]Department of Plant Sciences, Quaid-I-Azam University, Islamabad 45320, Pakistan.

Doyel Roy[2], Ayan Adhikari[2] and Zahed Hossain*[2]
[2]Plant Stress and Molecular Biology Laboratory, Department of Botany, University of Kalyani, Kalyani 741235, West Bengal, India

18.1 INTRODUCTION

Nanotechnology is an emerging field that deals with the synthesis and applications of nanomaterials (NMs). Recent developments in nanotechnology have revolutionized several domains of biology and medicines including biomedical industries, theragnostics, tissue engineering, cell manipulation and electronics (Biswas and Wu 2005). Increasing innovations in the field of nano-sciences have led to the identification of novel approaches for the synthesis of different kinds of NPs. Four major categories are metal or metal oxide NPs, carbon-based, dendrimers, and bio-inorganic complex NPs (Jha and Pudake 2016). NPs are synthesized through physical, chemical, biological, and hybrid methods (Mustafa et al. 2020). Along with chemical synthesis, NPs are also synthesized through biological ways with the involvement of plants, microorganisms, or their byproducts (Shah and Belozerova 2009). Nanoparticles synthesized through biological approaches are less toxic as compared to chemically synthesized NPs (Mustafa et al. 2020, Hasan et al. 2020, 2021a, b).

Plants are at continuous exposure to the NPs, due to their improper discharge into the environment. The unintended exposure poses biological risks to different life forms. NPs are ultrafine particles with a size range between 1–100 nm and their various applications in plant biotechnology and genetic engineering permit extensive developments in crop sciences (Wang et al. 2019, Khan and Upadhyay 2019, Kwak et al. 2019). Therefore, understanding the NPs-plant interaction mechanism is a topic of recent attention. Biological methods of NPs synthesis

*Corresponding Author: zahedkly@gmail.com

are preferred over conventional methods. NPs synthesized by conventional methods are ecotoxic while those produced by biological approaches are environment-friendly (Saxena et al. 2016). Furthermore, NPs synthesized through such methods are being effectively used for crop improvements and tolerance against various stress conditions.

Reactive properties and biochemical activities of NPs depend on their increased surface-to-volume ratio. These characteristics make them suitable entities for influencing multiple biological activities (Dubchak et al. 2010). A vast array of research has revealed the interaction pattern and transport of NPs in living systems. Different NPs are produced and successfully applied to plants for growth enhancement and stress tolerance (Saxena et al. 2016). Plant species exposed to NPs exhibit better growth and metabolism under stress conditions (Giraldo et al. 2014). Usage of chemically or biologically synthesized NPs in different products results in straight release into the environment leading to environmental hazards (Nair et al. 2010, Singh et al. 2015a). Plenty of research has been carried out on the impact of various metallic and metal oxide NPs on plants including silver (Mustafa et al. 2016, Hossain et al. 2016), aluminum oxide (Mustafa et al. 2015, Hossain et al. 2016), cerium oxide (Salehi et al. 2018), titanium oxide (Poscic et al. 2016, Hu et al. 2020), zinc oxide (Hossain et al. 2016, Taran et al. 2017, Faizan et al. 2020, Mustafa et al. 2022), and iron oxide (Adrees et al. 2020, Manzoor et al. 2021). Notably, impact of these NPs can be either positive or negative depending on the plant species under investigation and the type, size and concentration of nanoparticles used.

18.2 INSIGHTS INTO THE CELLULAR RESPONSES OF PLANTS TO NANOPARTICLE STRESS

In the modern era of agricultural technology, chemical fertilizers are extensively being used for crop improvement. Applications of conventional fertilizers have fulfilled the food demands of growing population; however, their discharge into the environment causes serious problems. Several physiological processes such as leaching, photolytic degradation, and decomposition, might restrict the complete internalization of conventional fertilizers by plants. To cope with this problem, the use of nanofertilizers is considered an emerging approach (Singh et al. 2015a). Nanofertilizers provide great possibility of minimizing nutrient losses leading to crop betterment (Solanki et al. 2015). Application of nanofertilizers is a potential source of NPs introduction into the environment due to their frequent use in agricultural industry. As primary producers, plants are at the forefront of NPs exposure which further leads to their accumulation in higher organisms (Zhang et al. 2012). Thus, understanding plant-NPs interactions is a major topic of concern. These NPs interactions at physiological, biochemical, and genetic levels depend upon the nature and extent of NPs internalization (Khodakovskaya et al. 2011, Laware and Rasker 2014). The nature of these interactions depends upon the unique physiochemical features of NPs such as surface charge, size, reactivity, and chemical composition (Spielman-Sun et al. 2017). Nanoparticles interactions with plants can either be positive or negative. It has been widely established that NPs, such as Co_3O_4, CuO, ZnO, and others, affect seedling germination and growth (Ghodake et al. 2011, Wang et al. 2012, Du et al. 2011). Seeds can efficiently uptake NPs which causes variation in germination rate. Silver nanoparticles (Ag NPs) have been known to enhance the germination rate in *Brassica* (Vishwakarma et al. 2017). Similarly, silicon nanoparticles (Si NPs) exposure was found to enhance growth and yield in wheat (Hussain et al. 2019). *Zea mays* and *Lycopersicon esculentum* seeds when exposed to SiO_2 NPs showed an enhanced germination rate (Suriyaprabha et al. 2012, Siddiqui and Al-whaibi 2014). Plants are exposed to NPs either from the atmosphere or soil. Aerially exposed NPs get deposited on leaves and are internalized by stomatal pathways while those released into the soil or water, interact with roots (Navarro et al. 2008, Klaine et al. 2008, Zhu et al. 2008). Once entered, NPs are either translocated through vascular pathways, bioaccumulated or transformed (Zhang et al. 2012). Physiochemistry of NPs allows them to cross initial barriers of the plant body for proper internalization. However, the adherence may sometimes end in the surface aggregation of NPs rather than internalization. Nature of the interactions between plants and NPs depends upon the properties of both nanomaterials as well as plants.

Nanoparticles are translocated symplastically to aerial parts through vascular system (Tripathi et al. 2017a). Advanced microscopy techniques have been used to investigate the translocation of NPs within the plant body. Uptake and distribution of CeO_2 NPs in cucumber roots were detected by electron microscopy (Corredor et al. 2009). Wang et al. (2012) used TEM and EDS for detecting CuO in xylem sap of *Zea mays*. Lin et al. (2009) reported the translocation of C70 and multiwalled carbon nanotubes through the xylem in rice plants. Previous studies indicate that after internalization into root, CeO_2, ZnO, and TiO_2 NPs are not further translocated towards the shoot (Lin and Xing 2008, Kurepa et al. 2010). NPs interact with plasma membrane pores to translocate across cellular boundaries (Rico et al. 2011). The phytotoxicity of NPs depends upon their size as well as the concentration (Barrena et al. 2009). Once inside the cell, NPs interact with metabolic machinery thereby modifying its physiology.

Oxidative stress leading to restricted plant growth and biomass is an inevitable consequence of NPs exposure (Ma et al. 2010, Miralles et al. 2012). Moreover, exposure to NPs causes changes in membrane permeability (metallic NPs), blockage of transpiration (TiO_2 NPs), and reduction in chlorophyll content (CeO_2 NPs and ZnO NPs) (Asli and Neuman 2009, Mukherjee et al. 2014). Quantum dots, much smaller than NPs are able to reach sub-cellular organelles causing oxidative burst. Santos et al. (2010) reported that Cd-Se/Zn-S QD (quantum dots) accumulate inside the nucleus causing oxidative cell damages. Single walled and functionalized carbon nanotubes induce phytotoxicity in *Oryza sativa*, *Lycopersicon esculentum*, and *Brassica oleracea* (Cañas et al. 2008). De La Torre-Roche et al. (2013) reported reduction of plant biomass in *Glycine max* and *Zea mays* under C60 fullerenes and carbon nanotubes exposure. Moreover, NPs cause mechanical damage to cells by inducing pore clogging (Jha and Pudaka 2016). Nevertheless, positive effect of NPs on plant growth and development has also been reported. For instance, PVP-coated Ag NPs induced oxidative damage in the model plant *Arabidopsis* (Kaveh et al. 2013), while enhanced growth in *Glycine max* (Mustafa et al. 2016). A lower concentration of SiO_2 NPs was found to enhance seed germination in tomatoes (Siddiqui and Al-Whaibi 2014). Other metallic NPs have also been reported to be less damaging for plant growth. These include iron, ZnO, copper, CeO, SiO and silver (Wang et al. 2012, Jayarambabu et al. 2015, Mustafa et al. 2016, Yasmeen et al. 2017, Jalil and Ansari 2019, Manzoor et al. 2021). Taken together all these data, it becomes apparent that NPs-plant interactions vary depending upon type of plant species, physiochemical nature of NPs and environmental conditions. Wide ranges of studies have identified these interactions at physiological, biochemical, and genetic levels. Still, the exact interaction mechanisms need to be elaborated in further detail.

18.3 NANOPARTICLE STRESS-INDUCED CHANGES IN PROTEOME COMPOSITION

Tremendous development in the fields of nanotechnology and proteomics has initiated the era of modern research in systems biology. Advanced tools of proteomics facilitate scientists to understand the molecular mechanisms of complex biological systems. Besides this, NPs have gained a lot of attention in scientific developments owing to their dynamic physical, chemical, and organic properties (Mondal et al. 2011). NPs such as metal oxide NPs, carbon nanotubes, quantum dots, and fullerenes are continuously introduced into the environment in one way or another (Navarro et al. 2008). Such contamination adversely affects the ecosystem causing water and soil pollution. NPs are being used to improve crop production, seed germination, and root/shoot development under stress and non-stress conditions (Saxena et al. 2016, Budhani et al. 2019). Thus, proteomic strategies significantly contribute to understand NPs mediated responses in plants by correlating the cellular metabolic processes with protein expression pattern. Employment of multi-omics approaches provides deep insights to understand plant response mechanisms under environmental stresses (Wang et al. 2013a). Notably, proteomics offers several benefits over genomics and transcriptomics techniques by directly dealing with proteins rather than DNA or mRNA (Barkla et al. 2013). Protein biomarkers, protein microarrays/chips, and gel-based/gel-free proteomic techniques are used for the quantification and identification of proteins (Abdulhamid et al. 2015). Furthermore, organelle proteomic shed light on the mechanisms of plant responses to NPs (Hossain et al. 2020). Nano-proteomic studies revealed that NPs interact with plants both physically and chemically. For instance, carbon nanotubes pierce the seed coat for water procurement, which is needed for seed germination in *Lycopersicon esculentum* (Khodakovskaya et al. 2009). C70 NMs in combination with carbon-based matter hindered the flowering period in *Oryza sativa* (Lin et al. 2009). Exposure of plant cells to metal-based NPs leads to oxidative damage (Foley et al. 2002), formation of ROS, lipid peroxidation (Kamat et al. 2000), and disturbed transport across the cell membrane (Auffan et al. 2008). Once NPs enter plant cells, they act as metal ions by countering with carboxyl and sulfhydryl groups, altering the protein activities (Hossain et al. 2015). Precise datasets generated by proteomic methods have assisted in protein functionalization in response to NPs. Identification of proteins, PTMs, and PPIs are very helpful in understanding the dynamics of nanobiotechnology in precision agriculture. Significant proteomic studies performed so far to understand the effects of NPs on cellular proteomes are summarized in Table 18.1.

Table 18.1 Summary of research works highlighting proteomic responses of plants to various NPs stresses

NPs (Size)	Plant species (Cultivar)	Exposed plant organ	Proteomic technique used	Major findings	References
Ag NPs (50 nm)	*Nicotiana tabacum* L. cv. Burley	Root	Gel based (2-DE), MALDI/TOF MS	Lower abundance of proteins involved in amino acid metabolic process in mitochondrion and in glutamine biosynthetic process in cytosol.	Peharec Štefanić et al. 2019
Ag NPs (15, 16 nm)	*Glycine max* L. cv. Enrei	2d old seedling	Immunoblot Label-free nanoLC/MS	Proteins related to protein degradation and synthesis, stress signaling were found to be differentially modulated, while no change was recorded in ascorbate peroxidase (APX), glutathione reductase (GR), and peroxiredoxin pattern.	Mustafa et al. 2020
Ag NPs (2–80 nm)	*Glycine max* L. cv. Enrei	Root	Gel free	Protein metabolism, amino acid metabolism, cell division and organization related proteins were changed. Growth was enhanced under flooding stress.	Mustafa et al. 2016
Ag NPs (15 nm)	*Glycine max* L. *cv. Enrei*	Root/ cotyledon	Gel free nanoLC- ESI-MS/MS	Aspartyl protease family protein, glyoxalase II 3, expansin-like B1 up-accumulated in roots; beta-ketoacyl reductase up-accumulated in cotyledons.	Mustafa et al. 2015

NPs (Size)	Plant species (Cultivar)	Exposed plant organ	Proteomic technique used	Major findings	References
Ag NPs (60 nm)	*Glycine max* L.	21 d old plant	Gel based (2D-DIGE), nano EI-LC-MS/MS	Chlorophyll a-b binding protein 6A, SOD (superoxide dismutase), nucleoside diphosphate kinase (NDK), and gamma-glutamyl hydrolase were significantly modulated.	Galazzi et al. 2019
Ag NPs (10 nm)	*Triticum aestivum* L.	Seed exposure and 5 d old seedlings	2-DE IEF/SDS-PAGE, LC–ESI-MS/MS	Cytoplasmic malate dehydrogenase (MDH) increased in roots along with proteins related to primary metabolism and defense were significantly regulated	Vannini et al. 2014
AgNPs (15-20nm)	*Triticum aestivum* L. var Punjab-2011	Root exposure	Gel free analysis	Glycolytic proteins (G3PH, G6P and PEPC) were down accumulated along with proteins of redox homeostasis.	Jhanzab et al. 2019
Ag NPs (18.34 nm)	*Oryza sativa* L.	Root/soil irrigation	Gel-based (2-DE, nano LC/FT-ICR MS)	Proteins linked to oxidative stress regulation, calcium dependent signaling, cell division and apoptosis were differentially regulated.	Mirzajani et al. 2014
Ag NPs (10 nm)	*Eruca sativa*	Seedling	Gel based (2DE) combined with nano-LC–ESI–MS/MS	Proteins involved in metabolism, and defense were differently regulated. V-ATPase subunits were downregulated.	Vannini et al. 2013
C NPs (28-77nm)	*Nicotiana tabacum* L.	Tobacco callus	Gel based (2D-LC MS/MS, iTRAQ	Mitochondrial proteins including CytC oxidase, membrane transferase proteins were significantly altered. Other up regulated proteins included calmodulin (CaM).	Zhao et al. 2022
Al_2O_3 NPs (30-60nm)	*Glycine max* L. *cv. Enrei*	Root	Gel free (nanospray) LTQ XL Orbitrap MS	Glycolysis related proteins were amplified. S-adenosyl-l-methionine dependent methyltransferases protein synthesis and stress signaling were significantly regulated.	Yasmeen et al. 2016
Al_2O_3 NPs (30-60 nm)	*Glycine max* L. *cv. Enrei*	Root	Gel-free	Upregulation of NmrA-like negative transcriptional regulator, and downregulation of flavodoxin-like quinone reductase, regulating energy metabolism and cell death.	Mustafa et al. 2015

Table 18.1 *Contd.*

Table 18.1 (*Contd.*) Summary of research works highlighting proteomic responses of plants to various NPs stresses

NPs (Size)	Plant species (Cultivar)	Exposed plant organ	Proteomic technique used	Major findings	References
Al_2O_3 NPs (30-60 nm)	*Glycine max* L. *cv. Enrei*	Root mitochondria	Gel-based/ LCMS	Suppression of glycolytic proteins; higher abundance of voltage-dependent anion channel protein indicating enhanced membrane permeability.	Mustafa and Komatsu 2016
Al_2O_3 NPs 30-60 nm+ ZnO and silver NPs	*Glycine max* L. *cv. Enrei*	Root/ leaf	Gel free (nanospray) LTQ XL orbitrap MS	Significant downregulation of methionine gamma lyase and upregulation of quinone reductase were observed; redox homeostasis related proteins such as thioredoxin family protein and NADH-CytB5 reductase were specifically regulated in response to Al_2O_3 NPs.	Hossain et al. 2016
CeO_2 NPs (10–30 nm)	*Phaseolus vulgaris* var. pinto bean	Root /foliar spray	Gel free Q-TOF tandem MS/MS	Proteins associated with photosynthetic ETC, protein turnover and folding and antioxidant defense (POD) were significantly regulated.	Salehi et al. 2018
FeNPs (20 nm/ 1–50 ppm)	*Triticum aestivum* L. Var. Pakistan-13 & NARC-11	Shoot	Gel free nanoLC/MS Immunoblot	Proteins of major energy regulatory pathways (glycolysis, electron transport chain etc.) were found to be significantly regulated.	Yasmeen et al. 2016, 2017b
Cu (15-30 nm) & Fe (20–30 nm) NPs (20-40 ppm)	*Triticum aestivum* L. Var. Pakistan-13 & NARC-11	Seed	Gel free, Nanospray LTQ XL Orbitrap MS/MS	Proteins of TCA cycle and glycolysis were upregulated; proteins of starch degradation were significantly changed in abundance	Yasmeen et al. 2017a
Copper NPs (10 nm)	*Triticum aestivum* L. Var. Pakistan-13 & NARC-11	Shoot	nanoLC/MS Gel free approach	Proteins of lipid metabolism were increased in Pakistan-13 while found to be decreased in NARC-11	Yasmeen et al. 2018

18.4 COPPER NANOPARTICLES

Copper (Cu), an essential micronutrient plays a key role in the growth and development of plants. Important cellular functions of Cu include photosynthetic electron transport, mitochondrial respiration, cell wall metabolism, hormone signalling, protein trafficking, and iron mobilization (Raven et al. 1999, Yruela 2005, 2009). Copper becomes harmful to plants at high concentrations, causing chlorosis, necrosis, stunting, and root growth inhibition (Yruela 2005). Cu^{1+} and Cu^{2+} are the two oxidation levels of copper that are available. Owing to this property, Cu can act

as an oxidising or reducing agent in biological processes. However, this feature also makes Cu potentially dangerous because Cu ions have the ability to stimulate the generation of free radicals (Hänsch and Mendel 2009, Ivask et al. 2010), causing oxidative stress (Valko et al. 2005), and transform into genotoxic materials. The malfunctioning of proteins caused by the interaction of too many copper ions with their sulfhydryl groups results in the inactivation of enzymes. Copper being a component of regulatory proteins, takes part in electron transport in the photosynthetic and respiratory chains, and functions as a cofactor in several enzymes, including cytochrome c oxidase, ascorbate oxidase, amino oxidase, laccase, plastocyanin and polyphenol oxidase, and Cu/Zn-superoxide dismutase (Chernavina 1970, Yruela 2005). Being a transition metal with redox activity, Cu triggers ROS (reactive oxygen species) generation (Halliwell and Gutteridge 1989).

Copper (Cu) NPs are the most widely studied among the metal-based NPs and their toxicity has been assessed in many crops. This form of metal oxide nanomaterial is widely used in a variety of industries including those related to agriculture, cosmetics, food, chemicals, fuel additives, textiles, medicine, paints, plastics, wastewater treatment, and electronics (Rajput et al. 2018). Due to anthropogenic activities, CuO NPs reach many environmental compartments, such as water, soil, and soil-exposed organisms, outside the scope of their intended usage.

CuO NPs have a negative impact on seed germination, photosynthesis, transpiration rate, root and shoot length, biomass and yield (Rajput et al. 2018). Shaw and Hossain (2013) took the initiative to explore the cellular responses of rice plants against nano-CuO stress to unravel the underlying mechanism of NPs-induced oxidative stress adaptation. The study revealed that nano-CuO treatment exerts oxidative damages to rice seedlings with enhanced ROS scavenging antioxidant enzymes activity and elevated MDA level. Similar study on the nano-CuO stress mediated changes of antioxidative defense and photosynthetic activity in barley (*Hordeum vulgare* L.) showed dose dependent reduction in shoot and root growth with decreased photosynthetic performance index (Shaw et al. 2014). Usually, NPs enter the plant system via root and translocate to the aboveground tissues. The absence of CuO NPs particles in the centre cylinder of the rice plant, however, may suggest that the casparian strip serves as the last apoplastic barrier separating the outer part from the vascular tissue (Rico et al. 2011).

When diverse group of plants were studied together, the influence of CuO NPs was found to be different (Lee et al. 2008, Adhikari et al. 2012, Atha et al. 2012, Van et al. 2016, Wang et al. 2016). It has also been demonstrated that in mung bean and wheat, copper nanoparticles can penetrate the cell directly to exercise their harmful effects including oxidative damages to the cellular components (Lee et al. 2008). Several proteins concerned with tricarboxylic acid cycle and glycolysis were found to be increased, whereas proteins associated with photosynthesis and tetrapyrrole decreased in wheat seedlings under Cu NPs stress (Yasmeen et al. 2018). Previous study with gel-free proteomic analysis revealed upregulation of glycolysis and starch degradation proteins in wheat grains exposed to Cu NPs (Yasmeen et al. 2017a).

It has been found that the concentration and time of exposure affect the accumulation and uptake of NPs in plants (Karlsson et al. 2009). Jain et al. (2017) reported that NPs phytotoxicity varied depending on the size and surface architecture of seeds and plant species. Notably, inadequate information on the particle size effect of CuO on plants is presently available (Shi et al. 2011, Koce 2017). Our recent comparative phytotoxicity study on maize seedlings has revealed that toxicity imparted by CuO bulk particles (BPs) was much higher as compared to CuO nanoparticles (CuO NPs) and the negative effects of CuO BPs were found to be due to higher dissolution of Cu ions (Roy et al. 2022).

18.5 ALUMINUM OXIDE NANOPARTICLES

Aluminum oxide is the key material used in products like high-temperature electrical insulators, high-voltage insulators, thermometry sensors, wear pads, ballistic armour, and grinding media because of its unique properties like extremely wear-resistant, good thermal conductivity, resist strong acids and alkali. Aluminum oxide (Al_2O_3) nanoparticles are nowadays used in agricultural products like herbicides and fertilizers as well as an energy substance (Novrotsky 2003, Vernikov et al. 2009, Stadler et al. 2011). Al_2O_3 NPs are also a source of exceptional fuel for various rockets and missiles as well as for space launch vehicles (Doshi et al. 2008). Although the mechanism of aluminium (Al^{3+}) toxicity is reasonably well understood and has been well documented in copious publications (Dogan et al. 2014, Matsumoto 2000), the mechanism of interaction and the fate of alpha aluminum oxide nanoparticles (Al_2O_3 NPs) after release into ecosystems have not been fully understood (Riahi Madvar et al. 2012). Even while Al_2O_3 NPs' effects on animals have been studied (Di Virgilio et al. 2010), there has been little in depth research into how hazardous these materials are to autotrophic organisms like plants. Al_2O_3 NPs were found to be less harmful than zinc oxide, iron oxide, and silicon oxide NPs as revealed in the comparative investigation of the effects of several types of nanoparticles on Arabidopsis (Lee et al. 2010). Although Al_2O_3 nanoparticles promote root elongation in radish, rape, ryegrass, and lettuce (Lin and Xing 2007), they had a negative effect on the root growth in corn, cucumber, soybean, cabbage, and carrot (Yang and Watts 2005). Al_2O_3 NPs also have injurious impact on wheat and tobacco growth and development (Burklew et al. 2012, Riahi-Madvar et al. 2012). Notably, the negative effect of Al_2O_3 NPs on wheat was reported to be alleviated by enhanced scavenging of reactive oxygen species by antioxidant enzymes system (Riahi-Madvar et al. 2012).

Proteomic study revealed that proteins associated with transport and development were increased during the post-flooding recovery phase under Al_2O_3 NPs exposure in *G. max* root (Yasmeen et al. 2016). Furthermore, Al_2O_3 NPs exposure found to enhance ribosomal protein abundance and regulated the ascorbate-glutathione pathway in *G. max* (Mustafa and Komatsu 2016).

A comparative proteome study of Al_2O_3 NPs with other two metal-based nanoparticles (ZnO and Ag) revealed less toxicity of Al_2O_3 NPs with no adverse effect on soybean (*G. max*) seedling growth (Hossain et al. 2016). Gel-free proteomic analysis of NPs challenged soybean roots showed that commonly expressed proteins are predominantly linked with cell organization, secondary metabolism, and hormone metabolism. Mustafa et al. (2015) reported that Al_2O_3 NPs treatment during flooding could enhance the root growth of *G. max* by modulating cell death and energy metabolisms. In a nutshell, nano-proteomic investigations reveal the beneficial role of Al_2O_3 NPs on wide range of economically important crops in agriculture sector.

18.6 ZINC NANOPARTICLES

Zinc (Zn) is an essential trace element, important for plant growth and development. In the absence of Zn, several enzymes, including carbonic anhydrase, carboxypeptidase, and alcohol dehydrogenase may lose their functional ability, but cadmium and mercury, two other members of the same group of elements with the same electrical configuration, are hazardous to plants when present in excess. Notably, ZnO NPs are the most synthetically produced nanomaterials in the world due to their extensive uses in batteries, semiconductors, paints, pigments, ceramics, biosensors, and in solar cells, thus causing environmental pollution. ZnO NPs have been found to be hazardous to a variety of plant species, including Arabidopsis (Lee et al. 2010), duckweed (Lee et al. 2013), wheat (Du et al. 2011), cucumber (Zhao et al. 2013), rapeseed (Kouhi et al. 2014), alfalfa (Wang et al. 2013b), etc. The majority of the investigations demonstrate that negative impact of ZnO NPs on plant growth and development is likely caused by the NPs induced oxidative stress (Hernandez-Viezcas et al. 2011, Thwala et al. 2013, Bandyopadhyay et al. 2015). In contrast, ZnO NPs were found to enhance the growth of *G. max* and affect root cell viability and rigidity (Hossain et al. 2016). Furthermore, they also downregulate proteins associated with the oxidation-reduction cascade. ZnO NPs promote growth in a concentration-dependent manner (Faizan et al. 2018).

Biologically synthesized ZnO nano-flowers exhibited potential antibacterial activity and maximum algae harvesting capacity (Hasan et al. 2021a). Zn NPs are less phytotoxic and have enhancing effects on plants. Biologically synthesized Zn-nanoflowers improve the soybean growth while chemically synthesized nanowires increased oxidative stress (Mustafa et al. 2022).

Recently, our research group has assessed the phytotoxicity of ZnO NPs on maize seedlings exploiting the microRNAomic approach (Adhikari et al. 2020). A large number of stress responsive miRNAs were found to be downregulated in ZnO NPs challenged maize leaves. Potential target genes of these miRNAs primarily encode transcription factors and proteins linked with varied biological processes including plant growth and development, chlorophyll biosynthesis, cell wall modification, redox homeostasis, and heavy metal detoxification, etc. Taken together, our findings highlight the roles of stress-responsive miRNAs in adaptive mechanisms of maize plants towards ZnO-NPs stress.

18.7 SILVER NANOPARTICLES

Silver nanoparticles (Ag NPs) are well-studied and have antifungal and antibacterial properties (Weir et al. 2008). There are various methods including biological, physical, and chemical to synthesize these NPs under definite constraints of shape and size (Iravani and Golghar 2014). Various nano-proteomic studies have been performed to understand the effects of Ag NPs in different crops. These studies are mostly based on gel-based proteomic techniques combined with LC-MS. Interactions of plants with Ag NPs are of both positive and negative nature. In *G. max*, Ag NPs reduced seedling growth by downregulating proteins of cellular organization, and secondary metabolisms. However, it was reported that a number of proteins associated with primary metabolisms were upregulated under Ag NPs treatment (Hossain et al. 2016). Interestingly, Ag NPs also have a mitigative role against abiotic stresses. Under flooding stress, Ag NPs promote growth of *G. max* by improving proteins related to the synthesis of amino acids (Mustafa et al. 2016). Exogenous application of Ag-NPs in *G. max* root causes an increase in proteins linked with degradation while decreasing proteins related to synthesis. Moreover, Ag NPs were found to improve the growth of *G. max* under flooding stress as revealed by gel-free proteomic technique (Hashimoto et al. 2020).

Jhanzab et al. (2019) reported that proteins linked to cell wall, signaling, and glycolysis decreased in wheat under Ag NPs stress using gel-free proteomics approach. Changes in expression level of metabolic and cell defense related proteins in wheat root have also been reported, using gel-based proteomic techniques (Vannini et al. 2014). Proteins associated with cell division, transcription, degradation, apoptosis, and oxidative stress were found to enhance in *Oryza sativa*, when exposed to Ag NPs (Mirzajani et al. 2014). Vacuole and ER-related proteins linked with sulfur metabolism were modulated in *Eruca sativa* roots under Ag NPs exposure (Vannini et al. 2013). These studies suggest that Ag NPs modulate several metabolic processes and protein regulation ensuring improved plant growth.

18.8 IRON NANOPARTICLES

Iron NPs have wide biomedical, industrial, and commercial applications. These NPs are considered as bio-remediation agents due to their highly reactive and magnetic nature (Yan et al. 2013). Fe NPs are reported to enhance plant growth and seed germination in wheat (Yasmeen et al. 2017b). Only a few proteomic studies have been conducted to understand the interactions of Fe NPs with plants. Gel-free proteomic study revealed that salinity and drought-resistant wheat varieties are influenced by Fe NPs. The study reported differentially expressed photosynthesis-related proteins in both wheat cultivars. Proteins associated with light reactions were found to be improved in salt tolerant cultivars as compared to drought tolerant cultivars. Wheat grain analysis elucidated upregulation in proteins linked to glycolysis, TCA cycle, and starch degradation (Yasmeen et al. 2017b). Proteomics analyses suggest that Fe NPs have more prominent role in changing expression level of photosynthetic proteins. However, further nano-proteomic studies need to be conducted to better understand the specific plant-Fe NPs interactions.

18.9 OTHER NANOPARTICLES

TiO_2 NPs decrease cell growth and photosynthetic efficiency while increasing chlorophyll b and carotenoids contents in green alga (Chen et al. 2012). CeO_2 NPs increase heat-shock proteins and enzymatic activities of catalase and ascorbate peroxidase in *Zea mays* (Zhao et al. 2012). In agriculture sector, Au NPs are found to be important. The proteome study of *Arabidopsis thaliana* exposed to Au confirmed that glutathione-S-transferases expressed in the shoot and played an important role in the biosynthesis of Au NPs (Tiwari et al. 2016). Metallic NPs interact with plants at morphophysiological as well as molecular levels and interfere with major cellular pathways.

18.10 NANOPARTICLE-MEDIATED REGULATION OF OXIDATIVE STRESS

Lipid peroxidation, H_2O_2 induced membrane damage, and protein degradation are known potential impacts imposed by oxidative damage in plants (Venkatachalam et. al. 2017). Excessive production of reactive oxygen species (ROS) particularly in response to biotic and abiotic stresses is associated with oxidative damages in plants. Moreover, ROS are continuously produced inside the plant organelles such as chloroplast, mitochondria, peroxisomes, ER, and plasma membranes (Karuppanapandian et al. 2011). The cascade of electron transfer reactions occur in plant cells which generate ROS. These can also be generated indirectly through energy transfer mechanism of oxygen. Studies have highlighted subcellular production of ROS that have the ability to induce cytotoxicity through certain modes of action. For instance, some ROS are target specific like $O_2^{-\bullet}$, which only interacts with compounds that have H-bonds in structural conformity (Gill and Tuteja 2010, Sharma et al. 2012). The target proteins for these reactive oxygen species could be the proteins that contain nitric acid or Fe–S functional groups. The modulation of oxidative stress-related proteins identified through nano-proteomic approaches are summarized in Figure 18.1.

Redox recycling or photo-oxidation reactions lead to ROS generation inside the chloroplast and plasma membranes. ROS become toxic when they are produced in excess inside plant cell and hence induce various damages to lipid bilayers and degrade essential biomolecules (Shen et al. 2010, Yadav et al. 2014). While mentioning ROS toxicity, enzymatic and non-enzymatic antioxidant defense system is often taken into consideration which comprises of superoxide dismutase (SOD) enzyme which functions as central scavenging system targeting superoxide radicals (Gill and Tuteja 2010, Sharma et al. 2012), along with other antioxidants such as POD, CAT, APX etc. (Laware and Raksar, 2014). Various kinds of NPs have potential effects on the antioxidant defense system. Studies have elaborated that nanoparticles can mediate ROS-dependent cytotoxicity by enhancing antioxidant enzyme activities under stress conditions (Laware and Raksar 2014). Rico et al. (2015) have suggested that NPs induced oxidative stress regulation in plants is one of the toxicological mechanisms which might be directly estimated by ROS production. However, many nanoparticles have been reported as having active roles in the regulation of oxidative damages in plants. Tan et al. (2009) observed that multiwalled carbon nanotubes can affect the rate of cell proliferation and cause plant cell death by inducing ROS generation. Similar effects of multi-walled nanotubes were observed in lettuce cells in which apoptotic responses were observed in cells (Lin and Xing

2008). Phytotoxicity induced by MWCNTs was also reported in *Oryza sativa* (Lin et al. 2009). Further investigation revealed that these multiwalled carbon nanotubes can penetrate cell walls thereby entering inside the cell and increasing antioxidant production (Smirnova et al. 2011). Shen et al. (2010) observed that single walled carbon NPs (1–2 nm size) applied at a concentration of 25 mg/L have the same effects. Besides this, exposure of plants to engineered nanoparticles demonstrated several hypersensitive responses, apoptosis, and necrosis (Nair et al. 2010).

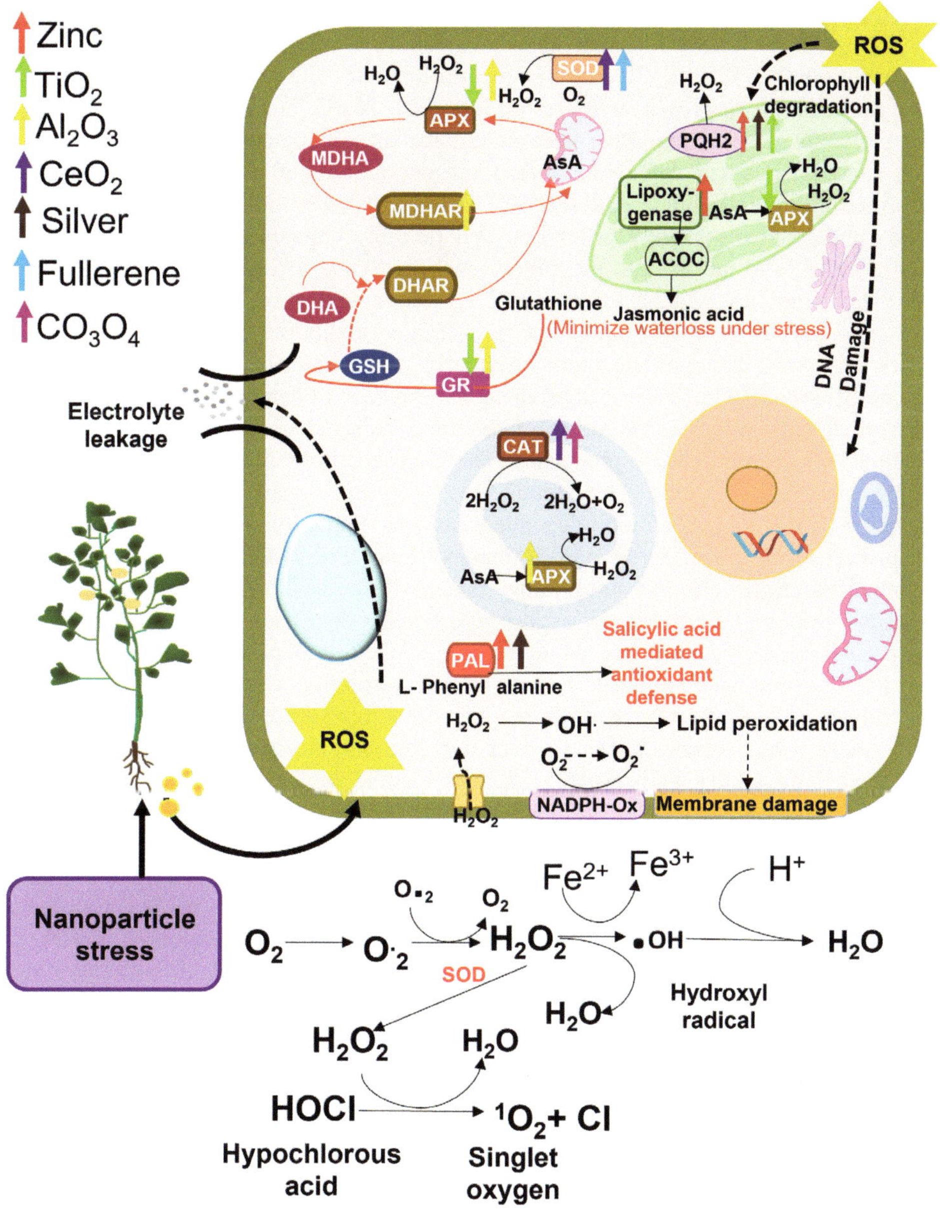

Figure 18.1 Schematic representation of NPs-mediated oxidative damage to plant cell. Effects of different NPs (colored arrows) on antioxidant enzymes and major redox homeostatic pathways (glutathione cycle: red arrows) has been shown. Upwards arrow: Upregulation of proteins; Downwards arrow: Downregulation of protein. Abbreviations: PAL, phenyl ammonia lyase; PQH2, plastoquinol; SOD, superoxide dismutase; CAT, catalase; APX, ascorbate peroxidase; DHAR, dehydroascorbate reductase; GR, glutathione reductase; MDHA, monodehydroascorbate; MDHAR, monodehydroascorbate reductase.

ROS estimation helps in studying the toxicological mechanisms (Rico et al. 2013a, b) induced due to NPs and heavy metal exposure (Nair et al. 2010, Begum and Fugetsu 2012, Sharma et al. 2012). As previously documented, lipid peroxidation and membrane damages are the known underlying symptoms of the oxidative damages in plants thus these can be used for assessing the impact of NPs induced oxidative stress regulation. In this context various studies have been performed. For instance, Rico et al. (2013a, b) and Siddiqui et al. (2015) reported that nano cerium dioxide nanoparticles at different exposure concentrations generated ROS in *Oryza sativa*. Yin et al. (2012) suggested that TiO_2 NPs induce phytotoxic effects, when exposed to an array of cell systems, were led by ROS which participates in stress signaling to alleviate the free radical toxicity. Additionally, 10 and 30 ppm concentrations of TiO_2 NPs also modify antioxidant enzymatic activities in *Phaseolus vulgaris* (Jacob et al. 2013), but very low concentrations i.e., 0.25% of TiO_2 NPs generate oxidative stress in chloroplasts of spinach seedlings (Lei et al. 2008). Variety of nanoparticles have been known for their ameliorating effects on plants. One such study was performed by Pérez-Labrada et al. (2019). These authors found that foliar application of Cu NPs makes tomato plants salinity tolerant which possibly appears due to stimulating enzymes associated with cellular antioxidant machinery such as SODs, PAL, CAT, APX, GPOX in leaf tissue. Another study by Mukherjee et al. (2014) documented on ZnO NPs (10 nm in size) suggested exposure in *Pisum sativum* at 500 ppm triggers oxidative stress.

Studies justify that nanoparticles induce oxidative stress in plants and have mitigative effects against toxic ROS. In contrary, NPs can produce ROS instead of detoxifying them. Jalil and Ansari (2019) reported that higher concentration of selenium NPs induces oxidative stress while lower concentrations show antioxidative properties. These observations are consistent with the studies reported by Rico et al. (2013a). They found that exposure to nano CeO_2 modifies antioxidant status in rice plants in a concentration dependent manner. Some other studies have even indicated potential protective effects of some specific nanoparticles. For instance, iron nanoparticles seem to have enhanced effects on plant morphology and photosynthesis when applied in certain doses (5, 10, 15, and 20 mg/L). Furthermore, exposure to Fe NPs decreases the activities of peroxidases, SOD and protect along electrolyte loss when exposed to plants under cadmium stress. Thus, it seems clear that ROS generation may indeed be one of the symptoms of nanoparticles induced toxicity in plants. However, in-depth investigations are still needed to further understand this phenomenon (Paunovic 2020). It has become evidently clear that nanoparticles (either metallic or nonmetallic) cause disturbance in cellular redox balance and hence could be considered as phytotoxic entities.

Although investigation provides sufficient evidence about ROS-mediated cellular responses to NPs, but information about NPs interaction at sub-cellular level and how underlying cascades of oxidative defense function inside the cell, is still lacking (Rico et al. 2015, Tripathi et al. 2017a). It is already known that plants' antioxidant defense comprises of non-enzymatic such as phenolics, thiols, ascorbate, etc. as well as enzymatic agents like superoxide dismutase (SOD), catalase (CAT), ascorbate peroxidase (APX), guaiacol peroxidase (GPX) dehydroascorbate reductase (DHAR) and glutathione reductase (GR) (Singh et al. 2015b, Rico et al. 2015). Both these enzymatic and non-enzymatic antioxidants are found to be engaged in highly specific and complex redox homeostatic reactions. Among them, APX, DHAR, and GR, the enzymatic components, are employed in the ascorbate-glutathione cycle (AsA-GSH cycle or Halliwell-Asada pathway) that regulate oxidative status of plants. Proposed mechanism of NPs induced ROS generation suggests that SOD synthesizes H_2O_2 which is further reduced by APX into water molecule. For this reduction, APX utilizes ascorbate which is regenerated by DHAR. The GR enzyme generates glutathione in reduced form which further acts as reducing agent for ascorbate synthesis (Rico et al. 2015). Studies have identified number of nanoparticles that mimic antioxidant properties in plants such as CeO_2, MnO_2, Fe_3O_4, fullerene, Co_3O_4, CuO and Au (Karakoti et al. 2010, Celardo et al. 2011, Wei and Wang 2013). Furthermore, studies revealed that antioxidant enzymes are induced by these nanoparticles. For instance, catalase in induced by Co_3O_4, CeO_2 and Fe_3O_4 while GPX was found to be induced by CeO_2, MnO_2, Fe_3O_4, Co_3O_4, Au, and SOD is induced by

Pt, CuO, fullerene and CeO_2 (Wei and Wang 2013). It must be highlighted that these interrelated mechanisms are very difficult to distinguish and to identify specific reaction pathways triggered in response to individual NPs. However, correlation could be established among these reactions up to some extent.

Often it has been observed that, in certain conditions, uneven effects of nanoparticles appear on enzymes activities of antioxidant defense system. Certain studies have also demonstrated this phenomenon contrarily. For instance, Lei et al. (2008) and Almutairi (2019) found that titanium dioxide NPs enhanced the enzyme activities of catalase, ascorbate peroxidase, superoxide dismutase, and glutathione peroxidase in spinach and *Arabidopsis thaliana*. While the same nanoparticle ($nTiO_2$) reduced GR and APX activities in *Vicia faba* (Foltête et al. 2011). Thus, it seems difficult to demonstrate the exact positive or negative interactions of individual nanoparticles within the cellular oxidative stress enzymes. Although there is contradiction between these investigations, still some studies highlight the direct influence of nanoparticles on antioxidant enzymes (Rico et al. 2015, Tripathi et al. 2016, 2017b). It seems that differential responses of nanoparticles mediated oxidative stress regulation in plants as well as their own antioxidant properties widely depend upon their certain physiochemical properties such as morphology, composition and concentration (Wei and Wang 2013, Tripathi et al. 2017a). Rico et al. (2013a) found that CeO_2–NPs modified rice antioxidant system while the same NPs induced toxicity in *Phaseolus vulgaris* estimated by disturbed antioxidative defense system and also induced oxidative stress in wheat plants (Majumdar et al. 2014, Iftikhar et al. 2020). In this scenario, it seems that interrelated phenomenon of nanoparticles toxicity and respective oxidative stress regulation in plants is not yet clear. Furthermore, evidences to correlate these phenomena with the chemical properties of nanoparticles are still a prerequisite.

18.11 CONCLUSION AND FUTURE PERSPECTIVES

This chapter outlines the response mechanisms of plants to various NPs stresses with special reference to the modulation of proteome composition. In recent years, nanotechnology has achieved enormous impetus in recent times owing to the wide applications of nanoparticles in agriculture, cellular imaging, cosmetic industry, biosensing, diagnosis, and cancer therapy. However, unintentional release of these commercially produced nanoparticles into the environment caused widespread alarm. Substantial consideration is now given to the nanoparticle synthesis methods, their proper discharge into the environment, and their long-term fate. As linked to conventional physicochemical methods, biosynthesis of nanoparticles with the help of plants and microorganisms is an environment-friendly approach. Silver, aluminum, iron, and zinc nanoparticles differentially modulate the various metabolic processes leading to plant's stress tolerance. Most of the research works were focused on the impact of single type of NPs on growth and development of various crop species. Further proteomic studies on plant responses to multiple NPs stress need to be encouraged to shed light on the cross-talk between different NPs stress signaling pathways. Moreover, in-depth organelle proteomic study will give a more precise picture about the protein networks essential for accurate dissection of the underlying NPs stress adaptive mechanisms. On a summative note, more comprehensive investigations integrating multi-omics approaches including genomics, proteomics, transcriptomics, and metabolomics, need to be undertaken to further dissect the plant-nanoparticle interaction mechanisms in details.

ACKNOWLEDGEMENTS

This work was supported by research grant [38(1440)/17/EMR-II] from Council of Scientific and Industrial Research (CSIR), Govt. of India. The authors D.R., A.A. thankfully acknowledge the economic support extended by CSIR as CSIR-SRF and University of Kalyani as URS respectively. The authors also express gratitude to DST-PURSE II for financial assistance.

References

Abdulhamid, H.N. and Wu, H.F. 2015. Proteomics analysis of the mode of antibacterial action of nanoparticles and their interactions with proteins. Trends Anal. Chem. 65: 30–46.

Adhikari, T., Kundu, S., Biswas, A.K., Tarafdar, J.C. and Rao, A.S. 2012. Effect of copper oxide nano particle on seed germination of selected crops. J. Agric. Sci. Technol. A. 2: 815–823. doi: 10.17265/2161-6256/2012.06A.009.

Adhikari, S., Adhikari, A., Ghosh, S., Roy, D., Azahar, I., Basuli, D. and Hossain, Z. 2020. Assessment of ZnO–NPs toxicity in maize: an integrative microRNAomic approach. Chemosphere. 12: 249: 126197.

Adrees, M., Khan, Z.S., Ali, S., Hafeez, M., Khalid, S., Ur Rehman, M.Z., Hussain, A., et al. 2020. Simultaneous mitigation of cadmium and drought stress in wheat by soil application of iron nanoparticles. Chemosphere. 238: 124681.

Almutairi, Z.M.T. 2019. Plant molecular defense mechanisms promoted by nano-particles against environmental stress. Int. J. Agric. Biol. 21: 259.

Asli, S. and Neumann, P.M. 2009. Colloidal suspensions of clay or titanium dioxide nanoparticles can inhibit leaf growth and transpiration via physical effects on root water transport. Plant Cell Environ. 32: 577–584.

Atha, D.H., Wang, H., Petersen, E.J., Cleveland, D., Holbrook, R.D., Jaruga, P., et al. 2012. Copper oxide nanoparticle mediated DNA damage in terrestrial plant models. Environ. Sci. Technol. 46: 1819–1827. doi: 10.1021/es202660k.

Auffan, M., Achouak, W., Rose, J., Roncato, M.A., Chanéac, C., Waite, D.T., Masion, A., et al. 2008. Relation between the redox state of iron-based nanoparticles and their cytotoxicity toward *Escherichia coli*. Environ. Sci. Technol. 42: 6730–6735.

Bandyopadhyay, S., Plascencia-Villa, G., Mukherjee, A., Rico, C.M., José-Yacamán, M., Peralta-Videa, J.R., et al. 2015. Comparative phytotoxicity of ZnO NPs, bulk ZnO, and ionic zinc onto the alfalfa plants symbiotically associated with *Sinorhizobium meliloti* in soil. Sci. Total Environ. 515–516: 60–69. doi: 10.1016/j.scitotenv.2015.02.014.

Barkla, B.J., Vera-Estrella, R. and Pantoja, O. 2013. Progress and challenges for abiotic stress proteomics of crop plants. Proteomics. 13: 1801–1815.

Barrena, R., Casals, E., Colon, J., Font, X. and Sanchez, A., et al. 2009. Evaluation of the ecotoxicity of model nanoparticles. Chemosphere 75: 850–857.

Begum, P. and Fugetsu, B. 2012. Phytotoxicity of multi-walled carbon nanotubes on red spinach (*Amaranthus tricolor* L.) and the role of ascorbic acid as an antioxidant. J. Hazard. Mater. 243: 212–222.

Biswas, P. and Wu, C.Y. 2005. Nanoparticles and the environment. J. Air Waste Manag. Assoc. 55(6): 708–746.

Budhani, S., Egboluche, N.P., Arslan, Z., Yu, H. and Deng, H. 2019. Phytotoxic effect of silver nanoparticles on seed germination and growth of terrestrial plants. J. Environ. Sci. Health. Part C. Environ. Carcinog. Ecotoxicol. Rev. 37(4): 330–355.

Burklew, C.E., Ashlock, J., Winfrey, W.B. and Zhang, B. 2012. Effects of aluminum oxide nanoparticles on the growth, development, and micro RNA expression of tobacco (*Nicotiana tabacum*). PLoS ONE. 7: e34783.

Cañas, J.E., Long, M., Nations, S., Vadan, R., Dai, L., Luo, M., Ambikapathi, R., et al. 2008. Effects of functionalized and nonfunctionalized single-walled carbon nanotubes on root elongation of select crop species. Environ. Toxicol. Chem. 27(9): 1922–1931.

Celardo, I., Pedersen, J.Z., Traversa, E. and Ghibelli, L. 2011. Pharmacological potential of cerium oxide nanoparticles. Nanoscale. 3(4): 1411–1420.

Chen, L.Z., Zhou, L.N., Liu, Y.D., Deng, S.Q., Wu, H. and Wang, G.H. 2012. Toxico-logical effects of nanometer titanium dioxide (nano-TiO_2) on *Chlamydomonas reinhardtii*. Ecotoxicol. Environ. Saf. 84: 155–162.

Chernavina, I.A. 1970. The Physiology and Biochemistry of Microelements. Moscow: Vysshaya Shkola 1970.

Corredor, E., Testillano, P.S., Coronado, M.J., González-Melendi, P., Fernán-dez-Pacheco, R., Marquina, C. and Risueño, M.C. 2009. Nanoparticle penetration and transport in living pumpkin plants: *in situ* subcellular identification. BMC Plant Biol. 9(1): 1–11.

De La Torre-Roche, R., Hawthorne, J., Deng, Y., Xing, B., Cai, W., Newman, L.A., Wang, Q., et al. 2013. Multiwalled carbon nanotubes and c60 fullerenes differentially impact the accumulation of weathered pesticides in four agricultural plants. Environ. Sci. Technol. 47(21): 12539–12547.

Di,Virgilio, A.L., Reigosa, M., Arnal, P.M., Lorenzo, F and, de Mele, M. 2010. Comparative study of the cytotoxic and genotoxic effects of titanium oxide and aluminium oxide nanoparticles in Chinese hamster ovary (CHO–K1) cells. J. Hazard Mater. 177: 711–718.

Dogan, I., Ozyigit, I.I. and Demir, G. 2014. Influence of aluminum on mineral nutrient uptake and accumulation in *Urtica pilulifera* L. J. Plant Nutr. 37: 469–481.

Doshi, R., Braida, W., Christodoulatos, C., Wazne, M. and O'Connor, G. 2008. Nano-aluminum: transport through sand columns and environmental effects on plant and soil communication. Environ. Res. 106: 296–303.

Du, W., Sun, Y., Ji, R., Zhu, J., Wu, J. and Guo, H. 2011. TiO_2 and ZnO nanoparticles negatively affect wheat growth and soil enzyme activities in agricultural soil. J. Environ. Monit. 13: 822–828. doi: 10.1039/c0em00611d.

Dubchak, S., Ogar, A., Mietelski, J.W. and Turnau, K. 2010. Influence of silver and titanium nanoparticles on arbuscularmycorrhiza colonization and accumulation of radiocaesium in *Helianthus annuus*. Span. J. Agric. Res. (1): 103–108.

Faizan, M., Faraz, A., Yusuf, M., Khan, S.T. and Hayat, S. 2018. Zinc oxide nanoparticle-mediated changes in photosynthetic efficiency and antioxidant system of tomato plants. Photosynthetica. 56(2): 678–686.

Faizan, M., Hayat, S. and Pichtel, J. 2020. Effects of zinc oxide nanoparticles on crop plants: a perspective analysis. Sustainable Agric. Rev. 41: 83–99.

Foley, S., Crowley, C., Smaihi, M., Bonfils, C., Erlanger, B.F., Seta, P. and Larroque, C. 2002. Cellular localization of a water-soluble fullerene derivative. Biochem. Biophys. Res. Commun. 294: 116–119.

Foltête, A.S., Masfaraud, J.F., Bigorgne, E., Nahmani, J., Chaurand, P., Botta, C., La-bille, J., et al. 2011. Environmental impact of sunscreen nanomaterials: ecotoxicity and genotoxicity of altered TiO_2 nanocomposites on *Vicia faba*. Environ. Pollut. 159: 2515–2522. doi: 10.1016/j.envpol.2011.06.020.

Galazzi, R.M., Lopes Júnior, C.A., de Lima, T.B., Gozzo, F.C. and Arruda, M.A.Z. 2019. Evaluation of some effects on plant metabolism through proteins and enzymes in transgenic and non-transgenic soybeans after cultivation with silver nanoparticles. J. Proteomics. 191: 88–106. doi: 10.1016/j.jprot.2018.03.024.

Ghodake, G., Seo, Y.D. and Lee, D.S. 2011. Hazardous phytotoxic nature of cobalt and zinc oxide nanoparticles assessed using *Allium cepa*. J. Hazard Mater. 186(1): 952–955.

Gill, S.S. and Tuteja, N. 2010. Reactive oxygen species and antioxidant machinery in abiotic stress tolerance in crop plants. Plant Physiol. Biochem. 48(12): 909–930.

Giraldo, J.P., Landry, M.P., Faltermeier, S.M., McNicholas, T.P., Iverson, N.M., Boghossian, A.A., Reuel, N.F., et al. 2014. Plant nanobionics approach to augment photosynthesis and biochemical sensing. Nat. Mater. 13(4): 400–408.

Halliwell, B. and Gutteridge, J.M.C. 1989. Free Radicals in Biology and Medicine, 2nd Ed. Clarendon Press, Oxford.

Hänsch, R. and Mendel, R.R. 2009. Physiological functions of mineral micronutrients (Cu, Zn, Mn, Fe, Ni, Mo, B, Cl). Curr. Opin. Plant Biol. 12: 259–266.

Hasan, M., Rafique, S., Zafar, A., Loomba, S., Khan, R., Hassan, S.G., Khan, M.W., et al. 2020. Physiological and anti-oxidative response of biologically and chemically synthesized iron oxide: *Zea mays* a case study. Heliyon. 6(8): e04595.

Hasan, M., Altaf, M., Zafar, A., Hassan, S.G., Ali, Z., Mustafa, G., Munawar, T., et al. 2021a. Bioinspired synthesis of zinc oxide nano-flowers: a surface enhanced antibacterial and harvesting efficiency. Materials Science and Engineering: C, Materials for Biological Applications. 119: 111280.

Hasan, M., Mehmood, K., Mustafa, G., Zafar, A., Tariq, T., Hassan, S.G., Loomba, S., et al. 2021b. Phytotoxic evaluation of phytosynthesized silver nanoparticles on lettuce. Coatings. 11(2): 225. https://doi.org/ 10.3390/coatings11020225.

Hashimoto, T., Mustafa, G., Nishiuchi, T. and Komatsu, S. 2020. Comparative analysis of the effect of inorganic and organic chemicals with silver nanoparticles on soybean under flooding stress. Int. J. Mol. Sci. 21: 1300.

Hernandez-Viezcas, J.A., Castillo-Michel, H., Servin, A.D., Peralta-Videa, J.R. and Gardea-Torresdey, J.L. 2011. Spectroscopic verification of zinc absorption and distribution in the desert plant *Prosopis juliflora-velutina* (velvet mesquite) treated with ZnO nanoparticles. Chem. Eng. J. 170: 346–352. doi: 10.1016/j.cej.2010.12.021.

Hossain, Z., Mustafa, G. and Komatsu, S. 2015. Plant responses to nanoparticle stress. Int. J. Mol. Sci. 16(11): 26644–26653.

Hossain, Z., Mustafa, G., Sakata, K. and Komatsu, S. 2016. Insights into the proteomic response of soybean towards Al_2O_3, ZnO, and Ag nanoparticles stress. J. Hazard. Mater. 304: 291–305.

Hossain, Z., Yasmeen, F. and Komatsu, S. 2020. Nanoparticles: synthesis, morphophysi-ological effects, and proteomic responses of crop plants. Int. J. Mol. Sci. 21(9): 3056.

Hu, J., Wu, X., Wu, F., Chen, W., White, J.C., Yang, Y., Wang, B., et al. 2020. Potential application of titanium dioxide nanoparticles to improve the nutritional quality of coriander (*Coriandrum sativum* L.). J Hazard Mater. 389: 121837.

Hussain, A., Rizwan, M., Ali, Q. and Ali, S. 2019. Seed priming with silicon nanoparticles improved the biomass and yield while reduced the oxidative stress and cadmium concentration in wheat grains. Environ. Sci. Pollut. Res. 26(8): 7579–7588.

Iftikhar, A., Rizwan, M., Adrees, M., Ali, S., urRehman, M.Z., Qayyum, M.F. and Hussain, A. 2020. Effect of gibberellic acid on growth, biomass, and antioxidant defense system of wheat (*Triticum aestivum* L.) under cerium oxide nanoparticle stress. Environ. Sci. Pollut. Res. 27(27): 33809–33820.

Iravani, S. and Golghar, B. 2014. Green synthesis of silver nanoparticles using *Pinus eldarica* bark extract. BioMed. Res. Int. 2013: 1–5.

Ivask, A., Bondarenko, O., Jepihhina, N. and Kahru, A. 2010. Profiling of the reactive oxygen species related ecotoxicity of CuO, ZnO, TiO_2, silver and fullerene nanoparticles using a set of recombinant luminescent *Escherichia coli* strains: differentiating the impact of particles and solubilised metals. Anal. Bioanal. Chem. 398: 701–716.

Jacob, D.L., Borchardt, J.D., Navaratnam, L., Otte, M.L., and Bezbaruah, A.N. 2013 Uptake and translocation of Ti from nanoparticles in crops and wetland plants. Int. J. Phytoremediat. 15: 142–153.

Jain, N., Bhargava, A., Pareek, V., Akhtar, M.S. and Panwar, J. 2017. Does seed size and surface anatomy play role in combating phytotoxicity of nanoparticles? Ecotoxicology. 26: 238–249. https://doi.org/10.1007/s10646-017-1758-7.

Jalil, S.U. and Ansari, M.I. 2019. Nanoparticles and abiotic stress tolerance in plants: synthesis, action, and signaling mechanisms. Plant Signaling Molecules. 549–561.

Jayarambabu, N., Kumari, B.S., Rao, K.V. and Prabhu, Y.T. 2015. Beneficial role of zinc oxide nanoparticles on green crop production. IJMART. 10: 273–282.

Jha, S. and Pudake, R.N. 2016. Molecular mechanism of plant–nanoparticle interactions. pp. 155–181. *In*: Kole, C., Kumar, D. and Khodakovskaya, M. (eds). Plant Nanotechnology: Principles and Practices. Springer, Cham. https://doi.org/10.1007/978-3-319-42154-4_7.

Jhanzab, H.M., Razzaq, A., Bibi, Y., Yasmeen, F., Yamaguchi, H., Hitachi, K., Tsuchida, K., et al. 2019. Proteomic analysis of the effect of inorganic and organic chemicals on silver nanoparticles in wheat. Int. J. Mol. Sci. 20: 4.

Kamat, J.P., Devasagayam, T.P., Priyadarsini, K.I. and Mohan, H. 2000. Reactive oxygen species mediated membrane damage induced by fullerene derivatives and its possible biological implications. Toxicology. 155: 55–61.

Karakoti, A., Singh, S., Dowding, J.M., Seal, S. and Self, W.T. 2010. Redox-active radical scavenging nanomaterials. Chem. Soc. Rev. 39(11): 4422–4432.

Karlsson, H.L., Gustafsson, J., Cronholm, P. and Möller, L. 2009. Size-dependent toxicity of metal oxide particles—a comparison between nano- and micrometer size. Toxicol. Lett. 188: 112–118.

Karuppanapandian, T., Moon, J.C., Kim, C., Manoharan, K. and Kim, W. 2011. Reactive oxygen species in plants: their generation, signal transduction, and scavenging mechanisms. Aust. J. Crop Sci. 5(6): 709–725.

Kaveh, R., Li, Y.S., Ranjbar, S., Tehrani, R., Brueck, C.L. and Van Aken, B. 2013. Changes in *Arabidopsis thaliana* gene expression in response to silver nanoparticles and silver ions. Environ. Sci. Technol. 47: 10637–10644.

Khan, Z. and Upadhyaya, H. 2019. Impact of nanoparticles on abiotic stress responses in plants: an overview. Nanomaterials in Plants, Algae and Microorganisms. 305–322.

Khodakovskaya, M., Dervishi, E., Mahmood, M., Xu, Y., Li, Z., Watanabe, F. and Biris, A.S. 2009. Carbon nanotubes are able to penetrate plant seed coat and dramatically affect seed germination and plant growth. ACS Nano. 3(10): 3221–3227.

Khodakovskaya, M.V., de Silva, K., Nedosekin, D.A., Dervishi, E., Biris, A.S., Shashkov, E.V., Galanzha, E.I., et al. 2011. Complex genetic, photothermal, and photoacoustic analysis of nanoparticle-plant interactions. Proc. Natl. Acad. Sci. U.S.A. 108(3): 1028–1033.

Klaine, S.J., Alvarez, P.J.J., Batley, G.E., Fernandes, T.F., Handy, R.D., Lyon, D.Y., Mahendra, S., et al. 2008. Nanomaterials in the environ-ment: behavior, fate, bioavailability, and effects. Environ. Toxicol. Chem. 27(9): 1825–1851.

Koce, J.D. 2017. Effects of exposure to nano and bulk sized TiO_2 and CuO in *Lemna minor*. Plant Physiol. Biochem. 119: 43–49.

Kouhi, S.M.M., Lahouti, M., Ganjeali, A. and Entezar, M.H. 2014. Comparative phytotoxicity of ZnO nanoparticles, ZnO microparticles, and Zn^{2+} on rapeseed (*Brassica napus* L.): investigating a wide range of concentrations. Toxico. Environ. Chem. 96: 861–868. doi: 10.1080/02772248.2014.994517.

Kurepa, J., Paunesku, T., Vogt, S., Arora, H., Rabatic, B.M., Lu, J., Wanzer, M.B., et al. 2010. Uptake and distribution of ultrasmallanatase TiO_2 Alizarin red S nanoconjugates in *Arabidopsis thaliana*. Nano Lett. 10(7): 2296–2302.

Kwak, S.Y., Lew, T., Sweeney, C.J., Koman, V.B., Wong, M.H., Bohmert-Tatarev, K., Snell, K.D., et al. 2019. Chloroplast-selective gene delivery and expression in planta using chitosan-complexed single-walled carbon nanotube carriers. Nat. Nanotechnol. 14(5): 447–455.

Laware, S.L. and Raskar, S. 2014. Effect of titanium dioxide nanoparticles on hydrolytic and antioxidant enzymes during seed germination in onion. Int. J. Curr. Microbiol. Appl. Sci. 3(7): 749–760.

Lee, W.M., An, Y.J., Yoon, H. and Kweon, H.S. 2008. Toxicity and bioavailability of copper nanoparticles to the terrestrial plants mung bean (*Phaseolus radiatus*) and wheat (*Triticum aestivum*): plant agar test for water-insoluble nanoparticles. Environ. Toxicol. Chem. 27: 1915–1921. doi: 10.1897/07-481.1.

Lee, C.W., Mahendra, S., Zodrow, K., Li, D., Tsai, Y.C., Braam, J., et al. 2010. Developmental phytotoxicity of metal oxide nanoparticles to *Arabidopsis thaliana*. Environ. Toxicol. Chem. 29: 669–675. doi: 10.1002/etc.58.

Lee, S., Kim, S., Kim, S. and Lee, I. 2013. Assessment of phytotoxicity of ZnO NPs on a medicinal plant, *Fagopyrum esculentum*. Environ. Sci. Pollut. Res. Int. 20: 848–854. doi: 10.1007/s11356-012-1069-8.

Lei, Z., Mingyu, S., Xiao, W., Chao, L., Chunxiang, Q., Liang, C., Hao, H., et al. 2008. Antioxidant stress is promoted by nano-anatase in spinach chloro-plasts under UV-B radiation. Biol. Trace Elem. Res. 121(1): 69–79.

Lin, D. and Xing, B. 2007. Phytotoxicity of nanoparticles: inhibition of seed germination and root growth. Environ. Pollut. 150(2): 243–250.

Lin, D. and Xing, B. 2008. Root uptake and phytotoxicity of ZnO nanoparticles. Environ. Sci. Technol. 42(15): 5580–5585.

Lin, S., Reppert, J., Hu, Q., Hudson, J.S., Reid, M.L., Ratnikova, T.A. and Ke, P.C. 2009. Uptake, translocation, and transmission of carbon nanomaterials in rice plants. Small. 5(10): 1128–1132.

Ma, X., Geiser-Lee, J., Deng, Y. and Kolmakov, A. 2010. Interactions between engineered nanoparticles (ENPs) and plants: phytotoxicity, uptake and accumulation. Sci. Total Environ. 408(16): 3053–3061.

Majumdar, S., Peralta-Videa, J.R., Bandyopadhyay, S., Castillo-Michel, H., Hernandez-Viezcas, J.A., Sahi, S. and Gardea-Torresdey, J.L. 2014. Exposure of cerium oxide nanoparticles to kidney bean shows disturbance in the plant defense mechanisms. J. Hazard Mater. 278: 279–287.

Manzoor, N., Ahmed, T., Noman, M., Shahid, M., Nazir, M.M., Ali, L., Alnusaire, T.S., et al. 2021. Iron oxide nanoparticles ameliorated the cadmium and salinity stresses in wheat plants, facilitating photosynthetic pigments and restricting cadmium uptake. Sci. Total Environ. 769: 145221.

Matsumoto, H. 2000. Cell biology of aluminum toxicity and tolerance in higher plants. Int. Rev. Cytol. 200: 1–46.

Miralles, P., Church, T.L. and Harris, A.T. 2012. Toxicity, uptake, and translocation of engineered nanomaterials in vascular plants. Environ. Sci. Technol. 46(17): 9224–9239.

Mirzajani, F., Askari, H., Hamzelou, H, Schober, Y., Römpp, A., Alireza Ghassempour. and Bernhard S. 2014. Proteomics study of silver nanoparticles toxicity on *Oryza sativa* L. Ecotoxicol. Environ. Saf. 108: 335–339.

Mondal, A., Basu, R., Das, S. and Nandy, P. 2011. Beneficial role of carbon nanotubes on mustard plant growth: an agricultural prospect. J. Nanopart. Res. 13(10): 4519–4528.

Mukherjee, A., Peralta-Videa, J.R., Bandyopadhyay, S., Rico, C.M., Zhao, L. and Gardea-Torresdey, J.L. 2014. Physiological effects of nanoparticulate ZnO in green peas (*Pisumsativum* L.) cultivated in soil. Metallomics. 6(1): 132–138.

Mustafa, G., Sakata, K. and Komatsu, S. 2015. Proteomic analysis of flooded soybean root exposed to aluminumoxide nanoparticles. J. Proteomics. 128: 280–297.

Mustafa, G. and Komatsu, S. 2016. Insights into the response of soybean mitochondrial proteins to various sizes of aluminum oxide nanoparticles under flooding stress. J. Proteome Res. 15(12): 4464–4475.

Mustafa, G., Sakata, K. and Komatsu, S. 2016. Proteomic analysis of soybean root exposed to varying sizes of silver nanoparticles under flooding stress. J. Proteomics. 148: 113–125.

Mustafa, G., Hasan, M., Yamaguchi, H., Hitachi, K., Tsuchida, K. and Komatsu, S. 2020. A comparative proteomic analysis of engineered and bio synthesized silver nanopar-ticles on soybean seedlings. J. Proteomics. 224: 103833.

Mustafa, G., Miyagawa, S., Hasan, M. Yamaguchi, H., Hitachi, K., Tsuchida K. and Komatsu, S. 2022. Bio-synthesized nanoflowers and chemically synthesized nanowires zinc-oxide induced changes in the redox and protein folding in soybean seedlings: a proteomic analysis. J. Plant Growth Regul. 42(5): 1–15. https://doi.org/10.1007/s00344-022-10728-9.

Nair, R., Varghese, S.H., Nair, B.G., Maekawa, T., Yoshida, Y. and Kumar, D.S. 2010. Nanoparticulate material delivery to plants. Plant Science. 179(3): 154–163.

Navarro, E., Baun, A., Behra, R., Hartmann, N.B., Filser, J., Miao, A.J., Quigg, A., et al. 2008. Environmental behavior and ecotoxicity of engineered nanoparticles to algae, plants, and fungi. Ecotoxicology (London, England). 17(5): 372–386.

Novrotsky, A. 2003. Energetics of nanoparticles oxides: interplay between surface energy and polymorphism. Genochem. Trans. 4: 34–37.

Paunovic, J., Vucevic, D., Radosavljevic, T., Mandić-Rajčević, S. and Pantic, I. 2020. Iron-based nanoparticles and their potential toxicity: focus on oxidative stress and apoptosis. Chem. Biol. Interact. 316: 108935.

Peharec Štefanić, P., Jarnević, M., Cvjetko, P., Biba, R., Šikić, S., Tkalec, M., Cindrić, M., et al. 2019. Comparative proteomic study of phytotoxic effects of silver nanoparticles and silver ions on tobacco plants. Environ. Sci. Pollut. Res. Int. 26(22): 22529–22550. doi: 10.1007/s11356-019-05552-w.

Pérez-Labrada, F., López-Vargas, E.R., Ortega-Ortiz, H., Cadenas-Pliego, G., Be-navides-Mendoza, A. and Juárez-Maldonado, A. 2019. Responses of tomato plants under saline stress to foliar application of copper nanoparticles. Plants (Basel, Switzerland). 8(6): 151.

Poščić, F., Mattiello, A., Fellet, G., Miceli, F. and Marchiol, L. 2016. Effects of cerium and titanium oxide nanoparticles in soil on the nutrient composition of barley (*Hordeum vulgare* L.) Kernels. Inter. J. Environ. Res. Public Health. 13(6): 577.

Rajput, V.D., Minkina, T., Suskova, S., Mandzhieva, S., Tsitsuashvili, V., Chapligin, V. and Fedorenko, A. 2018. Effect of copper nanoparticles (CuO NPs) on crop plants a mini review. Bio. Nano Sci. 8(1): 36–42.

Raven, J.A., Evans, M.C.W. and Korb, R.E. 1999. The role of trace metals in photosynthetic electron transport in O2-evolving organisms. Photosynth. Res. 60: 111–149.

Riahi-Madvar, A., Rezaee, F. and Jalali, V. 2012. Effects of alumina nanoparticles on morphological properties and antioxidant system of *Triticum aestivum*. Iran J. of Plant Physiol. 3: 595–603.

Rico, C.M., Majumdar, S., Duarte-Gardea, M., Peralta-Videa, J.R. and Gardea-Torresdey, J.L. 2011. Interaction of nanoparticles with edible plants and their possible implications in the food chain. J. Agric. Food Chem. 59(8): 3485–3498.

Rico, C.M., Morales, M.I., McCreary, R., Castillo-Michel, H., Barrios, A.C., Hong, J., Tafoya, A., et al. 2013a. Cerium oxide nanoparticles modify the antioxidative stress enzyme activities and macromolecule composition in rice seedlings. Environ. Sci. Technol. 47(24): 14110–14118.

Rico, C.M., Hong, J., Morales, M.I., Zhao, L., Barrios, A.C., Zhang, J.Y. and Gardea-Torresdey, J.L. 2013b. Effect of cerium oxide nanoparticles on rice: a study involving the antioxidant defense system and in vivo fluorescence imaging. Environ. Sci. Technol. 47(11): 5635–5642.

Rico, C.M., Peralta-Videa, J.R. and Gardea-Torresdey, J.L. 2015. Chemistry, bio-chemistry of nanoparticles, and their role in antioxidant defense system in plants. pp. 1–17. *In*: Siddiqui, M., Al-Whaibi, M. and Mohammad, F. (eds). Nanotechnology and Plant Sciences. Springer, Cham.

Rose, J., Férard, J.F. and Cotelle, S. 2011. Environmental impact of sunscreen nanomaterials: ecotoxicity and genotoxicity of altered TiO_2 nanocomposites on *Vicia faba*. Environ. Pollut. (Barking, Essex: 1987). 159(10): 2515–2522.

Roy, D., Adhikari, S., Adhikari, A., Ghosh, S., Azahar, I., Basuli, D. and Hossain, Z. 2022. Impact of CuO nanoparticles on maize: Comparison with CuO bulk particles with special reference to oxidative stress damages and antioxidant defense status. Chemosphere. 287: 131911.

Salehi, H., Chehregani, A., Lucini, L., Majd, A. and Gholami, M. 2018. Morphological, proteomic and metabolomic insight into the effect of cerium dioxide nanoparticles to *Phaseolus vulgaris* L. under soil or foliar application. Sci. Total Environ. 616: 1540–1551.

Santos, A.R., Miguel, A.S., Tomaz, L., Malhó, R., Maycock, C., Vaz Patto, M.C., Fevereiro, P., et al. 2010. The impact of CdSe/ZnS quantum dots in cells of *Medicago sativa* in suspension culture. J. Nanobiotechnol. 8(1): 24.

Saxena, R., Tomar, R.S. and Kumar, M. 2016. Exploring nanobiotechnology to mitigate abiotic stress in crop plants. J. Pharm. Sci. Res. 8(9): 974.

Shah, V. and Belozerova, I. 2009. Influence of metal nanoparticles on the soil microbial community and germination of lettuce seeds. Water Air Soil Pollut. 9: 143–148.

Sharma, P., Jha, A.B., Dubey, R.S. and Pessarakli, M. 2012. Reactive oxygen species, oxidative damage, and antioxidative defense mechanism in plants under stressful conditions. J. Bot. Article. ID. 217037.

Shaw, A.K. and Hossain, Z. 2013. Impact of nano-CuO stress on rice (*Oryza sativa* L.) seedlings. Chemosphere. 93(6): 906–915.

Shaw, A.K., Ghosh, S., Kalaji, H.M., Bosa, K., Brestic, M., Zivcak, M. and Hossain, Z. 2014. Nano-CuO stress induced modulation of antioxidative defense and photosynthetic performance of syrian barley (*Hordeum vulgare* L.). Environ. Exp. Bot. 102: 37–47.

Shen, C.X., Zhang, Q.F., Li, J., Bi, F.C. and Yao, N. 2010. Induction of programmed cell death in arabidopsis and rice by single wall carbon nanotubes. Am. J. Bot. 97(10): 1602–1609.

Shi, J.Y., Abid, A.D., Kennedy, I.M., Hristova, K.R. and Silk, W.K. 2011. To duckweeds (*Landoltia punctata*), nanoparticulate copper oxide is more inhibitory than the soluble copper in the bulk solution. Environ. Pollut. 159: 1277–1282.

Siddiqui, M.H. and Al-Whaibi, M.H. 2014. Role of nano-SiO_2 in germination of tomato (*Lycopersicum esculentum* seeds Mill.). Saudi J. Biol. Sci. 21(1): 13–17.

Siddiqui, M.H., Al-Whaibi, M.H. and Mohammad, F. 2015. Nanotechnology and Plant Sciences. Springer International Publishing. Switzerland.

Singh, A., Singh, N.B., Hussain, I., Singh, H. and Singh, S.C. 2015a. Plant-nanoparticle interaction: an approach to improve agricultural practices and plant productivity. Int. J. Pharm. Sci. Invent. 4(8): 25–40.

Singh, V.P., Singh, S., Kumar, J. and Prasad, S.M. 2015b. Investigating the roles of ascorbate-glutathione cycle and thiol metabolism in arsenate tolerance in ridged Luffa seedlings. Protoplasma. 252(5): 1217–1229.

Singh, P.K., Dubey, N.K., Pandey, A.C. and Chauhan, D.K. 2017. Nitric oxide alleviates silver nanoparticles (AgNps)-induced phytotoxicity in *Pisum sativum* seedlings. Plant Physiol. Biochem. PPB. 110: 167–177.

Smirnova, E.A., Gusev, A.A., Zaitseva, O.N., Lazareva, E.M., Onishchenko, G.E., Kuznetsova, E.V., Tkachev, A.G., et al. 2011. Multi-walled carbon nanotubes penetrate into plant cells and affect the growth of *Onobrychis arenaria* seedlings. Actanaturae. 3(1): 99–106.

Solanki, P., Bhargava, A., Chhipa, H., Jain, N. and Panwar, J. 2015. Nano-fertilizers and their smart delivery system. pp. 81–101. *In*: Rai, M., Ribeiro, C., Mattoso, L. and Duran, N. (eds). Nanotechnologies in Food and Agriculture. Springer, Cham.

Spielman-Sun, E., Lombi, E., Donner, E., Howard, D., Unrine, J.M. and Lowry, G.V. 2017. Impact of surface charge on cerium oxide nanoparticle uptake and translocation by wheat (*Triticum aestivum*). Environ. Sci. Technol. 51(13): 7361–7368.

Stadler, T., Buteler, M. and Weaver, D.K. 2011. Novel use of nanostructured alumina as an insecticide. Pest Manag. Sci. 66: 577–579.

Suriyaprabha, R., Karunakaran, G., Yuvakkumar, R., Rajendran, V. and Kannan, N. 2012. Silica nanoparticles for increased silica availability in maize (*Zea mays* L.) seeds under hydroponic conditions. Curr. Nanoscience. 8(6): 902–908.

Tan, X.M., Lin, C. and Fugetsu, B. 2009 Studies on toxicity of multi-walled carbon nanotubes on suspension rice cells. Carbon. 47(15): 3479–3487.

Taran, N., Storozhenko, V., Svietlova, N., Batsmanova, L., Shvartau, V. and Kovalenko, M. 2017. Effect of zinc and copper nanoparticles on drought resistance of wheat seedlings. Nanoscale Res. Lett. 12(1): 1–6.

Thwala, M., Musee, N., Sikhwivhilu, L. and Wepener, V. 2013. The oxidative toxicity of Ag and ZnO nanoparticles towards the aquatic plant *Spirodela punctuta* and the role of testing media parameters. Environ. Sci. Proc. Impacts 15: 1830–1843.

Tiwari, M., Krishnamurthy, S., Shukla, D., Kiiskila, J., Jain, A., Datta, R., Sharma, N., et al. 2016. Comparative transcriptome and proteome analysis to reveal the biosynthesis of gold nanoparticles in arabidopsis. Sci. Rep. 6: 21733.

Tripathi, D.K., Singh, S., Singh, S., Srivastava, P.K., Singh, V.P., Singh, S., Prasad, S.M., et al. 2011. To duckweeds (*Landoltia punctata*), nanoparticulate copper oxide is more inhibitory than the soluble copper in the bulk solution. Environ. Pollut. 159: 1277–1282.

Tripathi, D.K., Singh, S., Singh, V.P., Prasad, S.M., Chauhan, D.K. and Dubey, N.K. 2016.. Silicon nanoparticles more efficiently alleviate arsenate toxicity than silicon in maize cultivar and hybrid differing in arsenate tolerance. Front. Environ. Sci. 4: 46.

Tripathi, D.K., Shweta, Singh, S., Singh, S., Pandey, R., Singh, V.P., Sharma, N.C., et al. 2017a. An overview on manufactured nanoparticles in plants: uptake, translocation, accumulation and phytotoxicity. Plant Physiol. Biochem.: PPB. 110: 2–12.

Valko, M., Morris, H. and Cronin, M.T. 2005. Metals, toxicity and oxidative stress. Curr. Med. Chem. 12: 1161–1208.

Van, N.L., Ma, C., Shang, J., Rui, Y., Liu, S. and Xing, B. 2016. Effects of CuO nanoparticles on insecticidal activity and phytotoxicity in conventional and transgenic cotton. Chemosphere. 144: 661–670.

Vannini, C., Domingo, G., Onelli, E., Prinsi, B., Marsoni, M., Espen, L. and Bracale, M. 2013. Morphological and proteomic responses of Erucasativa exposed to silver nanoparticles or silver nitrate. PLoS One. 8(7): e68752.

Vannini, C., Domingo, G., Onelli, E., De Mattia, F., Bruni, I., Marsoni, M. and Bracale, M. 2014. Phytotoxic and genotoxic effects of silver nanoparticles exposure on germinating wheat seedlings. J. plant Physiol. 171(13): 1142–1148.

Venkatachalam, P., Jayaraj, M., Manikandan, R., Geetha, N., Rene, E.R., Sharma, N.C. and Sahi, S.V. 2017. Zinc oxide nanoparticles (ZnONPs) alleviate heavy metal-induced toxicity in *Leucaena leucocephala* seedlings: a physiochemical analysis. Plant Physiol. Biochem. 110: 59–69.

Vernikov, V.M., Arianova, E.A., Gmoshinskii, I.V., Khotimchenko, S.A. and Tutelian, V.A. 2009. Nanotechnology in food production: advances and problems. Vopr. Pitan.78: 4–17.

Vishwakarma, K., Shweta, Upadhyay, N., Singh, J., Liu, S., Singh, V.P., Prasad, S.M., et al. 2017. Differential phytotoxic impact of plant mediated silver nanoparticles (AgNPs) and silver nitrate ($AgNO_3$) on *Brassica* sp. Front. Plant Sci. 8: 1501.

Wang, Z., Xie, X., Zhao, J., Liu, X., Feng, W., White, J.C. and Xing, B. 2012. Xylem-and phloem-based transport of CuO nanoparticles in maize (*Zea mays* L.). Environ. Sci. Technol. 46(8): 4434–4441.

Wang, W., Luo, M., Fu, Y., Wang, S., Efferth, T. and Zu, Y. 2013a. Glycyrrhizic acid nano-particles inhibit LPS-induced inflammatory mediators in 264.7 mouse macrophages compared with unprocessed glycyrrhizic acid. Int. J. Nanomed. 8: 1377.

Wang, P., Menzies, N.W., Lombi, E., McKenna, B.A., Johannessen, B., Glover, C.J., Kappen, P., et al. 2013b. Fate of ZnO nanoparticles in soils and cowpea (*Vigna unguiculata*). Environ. Sci. Technol. 47: 13822–13830. doi: 10.1021/es403466p.

Wang, P., Lombi, E., Zhao, F.J. and Kopittke, P.M. 2016. Nanotechnology: a new opportunity in plant sciences. Trends Plant Sci. 21: 699–712.

Wang, J.W., Grandio, E.G., Newkirk, G.M., Demirer, G.S., Butrus, S., Giraldo, J.P. and Landry, M.P. 2019. Nanoparticle-mediated genetic engineering of plants. Mol. Plant. 12(8): 1037–1040.

Wei, H. and Wang, E. 2013. Nanomaterials with enzyme-like characteristics (nanozymes): next-generation artificial enzymes. Chem. Soc. Rev. 42(14): 6060–6093.

Weir, E., Lawlor, A., Whelan, A. and Regan, F. 2008. The use of nanoparticles in antimi-crobial materials and their characterization. Analyst. 133: 835–845.

Yadav, T., Mungray, A.A. and Mungray, A.K. 2014. Fabricated nanoparticles: current status and potential phytotoxic threats. pp. 83–110. *In*: Whitacre, D. (eds). Reviews of Environmental Contamination and Toxicology volume. Reviews of Environmental Contamination and Toxicology, Vol. 230. Springer, Cham.

Yan, W., Lien, H.I., Koel, B.E., and Zhang, W. 2013. Iron nanoparticles for environmental clean-up: recent development and future outlook. Environ. Sci. Processes Impacts. 15: 63–77.

Yang, L. and Watts, D.J. 2005. Particle surface characteristics may play an important role in phytotoxicity of alumina nanoparticles. Toxicol. Lett. 158: 122–132.

Yasmeen, F., Raja, N.I., Mustafa, G., Sakata, K. and Komatsu, S. 2016. Quantitative proteomic analysis of post-flooding recovery in soybean root exposed to aluminum oxide nanoparticles. J. Proteomics. 143: 136–150.

Yasmeen, F., Raja, N.I., Razzaq, A. and Komatsu, S. 2017a. Proteomic and physiological analyses of wheat seeds exposed to copper and iron nanoparticles. Biochim. Biophys. Acta, Proteins Proteomics. 1865(1): 28–42.

Yasmeen, F., Raja, N.I., Razzaq, A. and Komatsu, S. 2017b. Gel-free/label-free proteomic analysis of wheat shoot in stress tolerant varieties under iron nanoparticles exposure. Biochim. Biophys. Acta, Proteins Proteomics. 1864(11): 1586–1598.

Yasmeen, F., Raja, N.I., Ilyas, N. and Komatsu, S. 2018. Quantitative proteomic analysis of shoot in stress tolerant wheat varieties on copper nanoparticle exposure. Plant Mol. Biol. Rep. 36(2): 326–340.

Yin, L., Colman, B.P., McGill, B.M., Wright, J.P. and Bernhardt, E.S. 2012. Effects of silver nanoparticle exposure on germination and early growth of eleven wetland plants. PLoS One. 7(10): e47674.

Yruela, I. 2005. Copper in plants. Braz. J. Plant Physiol. 17(1): 145–156.

Yruela, I. 2009. Copper in plants: acquisition, transport and interactions. Funct. Plant Biol. 36: 409–430.

Zhang, P., Ma, Y., Zhang, Z., He, X., Zhang, J., Guo, Z. and Chai, Z. 2012. Biotrans-formation of ceria nanoparticles in cucumber plants. ACS Nano. 6(11): 9943–9950.

Zhao, L., Peng, B., Hernandez-Viezcas, J.A., Rico, C., Sun, Y., Peralta-Videa, J.R., Tang, X., et al. 2012. Stress response and tolerance of *Zea mays* to CeO_2 nanoparticles: cross talk among H_2O_2, heat shock protein, and lipid peroxidation. ACS Nano. 6: 9615–9622.

Zhao, L., Sun, Y., Hernandez-Viezcas, J.A., Servin, A.D., Hong, J., Niu, G., et al. 2013. Influence of CeO_2 and ZnO nanoparticles on cucumber physiological markers and bioaccumulation of Ce and Zn: a life cycle study. J. Agric. Food Chem. 61: 11945–11951.

Zhao, Z., Hu, L., Chen, Q., Dai, H., Meng, X., Yin, Q. and Liang, T. 2022. iTRAQ-based comparative proteomic analysis provides insights into tobacco callus response to carbon nanoparticles. Plant. Mol. Biol. Rep. 40: 556–565. https://doi.org/10.1007/s11105-022-01337-7.

Zhu, H., Han, J., Xiao, J.Q. and Jin, Y. 2008. Uptake, translocation, and accumulation of manufactured iron oxide nanoparticles by pumpkin plants. J. Environ. Monit. 10(6): 713–717.

Chapter 19

Applications of Nanomaterials for Environment Management: Toxicity, Safety, and Regulations

Patrycja Golińska*[1], Magdalena Wypij[1] and Joanna Trzcińska-Wencel[1]

[1]Department of Microbiology, Nicolaus Copernicus University, Toruń, Poland

Mahendra Rai[1,2]

[2]Nanobiotechnology Laboratory, Department of Biotechnology, Sant Gadge Baba Amravati University, Amravati, Maharashtra, India

19.1 INTRODUCTION

Nanotechnology is recognized as applied technology in various areas of life sciences, industry, and medicine (Taran et al. 2021, Malik et al. 2023) which focuses on the design, synthesis, characterization, and application of materials and devices on the nanoscale (Chen et al. 2016). As nanomaterials, we consider incidental, natural, and also fabricated materials in which one or more external dimensions are in the size range of 1–100 nm (European Commission 2016). To date, variable nanomaterials, both organic (polymeric nanoparticles, liposomes, micelles) and inorganic (metal and non-metal nanoparticles, carbon nanomaterials) can be used for environment management including health improvement, pathogens monitoring, bioremediation, and delivery of bioactive substances (Chen et al. 2016).

However, the widespread use of nanomaterials (NMs), not only for environment management but also in various industries, such as pharmaceuticals, chemicals, food, forestry, and agriculture, results in the inadvertent or intentional release of nanomaterials into the environment (Loureiro et al. 2018, Taran et al. 2021, Avila-Quezada et al. 2022).

For instance, to improve agricultural production a variety of nanomaterials have found applications as nano-fertilizers, nanopesticides, nanofungicides, nanosensors, carriers of growth stimulators or many others (Zhao et al. 2020, Avila-Quezada et al. 2022). Most of the nanomaterials will be at some stage get released into the environment, and there is a high risk

*Corresponding Author: golinska@umk.pl

of their dispersibility and acting as pollutants. Nanomaterials released in the environment can be affected by both their own physico-chemical properties (shape, size, surface modification, elemental composition or crystalline structure), but also the exposure conditions that are specific to the environment (Ahlbom et al. 2009, Lead et al. 2018, Consiglia Esposito 2021). Therefore, the physico-chemical characterization of nanomaterials is key to understanding the behavior of nanomaterials in the environment (Consiglia Esposito 2021). The mobility of nanomaterials is influenced by their stability. Destabilization of nanomaterials very often leads to aggregation and subsequent deposition in the environment (Gimbert et al. 2007, Rana and Kalaichelvan 2013). In case of effect of exposure conditions of the environment on nanomaterials behavior, for example, complex structure and high biodiversity of soil affect the ability of NMs to aggregate, release toxic ions, attach to and interact with molecules, compounds or surfaces (Lead et al. 2018). The behavior and fate of nanoparticles in the air are determined by three factors, namely the duration of NPs in the air, the distance travelled in the air, and the interaction with other particles or biomacromolecules (Aitken et al. 2009). The nanoparticles suspended in the air may uncontrollably endanger human health as well as aquatic and terrestrial organisms, including microorganisms, as they spread over long distances (Gupta and Xie 2018).

It is claimed that many analyzed nanomaterials show adverse effects on the environment, aquatic and soil organisms, human beings, flora, and fauna. Consequently, any methods for assessment of the fate of nanomaterials in the environment and their toxicological properties to ecologically relevant organisms are essential to assess their risk (Hegde 2015). Unfortunately, efficient analytical methods for monitoring nanomaterials in the environment are still lacking.

The assessment of the impact of nanoparticles as a pollutant entering the environment is influenced by several aspects such as ecotoxicity, movement from place to place and modifications (Joner et al. 2008, Rana and Kalaichelvan 2013).

The environmental risk assessment addresses changes in environment parameters, including pH, the bioavailability of nutrients, micro- and macroelements, natural organic matter, and effects on the diversity of the microbiome (Ben-Moshe et al. 2013). Therefore, more and more attention is being paid to the risks that are associated with the application of nanomaterials and are a matter of concern for environmental safety (Laux et al. 2018).

In order to properly care for the environment and human health, when using various nanomaterials, appropriate strategies such as prevention, care, and treatment should be implemented (Rezaei et al. 2015, Kaur et al. 2019).

It seems to be most important to inform the public about the amount of nanomaterials produced and placed on the market in order to assess the possible exposure of the environment to this risk. Unfortunately, the provision of such information is limited and negligible, and it can even be emphasized that it is impossible. There are some regulations that define the risk management of nanomaterials, nanoobjects, aggregates, and agglomerates (NOAA) e.g., international standard ISO 31000.

This chapter is focused on application of variable organic and inorganic nanomaterials in the environment for its management, namely environment remediation, wastewater treatment, and sustainable delivery of bioactive compounds. Aspects of toxicity, safety, and regulations of the applications of nanomaterials in aquatic and terrestrial environments are also comprehensively discussed.

19.2 APPLICATION OF NANOMATERIALS FOR ENVIRONMENT MANAGEMENT

Environmental management is mainly associated with minimizing the negative factors affecting the environment (Kazancoglu et al. 2020). Therefore, an increasing number of researchers have focused on the application of nanotechnology in removing pollutants, such as heavy metals, organic and inorganic pollutants, and also pathogens from contaminated surface waters, groundwater, sediments, soil, and air (Khin et al. 2012) (Figure 19.1). The nanomaterials such as nanotubes, nanocomposites, nanofilters and also nanoparticles are used to build systems and significant applications in environmental issues (Fang and Guo 2018).

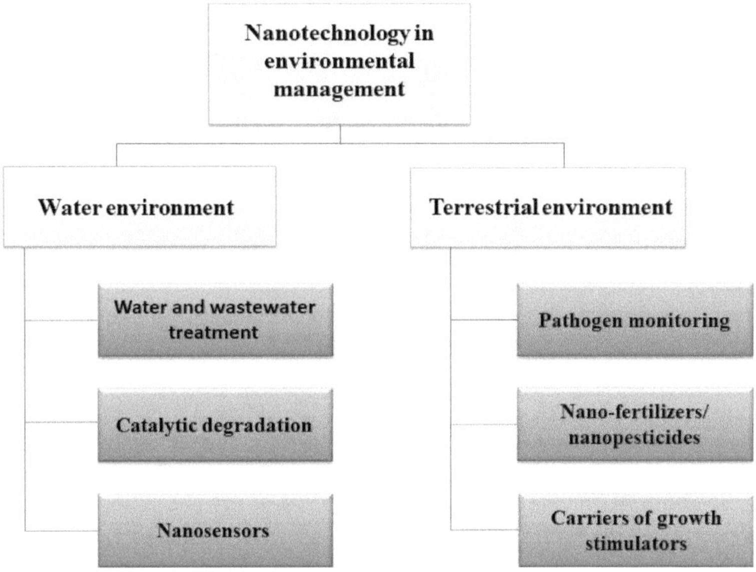

Figure 19.1 Environmental management by using nanotechnology.

19.3 HEAVY METAL POLLUTION

Heavy metal pollution is one of the most important areas of environmental study. Therefore, the presence and concentration of heavy metals such as Cd, As, Fe, Cr, Zn, Cu, Mn, Pb and Ni are constantly and tightly controlled in the soil, surface, and groundwater around the world (Kaonga et al. 2017). Over decades, many physical, chemical, and biological methods have been developed to deal with heavy metal contamination. Adsorption turns out to be the most proven and efficient method, and good adsorbents include activated carbon, carbon film, biopolymers, clay materials, and nanomaterials (Zou et al. 2016). Nanomaterials in environmental remediation arouse great interest because they show good chemical reactivity, active atomicity, large adsorption surface, and low-temperature modification (Yu et al. 2021). Nanomaterials are used for adsorption of organic ions as well as metals and in the remediation of water pollutants (Saleem and Zaidi 2020). Moreover, nanophotocatalysts increase the oxidation capacity of compounds, which contributes to better degradation of pollutants in the aquatic environment (Gómez-Pastora et al. 2017). Nanomaterials such as carbon-based nanomaterials and inorganic nanomaterials are widely used in environmental remediation (Stone et al. 2010), especially nano zero-valent iron, carbon nanotubes, and titanium dioxide nanoparticles (Yu et al. 2021). Moreover, sensors have been developed that are easily able to detect heavy metal ions before their concentrations reach dangerous levels (Kharwar and Singh 2021).

A brief description of nanomaterials and nanosensors and their application for heavy metal pollution is presented below.

Nano Zero-Valent Iron is an effective and efficient adsorbent, reducing agent, and catalyst for heavy metal ions, pharmaceuticals or organic compounds. It also demonstrates the ability of surface sorption and precipitation of various impurities (Cao et al. 2018) and works effectively as an adsorbent to precipitate cadmium from contaminated water (Boparai et al. 2011).

Carbon Nanotubes are characterized by a large specific surface area, unique morphology and high reactivity. Therefore, they are used to remove both organic and inorganic pollutants (Lv et al. 2011). Carbon nanotubes could be split into single-wall CNTs (SWCNTs) and multi-wall CNTs (MWCNTs). The latter are excellent at removing Zn(II) from wastewater (Mubarak et al. 2015, Yu et al. 2021).

Titanium Dioxide Nanoparticles are easily synthesized and used for photocatalysis, environmental remediation, and the assessment of the toxicity of heavy metals in water, soil, and sediments (Taghipour et al. 2019). These nanoparticles were found to be efficient in the remediation of carbonate, and iron-manganese oxide of As and Pb in the sediments (Fan et al. 2018).

Quantum Dots are very advanced sensors that detect heavy metal contamination in the environment (Wang et al. 2021). The Quantum dots based on carbon obtained from green algae waste are used to detect Fe (III) in effluents (Liu et al. 2020) while zero-dimensional graphene dots exhibit interesting physico-chemical properties and are used as nanodetectors for the heavy metal ions (Zhang et al. 2018).

Nanocomposites with a core-shell structure of gold-silica show strong binding of surface plasmons, therefore they are able to detect trace amounts of heavy metals in drinking water (Meng et al. 2018).

19.4 WATER AND WASTEWATER TREATMENT

Nowadays, an important application of nanotechnology in the environment is the use of nanofilters for water and wastewater treatment (Mathew et al. 2019). The application of various polymer membranes in nanofilters allows purifying water from bacteria, viruses, pesticides, organic pollutants and calcium and magnesium salts (Mathew et al. 2019). These membranes include ceramic membranes coated with catalytic and zeolite nanoparticles, organic-inorganic hybrid nanocomposite membranes, and biomimetic membranes. The latter include biopolymer membranes containing protein molecules, carbon nanotube membranes, and block copolymer membranes with equivalent porosity (Beni and Jabbari 2022). Polymer membranes are currently the most used materials in water treatment as they exhibit interesting mechanical, physical and chemical properties (Boffa and Marino 2020). Nanomaterial-based membranes exhibit antifouling, antibacterial, and photodegradable properties (Ying et al. 2017). Membranes made of inorganic materials (zeolites, ceramics, glass, and metals) deserve special attention, as they are effective during water filtration (Boffa and Marino 2020). Purification of drinking water is crucial for the safety of consumers. For this purpose, block copolymer membranes (BCP) (John et al. 2021), chlorine-tolerant TFC membranes (thin-film composite membranes) (Lau et al. 2012), electrospun nanofibrous membranes (ENMs) (Selatile et al. 2018) can be effectively used. Also, the demand for capacitive deionization (CDI) for water treatment is an area of interest to researchers (Saleem and Zaidi 2020).

Nanotechnology has improved the water treatment and purification systems, due to introduction of polymer, nanofiber membranes with the addition of nanoparticles, shedding new light on nanoscale materials (Xu et al. 2022).

19.5 ADSORPTION OF POLLUTANTS

Pollutants such as carbon monoxide (CO), unburned hydrocarbons (HC), nitrogen oxides (NO_x), dioxin are persistent and highly toxic to the environment. Mubeen et al. (2021) reported that carbon nanotubes are more efficient and effective for the degradation of dioxin compared to activated carbon and Al_2O_3–γ. Moreover, carbon nanotubes could be used to adsorb NOx (Dai et al. 2009). Recently, membranes enriched in zeolite or nanosilica have been manufactured and placed on factory chimneys to collect pollutants such as carbon dioxide and reduce the greenhouse effect (Baghery et al. 2019). In turn, manganese oxide-based catalysts with nanoparticles, namely gold NPs have been proposed as materials for removing harmful compounds such as acetaldehyde, toluene and hexane from room air (Hosseini et al. 2016).

19.6 NANOMATERIALS FOR PLANT GROWTH PROMOTION AND PROTECTION

Chemicals used for environmental management might change under physico-chemical conditions, but also easily interact with other compounds. In order to ensure their stability, less susceptibility to adverse environmental factors, controlled release, and enhanced bioavailability of components, they are coated with nanomaterials (Bazana et al. 2019). Among the nanocarriers are those based on proteins (Pérez-Masiá et al. 2015, Fuciños et al. 2017), lipids (Wang et al. 2017a, Suzuki et al. 2021) or polysaccharides (Chin et al. 2014, Ntoutoume et al. 2016, Hadidi et al. 2021).

19.7 NANOFERTILIZERS

Fertilizers are responsible for providing one or more nutrients to plants, promoting their growth, and enhancing biomass production. The application of traditional NPK fertilizers generally results in low nutrient use effectiveness and water eutrophication (Preetha and Balakrishnan 2017). Yoon et al. (2020) reported a multifunctional nanofertilizer based on hydroxyapatite (HA) nanoparticles combined with humic substances (HS) for the co-release of phosphate ions and plant growth stimulants. Applied nonfertilizer increased phosphorus uptake by maize and improved biomass productivity. Moreover, a combination of HA-HS improved the stability of nanoparticles in water, which may increase the potential for use in the form of water sprays to improve crop quality (Yoon et al. 2020). Similarly, application of nanohybrid of HA and urea for slow release of nitrogen, which allowed to reduce two times the dose of fertilizer and obtain higher yields was reported by Kottegoda et al. (2017). Fe-loaded nanoliposomes were successfully used for basil (*Ocimum basilicum* L.) growth promotion. An enhanced effect was observed at 20 mM of Fe-liposomes, compared with $FeSO_4$ and empty liposomes. This indicates the role of liposomes in facilitating Fe delivery to plant cells, thereby further promoting chlorophyll synthesis and biomass growth (Farshchi et al. 2021).

19.8 NANOMATERIALS FOR PLANT PROTECTION

The critical challenges in delivering substances with plant-protective properties to the environment are related to their stability in transport and the release of compounds in a time-appropriate manner, in a safe and reproducible way, and usually at a specific destination (Vega-Vásquez et al. 2020). Pesticides are defined as the chemical substances (natural/engineered, organic/inorganic) applied to eliminate or suppress the reproduction of pests (weeds, insects, nematodes, microbial pathogens, etc.). However, the beneficial effects of pesticide use in crop production, carry the risk of harming the environment and biodiversity, and impacting non-target species (Kumar and Kumar 2019). The use of nanomaterials to formulate pesticides may improve their properties and behavior, such as stability, mobility, and targeted delivery. The higher efficacy, extended duration of action, and reduced dosages affect safety, cost-effectiveness, and reduction of toxic residues in the environment (Dangi and Verma 2021). Recently, antifungal, antibacterial, and antinematode effects of nanoparticles were reported for Ag, Cu, Zn, and Mg nanoparticles, as presented in Table 19.1.

Table 19.1 Nanomaterials for plant pathogen management

Nanomaterials	Activity of NMs	Targeted pathogen	References
Silver-based			
AgNPs	antibacterial	*Pectobacterium carotovorum,*	Pineda et al. 2021
AgNPs	antibacterial	*Ralstonia solanacearum*	Khan et al. 2021
AgNPs	antibacterial	*Pectobacterium Carotovorum* subsp. *carotovorum*	Abdelghany et al. 2022
AgNPs	antifungal	*Fusarium oxysporum f.* sp. *radices-lycopersici*	Akpinar et al. 2021
AgNPs	antifungal	*Fusarium* spp.	Dawoud et al. 2021
AgNPs	antifungal	*Pestalotiopsis mangiferae*	Jenish et al. 2022
AgNPs	antinematode	*Meloidogyne incognita*	Khan et al. 2021
Copper-based			
CuONPs	antifungal	*Bipolaris maydis*	Huang et al. 2021
CuONPs	antibacterial	*Xanthomonas oryzae*	Singh et al. 2021
CuSNPs	antibacterial	*Pectobacterium carotovorum* subsp. *carotovorum*	Yan et al. 2021b
Cu-based nanocomposites	antibacterial	*Xanthomonas euvesicatoria*	Fan et al. 2021
CuNP - alginate nanogel	antibacterial	*Pseudomonas syringae* pv. *tabaci*	Ma et al. 2022
Cu-NPs	antifungal	*Botrytis cinerea, Sclerotinia sclerotiorum*	Sadek et al. 2022
CuONPs-starch nanocomposite	antifungal	*Penicillium expansum, Fusarium oxysporum, Rhizoctonia solani,*	Hasanin et al. 2021
Magnesium-based			
Chitosan-MgNPs nanocomposite	antibacterial	*Acidovorax oryzae*	Ahmed et al. 2021a
MgNPs	antifungal	*Fusarium oxysporum*	Pathania et al. 2022
MgNPs	antifungal	*Fusarium solani, F. oxysporum*	Abdallah et al. 2022
Chitosan-MgNPs nanocomposite	antifungal	*Rhizoctonia solani*	Ahmed et al. 2021a
Zinc-based			
ZnONPs	antibacterial	*Pectobacterium carotovorum* subsp. *carotovorum*	Abdelghany et al. 2022
ZnONPs	antibacterial	*Pectobacterium betavasculorum,*	Khan and Siddiqui 2021
ZnONPs	antifungal	*Rhizoctonia solani*	Khan and Siddiqui 2021
ZnONPs	antifungal	*Alternaria alternata CGJM3078, A. alternata CGJM3006, Fusarium verticilliodes CGJM3823.*	Akpomie et al. 2021
PEGylated ZnNPs	antifungal	*Fusarium solani, F. sporotrichioides, F. oxysporum, Rhizoctonia solani, Macrophomina phaseolina.*	Ammar et al. 2021
ZnONPs	antinematode	*Meloidogyne incognita*	Khan and Siddiqui 2021
Other ones			
Bi_2O_3/TiO_2@ reduced graphene oxide (rGO)	antibacterial	*Pseudomonas syringae* pv. *tomato* DC3000	Yu et al. 2021
Cr_2O_3NPs,	antibacterial	*Pectobacterium carotovorum* subsp. *carotovorum*	Abdelghany et al. 2022
SeNPs	antifungal	*Rhizoctonia solani*	Hashem et al. 2021
sulfur NPs (S-NPs)	antifungal	*Botrytis cinerea, Sclerotinia sclerotiorum*	Sadek et al. 2022
Fe-chitosan nanocomposite	antibacterial	*Xanthomonas oryzae* pv. *oryzae*	Ahmed et al. 2022

The majority of research interest is paid to antimicrobial activity of silver nanoparticles (AgNPs) that has been showed against numerous bacterial and fungal strains (Huq et al. 2022). Out of diverse types of AgNPs, these biologically synthesized, and prepared from plants, algae, bacteria or fungi were found to be efficient agents against phytopathogens (Mustapha et al. 2022). For instance, plant growth-promoting bacteria (PGPR), *Burkholderia* sp. 15B strain, was used to synthesize AgNPs with antifungal activity against phytopathogenic fungi, namely *Fusarium oxysporum, Aspergillus niger, A. fumigatus, Pythium* sp., and *Rosellinia* sp. (Kumar et al. 2022). Derakhsh Ahmadi et al. (2021) used nanocomposite with silver nanoparticles coated with polyethylene glycol (PEG) and chitosan (Ch) to improve chlorpyrifos (CPS) action against termites. Slow release of insecticides was observed for each tested dose (5, 10, 50, 100 ppm) after 72 h and allowed to kill new generations of termites. In turn, Se-NPs biosynthesized from *Bacillus megaterium* showed antifungal activity against the pathogen of *Vicia faba,* which was *Rhizoctonia solani.* Experiments *in vitro* and *in vivo* confirmed inhibitory effects on fungal growth and promoting effect on plant development, metabolic parameters, and defense systems (Hashem et al. 2021). Interestingly, Khan et al. (2021a) found that foliar spraying of zinc oxide nanoparticles (ZnONPs) on beetroot co-infected with nematode of *Meloidogyne incognita, Pectobacterium betavasculorum,* and *Rhizoctonia solani*) reduced nematode population, root galling, and symptoms of soft and dry root rot. Moreover, chitosan-based CuNPs at a concentration of 0.2% revealed fungicidal activity against *Rhizoctonia solani* and *Pythium aphanidermatum* without phytotoxic effect on chili, tomato and cowpea seedlings (Vanti et al. 2020). Similarly, studies by Choudhary et al. (2017) proved that Cu-chitosan nanoforms might act as efficient antifungal agents without toxic effects on plants. In addition, studies showed increased growth of infected plants treated with this type of nanoformulation and indicated that the applied dose and gradual release of Cu are crucial to enhancing the plant defense response against biotic stress (Choudhary et al. 2017).

There is a potential for applying nanocomposites for controlled delivery of other pesticides (Meyer et al. 2015, Ihegwuagu et al. 2016, Yan et al. 2021a). Besides, chitosan nanocomposites as delivery systems have been used in combination with other bioactive compounds such as lysozyme (Giacomello et al. 2020) or hormones (Rather et al. 2013). In addition, chitosan nanoencapsulation enhanced insecticidal and antifungal activities of essential oils (Ibrahim et al. 2022, Mohammed and Attia 2022, Roshan et al. 2022, Upadhyay et al. 2022), which might be used for forest and crop protection (Bahmani and Schmidt 2018, Alonso-Gato et al. 2021).

19.9 OTHER NANOMATERIALS IN ENVIRONMENTAL MANAGEMENT

Nanomaterials can be used for the control of plant pathogens in aquacultures, as exemplified in studies by Ali et al. (2021). Authors reported application of chitosan nanoparticles loaded with oxytetracycline hydrochloride which significantly increased the survival rate of carps infected with *Flavobacterium columnare* compared to antibiotic used alone. Analysis of protein levels (total, albumin, globulin) suggested an improvement of innate immunity in fish after treatment with chitosan nanocomposites conjugated with oxytetracycline (Ali et al. 2021). Chitosan nanocomposite has been also used for vitamin E supplementation in Nile tilapia (*Oreochromis niloticus*), improving their growth, intestinal morphology, and immune status. The results indicated that vitamin E supplementation, especially using nanocomposites, has the potential to alleviate stressful conditions created by raising fish in high-density cultures (Ahmed et al. 2021b).

19.10 TOXICITY OF NANOMATERIALS IN THE ENVIRONMENTS

Over the years, a huge number of nanomaterials have been released into the environment (water, soil, and air) due to the development of nanotechnology (Rana and Kalaichelvan 2013). When assessing the toxicity of nanomaterials, both short-term and long-term tests are used in practice. However, determination of the exposure concentration of nanomaterials is extremely difficult as they aggregate very quickly (Handy et al. 2012). It is important to analyze nanomaterials in their pure form, but also to control and study their fate in the environment (Wiesner et al. 2009, Nowack et al. 2012). An engineered nanoparticle based on carbon might reduce the freely dissolved concentration or bioavailability of organic toxicants and affect the biota (Wiesner et al. 2009, Nowack et al. 2012). While inorganic nanoparticles could release ions, and show various chemical and biological transformations (Farre et al. 2009). Nanomaterials as Ag, Cu, ZnO, and Fe_2O_3 might dissolve very fast, but others as TiO_2, SiO_2, carbon nanotubes, and graphene are much more durable (Valsami-Jones and Lynch 2015). Zhao et al. (2021) studied environmental concentrations of nanomaterials such as TiO_2, Ag, CuO and graphene. The research has shown the high concentration of these nanoparticles in sewage sludges, wastewater effluents, and sediments but significantly low in soil and surface water. Also, authors assessed ecotoxicity of the above-mentioned materials. In surface water and soil, the ecotoxicity was low but their high toxicity in the wastewater and sludge was of concern.

After reaching the natural environment, nanomaterials can have a direct effect on organisms. Thus, fungi, bacteria, algae, protozoa, and also many other organisms from aquatic environments can be used in the toxicity evaluation of different nanomaterials (Joner et al. 2008). The toxicity of nanomaterials to the biota depends on the type of base material, size, substituents, and also coatings (Hallock et al. 2009) and includes mainly disturbances in membrane transport and strong adsorption to the cell surface (Gupta and Xie 2018). It can be monitored by assessment of biological effects such as metabolism, excretion, and adduction to biological molecules, and environmental effects like temperature, pH, presence of microbes, salinity, ionic strength on nanomaterials exposure (Scown et al. 2010, Roberto and Christofoletti 2019). Such assays also allow to learn about the mechanism of action on living organisms, distribution in organisms and interaction of nanomaterials with pollutants. It was reported that ultrafine particles had greater toxicity than fine particulates (Hallock et al. 2009). The toxicity of nanoparticles is related to the number of functional groups on their surface (Warheit et al. 2007a, b, Warheit 2008).

There is no doubt that there is a lack of consistency in the risk assessment of nanomaterials in the environment (OECD 2012). There is a need to develop standardized test methods to assess the toxicity and safety of nanomaterials. Hence, multiple government organizations are working to fund and assemble toxicology information on nanomaterials (Hallock et al. 2009). The time-related effects of chronic dose-response toxicity are crucial to improve the safety approach of nanotechnological development (Silva et al. 2019).

19.11 TOXICITY IN THE AQUATIC ENVIRONMENT

The aquatic ecosystem is most exposed to the dangers of nanomaterials because this environment is where they usually end up (Griffitt et al. 2008). In the aquatic environment, nanomaterials can easily interact with other particles, colloids, pollutants or solids (Wiesner et al. 2009, Nowack et al. 2012). In addition to the benefits of nanomaterials, there is also a disadvantageous aspect of their use. The interactions between nanomaterials (obtained from different sources) and aquatic ecosystem might affect their toxicity (Sahu and Hayes 2017). Therefore, many studies have been conducted to evaluate the toxicity of nanomaterials in aquatic environments (Hansen et al. 2007, Lu et al. 2016, Zhao et al. 2021). Wong and coauthors (2020) evaluated the toxicity of sunscreen (contained nZnO) on copepod *Tigriopus japonicus*. The median lethal concentration (LC_{50}) was in the range of 22.4–230 mg L^{-1}. Moreover, Gao et al. (2015) investigated the effect of socks PR–nAg-containing on zebrafish (*Danio rerio*) embryos. The embryos were exposed to the undiluted effluent directly from the socks and the centrifuged effluent. Mortality was recorded after

24 h with LC_{50} values at 0.4 and 0.26 mg L^{-1} for the undiluted effluent and the centrifuged effluent, respectively. Bai et al. (2010) studied effect of ZnONPs to zebrafish embryos and showed that nano-ZnO at 50 and 100 mg L^{-1} killed the zebrafish embryos while concentrations at 1–25 mg L^{-1} caused body malformation. Similarly, Syafiuddin et al. (2019) and Kanwal and coauthors (2019) studied exposition of 25 mg L^{-1} of AgNPs and nickel nanoparticles (NiNPs) on freshwater fish *Labeo rohita* for 21 days and showed that these nanoparticles did not affect fish survival. Hardman and coauthors (2006) informed that some quantum dots have a strong cytotoxic effect but data on their ecotoxicity are lacking. The mechanism of interaction of nanomaterials on cellular organelles is poorly understood to date while probably oxidative stress is one of the main mechanisms that lead to mitochondrial dysfunction and ultimately lead to cell death (Sahu and Hayes 2017).

Moreover, Muller and Nowack (2008) presented the first environmental research related to the risk assessment of nanoparticles and calculated the PNEC-values of AgNPs, TiO_2 and CNTs in water, which were 0.04 mg L^{-1}, <0.001 mg L^{-1} and <0.0001 mg L^{-1}, respectively. Also, the researcher calculated RQs values for these materials and the level of AgNPs and CNTs were low (below 0.001) therefore these materials did not present any risk to aquatic organisms while exposure to TiO_2 caused risks since its RQs ranged from 0.7 to 16.

Unfortunately, to date research on the ecotoxicity of nanomaterials in aquatic environment is still limited. There is a need for a broader perspective of research related to the absorption, distribution and localization of nanomaterials in aquatic organisms (Valavanidis and Vlachogianni 2010). Therefore, it should be emphasized that the assessment of the risk of nanomaterials in the environment and their impact on living creatures is a huge challenge (Sahu and Hayes 2017). The toxicity of nanomaterials depends on many factors including their solubility and other physico-chemical properties (Krysanov et al. 2010). Also, authors emphasize that the literature data on the toxicity of nanoparticles are inconclusive and there is no data on the chronic effects of nanomaterials (Krysanov et al. 2010).

19.12 TOXICITY OF NANOMATERIALS IN SOIL

Some studies indicated that in the number of nanomaterials, the metal oxide nanoparticles (MONPs) reveal the highest toxicity to the soil by affecting its fertility as well as microorganisms, invertebrates and plants (Frenk et al. 2013, Xu et al. 2015, Simonin et al. 2018). Toxic effects can also be attributed to the chemical composition of these particles. The harmful effect of inorganic, organic and carbon-based nanomaterials on soil organisms are shown in Table 19.2. For example, the uptake of Cu and Zn by plant roots after exposure to metal oxide nanoparticles was relatively high, however, adsorption and accumulation of Fe_2O_3NPs were limited (Van et al. 2016, Chen et al. 2018, Rui et al. 2018, Wei et al. 2021). Nevertheless, Wei et al. (2021) observed varying effects of ZnO, CuO and γ-Fe_2O_3 NPs applied in soil on *Salvia miltiorrhiza* growth. The results showed that depending on the type and concentration of nanoparticles, different biomass growth was observed. Plant biomass increased after treatment with various concentrations of Fe_2O_3NPs, but not with CuONPs. In contrast, a concentration-dependent effect was observed for ZnONPs, where lower doses promoted biomass growth, but higher doses reduced above ground biomass. Moreover, analysis of the rhizosphere microbiome after exposure to MONPs revealed an expansion in the structure and composition of bacterial and fungal communities. It was confirmed by an increase in the relative abundance of specific genera such as plant growth promoting bacteria of *Sphingomonas*, superoxide dismutase producers like *Aminobacter*, or metal tolerant bacteria of *Thiobacillus* and *Metarhizium* (Wei et al. 2021). Nevertheless, polyaniline nanorods (PANRs) decreased soil enzyme activity, namely sucrase, neutral phosphatase, protease, and urease. In addition, PANRs significantly inhibited the growth of earthworms, and caused oxidative stress and lipid damage. However, introduction of earthworms positively affected activity of soil enzymes in dose dependent manner, indicating that they can be used to improve quality and microbial functional diversity of soils (Shu et al. 2022). In turn, carbon-based nanomaterials (CNMs) (e.g., carbon nanotubes, fullerenes, graphene) have been found as sustainable solution for controlling environmental pollution through ability to adsorb or degrade harmful substances, namely heavy metals, antibiotic residues, crude oil etc. Carbon nanotubes (CNTs) were recognized as factors affecting bacterial diversity, associated nutrient cycle and also soil enzymes (Jin et al. 2013, He et al. 2015, Ge et al. 2016). Various effects on N_2 fixation in soil microorganisms were observed, in the presence of graphene nanoplatelets (Wang et al. 2017b) and graphene oxide (Yilihamu et al. 2020) that slightly lowered this process, while multi-walled carbon nanotubes (MWCNTs) significantly reduced it (Wang et al. 2017b). It has been suggested that carbon dots can accelerate electron transfer between active sites in nitrogenase enzyme through electron donating and accepting properties, thus improving the catalytic behavior of the enzyme and N_2 fixation (Wang et al. 2018).

Table 19.2 Toxicity of nanomaterial on soil organisms

Nanomaterials	Effect	References
	Inorganic	
AgNPs	Intestinal damage of earthworm *Pheretima guillemi*	Li et al. 2022
	Inhibition of enzymatic activity of urease, acid phosphatase, arylsulfatase, β-glucosidase	Shin et al. 2012
CuNPs	Reduction of catalase and polyphenol oxidase activity, enhancement of urease activity	Galaktionova et al. 2019
CuONPs	Reduction of microbial activity in the soil	Simonin et al. 2018
	Increased phytotoxicity to wheat (*Triticum aestivum*) after aging in soil	Gao et al. 2018
	Toxic effect on earthworm cells caused by released Cu^{2+} ions	Pacheco et al. 2021
γ-Fe_2O_3NPs	Promotion of *Salvia miltiorrhiza* growth	Wei et al. 2021
nZVI NPs	Enhancement of toxic effect of PCB77 on *Eisenia fetida*	Zhang et al. 2021a
ZnONPs	Reduction of dehydrogenase, acid phosphatase, and β-glucosidase activity	Kim et al. 2012
	Dose-dependent activity, 100 mg/kg promotion of growth, 700 mg/kg reduction of aboveground biomass of *Salvia miltiorrhiza*	Wei et al. 2021
NiONPs	Reduction of β-glucosidase activity	Avila-Arias et al. 2019
Ti_2ONPs	Genotoxic effect on rice, dose-depended reduction of rice biomass	Waani et al. 2021
	Organic	
PAMAM dendrimers,	Impaired photosynthesis in cyanobacterium *Anabaena*	Tamayo-Belda et al. 2019
Polyaniline nanorods	Inhibited earthworm growth	Shu et al. 2022
Carbon-based		
SWCNTs	Decreased soil microbial biomass	Jin et al. 2014
MWCNTs	Inhibited nitrogen removal process in long-term exposure	Yang et al. 2021
	Inhibition of N_2 fixation process	Wang et al. 2017b
Graphene oxide	Reduced enzyme activity of soil by15–50% in short term	Chung et al. 2015
	Inhibition of N_2 fixation process	Yilihamu et al. 2020
Nanodiamonds	Inhibition of photosynthesis in wheat	Chernysheva et al. 2018

19.13 EFFECT OF SOIL PROPERTIES ON THE TOXICITY OF NANOMATERIALS

Research efforts focus on the bioactivity and toxicity of NMs to various types of organisms (bacteria, fungi, animals and plants) (Calder et al. 2012, Bae et al. 2021, Sethulekshmi et al. 2022). However, nanomaterials undergo various types of modifications after being introduced into the soil environment (Dimpka 2018). Xin et al. (2020) analyzed the impact of soil properties on the toxicological behavior of four nanomaterials (AgNPs, TiO_2NPs, MWCNTs, and polysuccinimide (PSI) NPs). The results indicated that lower pH, soil organic matter and clay content suppressed the toxic effects of the nanomaterials on microbial activity and biochemical properties of soil. Only the highest application rates of NMs led to a significant reduction in microbial biomass and disruption of an ecosystem by inhibiting enzymes responsible for the decomposition of substrates and accessibility of assimilation products, along with the deceleration of nutrient cycling (Xin et al. 2020). Another study on nanodiamonds (NDs) showed a relationship between the content of various humic substances in soil and their uptake by wheat. NDs alone adversely affected the photosynthesis parameters of plants while further analysis pointed out that NDs inhibited electron transfer and had a negative impact on the PSII system. Humic acid (HA) reduced accumulation of NDs and enhanced photosynthesis while fulvic acid (FA) slightly improved photosynthesis and had no impact on the level of NDs in seedling roots (Chernysheva et al. 2018).

19.14 SELF-ADAPTATION OF SOIL TO NANOMATERIALS

All nanomaterials show different effects on the soil microbiome, depending on the time of exposure and concentration. When the level of the environmental factor exceeds the relevant threshold, it may generate adverse reactions. The potential toxic effect of zinc oxide nanoparticles used for improvement of crop production was evaluated by Mahmood et al. (2021). They studied the impact of biologically synthesized ZnONPs on calcareous soil. Treatment with ZnONPs (concentration 1mg kg^{-1} of soil) increased the pH and level of C mineralization of soil samples, while microbial biomass was significantly reduced after 24 hours of incubation when compared to control. However, after 9 weeks of incubation, no differences in microbial biomass between the test (ZnO) and control samples were observed. These results suggested that the toxic effect of ZnONPs on microbial biomass was impermanent due to the self-adaptation of soil. Moreover, authors suggested that the transient effect of ZnONPs on bacterial community may be due to the high tendency of nanoparticles to agglomerate. Limited toxicity of the agglomerated nanoparticles was explained as a result of the reduction of the volume ratio and surface area in the agglomerates (Mahmood et al. 2021).

In other studies, by García-Gómez et al. (2018) the impact of ZnONPs on microbiomes of two different soils, acidic and alkaline was compared. Results indicated that only at higher concentrations (1 mg kg^{-1}) ZnONPs had a pronounced negative effect on respiration potential or nitrogen transformation. At lower concentrations of ZnONPs, the activity of dehydrogenase (DH) and alkaline phosphatase (ALP) were slightly inhibited in both types of soil, besides the activity of acidic phosphatase (ACP) which was also decreased in calcareous soil. Moreover, based on enzymatic activity authors suggested self-adaptation of soils microbiome in both types of soil over 180 days (García-Gómez et al. 2018) that could be associated with changes in the properties and reactivity of NPs (Cupi et al. 2015, Mitrano et al. 2015) and/or adaptation to metal by the development of resistance in soil microbes and unrelated to changes in the abundance of microorganisms, but their adaptation (Mitchell et al. 2019). Auger et al. (2019) suggested that Mg-doped ZnONPs induce the synthesis of proteins associated with ROS detoxification and biofilm formation by soil bacteria of *Bacillus subtilis*. Exposition to NPs can lead to adaptation and development of new resistant phenotypes which could adversely affect the environment, and health of plants, animals, and people (Auger et al. 2019). It was suggested that carbon-based nanomaterials (SWCNT and fullerenes) affected adaptation of soil Gram-positive bacteria by changes in the lipid composition of cell membranes (Fang et al. 2007, Jin et al. 2014). Decreased level of saturated fatty acids (FAs) and increased levels of *iso*-branched FAs were observed after treatment with a higher concentration of SWCNT. These changes may be a physiological adaptive response of Gram-positive bacteria to stress triggered by exposure to CNs. The analysis of the above alterations can be used in determining environmental contamination by nanostructures (Jin et al. 2014).

19.15 REGULATIONS AND SAFETY

The sustainable development of nanotechnology requires an assessment of the hazards and risks arising from the introduction of nanomaterials into the environment. Detailed analysis, control, and monitoring of nanomaterials in the environment will allow the implementation of nanotechnological applications in industrial sectors and at the same time minimize the risk (Fajardo et al. 2022).

Based on published reports it can be seen that nanotechnology brings many benefits as well as threats to the environments and human life. However, a number of papers on the advantages of nanotechnology are much bigger than on its dangers and consequences for the environment (Suppan 2013, Suppan 2017, Malakar et al. 2021, Martinez et al. 2021, Wu et al. 2021). Therefore global authorities should consider the large-scale use of nanomaterials and their control and regulations to protect the environment and human health (Martinez et al. 2021). Unfortunately, the regulations for the proper evaluation and control of nanotechnology are not fully established. There is a fine line between overly restrictive regulations that may weaken the further boom of nanotechnology, and too weak ones that may lead to environmental pollution, threats to fauna and flora, and thus the extinction of some species (Wheeler 2020).

With the emergence of nanoproducts in commercial use, an important aspect is to control the production and use of NMs by appropriate regulations, recommendations or guidelines. Different approaches and strategies have been used by various governments and international organizations to properly manage the risks associated with the use of nanomaterials in various sectors (Amenta et al. 2015). In 2007, the European Commission (EC) included engineered nanomaterials (ENM) in the register of chemical compounds called Registration, Evaluation, Authorization and Restriction of Chemicals (REACH). It provides the principles of registration, use and control of chemicals considering their significance for human health and environmental safety (Esposito et al. 2021). The EC assumed that the assessment of their toxicity and safety for the environment should be related to the standards adopted for conventional chemical compounds (Esposito et al. 2021). Annex II, Section 12 (*Ecological information*) of this regulation clearly indicates, that all toxicity data on soil organisms (including microorganisms), as well as information on mobility in soil, should be provided during new substances registration to compile safety data sheets (EC 2006). In 2009, European Parliament's Resolution was imposed to include nanomaterials in regulations, standards on waste, and their emissions to the aquatic environment and air (Vijaya Lakshmi et al. 2011). However, in 2011, an analysis of European legislation, showed that current regulations were too general and revealed the enormous difficulties in monitoring nanoparticles in the environment (Hansen et al. 2011). The organizations such as European Research Council (ERC) and the Danish Consumer Council support the Nanodatabase (Nanodatabase 2020) project that groups nano-products in terms of their toxicity to the environment (Fajardo et al. 2022). Moreover, the European Chemicals Agency (ECHA) is responsible for the technical and administrative aspects of the implementation of chemical substances, including nanomaterials in the European Union (EU). Detailed regulations on the required physico-chemical characterization, depending on

Table 19.3 Directives involved for the regulatory control of nanomaterials (Final Report 2011)

Regulations/Directives	General scope
Waste Framework Directive 2008/98/EC	Waste, recovery, and disposal and key requirements for waste management
Decision 2000/532/EC on the List of Waste	Coverage of nanomaterials under the list of waste
Landfill Directive 1999/31/EC	Technical and operational requirements for the dumping of waste in landfills with the aim of preventing or reducing negative effects on the environment, in particular the pollution of surface water, groundwater, soil, and air
WEEE Directive 2002/96/EC	The prevention of waste electrical and electronic equipment (WEEE)/ Nanomaterials in electrical and electronic equipment
Directive 2002/95/EC on RoHS	The restriction of the use of hazardous substances in electrical and electronic equipment (EEE) with a view to contributing to the protection of human health and the environment
Packaging and Packaging Waste Directive 1994/62/EC	Preventing the production of packaging waste/ Nanomaterials in packaging
Directive 86/278/EEC on the protection of the environment, and in particular of the soil, when sewage sludge is used in agriculture (Sewage Sludge Directive)	Preventing harmful effects on soil, vegetation, animals and people of sewage sludge from wastewater
Water Framework Directive 2000/60/EC	The Water Framework Directive, sets the legal framework for the protection and restoration of clean water across Europe, with the aim of ensuring its long-term sustainable use
Directive 2008/105/EC	The environmental quality standards in the field of water policy
Directive 2000/60/EC introduces in Article 16(2)	Methodology for selecting priority substances on the basis of their significant risk to or via the aquatic environment
Directive 2008/105/EC on Environmental Quality Standards in the Field of Water Pollution	Environmental quality standards (EQS) for priority substances and certain other pollutants as required under Article 16 of the Water Framework Directive
Directive 2006/118/EC on the protection of groundwater against pollution and deterioration	Strategies to prevent and control the pollution of groundwater
Urban Waste Water Directive 91/271/EEC	Regulates the collection, treatment, and discharge of urban wastewater and the treatment and discharge of waste water from certain industrial sectors
Drinking Water Directive 98/83/EC	The quality of water intended for human consumption
Directive 96/82/EC on the control of major-accident hazards involving dangerous substances (Seveso II Directive)	Prevention of major-accident hazards involving dangerous substances
Air Quality Directive 2008/50/EC	The air quality assessment and prevention or reduce harmful effects on human health and the environment
Directive 2006/11/EC	The pollution caused by certain dangerous substances discharged into the aquatic environment
Directive 2006/118/EC	The protection of groundwater against pollution and deterioration
Directive 2000/60/EC	Water Framework
Directive 2008/105/EC	The environmental quality standards in the field of water policy

the form and application of nanomaterials (e.g., determination of redox potential is essential for inorganic NMs) are described in the European Food Safety Authority (EFSA) guidance (EFSA 2021). While according to the National Environmental Policy Act (NEPA), the approval of a drug or biologically active substance requires an environmental assessment.

In turn, the Food and Drug Administration in the USA (FDA) is responsible for the regulations related to nanomaterials in food products, the pharmaceutical industry and cosmetics while Environmental Protection Agency of the USA is for the regulations related to the emission of nanomaterials in water, soil or air (Resnik 2019, FDA 2020). The FDA regulations include a customized approach to specific substances, and their possible environmental impact is analyzed on a case-by-case basis (FDA 2022). The FDA has introduced inorganic metal oxides with the designations such as E171, E551 and E530 for direct use and consumption (EFSA 2021). Moreover, the important regulations that entered into force recently were Guidance document: *Aquatic and sediment ecotoxicity testing* and *Physical-chemical decision framework to inform decisions for risk assessment of manufactured nanomaterials* issued by Organization for Economic Co-operation and Development (OECD) (OECD 2020, OECD 2019).

To combat microbial pathogens, insects or weeds, a lot of chemical substances are widely used and pass directly into the soil. Thus, it is important to recognize the risks and reduce harmful effects on the soil microbiome, animals, and plants (Degrendele et al. 2022, Rasool et al. 2022). European Parliament (EP) and EC established Plant Protection Products (PPP) Regulation 1107/2009 (EC 2009). The document provides rules for the registration, commercial use, and control of plant protection products to ensure human and environment safety. For this purpose, approval criteria for products include effectiveness and any unacceptable toxic effects on plants and environment. If possible, harmful effects on human health, potential contamination of ecosystems and impact on biodiversity in ecosystems should be assessed before authorization and introduction to commercial use (EC 2009). Boxall et al. (2007) proposed a framework of simple algorithms that can be used to estimate the level of ENM released into the environment, including ENMs from sewage sludge on land, through the use of remediation technologies and plant protection products introduced to soil or water. However, the development and application of artificial intelligence and machine learning may be successfully used to simplify environmental risk assessment (Cohen et al. 2013, Singh et al. 2020). These include predictions of further modifications, transformations, transport or uptake by plants (Winkler et al. 2020, Zhang et al. 2021b). Nonetheless, continuous generation and storage of high-quality data is necessary to train and validate models, as well as solutions to incorporate the heterogenic and dynamic nature of the various biological environments, which is essential for designing safer ENMs (Singh et al. 2020).

Selected directives involved in the regulatory control of nanomaterials are presented in Table 19.3

19.16 CONCLUSION

The global application of various nanomaterials has revolutionized all fields of life including agriculture, medicine, environment, textiles, and so on. The applications of nanomaterials in different fields emit the nanomaterials as waste that enters an aquatic ecosystem or is suspended in the air. In this context, it is worth mentioning that physicochemical parameters such as shape, size, surface charge, and agglomeration determine the activity of the nanomaterials. Studies have already demonstrated that nanomaterials present in the environment are toxic to all organisms. However, more thorough studies related to toxicity are desired as the presently available data is not enough to explain the toxicity of nanomaterials. The remediation of the nanomaterials and treatment of the wastewater is mainly required for the sustainable management of the environment. Moreover, for managing nanomaterials filtration, recycling, and incineration methods are used but the studies related to the presence of nanomaterials in recycled materials are still scanty and thus warrant further in-depth studies.

19.17 ACKNOWLEDGEMENT

Financial support for JTW from Nicolaus Copernicus University (No. 38/2021) and National Science Centre, Poland (grant No. 2022/45/N/NZ9/01483) is acknowledged.

References

Abdallah, Y., Hussien, M., Omar, M.O., Elashmony, R., Alkhalifah, D.H.M. and Hozzein, W.N. 2022. Mung bean (*Vigna radiata*) treated with magnesium nanoparticles and its impact on soilborne *Fusarium solani* and *Fusarium oxysporum* in Clay Soil. Plants. 11: 1514.

Abdelghany, W.A., Mohamedin, A.H., Abo-Elyousr, K.A. and Hussein, M.A. 2022. Control of bacterial soft rot disease of potato caused by *Pectobacterium carotovorum* subsp. *carotovorum* using different nanoparticles. Arch. Phytopathol. 1–23.

Ahlbom, A., Bridges, J., De Jong, W., Jung, T., Mattsson, O., Pagès, J.M., Rydzynski, K., et al. 2009. Scientific committee on emerging and newly identified health risks SCENIHR risk assessment of products of nanotechnologies. Nanotech. 2: 15.

Ahmed, T., Noman, M., Luo, J., Muhammad, S., Shahid, M., Ali, M.A. and Li, B. 2021a. Bioengineered chitosan-magnesium nanocomposite: A novel agricultural antimicrobial agent against *Acidovorax oryzae* and *Rhizoctonia solani* for sustainable rice production. Int. J. Biol. Macromol. 168: 834–845.

Ahmed, S.A., Ibrahim, R.E., Farroh, K.Y., Moustafa, A.A., Al-Gabri, N.A., Alkafafy, M. and Amer, S.A. 2021b. Chitosan vitamin E nanocomposite ameliorates the growth, redox, and immune status of Nile tilapia (*Oreochromis niloticus*) reared under different stocking densities. Aquacult. 541: 736804.

Ahmed, T., Noman, M., Jiang, H., Shahid, M., Ma, C., Wu, Z. and Li, B. 2022. Bioengineered chitosan-iron nanocomposite controls bacterial leaf blight disease by modulating plant defense response and nutritional status of rice (*Oryza sativa* L.). Nano Today. 45: 101547.

Aitken, R. 2009. EMERGNANO: a review of completed and near completed environment, health and safety research on nanomaterials and nanotechnology. Report TM/09/01. Defra Project CB0409.

Akpinar, I., Unal, M. and Sar, T. 2021. Potential antifungal effects of silver nanoparticles (AgNPs) of different sizes against phytopathogenic *Fusarium oxysporum* f. sp. *radicis-lycopersici* (FORL) strains. SN App. Sci. 3: 1–9.

Akpomie, K.G., Ghosh, S., Gryzenhout, M. and Conradie, J. 2021. One-pot synthesis of zinc oxide nanoparticles via chemical precipitation for bromophenol blue adsorption and the antifungal activity against filamentous fungi. Sci. Rep. 11: 1–17.

Ali, M.N., Al-Saadi, T.M. and AL-Faragi, J.K. 2021. Effect of chitosan nanoparticles loaded oxytetracycline hydrochloride on health status of common carp (*Cyprinus carpio* L.) infected with columnaris disease. J. Physics: Conf. Series. 1879: 032075

Alonso-Gato, M., Astray, G., Mejuto, J.C. and Simal-Gandara, J. 2021. Essential oils as antimicrobials in crop protection. Antibiot. 10: 34.

Amenta, V., Aschberger, K., Arena, M., Bouwmeester, H., Moniz, F.B., Brandhoff P. and Peters, R.J. 2015. Regulatory aspects of nanotechnology in the agri/feed/food sector in EU and non-EU countries. Regul. Toxicol. Pharmacol.73: 463–476.

Ammar, H.A., Alghazaly, M.S., Assem, Y. and Abou Zeid, A.A. 2021. Bioengineering and optimization of PEGylated zinc nanoparticles by simple green method using *Monascus purpureus*, and their powerful antifungal activity against the most famous plant pathogenic fungi, causing food spoilage. Environ. Nanotechnol. Monit. Manag. 16: 100543.

Auger, S., Henry, C., Péchaux, C., Lejal, N., Zanet, V., Nikolic, M.V. and Vidic, J. 2019. Exploring the impact of Mg-doped ZnO nanoparticles on a model soil microorganism *Bacillus subtilis*. Ecotoxicol. Environ. Saf. 182: 109421

Avila-Arias, H., Nies, L.F., Gray, M.B. and Turco, R.F. 2019. Impacts of molybdenum-, nickel-, and lithium-oxide nanomaterials on soil activity and microbial community structure. Sci. Total Environ. 652: 202–211.

Avila-Quezada, G.D., Ingle, A.P., Golińska, P. and Rai, M. 2022. Strategic applications of nano-fertilizers for sustainable agriculture: benefits and bottlenecks. Nanotechnol. Rev. 11: 2022.

Bae, M., Oh, J.K., Liu, S., Nagabandi, N., Yegin, Y., DeFlorio, W. and Scholar, E. 2021. Nanotoxicity of 2D molybdenum disulfide, MoS_2, nanosheets on beneficial soil bacteria, *Bacillus cereus* and *Pseudomonas aeruginosa*. Nanomaterials. 11: 1453.

Baghery, R., Riahi, S., Abbasi, M. and Mohammadi-Khanaposhtani, M. 2019. Investigation of the CO_2 absorption in pure water and MDEA aqueous solution including amine functionalized multi-wall carbon nano tubes. J. Mol. Liq. 293: 111431.

Bahmani, M. and Schmidt, O. 2018. Plant essential oils for environment-friendly protection of wood objects against fungi. Maderas. Ciencia Technol. 20: 325–332.

Bai, W., Zhang, Z., Tian, W., He, X., Ma, Y., Zhao, Y. and Chai, Z. 2010. Toxicity of zinc oxide nanoparticles to zebrafish embryo: a physicochemical study of toxicity mechanism. J. Nanopart. Res. 12: 1645–1654.

Bazana, M.T., Codevilla, C.F. and de Menezes, C.R. 2019. Nanoencapsulation of bioactive compounds: challenges and perspectives. Current Opin. Food Scien. 26: 47–56.

Beni, A.A. and Jabbari, H. 2022. Nanomaterials for environmental applications. Results Eng. 15: 100467.

Ben-Moshe, T., Frenk, S., Dror, I., Minz, D. and Berkowitz, B. 2013. Effects of metal oxide nanoparticles on soil properties. Chemosph. 90: 640–646.

Boffa, V. and Marino, E. 2020. Inorganic materials for upcoming water purification membranes. pp. 117–140. *In*: Basile A. and Comite A. (eds). Current Trends and Future Developments on (Bio-) Membranes: Membrane Technology for Water and Wastewater Treatment—Advances and Emerging Processes. Elsevier, Cambridge, MA, US.

Boparai, H.K., Meera J.M. and O'Carroll, D.M. 2011. Kinetics and thermodynamics of cadmium ion removal by adsorption onto nano zerovalent iron particles. J. Hazard. Mater. 186: 458–465.

Boxall, A.B., Chaudhry, Q., Sinclair, C., Jones, A., Aitken, R., Jefferson, B. and Watts, C. 2007. Current and future predicted environmental exposure to engineered nanoparticles. Report by Central Science Laboratory. Department of the Environment and Rural Affairs. London. UK.

Calder, A.J., Dimkpa, C.O., McLean, J.E., Britt, D.W., Johnson, W. and Anderson, A.J. 2012. Soil components mitigate the antimicrobial effects of silver nanoparticles towards a beneficial soil bacterium, *Pseudomonas chlororaphis* O6. Sci. Total Environ. 429: 215–222.

Cao, Y., Zhang, S., Zhong, Q., Wang, G., Xu, X., Li, T., Wang, L., et al. 2018. Feasibility of nanoscale zero-valent iron to enhance the removal efficiencies of heavy metals from polluted soils by organic acids. Ecotoxicol. Environ. Saf. 162: 464–473.

Chen, D., Zhu, H., Yang, S., Li, N., Xu, Q., Li, H. and Lu, J. 2016. Micronanocomposites in environmental management. Adv. Mat. 28: 10443–10458.

Chen, J., Dou, R., Yang, Z., You, T., Gao, X. and Wang, L. 2018. Phytotoxicity and bioaccumulation of zinc oxide nanoparticles in rice (*Oryza sativa* L.). Plant Physiol. Biochem. 130: 604–612.

Chernysheva, M.G., Myasnikov, I.Y., Badun, G.A., Matorin, D.N., Gabbasova, D.T., Konstantinov, A.I. and Kulikova, N.A. 2018. Humic substances alter the uptake and toxicity of nanodiamonds in wheat seedlings. J. Soils Sediments. 18: 1335–1346.

Chin, S.F., Mohd Yazid, S.N.A. and Pang, S.C. 2014. Preparation and characterization of starch nanoparticles for controlled release of curcumin. Int. J. Polymer Sci. 4: 340121.

Choudhary, R.C., Kumaraswamy, R.V., Kumari, S., Sharma, S.S., Pal, A., Raliya, R. and Saharan, V. 2017. Cu-chitosan nanoparticle boost defense responses and plant growth in maize (*Zea mays* L.). Sci. Rep. 7: 1–11.

Chung, H., Kim, M.J., Ko, K., Kim, J.H., Kwon, H.A., Hong, I. and Kim, W. 2015. Effects of graphene oxides on soil enzyme activity and microbial biomass. Sci. Total Environ. 514: 307–313.

Cohen, Y., Rallo, R., Liu, R. and Liu, H.H. 2013. *In silico* analysis of nanomaterials hazard and risk. Acc. Chem. Res. 46: 802–812.

Consiglia Esposito, M., Corsi, I., Russo, G.L., Punta, P., Tosti, E. and Gallo, A. 2021. The era of nanomaterials: a safe solution or a risk for marine environmental pollution. Biomol. 11 :441.

Cupi, D., Hartmann, N.B. and Baun, A. 2015. The influence of natural organic matter and aging on suspension stability in guideline toxicity testing of silver, zinc oxide, and titanium dioxide nanoparticles with *Daphnia magna*. Environ. Toxicol. Chem. 34: 497–506.

Dai, J., Giannozzi, P. and Yuan, J. 2009. Surface science adsorption of pairs of NO_x molecules on single-walled carbon nanotubes and formation of NO^+ NO_3 from NO_2. Surf. Sci. 603: 3234–3238.

Dangi, K. and Verma, A.K. 2021. Efficient and eco-friendly smart nano-pesticides: emerging prospects for agriculture. Mat. Today: Proceed. 45: 3819–3824.

Dawoud, T.M., Yassin, M.A., El-Samawaty, A.R.M. and Elgorban, A.M. 2021. Silver nanoparticles synthesized by *Nigrospora oryzae* showed antifungal activity. Saudi J. Biol. Sci. 28: 1847–1852.

Degrendele, C., Klánová, J., Prokeš, R., Příbylová, P., Šenk, P., Šudoma, M. and Fuhrimann, S. 2022. Current use pesticides in soil and air from two agricultural sites in South Africa: Implications for environmental fate and human exposure. Sci. Total Environ. 807: 150455.

Derakhsh Ahmadi, V., Rafiei-Karahroudi, Z., Sanatgar, E., H Alizadeh, B. and Goldasteh, S. 2021. Chlorpyrifos-loaded silver/polyethylene glycol/chitosan nanocomposite: improved termiticidal activity against microcerotermes diversus *Isoptera termitidae*. J. Sci. Islamic Rep. Iran. 32: 43–51.

Dimkpa, C.O. 2018. Soil properties influence the response of terrestrial plants to metallic nanoparticles exposure. Curr. Opin. Environ. Sci. Health. 6: 1–8.

EC. 2006. Regulation No 1907/2006 of the European Parliament and of the Council of 18 december 2006. Available online: https://eur-lex.europa.eu/legal-content/EN/TXT/?uri=CELEX%3A02006R1907-20220501.

EC. 2009. Regulation No 1107/2009 of the European Parliament and of the Council of 21 October 2009. Available online: https://eur-lex.europa.eu/legal-content/EN/TXT/?uri=celex%3A32009R1107.

EFSA. 2021. European Food Safety Authority. Scientific Committee. Guidance on the Risk Assessment of the Application of Nanoscience and Nanotechnologies in the Food and Feed Chain. Available online: http://www.efsa.europa.eu/en/efsajournal/doc/2140.pdf.

Esposito, M.C., Corsi, I., Russo, G.L., Punta, C., Tosti, E. and Gallo, A. 2021. The era of nanomaterials: a safe solution or a risk for marine environmental pollution? Biomolecules. 11: 441.

European Commission. 2016. Recommendation on the definition of a nanomaterial. Available online: http://data.europa.eu/eli/reco/2011/696/oj.

Fajardo, C., Martinez-Rodriguez, G., Blasco, J., Mancera, J.M., Thomas, B. and De Donato, M. 2022. Nanotechnology in aquaculture: applications, perspectives and regulatory challenges. Aquac. Fish. 7: 185–200.

Fan, X., Wang, C., Wang, P., Hu, B. and Wang, X. 2018. TiO_2 nanoparticles in sediments: effect on the bioavailability of heavy metals in the freshwater bivalve *Corbicula fluminea*. J. Hazard. Mater. 342: 41–50.

Fan, Q., Liao, Y.Y., Kunwar, S., Da Silva, S., Young, M., Santra, S. and Paret, M.L. 2021. Antibacterial effect of copper composites against *Xanthomonas euvesicatoria.* Crop Prot.139: 105366.

Fang, J., Lyon, D.Y., Wiesner, M.R., Dong, J. and Alvarez, P.J. 2007. Effect of a fullerene water suspension on bacterial phospholipids and membrane phase behavior. Environ. Sci. Technol. 41: 2636–2642.

Fang, Y. and Guo, Y. 2018. Copper-based non-precious metal heterogeneous catalysts for environmental remediation. J. Catal. 39: 566–582.

Farre, M., Gajda-Schrantz, K., Kantiani, L. and Barcelo, D. 2009. Ecotoxicity and analysis of nanomaterials in the aquatic environment. Anal. Bioanal. Chem. 393: 81– 95.

Farshchi, H.K., Azizi, M., Teymouri, M., Nikpoor, A.R. and Jaafari, M.R. 2021. Synthesis and characterization of nanoliposome containing Fe^{2+} element: a superior nano-fertilizer for ferrous iron delivery to sweet basil. Sci. Horticul. 283: 110110.

FDA. 2020. Food and Drug Administration. Report: Nanotechnology—Over a Decade of Progress and Innovation. Available online: https://www.fda.gov/media/140395/download.

FDA. 2022. Food and Drug Administration. Guidance document. Drug Products, Including Biological Products, that Contain Nanomaterials—Guidance for Industry. Available online: https://www.fda.gov/regulatory-information/search-fda-guidancedocuments/drug-products-including-biological-products-contain-nanomaterials-guidance-industry.

Final Report Contract No 070307/2010/580540/SER/D. 2011. Review of Environmental Legislation for the Regulatory Control of Nanomaterials.

Frenk, S., Ben-Moshe, T., Dror, I., Berkowitz, B. and Minz, D. 2013. Effect of metal oxide nanoparticles on microbial community structure and function in two different soil types. PloS. one. 8: e84441.

Fuciños, C., Míguez, M., Fuciños, P., Pastrana, L.M., Rúa, M.L. and Vicente, A.A. 2017. Creating functional nanostructures: encapsulation of caffeine into α-lactalbumin nanotubes. Innovat. Food Sci. Emerg. Technol. 40: 10–17.

Galaktionova, L., Gavrish, I. and Lebedev, S. 2019. Bioeffects of Zn and Cu nanoparticles in soil systems. Toxicol. Environ. Health. Sci. 11: 259–270.

Gao, J., Sepúlveda, M.S., Klinkhamer, C., Wei, A., Gao, Y. and Mahapatra, C.T. 2015. Nanosilver-coated socks and their toxicity to zebrafish (*Danio rerio*) embryos. Chemosphere. 119: 948–952.

Gao, X., Avellan, A., Laughton, S., Vaidya, R., Rodrigues, S.M., Casman, E.A. and Lowry, G.V. 2018. CuO nanoparticle dissolution and toxicity to wheat (*Triticum aestivum*) in rhizosphere soil. Environ. Sci. Technol. 52: 2888–2897.

García-Gómez, C., Fernández, M.D., García, S., Obrador, A.F., Letón, M. and Babín, M. 2018. Soil pH effects on the toxicity of zinc oxide nanoparticles to soil microbial community. nEnviron. Sci. Pollut. Res. 25: 28140–28152.

Ge, Y., Priester, J.H., Mortimer, M., Chang, C.H., Ji, Z., Schimel, J.P. and Holden, P.A. 2016. Long-term effects of multiwalled carbon nanotubes and graphene on microbial communities in dry soil. Environ. Sci. Technol. 50: 3965–3974.

Giacomello, E., Sava, G., Vita, F., Delhom, N., Mahl, P. and Bergamo, A. 2020. Chitosan-coated alginate micro-particles delivery of active principles through conventional pelleted food—a study in Tilapia (*Oreochromis niloticus*). Int. J. Biol. Macromol. 165: 82–92.

Gimbert, L.J., Hamon, R.E., Casey, P.S. and Worsfold, P.J. 2007. Partitioning and stability of engineered ZnO nanoparticles in soil suspensions using flow field-flow fractionation. Environment. Chem. 4: 8–10.

Gómez-Pastora, J., Dominguez, S., Bringas, E., Rivero, M.J., Ortiz, I. and Dionysiou, D.D. 2017. Review and perspectives on the use of magnetic nanophotocatalysts (MNPCs) in water treatment. Chem. Eng. J. 310: 407–427.

Griffitt, R.J., Luo, J., Gao, J., Bonzongo, J.C. and Barber, D.S. 2008. Effects of particle composition and species on toxicity of metallic nanomaterials in aquatic organisms. Environ. Toxicol. Chem. 27: 1972–1978.

Gupta, R. and Xie, H. 2018. Nanoparticles in daily life: applications, toxicity and regulations. J. Environ. Pathol. Toxicol. Oncol. 37: 209–230.

Hadidi, M., Motamedzadegan, A., Jelyani, A.Z. and Khashadeh, S. 2021. Nanoencapsulation of hyssop essential oil in chitosan-pea protein isolate nanocomplex. LWT. 144: 111254.

Hallock, M.F., Greenley, P., DiBerardinis, L. and Kallin, D. 2009. Potential risks of nanomaterials and how to safely handle materials of uncertain toxicity. J. Chem. Health Saf. 16: 16–23.

Handy, R.D., Cornelis, G., Fernandes, T., Tsyusko, O., Decho, A., Sabo-Attwood, T., Metcalfe, C., et al. 2012. Ecotoxicity test methods for engineered nanomaterials: practical experiences and recommendations from the bench. Environ. Toxicol. Chem . 31: 15–31.

Hansen, S.F., Larsen, B.H., Olsen, S.I. and Baun, A. 2007. Categorization framework to aid hazard identification of nanomaterials. Nanotoxicology. 1: 243–250.

Hansen, S.F., Baun, A. and Ganzleben, C. 2011. Nanomaterials and the european water framework directive. Eur. J. Law Technol. 2: 1–11.

Hardman, R. 2006. A toxicologic review of quantum dots: toxicity depends on physicochemical and environmental factors. Environ. Health Perspect. 114: 165–172.

Hasanin, M., Hashem, A.H., Lashin, I. and Hassan, S.A. 2021. *In vitro* improvement and rooting of banana plantlets using antifungal nanocomposite based on myco-synthesized copper oxide nanoparticles and starch. Biomass Convers. Biorefin. 1–11.

Hashem, A.H., Abdelaziz, A.M., Askar, A.A., Fouda, H.M., Khalil, A.M., Abd-Elsalam, K.A. and Khaleil, M.M. 2021. Bacillus megaterium-mediated synthesis of selenium nanoparticles and their antifungal activity against *Rhizoctonia solani* in Faba Bean Plants. J. Fungi 7: 195.

He, F., Wang, H., Chen, Q., Yang, B., Gao, Y. and Wang, L. 2015. Short-term response of soil enzyme activity and soil respiration to repeated carbon nanotubes exposure. Soil. Sediment Contam. 24: 250–261.

Hegde, K. 2015. Environmental hazards and risks of nanomaterials. pp. 357–382. *In*: Rao, Y. (ed.). Nanomaterials in the Environment. American Society of Civil Engineers. Reston. VA.

Hosseini, M., Haghighi, M. and Kahforoushan, D. 2016. Sono-dispersion of ceria and palladium in preparation and characterization of Pd/Al_2O_3-clinoptilolite-CeO_2 nanocatalyst for treatment of polluted air via low temperature VOC oxidation. Process Saf. Environ. Protect. 106: 284–293.

Huang, W., Fang, H., Zhang, S. and Yu, H. 2021. Optimized green synthesis of copper oxide nanoparticles and their antifungal activity. Micro. Nano. Lett. 16 374–380.

Huq, M.A., Ashrafudoulla, M., Rahman, M.M., Balusamy, S.R. and Akter, S. 2022. Green synthesis and potential antibacterial applications of bioactive silver nanoparticles: a review. Polymers. 14: 742.

Ibrahim, S.S., Abou-Elseoud, W.S., Elbehery, H.H. and Hassan, M.L. 2022. Chitosan-cellulose nanoencapsulation systems for enhancing the insecticidal activity of citronella essential oil against the cotton leafworm *Spodoptera littoralis*. Indust. Crops Prod. 184: 115089.

Ihegwuagu, N.E., Sha'Ato, R., Tor-Anyiin, T.A., Nnamonu, L.A., Buekes, P., Sone, B. and Maaza, M. 2016. Facile formulation of starch–silver-nanoparticle encapsulated dichlorvos and chlorpyrifos for enhanced insecticide delivery. New J. Chem. 40: 1777–1784.

ISO 31000:2018. Second Edition: Risk management—Guidelines: 8.2.

Jenish, A., Ranjani, S. and Hemalatha, S. 2022. Moringa oleifera nanoparticles demonstrate antifungal activity against plant pathogenic fungi. Appl. Biochem. Biotechnol. 1–12.

Jin, L., Son, Y., Yoon, T.K., Kang, Y.J., Kim, W. and Chung, H. 2013. High concentrations of single-walled carbon nanotubes lower soil enzyme activity and microbial biomass. Ecotoxicol. Environ. Saf. 88: 9–15.

Jin, L., Son, Y., DeForest, J.L., Kang, Y.J., Kim, W. and Chung, H. 2014. Single-walled carbon nanotubes alter soil microbial community composition. Sci. Total Environ. 466: 533–538.

John, R., Pal, K., Jayan, J.S., Appukuttan, S. and Joseph, K. 2021. New emerging review on advances in block copolymer based water purification membranes. J. Mol. Struct. 5: 1231.

Joner, E.J., Hartnik, T. and Amundsen, C.E. 2008. Environmental fate and ecotoxicity of engineered nanoparticles. pp. 1–64. *In*: Hartnik, T. (ed.). Norwegian Pollution Control Authority Report no. TA 2304/2007. Bioforsk, As, Norway.

Kanwal, Z., Raza, M.A., Manzoor, F., Riaz, S., Jabeen, G., Fatima, S. and Naseem, S. 2019. A comparative assessment of nanotoxicity induced by metal (silver, nickel) and metal oxide (cobalt, chromium) nanoparticles in *Labeo rohita*. Nanomaterials. 9: 309.

Kaonga, C.C., Kosamu, I.B., Lakudzala, D.D., Mbewe, R., Thole, B. and Monjerezi, M. 2017. A review of heavy metals in soil and aquatic systems of urban and semi-urban areas in Malawi with comparisons to other selected countries. Afr. J. Environ. Sci. Technol. 11: 448–460.

Kaur, R., Sharma, S.K. and Tripathy, S.K. 2019. Advantages and limitations of environmental nanosensors. pp. 119–132. *In*: Sharma, S.K (ed.). Advances in Nanosensors for Biological and Environmental Analysis. Elsevier.

Kazancoglu, I., Sagnak, M., Mangla, S.K. and Kazancoglu, Y. 2021. Circular economy and the policy: a framework for improvingthe corporate environmental management in supply chains. Bus Strat. Env. 30: 590–608.

Khan, M.R. and Siddiqui, Z.A. 2021. Role of zinc oxide nanoparticles in the management of disease complex of beetroot (*Beta vulgaris* L.) caused by *Pectobacterium betavasculorum*, *Meloidogyne incognita* and *Rhizoctonia solani*. Hortic. Environ. Biotechnol. 62: 225–241.

Khan, M., Khan, A.U., Bogdanchikova, N. and Garibo, D. 2021. Antibacterial and antifungal studies of biosynthesized silver nanoparticles against plant parasitic nematode *Meloidogyne incognita*, plant pathogens *Ralstonia solanacearum* and *Fusarium oxysporum*. Molecules. 26: 2462.

Kharwar, S. and Singh, S. 2021. First-principles investigation of zigzag graphene nanoribbons based nanosensor for heavy metal detector. Mater. Today Proc. 15: 17.

Khin, M.M., Nair, A.S., Babu, V.J., Murugan, R. and Ramakrishna, S. 2012. A review on nanomaterials for environmental remediation. Energy Environ. Sci. 5: 8075–8109.

Kim, S., Kim, J. and Lee, I. 2012. Effects of Zn and ZnO nanoparticles and Zn^{2+} on soil enzyme activity and bioaccumulation of Zn in *Cucumis sativus*. Chem. Ecol. 27: 49–55.

Kottegoda, N., Sandaruwan, C., Priyadarshana, G., Siriwardhana, A., Rathnayake, U.A., Berugoda Arachchige, D.M. and Amaratunga, G.A. 2017. Urea-hydroxyapatite nanohybrids for slow release of nitrogen. ACS Nano. 11: 1214–1221.

Krysanov, E.Y., Pavlov, D.S., Demidova, T.B. and Dgebuadze, Y. 2010. Effect of nanoparticles on aquatic organisms. Biol. Bull. 37: 406–412.

Kumar, V. and Kumar, P. 2019. Pesticides in agriculture and environment: impacts on human health. pp. 76–95. *In:* Kumar, P. (ed.). Contaminants in Agriculture and Environment: Health Risks and Remediation. Haridwar. India.

Kumar, A., Sharma, D., Balasubramaniam, B., Thakur, R., Saini, R.V., Gupta, R.K., Mittal, D., et al. 2022. Application of novel biogenic nanoparticles for antimicrobial traits. Biomat. Polym. Horizon. 1(2).

Lau, W.J., Ismail, A.F., Misdan, N. and Kassim, M.A. 2012. A recent progress in thin film composite membrane: a review. DES. 287: 190–199.

Laux, P., Tentschert, J., Riebeling, C., Braeuning, A., Creutzenberg, O., Epp, A. and Luch, A. 2018. Nanomaterials: certain aspects of application, risk assessment and risk communication. Arch. Toxicol. 92: 121–141.

Le Van, N., Ma, C., Shang, J., Rui, Y., Liu, S. and Xing, B. 2016. Effects of CuO nanoparticles on insecticidal activity and phytotoxicity in conventional and transgenic cotton. Chemosphere. 144: 661–670.

Lead, J.R., Batley, G.E., Alvarez, P.J., Croteau, M.N., Handy, R.D., McLaughlin, M.J. and Schirmer, K. 2018. Nanomaterials in the environment: behavior, fate, bioavailability, and effects—an updated review. Environment. Toxicol. Chem. 37: 2029–2063.

Li, M., Ruan, L.Y., Dang, F., Liu, H.L., Zhou, D.M., Yin, B. and Wang, J.S. 2022. Metabolic response of earthworms (*Pheretima guillemi*) to silver nanoparticles in sludge-amended soil. Environ. Pollut. 300: 118954.

Liu, F., Zhu, S., Li, D., Chen, G. and Ho, S.H. 2020. Detecting ferric iron by microalgal residuederived fluorescent nanosensor with an advanced kinetic model. IScien. 23: 101174.

Loureiro, S., Tourinho, P.S., Cornelis, G., Van Den Brink, N.W., Díez-Ortiz, M., Vázquez-Campos, S. and Van Gestel, C.A. 2018. Nanomaterials as soil pollutants. pp. 161–190. *In*: Duarte, A.C., Cachada, A. and Rocha-Santos, T. (eds). Soil Pollution. Academic Press. Cambridge. MA. USA.

Lu, H., Wang, J., Stoller, M., Wang, T., Bao, Y. and Hao, H. 2016. An overview of nanomaterials for water and wastewater treatment. Adv. Mater. Sci. Eng. 2016

Lv, X., Xu, J., Jiang, G. and Xu, X. 2011. Removal of chromium (VI) from wastewater by nanoscale zero-valent iron particles supported on multiwalled carbon nanotubes. Chemosph. 85: 1204–1209.

Ma, X., Zhu, X., Qu, S., Cai, L., Ma, G., Fan, G. and Sun, X. 2022. Fabrication of copper nanoparticle composite nanogel for high-efficiency management of *Pseudomonas syringae* pv. *tabaci* on tobacco. Pest Manag. Sci. 78: 2074–2085.

Mahmood, A., Hussain, S., Mahmood, F., Iqbal, M., Shahid, M., Ibrahim, M. and Shahzad, T. 2021. Toxicity of biogenic zinc oxide nanoparticles to soil organic matter cycling and their interaction with rice-straw derived biochar. Sci. Rep. 11: 1–12.

Malakar, A., Kanel, S.R., Ray, C., Snow, D.D. and Nadagouda, M.N. 2021. Nanomaterials in the environment, human exposure pathway, and health effects: a review. Sci. Total Environ. 759: 143470.

Malik, S., Muhammad, K. and Waheed, Y. 2023. Nanotechnology: a revolution in modern industry. Molecules. 28: 661.

Martínez, G., Merinero, M., Pérez-Aranda, M., Pérez-Soriano, E.M., Ortiz, T., Begines, B. and Alcudia, A. 2021. Environmental impact of nanoparticles' application as an emerging technology: a review. Materials. 14: 166.

Mathew, J., Joy, J. and George, S.C. 2019. Potential applications of nanotechnology in transportation: a review. J. King Saud Univ. Sci. 31: 586–594.

Meng, C., Zhikun, W., Qiang, L., Chunling, L., Shuangqing, S. and Songqing, H. 2018. Preparation of amino-functionalized Fe_3O_4@mSiO_2core-shell magnetic nanoparticles and their application for aqueous Fe^{3+} removal. J. Hazard Mater. 341: 198–206.

Meyer, W.L., Gurman, P., Stelinski, L.L. and Elman, N.M. 2015. Functional nanodispensers (FNDs) for delivery of insecticides against phytopathogen vectors. Green Chem. 17: 4173–4177.

Mitchell, S.L., Hudson-Smith, N.V., Cahill, M.S., Reynolds, B.N., Frand, S.D., Green, C.M. and Carlson, E.E. 2019. Chronic exposure to complex metal oxide nanoparticles elicits rapid resistance in *Shewanella oneidensis* MR-1. Chem. Sci. 10: 9768–9781.

Mitrano, D.M., Motellier, S., Clavaguera, S. and Nowack, B. 2015. Review of nanomaterial aging and transformations through the life cycle of nano-enhanced products. Environ. Int. 77: 132–147.

Mohammed, A.N. and Attia, A.S. 2022. Control of biofilm-producing *Aeromonas* bacteria in the water tanks and drinkers of broiler poultry farms using chitosan nanoparticle-based coating thyme oil. Iraqi J. Vet. Sci. 36: 659–669.

Mubarak, N.M., Sahu, J.N., Abdullah, E.C., Jayakumar, N.S. and Ganesan, P. 2015. Microwave-assisted synthesis of multi-walled carbon nanotubes for enhanced removal of Zn(II) from wastewater. Res. Chem. Intermed. 42: 1–25.

Mubeen, I., Tulaphol, S., Shengyong, L., Pan, D., Zhang, P., Md. Sajid Khan, Yan, M., et al., 2021. Online measurement of 1,2,4-trichlorobenzene as dioxin indicator on multi-walled carbon nanotubes. Environ. Pollut. A. 268: 115329.

Mueller, N.C. and Nowack, B. 2008. Exposure modeling of engineered nanoparticles in the environment. Environ. Sci. Technol. 42: 4447–4453.

Mustapha, T., Misni, N., Ithnin, N.R., Daskum, A.M. and Unyah, N.Z. 2022. A review on plants and microorganisms mediated synthesis of silver nanoparticles, role of plants metabolites and applications. Int. J. Environ. Res. Public Health. 19: 674.

Nanodatabase. Available online: https://nanodb.dk/en/ (accessed on 11 December 2020).

Nowack, B., Ranville, J.F., Diamond, S., Gallego-Urrea, J.A., Metcalfe, C., Rose, J. and Klaine, S.J. 2012. Potential scenarios for nanomaterial release and subsequent alteration in the environment. Environ. Toxicol. Chem. 31: 50–59.

Ntoutoume, G.M.N., Granet, R., Mbakidi, J.P., Brégier, F., Léger, D.Y., Fidanzi-Dugas, C. and Sol, V. 2016. Development of curcumin–cyclodextrin/cellulose nanocrystals complexes: new anticancer drug delivery systems. Bioorg. Med. Chem. Lett. 26: 941–945.

OECD. 2012. Development Co-operation Report: Lessons in linking sustainability and development. Available online: https://www.oecd.org/development/dcr2012.htm

OECD. 2019. Physical-chemical decision framework to inform decisions for risk assessment of manufactured nanomaterials. Paris: OECD Publishing. Available online: https://www.oecd.org/officialdocuments/publicdisplaydocumentpdf/?cote=env/jm/mono(2019)12&do clanguage=en.

OECD. 2020. Guidance document on aquatic and sediment toxicological testing of nanomaterials. Paris: OECD Publishing. Available online: https://www.oecd.org/officialdocuments/publicdisplaydocumentpdf/?cote=env/jm/mono(2020)8&doclanguage=en.

Pacheco, N.I.N., Roubalova, R., Dvorak, J., Benada, O., Pinkas, D., Kofronova, O. and Prochazkova, P. 2021. Understanding the toxicity mechanism of CuO nanoparticles: the intracellular view of exposed earthworm cells. Environ. Sci. Nano. 8: 2464–2477.

Pathania, D., Kumar, S., Thakur, P., Chaudhary, V., Kaushik, A., Varma, R.S. and Khosla, A. 2022. Essential oil-mediated biocompatible magnesium nanoparticles with enhanced antibacterial, antifungal, and photocatalytic efficacies. Sci. Rep. 12: 1–13.

Pérez-Masiá, R., López-Nicolás, R., Periago, M.J., Ros, G., Lagaron, J.M. and López-Rubio, A. 2015. Encapsulation of folic acid in food hydrocolloids through nanospray drying and electrospraying for nutraceutical applications. Food Chem. 168: 124–133.

Pineda, M.E.B., Lizarazo Forero, L.M. and Sierra Avila, C.A. 2022. Antibacterial activity of biosynthesized silver nanoparticles (AgNps) against *Pectobacterium carotovorum.* Braz. J. Microbiol. 1–12.

Preetha, P.S. and Balakrishnan, N. 2017. A review of nano fertilizers and their use and functions in soil. Int. J. Curr. Microbiol. Appl. Sci. 6: 3117–3133.

Rana, S. and Kalaichelvan, P.T. 2013. Ecotoxicity of Nanoparticles. ISRN Toxicol. 2013: 574648.

Rasool, S., Rasool, T. and Gani, K.M. 2022. A review of interactions of pesticides within various interfaces of intrinsic and organic residue amended soil environment. Adv. Chem. Eng. 100301.

Rather, M.A., Sharma, R., Gupta, S., Ferosekhan, S., Ramya, V.L. and Jadhao, S.B. 2013. Chitosan-nanoconjugated hormone nanoparticles for sustained surge of gonadotropins and enhanced reproductive output in female fish. PloS One. 8: e57094.

Resnik, D.B. 2019. How should engineered nanomaterials be regulated for public and environmental health? AMA J. Ethics. 21: 363–369.

Rezaei, F., Safarzadeh, M., Mozafari, H. and Tavakoli, P. 2015. Prevalence of geographic tongue and related predisposing factors in 7–18 year-old students in Kermanshah. Glob. J. Health Sci. 7: 91–95.

Roberto, M.M. and Christofoletti, C.A. 2019. How to assess nanomaterial toxicity? An environmental and human health approach. *In:* Clichici, S., Filip, G.A. and Do Nascimento, G.M. (eds). Nanomaterials-Toxicity, Human Health and Environment. London. UK. Intech. Open.

Roshan, A.B., Venkatesh, H.N., Dubey, N.K. and Mohana, D.C. 2022. Chitosan-based nanoencapsulation of *Toddalia asiatica* (L.) Lam. essential oil to enhance antifungal and aflatoxin B1 inhibitory activities for safe storage of maize. Int. J. Biol. Macromol. 204: 476–484.

Rui, M., Ma, C., White, J.C., Hao, Y., Wang, Y., Tang, X. and Xing, B. 2018. Metal oxide nanoparticles alter peanut (*Arachis hypogaea* L.) physiological response and reduce nutritional quality: a life cycle study. Environ. Sci. Nano 5: 2088–2102.

Sadek, M.E., Shabana, Y.M., Sayed-Ahmed, K. and Abou Tabl, A.H. 2022. Antifungal activities of sulfur and copper nanoparticles against cucumber postharvest diseases caused by botrytis cinerea and sclerotinia sclerotiorum. J. Fungi. 8: 412.

Sahu, S.C. and Hayes, AW. 2017. Toxicity of nanomaterials found in human environment: a literature review. Toxicol. Res. Appl.1: 1–13.

Saleem, H. and Zaidi, S.J. 2020. Developments in the application of nanomaterials for water treatment and their impact on the environment. Nanomat. 10: 1764.

Scown, T.M., Aerle, R.V. and Tyler, C.R. 2010. Do engineered nanoparticles pose a significant threat to the aquatic environment. Crit. Rev. Toxicol. 40: 653–670.

Selatile, M.K., Ray, S.S., Ojijo, V. and Sadiku, R. 2018. Recent developments in polymeric electrospun nanofibrous membranes for seawater desalination. RSC Adv. 8: 37915–37938.

Sethulekshmi, A.S., Saritha, A., Joseph, K., Aprem, A.S. and Sisupal, S.B. 2022. MoS_2 based nanomaterials: advanced antibacterial agents for future. J. Control. Release. 348: 158–185.

Shin, Y.J., Kwak, J.I. and An, Y.J. 2012. Evidence for the inhibitory effects of silver nanoparticles on the activities of soil exoenzymes. Chemosphere. 88: 524–529.

Shu, W., Yang, Z., Xu, Z., Zhu, T., Tian, X. and Yang, Y. 2022. Effects of one-dimensional nanomaterial polyaniline nanorods on earthworm biomarkers and soil enzymes. Environ. Sci. Pollut. Res. 29: 35217–35229.

Silva, G.F.B.L., Viana, C., Domingues, D. and Vieira, F. 2019. Risk assessment and health, safety, and environmental management of carbon nanomaterials. *In:* Clichici, S., Filip, A. and Nascimento, G.M. (eds). Nanomaterials-Toxicity, Human Health and Environment. Intech. Open. London. UK.

Simonin, M., Cantarel, A.A., Crouzet, A., Gervaix, J., Martins, J.M. and Richaume, A. 2018. Negative effects of copper oxide nanoparticles on carbon and nitrogen cycle microbial activities in contrasting agricultural soils and in presence of plants. Front. Microbiol. 9: 3102.

Singh, A.V., Ansari, M.H.D., Rosenkranz, D., Maharjan, R.S., Kriegel, F.L., Gandhi, K. and Luch, A. 2020. Artificial intelligence and machine learning in computational nanotoxicology: unlocking and empowering nanomedicine. Adv. Healthc. Mater. 9: 1901862.

Singh, P., Singh, K.R., Singh, J., Das, S.N. and Singh, R.P. 2021. Tunable electrochemistry and efficient antibacterial activity of plant-mediated copper oxide nanoparticles synthesized by Annona squamosa seed extract for agricultural utility. RSC Advances. 11: 18050–18060.

Stone, V., Nowack, B., Baun, A., van den Brink, N., Kammer, F., Dusinska, M., Handy, R., et al. 2010. Nanomaterials for environmental studies: classification, reference material issues, and strategies for physicochemical characterisation. Sci. Total Environ. 408: 1745–1754.

Suppan, S. 2013. Institute for Agriculture and Trade Policy. Nanomaterials in Soil. Washington. DC. USA.

Suppan, S. 2017. Institute for Agriculture and Trade Policy. Applying Nanotechnology to Fertilizer: Rationales, Research, Risks and Regulatory Challenges. Washington. DC. USA.

Suzuki, Y., Onuma, H., Sato, R., Sato, Y., Hashiba, A., Maeki, M. and Harashima, H. 2021. Lipid nanoparticles loaded with ribonucleoprotein–oligonucleotide complexes synthesized using a microfluidic device exhibit robust genome editing and hepatitis B virus inhibition. J. Controll. Rel. 330: 61–71.

Syafiuddin, A., Salmiati, S., Hadibarata, T., Salim, M.R., Kueh A.B,H. and Suhartono, S. 2019. Removal of silver nanoparticles from water environment: Experimental, mathematical formulation, and cost analysis. Water Air Soil Pollut. 230: 102.

Taghipour, S., Hosseini, S.M. and Ataie-Ashtiani, B. 2019. Engineering nanomaterials for water and wastewater treatment: review of classifications, properties and applications. New J. Chem. 43: 7902–7927.

Tamayo-Belda, M., González-Pleiter, M., Pulido-Reyes, G., Martin-Betancor, K., Leganés, F., Rosal, R. and Fernández-Piñas, F. 2019. Mechanism of the toxic action of cationic G5 and G7 PAMAM dendrimers in the cyanobacterium *Anabaena* sp. PCC7120. Environ. Sci. Nano. 6: 863–878.

Taran, M., Safaei, M., Karimi, N. and Almasi, A. 2021. Benefits and application of nanotechnology in environmental science: an overview. Biointerf. Res. Appl. Chem. 11: 7860–7870.

Upadhyay, N., Singh, V.K., Dwivedy, A.K., Chaudhari, A.K. and Dubey, N.K. 2022. Assessment of nanoencapsulated *Cananga odorata* essential oil in chitosan nanopolymer as a green approach to boost the antifungal, antioxidant and in situ efficacy. Int. J. Biol. Macromol. 171: 480–490.

Valavanidis, A. and Vlachogianni, T. 2010. Nanomaterials and nanoparticles in the aquatic environment: toxicological and ecotoxicological risks. Sci. Adv. Environ. 2010.

Valsami-Jones, E. and Lynch, I. 2015. Nanosafety. How safe are nanomaterials? Science. 350: 388–389.

Van, N.L., Ma, C., Shang, J., Rui, Y., Liu, S. and Xing, B. 2016. Effects of CuO nanoparticles on insecticidal activity and phytotoxicity in conventional and transgenic cotton. Chemosphere. 144: 661–670.

Vanti, G.L., Masaphy, S., Kurjogi, M., Chakrasali, S. and Nargund, V.B. 2020. Synthesis and application of chitosan-copper nanoparticles on damping off causing plant pathogenic fungi. Int. J. Biol. Macromol. 156: 1387–1395.

Vega-Vásquez, P., Mosier, N.S. and Irudayaraj, J. 2020. Nanoscale drug delivery systems: from medicine to agriculture. Frontiers Bioengin. Biotech. 8: 79.

Vijaya Lakshmi, K., Nagrath, K. and Jha, A. 2011. Technical Report. Access to safe water: approaches for nanotechnology benefits to reach the bottom of the pyramid. New Delhi: Development Alternatives Group. Department for International Development, UK. Available online: https://www.gov.uk/research-for-development-outputs/access-to-safe-water-approaches-for-nanotechnology-benefits-to-reach-the-bottom-of-the-pyramid-final-technical-report-may-2011.

Waani, S.P.T., Irum, S., Gul, I., Yaqoob, K., Khalid, M.U., Ali, M.A. and Arshad, M. 2021. TiO_2 nanoparticles dose, application method and phosphorous levels influence genotoxicity in Rice (*Oryza sativa* L.), soil enzymatic activities and plant growth. Ecotoxicol. Environ. Saf. 213: 111977.

Wang, T., Xue, J., Hu, Q., Zhou, M. and Luo, Y. 2017a. Preparation of lipid nanoparticles with high loading capacity and exceptional gastrointestinal stability for potential oral delivery applications. J. Colloid Interf. Sci. 507: 119–130.

Wang, Y., Chang, C.H., Ji, Z., Bouchard, D.C., Nisbet, R.M., Schimel, J.P. and Holden, P.A. 2017b. Agglomeration determines effects of carbonaceous nanomaterials on soybean nodulation, dinitrogen fixation potential, and growth in soil. ACS Nano. 11: 5753–5765.

Wang, H., Li, H., Zhang, M., Song, Y., Huang, J., Huang, H. and Kang, Z. 2018. Carbon dots enhance the nitrogen fixation activity of *Azotobacter chroococcum.* ACS Appl. Mater. Interfaces. 10: 16308–16314.

Wang, X., Kong, L., Zhou, S., Ma, C., Lin, W., Sun, X., Kirsanov, D., et al. 2021. Development of QDs-based nanosensors for heavy metal detection: a review on transducer principles and in-situ detection. Talanta. 3: 34.

Warheit, D. 2008. How meaningful are the results of nanotoxicity studies in the absence of adequate material characterization? Toxicol. Sci. 101: 183–185.

Warheit, D.B., Webb, T.R., Colvin, V.L., Reed, K.L. and Sayes, C.M. 2007a. Pulmonary bioassay studies with nanoscale and fine-quartz particles in rats: toxicity is not dependent upon particle size but on surface characteristics. Toxicol. Sci. 95: 270–280.

Warheit, D.B., Webb, T.R., Reed, K.L., Frerichs, S. and Sayes, C.M. 2007b. Pulmonary toxicity study in rats with three forms of ultrafine-TiO_2 particles: differential responses related to surface properties. Toxicology. 230: 90–104.

Wheeler, A. 2020. United States Environmental Protection Agency. Available online: https://www.epa.gov.

Wei, X., Cao, P., Wang, G., Liu, Y., Song, J. and Han, J. 2021. CuO, ZnO, and γ-Fe_2O_3 nanoparticles modified the underground biomass and rhizosphere microbial community of *Salvia miltiorrhiza* (Bge.) after 165–day exposure. Ecotoxicol. Environ. Saf. 217: 112232.

Wiesner, M.R., Lowry, G.V., Jones, K.L., Hochella, M.F., Di Giulio, R.T., Casman, E. and Bernhardt, E.S. 2009. Decreasing uncertainties in assessing environmental exposure, risk, and ecological implications of nanomaterials. Environ. Sci. Technol. 43: 6458–6462.

Winkler, D.A. 2020. Role of artificial intelligence and machine learning in nanosafety. Small. 16: 2001883.

Wong, S.W.Y., Zhou, G.J., Leung, P.T.Y., Han, J., Lee, J.S., Kwok, K.W.H. and Leung, K.M.Y. 2020. Sunscreens containing zinc oxide nanoparticles can trigger oxidative stress and toxicity to the marine copepod *Tigriopus japonicus*. Mar. Pollut. Bull. 154: 111078.

Wu, S., Gaillard, J.F. and Gray, K.A. 2021. The impacts of metal-based engineered nanomaterial mixtures on microbial systems: a review. Sci. Total Environ. 780: 46496.

Xin, X., Zhao, F., Zhao, H., Goodrich, S.L., Hill, M.R., Sumerlin, B.S. and He, Z. 2020. Comparative assessment of polymeric and other nanoparticles impacts on soil microbial and biochemical properties. Geoderma. 367: 114278.

Xu, C., Peng, C., Sun, L., Zhang, S., Huang, H., Chen, Y. and Shi, J. 2015. Distinctive effects of TiO_2 and CuO nanoparticles on soil microbes and their community structures in flooded paddy soil. Soil Biol. Biochem. 86: 24–33.

Xu, X.,Yang, Y., Liu, T. and Chu, B. 2022. Cost-effective polymer-based membranes for drinking water purification. Giant. 10: 100099.

Yan, S., Hu, Q., Jiang, Q., Chen, H., Wei, J., Yin, M. and Shen, J. 2021a. Simple osthole/nanocarrier pesticide efficiently controls both pests and diseases fulfilling the need of green production of strawberry. ACS Appl. Mat. Interf. 13: 36350–36360.

Yan, W., Fu, X., Gao, Y., Shi, L., Liu, Q., Yang, W. and Feng, J. 2021b. Synthesis, antibacterial evaluation, and safety assessment of CuS NPs against *Pectobacterium carotovorum* subsp. carotovorum. Pest Management Sci. 78: 733–742.

Yang, X., Gong, J., Zhang, X., Huang, Y., Zhang, W., Yang, J. and Liu, J. 2021. Evaluation of the combined toxicity of multi-walled carbon nanotubes and cadmium on earthworms in soil using multi-level biomarkers. Ecotoxicol. Environ. Saf. 221: 112441.

Yilihamu, A., Ouyang, B., Ouyang, P., Bai, Y., Zhang, Q., Shi, M. and Yang, S.T. 2020. Interaction between graphene oxide and nitrogen-fixing bacterium *Azotobacter chroococcum*: transformation, toxicity and nitrogen fixation. Carbon. 160: 5–13.

Ying, Y., Ying, W., Li, Q., Meng, D., Ren, G., Yan, R. and Peng, X. 2017. Recent advances of nanomaterial-based membrane for water purification. Appl. Mater Today. 7: 144–158.

Yoon, H.Y., Lee, J.G., Esposti, L.D., Iafisco, M., Kim, P.J., Shin, S.G. and Adamiano, A. 2020. Synergistic release of crop nutrients and stimulants from hydroxyapatite nanoparticles functionalized with humic substances: toward a multifunctional nanofertilizer. ACS Omega. 5: 6598–6610.

Yu, G., Wang, X., Liu, J., Jiang, P., You, S., Ding, N., Guo, Q., et al. 2021. Applications of nanomaterials for heavy metal removal from water and soil: a review. Sustainability. 13: 713.

Yu, H., Wang, L., Qu, J., Wang, X., Huang, F., Jiao, Y. and Zhang, Y. 2022. Bi_2O_3/TiO_2@ reduced graphene oxide with enzyme-like properties efficiently inactivates *Pseudomonas syringae* pv. *tomato* DC3000 and enhances abiotic stress tolerance in tomato. Environ. Sci. Nano. 9: 118–132.

Zhang, L., Peng, D., Liang, R.P. and Qiu, J.D. 2018. Graphene-based optical nanosensors for detection of heavy metal ions TrAC. Trends Anal. Chem. 102: 280–289.

Zhang, F., He, M., Zhang, C., Lin, D. and Zhang, J. 2021a. Combined toxic effects of dioxin-like PCB77 with Fe-based nanoparticles in earthworm Eisenia fetida. Sci. Total Environ. 766: 144347.

Zhang, P., Guo, Z., Ullah, S., Melagraki, G., Afantitis, A. and Lynch, I. 2021b. Nanotechnology and artificial intelligence to enable sustainable and precision agriculture. Nature Plants. 7: 864–876.

Zhao, L., Lu, L., Wang, A., Zhang, H., Huang, M., Wu, H. and Ji, R. 2020. Nanobiotechnology in agriculture: use of nanomaterials to promote plant growth and stress tolerance. J. Agricul. Food Chem. 68: 1935–1947.

Zhao, J., Lin, M., Wang, Z., Cao, X. and Xing, B. 2021. Engineered nanomaterials in the environment: are they safe? Crit. Rev. Environ. Sci. Technol. 51: 1443–1478.

Zou, Y., Wang, X., Khan, A., Wang, P., Liu, Y.H., Alsaedi, A., Hayat, T., et al. 2016. Environmental remediation and application of nanoscale zero-valent iron and its composites for the removal of heavy metal ions: a review. Environ. Sci. Technol. 50: 7290–7304.

Index

Brief Biography

Professor (Dr.) Jayanta Kumar Biswas is a Professor & Ex-Head, Department of Ecological Studies and International Centre for Ecological Engineering, University of Kalyani, India. He received many awards and fellowships including Fellow, National Institute of Ecology (FNIE); Fellow, West Bengal Academy of Science and Technology (FAScT); Fellow, National Environmentalists Association (FNEA); Fellow, Zoological Society of India (FZSI); Fellow, Zoological Society, Kolkata (FZS); Senior Scientist Award & Gold Medal (ZSI); Best Research Paper Award (AESA); Shrimati Saroma Sanyal Memorial Prize (IEI); Outstanding Reviewer Award (Chemosphere (Elsevier) & Environmental Geochemistry and Health (Springer); SRF (ICAR); National Merit Scholarship (MHRD, GoI), etc. He is credited with publishing 7 books, 35 book chapters and 165 original research papers in reputed international journals. He has editorial affiliations to: as *Associate Editor*: Environmental Geochemistry and Health (Springer); Environmental Chemistry Letters (Springer); iScience (Cell Press/Elsevier); Ecotoxicology (Springer); Archives of Environmental Contamination and Toxicology (Springer); as *Advisory Board Member*: Ambio (Springer); Heliyon (Cell Press/Elsevier); and as *Editorial Board Member*: Science of the Total Environment (Elsevier); Environmental Technology & Innovation (Elsevier); Current Pollution Reports (Springer); Review of Environmental Contamination and Toxicology (Springer); Sustainable Environment (T&F), Eco-Environment & Health (Elsevier).

Professor (Dr.) Mahendra Rai is presently a visiting Professor at the Department of Microbiology, Nicolaus Copernicus University, Torun, Poland. Formerly he was Professor and Head of the Department of Biotechnology, Sant Gadge Baba Amravati University, Maharashtra State, India. He was a visiting scientist at the University of Geneva, Debrecen University, Hungary; the University of Campinas, Brazil; VSB Technical University of Ostrava, Czech Republic, National University of Rosario, Argentina, and the University of Sao Paulo. He has published more than 425 research papers in national and international journals, with 85 h-index. In addition, he has edited/authored more than 70 books and 6 patents. His area of interest includes microbial nanotechnology and nanobiotechnology with special reference to nanoantimicrobials. Recently, he has been featured in Stanford's list of the top 2% of scientists in Nanoscience and Nanotechnology. As per research.com ranking, he has been ranked #15 in India among the best scientists for 2023 in biology and biochemistry.